INTRODUCTION TO BIOPHOTONICS

INTRODUCTION TO BIOPHOTONICS

Paras N. Prasad

A JOHN WILEY & SONS, INC., PUBLICATION

For general information on our other products and services please contact our Customer Care Department within the U.S. at 877-762-2974, outside the U.S. at 317-572-3993 or fax 317-572-4002.

Wiley also publishes its books in a variety of electronic formats. Some content that appears in print, however, may not be available in electronic format.

Library of Congress Cataloging-in-Publication Data:

Prasad, Paras N.
  Introduction to biophotonics / Paras N. Prasad.
     p.   cm.
  ISBN 0-471-28770-9 (cloth)
  1. Photobiology.   2. Photonics.   3. Biosensors.   4. Nanotechnology.   I. Title.
  QH515.P73 2003
  571.4′55—dc21

                                                                    2003000578

Printed in the United States of America

10 9 8 7 6 5

■ SUMMARY OF CONTENTS

▰▰▰ CONTENTS

Biophotonics deals with interactions between light and biological matter. It is an exciting frontier which involves a fusion of photonics and biology. It offers great hope for the early detection of diseases and for new modalities of light-guided and light-activated therapies. Also, biology is advancing photonics, since biomaterials are showing promise in the development of new photonic media for technological applications.

Biophotonics creates many opportunities for chemists, physicists, engineers, health professionals and biomedical researchers. Also, producing trained healthcare personnel and new generations of researchers in biophotonics is of the utmost importance to keep up with the increasing worldwide demands.

Although several books and journals exist that cover selective aspects of biophotonics, there is a void for a monograph that provides a unified synthesis of this subject. This book provides such an overview of biophotonics which is intended for multidisciplinary readership. The objective is to provide a basic knowledge of a broad range of topics so that individuals in all disciplines can rapidly acquire the minimal necessary background for research and development in biophotonics. The author intends that this book serve both as a textbook for education and training as well as a reference book that aids research and development of those areas integrating light, photonics and biological systems. Another aim of the book is to stimulate the interest of researchers and healthcare professionals and to foster collaboration through multidisciplinary programs.

This book encompasses the fundamentals and various applications involving the integration of light, photonics and biology into biophotonics. Each chapter begins with an introduction describing what a reader will find in that chapter. Each chapter ends with highlights which are basically the take home message and may serve as a review of the materials presented.

In each of the chapters, a description of future directions of research and development is also provided, as well as a brief discussion of the current status, identifying some of areas of future opportunities. A few of the existing commercial sources of instrumentation and supplies relevant to the content of many of the applications chapters (7 and higher) are listed in the respective chapters.

In order to author a book such as this, which covers a very broad range of topics, I received help from a large number of individuals at the Institute for

Lasers, Photonics and Biophotonics and from elsewhere. This help has consisted of gathering technical content, making illustrations, providing critiques and preparing the manuscript. A separate Acknowledgement recognizes these individuals.

Here I would like to acknowledge the individuals whose broad-based support has been of paramount value in completing the book. I wish to express my sincere gratitude to my wife, Nadia Shahram who has been a constant source of inspiration, providing support and encouragement for this project, in spite of her own very busy professional schedule. I am also indebted to our daughters, Melanie and Natasha, for showing their understanding by sacrificing their quality time with me.

I express my sincere appreciation to my colleague, Professor Stanley Bruckenstein, for his endless support and encouragement. I thank Dr. E.J. Bergey for his valuable general support and technical help in bio-related areas. Valuable help was provided by Dr. Haridas Pudavar and it is very much appreciated. I owe thanks to my administrative assistant, Ms. Margie Weber, for assuming responsibility for many of the non-critical administrative issues at the Institute. Finally, I thank Ms. Barbara Raff, whose clerical help in manuscript preparation was invaluable.

Paras N. Prasad
Buffalo, NY

![] ACKNOWLEDGMENTS

Technical Contents:

Dr. E. James Bergey, Dr. Ryszard Burzynski, Dr. Aliaksandr Kachynski, Dr. Andrey Kuzmin, Dr. Paul Markowicz, Dr. Tymish Ohulchanskyy, Dr. Haridas Pudavar, Dr. Marek Samoc, Professor Brenda Spangler, Professor Carlton Stewart

Technical Illustrations and References:

Professor J.M.J. Frechet, Mr. Christopher Friend, Dr. Jeffrey Kingsbury, Professor R. Kopelman, Dr. Tzu Chau Lin, Mr. Emmanuel Nishanth, Mr. Hanifi Tiryaki, Dr. Indrajit Roy, Dr. Kaushik RoyChoudhury, Dr. Yudhisthira Sahoo, Dr. Yuzchen Shen, Professor Hiro Suga, Dr. Richard Vaia, Dr. Jeffrey Winiarz, Mr. QingDong Zheng, Mr. Gen Xu

Chapter Critiques:

Professor Frank Bright, Professor Stanley Bruckenstein, Professor Allan Cadenhead, Mr. Martin Casstevens, Dr. Joseph Cusker, Professor Michael Detty, Professor Sarah Gaffen, Professor Margaret Hollingsworth, Dr. David James, Mr. William Kirkey, Dr. Joydeep Lahiri, Dr. Raymond Lanzafame, Professor Antonia Monteiro, Dr. Janet Morgan, Dr. Allan Oseroff, Dr. Ammasi Periasamy, Dr. Anthony Prezyna, Dr. David Rodman, Professor Malcolm Slaughter, Professor Joseph J. Tufariello, Professor Charles Spangler

Manuscript Preparation:

Cindy Hennessey, Michelle Murray, Kristen Pfaff, Barbara Raff, Patricia Randall, Theresa Skurzewski, Marjorie Weber

CHAPTER 1

Introduction

1.1 BIOPHOTONICS—A NEW FRONTIER

We live in an era of technological revolutions that continue to impact our lives and constantly redefine the breadth of our social interactions. The past century has witnessed many technological breakthroughs, one of which is photonics. Photonics utilizes photons instead of electrons to transmit, process, and store information and thus provides a tremendous gain in capacity and speed in information technology. Photonics is an all-encompassing light-based optical technology that is being hailed as the dominant technology for this new millennium. The invention of lasers, a concentrated source of monochromatic and highly directed light, has revolutionized photonics. Since the demonstration of the first laser in 1960, laser light has touched all aspects of our lives, from home entertainment, to high-capacity information storage, to fiber-optic telecommunications, thus opening up numerous opportunities for photonics.

A new extension of photonics is biophotonics, which involves a fusion of photonics and biology. Biophotonics deals with interaction between light and biological matter. A general introduction to biophotonics is illustrated in Figure 1.1.

The use of photonics for optical diagnostics, as well as for light-activated and light-guided therapy, will have a major impact on health care. This is not surprising since Nature has used biophotonics as a basic principle of life from the beginning. Harnessing photons to achieve photosynthesis and conversion of photons through a series of complex steps to create vision are the best examples of biophotonics at work. Conversely, biology is also advancing photonics, since biomaterials are showing promise as new photonic media for technological applications.

As an increasingly aging world population presents unique health problems, biophotonics offers great hope for the early detection of diseases and for new modalities of light-guided and light-activated therapies. Lasers have already made a significant impact on general, plastic, and cosmetic surgeries. Two popular examples of cosmetic surgeries utilizing lasers are skin resurfacing

Introduction to Biophotonics, by Paras N. Prasad
ISBN: 0-471-28770-9   Copyright © 2003 John Wiley & Sons, Inc.

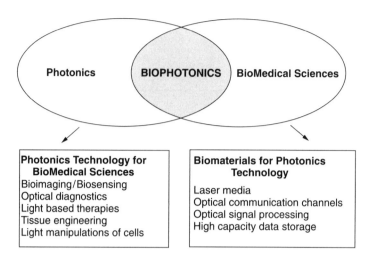

Figure 1.1. Biophotonics as defined by the fusion of photonics and biomedical sciences. The two broad aspects of biophotonics are also identified.

(most commonly known as wrinkle removal) and hair removal. Laser technology also allows one to administer a burst of ultrashort laser pulses that have shown promise for use in tissue engineering. Furthermore, biophotonics may produce retinal implants for restoring vision by reverse engineering Nature's methods.

This book provides an introduction to the exciting new field of biophotonics and is intended for multidisciplinary readership. The book focuses on its potential benefits to medicine. An overview of biophotonics for health care applications is presented in Figure 1.2. It illustrates the scope of biophotonics through multidisciplinary comprehensive research and development possibilities. The focus of the book is on optical probing, diagnostics, and light-activated therapies. However, biophotonics in a broad sense also includes the use of biology for photonics technology, such as biomaterials and development of bioinspired materials as photonic media. These topics are also briefly covered in this book.

1.2 AN INVITATION TO MULTIDISCIPLINARY EDUCATION, TRAINING, AND RESEARCH

In the 21st century, major technological breakthroughs are more likely to occur at the interfaces of disciplines. Biophotonics integrates four major technologies: lasers, photonics, nanotechnology, and biotechnology. Fusion of these technologies truly offers a new dimension for both diagnostics and therapy. Biophotonics creates many opportunities for chemists, physicists, engineers, physicians, dentists, health-care personnel, and biomedical researchers. The

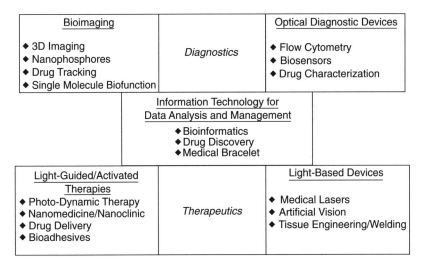

Figure 1.2. The comprehensive multidisciplinary scope of biophotonics for health care.

need for new materials and technologies to provide early detection of diseases, to produce more effective targeted therapies, and to restore impaired biological functions is constantly increasing. The world we live in has become more complex and increasingly dependent upon advanced technologies.

The benefits of lasers to health care are well recognized, even by the general population. Many light-based and spectroscopic techniques are already currently being used as optical probes in clinical laboratories as well as in medical and other health-care practices. Photodynamic therapy, which uses light to treat cancer and has a great potential for growth, is now being practiced.

Producing trained health-care personnel and new generations of researchers in biophotonics is of the utmost importance to keep up with the increasing worldwide demands. Undergraduate and graduate research training programs are needed to develop a skilled workforce and a future generation of researchers respectively for a rapidly growing biotechnology industrial sector. The number of conferences being organized in this field are rapidly increasing, as are the education and training programs at various institutions worldwide. The NSF sponsored integrative graduate education and training program (IGERT) on biophotonics at the University at Buffalo's Institute for Lasers, Photonics, and Biophotonics is a prime example of this trend. This IGERT program is developing multiple interdepartmental courses to provide the needed multidisciplinary education.

A monthly journal, *Biophotonics International*, has emerged as a major reference source. In the areas of research and development, many disciplines can contribute individually as well as collaboratively. Multidisciplinary interactions create unique opportunities that open new doors for the development and application of new technologies.

The author intends that this book serve both as a textbook for education and training as well as a reference book aiding research and development. An aim of the book is to stimulate the interest of researchers and to foster collaboration through multidisciplinary programs. This can lead to the creation of a common language among researchers of widely varying background. The inability to communicate effectively is a major hurdle in establishing any interdisciplinary program.

1.3 OPPORTUNITIES FOR BOTH BASIC RESEARCH AND BIOTECHNOLOGY DEVELOPMENT

Biophotonics offers tremendous opportunities for both biotechnology development and fundamental research. From a technological perspective, biophotonics, as described above, integrates four major technologies: lasers, photonics, nanotechnology, and biotechnology. These technologies have already established themselves in the global marketplace, collectively generating hundreds of billions of dollars per year. Biophotonics also impacts a wide range of industries including biotechnology companies, health care organizations (hospitals, clinics, and medical diagnostic laboratories), medical instrument suppliers, and pharmaceutical manufacturers, as well as those dealing with information technology and optical telecommunication. In the future, biophotonics will have a major impact both in generating new technologies and in offering huge commercial rewards worldwide.

Biophotonics offers challenging opportunities for researchers. A fundamental understanding of the light activation of biomolecules and bioassemblies, and the subsequent photoinduced processes, is a fundamental requirement in designing new probes and drug delivery systems. Also, an understanding of multiphoton processes utilizing ultrashort laser pulses is a necessity both for developing new probes and creating new modalities of light-activated therapy. Some of the opportunities, categorized by discipline, are listed below:

Chemists

- Development of new fluorescent tags
- Chemical probes for analyte detection and biosensing
- Nanoclinics for targeted therapy
- Nanochemistries for materials probes and nanodevices
- New structures for optical activation

Physicists

- Photoprocesses in biomolecules and bioassemblies
- New physical principles for imaging and biosensing

- Single-molecule biophysics
- Nonlinear optical processes for diagnostics and therapy

Engineers

- Efficient and compact integration of new generation lasers, delivery systems, detectors
- Device miniaturization, automation, and robotic control
- New approaches to noninvasive or minimally invasive light activation
- Optical engineering for *in vivo* imaging and optical biopsies
- Nanotechnologies for targeted detection and activation
- Optical BioMEMS (micro-electro-mechanical systems) and their nanoscale analogues.

Biomedical Researchers

- Bioimaging to probe molecular, cellular, and tissue functions
- Optical signature for early detection of infectious diseases and cancers
- Dynamic imaging for physiological response to therapy and drug delivery
- Cellular mechanisms of drug action
- Toxicity of photoactivatable materials
- Biocompatibility of implants and probes

Clinicians

- *In vivo* imaging studies using human subjects
- Development of optical *in vivo* probes for infections and cancers
- *In vivo* optical biopsy and optical mammography
- Tissue welding, contouring, and regeneration
- Real-time monitoring of drug delivery and action
- Long-term clinical studies of side effects

The opportunities for future research and development in a specific biophotonics area are provided in the respective chapter covering that area.

1.4 SCOPE OF THIS BOOK

This book is intended as an introduction to biophotonics, not as an in-depth and exhaustive treatise of this field. The objective is to provide a basic knowledge of a broad range of topics so that even a newcomer can rapidly acquire the minimal necessary background for research and development.

Although several books and journals exist that cover selective aspects of biophotonics, there is a clear need for a monograph that provides a unified synthesis of this subject. The need for such a book as this became apparent while teaching this topic as an interdisciplinary course available to students in many departments at the University at Buffalo. While offering tutorial courses at several professional society conferences such as BIOS of SPIE, the need became even more apparent. The makeup of the registrants for these tutorial courses has been multidisciplinary. Over the years, participants in these courses have constantly emphasized the need for a comprehensive and multi-disciplinary text in this field.

The book is written with the following readership in mind:

- • Researchers working in the area; it will provide useful information for them in areas outside their expertise and serve as a reference source.
- • Newcomers or researchers interested in exploring opportunities in this field; it will provide for them an appreciation and working knowledge of this field in a relatively short time.
- • Educators who provide training and tutorial courses at universities as well as at various professional society meetings; it will serve them as a textbook that elucidates basic principles of existing knowledge and multidisciplinary approaches.

This book encompasses the fundamentals and various applications of bio-photonics. Chapters 1 through 6 cover the fundamentals intended to provide the reader with background, which may be helpful in understanding biophotonics applications covered in subsequent chapters. Chapters 7 through 11 illustrate the use of light for optical diagnostics. Chapters 12 and 13 provide examples of light-based therapy and treatment. Chapters 14 and 15 present specialized topics dealing with micromanipulation of biological objects by light and the infusion of nanotechnology into biophotonics. Chapter 16 discusses the other aspect of biophotonics—that is, the use of biomaterials for photonics technology (see Figure 1.1).

Each chapter begins with an introduction describing what a reader will find in that chapter. In the case of Chapters 1–6, the introductory section also provides a guide to which parts may be skipped by a reader familiar with the content or less inclined to go through details. Each chapter ends with highlights of the material covered in it. The highlights are basically the take home message from the chapter and may serve as a review of the materials presented in the chapter. For an instructor, the highlights may be useful in the preparation of lecture notes or power point presentations. For researchers who may want to get a cursory glimpse, the highlights will provide a summary of what the chapter has covered.

In each of the chapters dealing with applications (Chapters 7–16), a description of future directions of research and development is also provided, along with a brief discussion of current status and the identification of some areas of future opportunities. Each of these application chapters also lists commercial sources of instrumentation and suppliers relevant to the content of the chapter. This list may be useful to a researcher new to this area and interested in acquiring the necessary equipment and supplies or to a researcher interested in upgrading an existing facility.

The book is organized to be adapted for various levels of teaching. Chapters 2–6 can be covered partly or completely, depending on the depth and length of the course and its intended audience. Chapters 7–13 are the various applications of photonics to life sciences and are somewhat interrelated. Chapters 14–16 can be optional, because they deal with specialized topics and do not necessarily require the detailed contents of preceding chapters. Chapters 8–16 are, to a great degree, independent of each other, which allows considerable freedom in the choice of areas to be covered in a course.

Chapter 2 begins with a discussion of the fundamentals of light and matter at a basic level, emphasizing concepts and avoiding mathematical details. For those readers with little exposure to the subject, the materials of this chapter will assist them in grasping the concepts. For those readers already familiar with the subject, the chapter will serve as a condensed review. The dual nature of light as electromagnetic waves and photon particles is described, along with manifestations derived from them. The section on matter introduces a simplified quantum description of atoms, molecules, and the nature of chemical bonding. The description of Π-bonding and the effect of conjugation are provided. The geometric effect derived from the shapes of molecules, along with intermolecular effects, is also covered.

Chapter 3 focuses on building a molecular understanding of biological structures and their relation to biological functions. It provides the basics of biology and introduces the necessary terminology and concepts of biology used in this book. The chapter is written primarily for those unfamiliar with biological concepts, or those wishing to refresh their background in this subject. The chapter describes the chemical makeup of a biological cell and the different organelles. The various cellular functions are also discussed. Then, assembling of cells to form a tissue structure is described, along with the nature of extracellular components in a tissue. The chapter ends with a brief description of tumors and cancers.

Various aspects of light and matter, which form the fundamental basis for biophotonics, are addressed in Chapter 4. This chapter, written for a multidisciplinary readership with varied backgrounds, provides knowledge of the necessary tools of optical interactions utilized in biophotonics applications. These are covered in Chapters 7–16. The emphasis again is on introducing concepts and terminologies, avoiding complex theoretical details. Various spectroscopic techniques useful for biology are covered.

Chapter 5 describes the principle of laser action, relying on simple diagrammatic descriptions. The various steps involved and components used in laser operation are briefly explained. The present status of the laser technology, useful for biophotonics, is described. The chapter also introduces the concepts of nonlinear optical interactions that take place under the action of an intense laser beam. These nonlinear optical interactions are increasingly recognized as useful for biophotonics. The chapter also provides a brief discussion of laser safety.

Chapter 6 discusses photobiology—that is, the interactions of various molecular, cellular, and tissue components with light. Light-induced radiative and nonradiative processes are described, along with a discussion of the photochemical processes at both the cellular and tissue levels. As important examples of biophotonics in Nature, the processes of vision and photosynthesis are presented. A fascinating topic in photobiology is *in vivo* photoexcitation in live specimens, which has opened up the new area of optical biopsy. Another exciting new area is the use of optical techniques to probe interactions and dynamics at the single-cell/single-biomolecule level.

Chapter 7 describes the basic principles and techniques used for optical bioimaging, a major thrust area of biophotonics applications. Although ultrasonic imaging and MRI are well established in the biomedical field, optical imaging offers a complementary approach. For example, it allows multidimensional imaging (multicolor, three-dimensional, time-resolved) and also covers application to all biological organisms, from microbes to humans. Topics discussed include spectral imaging, fluorescence resonance energy transfer (FRET), and lifetime imaging. Newer nonlinear optical imaging methods utilizing multiphoton absorption, harmonic generation, and coherent anti-Stokes scattering (CARS) are also presented. Various types of microscopies described in this chapter include differential interference contrast (DIC), confocal, two-photon laser scanning, optical coherence tomography (OCT), total internal reflection fluorescence (TIRF), and near-field microscopy (NSOM or SNOM).

Chapter 8 provides examples of the wide usage of optical bioimaging to investigate structures and functions of cells and tissues and also to profile diseases at cellular, tissue, and *in vivo* specimen levels. This chapter also discusses the various fluorophores used for fluorescence imaging. Cellular imaging to probe structures and functions of viruses, bacteria, and eukaryotic cells are presented. Then imaging at the tissue level is presented. Finally, *in vivo* imaging, for example optical mammography is discussed.

Chapter 9 on biosensors describes the basic optical principles and the various techniques utilized in biosensing. Biosensors are of especially great interest right now. They are important in combating the constant health danger posed by new strands of microbial organisms and spread of infectious diseases, by characterizing them rapidly. They will be effective tools in the worldwide struggle against chemical and bioterriosm. Chapter 9 provides a detailed coverage of the various existing optical biosensors and ongoing activities given in the literature. The biosensors covered in this chapter are fiber-optic biosensors, planar waveguide biosensors, evanescent wave biosensors, interferometric biosensors, and surface plasmon resonance biosensors. Some novel sensing methods are also described.

Chapter 10 covers microarray technology. It is a natural extension of biosensing. Microarray technology utilizes a micropatterned array of biosensing capture agents for rapid and simultaneous probing of a large number of DNA, proteins, cells, or tissue fragments. It provides a powerful tool for high-throughput, rapid analysis of a large number of samples. This capability has been of significant value in advancing the fields of genomics, proteomics, and bioinformatics, which are at the forefront of modern structural biology, molecular profiling of diseases, and drug discovery. Biophotonics has played an important role in the development of microarray technology, since optical methods are commonly used for detection and readout of microarrays. Four types of microarrays are covered here: DNA microarrays, protein microarrays, cell microarrays, and tissue microarrays.

Chapter 11 introduces the flow cytometer, an optical diagnostic device that currently is used in research and clinical laboratories for disease profiling by measuring the physical and/or chemical characteristics of cells. Flow cytometry is also suitable for rapid and sensitive screening of potential sources of deliberate contamination, an increasing source of concern in bioterrorism. Flow cytometer is also emerging as a powerful technique for agricultural research and livestock development. The chapter describes the steps involved in flow cytometry. The various components of a flow cytometer are described. Methods of data collection, analysis, and display are also discussed.

Chapters 12 and 13 treat the use of light for therapy and treatment, an important area of biophotonics. These chapters provide examples of the use of light for therapy and medical procedures. Chapter 12 covers light-activated therapy, specifically the use of light to activate an administered photosensitizer that causes the destruction of cancer or treats a diseased tissue. This procedure is called *photodynamic therapy* (PDT) and constitutes a multidisciplinary area that has witnessed considerable global growth. Treatment of certain types of cancer using photodynamic therapy is already approved in the United States by the Food and Drug Administration as well as by equivalent agencies in other countries. Therefore, this chapter can be useful not only for researchers but also for clinicians and practicing oncologists. Applications of photodynamic therapy to areas other than cancer—for example, to age-related macular degeneration—are also discussed.

Lasers have emerged as powerful tools for tissue engineering. Chapter 13 discusses tissue engineering with light, utilizing various types of light–tissue interactions. Chapter 13 also has sufficient medical focus to be useful to medical practitioners as well. The chapter covers three main types of laser-based tissue engineering: (i) tissue contouring and restructuring, (ii) tissue welding, and (iii) tissue regeneration. Specific examples of tissue contouring and restructuring are chosen from dermatology and ophthalmology. The section on laser welding of tissues discusses how lasers are used to join tissues. Laser tissue regeneration is a relatively new area; recent work suggests that laser treatment can effect tissue regeneration to repair tissue damage due to an injury. A major impetus to the area of laser-based tissue engineering has been provided by developments in femtosecond laser technology, giving rise to the emergence of "femtolaser surgery."

Chapter 14 covers the usage of a laser beam as a tool for micromanipulation of biological specimens. Two types of laser micromanipulation discussed are laser tweezers for optical trapping and laser scissors for microdissection. The principle of laser optical trapping using a laser is explained. Chapter 14 also provides a detailed discussion of the design of a laser tweezer for the benefit of readers interested in building their own laser tweezers. The use of pulsed laser beam for microdissection of a tissue is discussed. The applications covered are both fundamental, such as in the studies of single molecular biofunction, and applied, such as for reproductive medicine and in plant breeding.

Chapter 15 covers the subject of bionanophotonics, the merging of biomedical science and technology and nanophotonics. Nanophotonics is an emerging field that describes nanoscale optical science and technology. Specifically, this chapter discusses the use of nanoparticles for optical bioimaging, optical diagnostics and light guided and activated therapy. The chapter includes the use of a nanoparticles platform for intracellular diagnostics and targeted drug delivery. Specifically discussed are (a) the PEBBLE nanosensors approach for monitoring intracellular activities and (b) the nanoclinic approach with carrier groups to target specific biological sites for diagnostics and external activation of therapy.

Chapter 16 describes the application of biomaterials to photonics-based information technology, which utilizes light–matter interactions for information processing, transmission, data storage, and display. The continued development of photonics technology is crucially dependent on the availability of suitable optical materials. Biomaterials are emerging as an important class of materials for a variety of photonics applications. The various types of biomaterials being investigated for photonics are presented. Examples of photonics applications discussed in this chapter include efficient harvesting of solar energy, low-threshold lasing, high-density data storage, and efficient optical switching and filtering.

The author hopes that this book will inspire new ideas and stimulate new directions in biophotonics.

CHAPTER 2

Fundamentals of Light and Matter

An understanding of properties of light and matter forms the very fundamental basis to create an insight into the nature of interactions between light and biological systems. This chapter discusses the fundamentals of light and matter at a very basic level, avoiding mathematical details and emphasizing concepts. For those readers having little exposure to the subject, the materials of this chapter will assist them in grasping the important concepts. However, it is not crucial that they understand all the details provided in this chapter. For those readers familiar with the subject, the chapter will serve as a condensed review.

Section 2.1 covers the fundamentals of light. The emphasis is on introducing basic concepts: (a) Light as photons carries energy, and (b) Light as waves exhibits properties such as interference and diffraction. Light propagation in a medium is dependent on its optical characteristics. The biological applications of spectroscopy (Chapter 4) and fluorescence microscopy (Chapter 7) utilize the photons while the interference feature of the wave is used in a number of biophotonics applications such as phase-contrast microscopy and optical coherence tomography (Chapter 7) as well as in biosensing (Chapter 9). For further basic discussions of light and optics, a good reference is Feynman et al. (1963). A rigorous text on the subject is by Born and Wolf (1965).

Section 2.2 covers the fundamentals of matter. The emphasis of this section is on introducing the concept that the energies of electrons in atoms and molecules have only certain permissible discrete values, a condition called *quantization of energies*. This feature is derived from the wave-like behavior of matter. In the case of molecules, the total energy can be divided into four parts: electronic, vibrational, rotational, and translational. Only electronic, vibrational, and rotational energies are quantized (have discrete values). Of these, only electronic and vibrational energy levels are of significance to biophotonics, because they are an integral part of spectroscopy (Chapter 4), bioimaging (Chapter 7 and Chapter 8), biosensing (Chapter 9), and flow cytometry (Chapter 11).

Introduction to Biophotonics, by Paras N. Prasad
ISBN: 0-471-28770-9   Copyright © 2003 John Wiley & Sons, Inc.

These concepts are also useful for photodynamic therapy (Chapter 12) and biomaterials for photonics (Chapter 16). Use of a mathematical relation called the *Schrödinger equation* is qualitatively demonstrated here to obtain the energy levels and to provide a probabilistic description of electrons in atoms and molecules. The flow sheets shown in Tables 2.5 and 2.6 summarize the approaches used to determine the energy levels of atoms and molecules. Readers less mathematically oriented may skip these tables, because a full understanding of the contents discussed there is not crucial for understanding subsequent chapters.

The σ and π bonding in organic molecules and π-electron delocalization effect in conjugated structures are important in understanding spectroscopy and fluorescence behavior of fluorophores used in bioimaging and biosensing.

The final section of this chapter deals with imtermolecular effects on the energy levels of matter. The concepts developed here are important in showing how functions of a biological organization can be probed using effect of intermolecular interaction on its energy levels.

2.1   NATURE OF LIGHT

2.1.1   Dual Character of Light

The description of the nature of light provided here is at the very basic level. It serves to review the concepts that most readers already may be familiar with. Light is an electromagnetic field consisting of oscillating electric and magnetic disturbances that can propagate as a wave through a vacuum as well as through a medium. However, modern theory, quantum mechanics, also imparts a particle-like description of light as energy packets called *photons* or *quanta* (Atkins and dePaula, 2002). This dual-character description of light, which is also shared by matter as described in Section 2.2, is represented in Table 2.1. In Table 2.1, the symbol c represents the speed of electromagnetic waves, more commonly called the speed of light in a vacuum. All electromagnetic waves travel with the same speed in a vacuum. In a medium such as a glass or a biological material, the speed of an electromagnetic wave, often labeled as v, is different. The ratio of the two speeds c and v is called the refractive index, n, of the medium. In other words,

$$\frac{c}{v} = \frac{\text{speed of light in a vacuum}}{\text{speed of light in a medium}} = n \quad \text{or} \quad v = \frac{c}{n} \tag{2.1}$$

Therefore, n can be viewed as the resistance offered by the medium toward the propagation of light. The higher the refractive index, the lower the speed.

The electromagnetic spectrum is defined by the spread (distribution) of a series of electromagnetic waves as a function of the wavelength, frequency, or

TABLE 2.1. Dual Character of Light

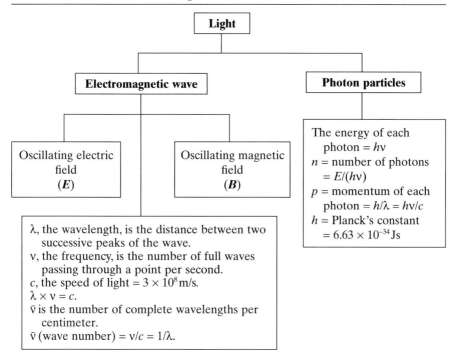

TABLE 2.2. Different Spectral Regions of Light

Region	Far-IR	Mid-IR	Near-IR	Visible	UV	Vacuum UV
Wavelength (nm)	$5000–10^6$	2500–5000	700–2500	400–700	200–400	100–200
Wave number (cm^{-1})	200–10	4000–200	$1.4 \times 10^4–4000$	$2.5 \times 10^4–1.4 \times 10^4$	$5 \times 10^4–2.5 \times 10^4$	$10^5–5 \times 10^4$

wave number. Different wavelength regions often carry a specific label such as radio frequency at the longer end and cosmic rays at the shorter wavelength end of the spectrum. The range from far-infrared (IR) to vacuum ultraviolet (UV) is called the *optic wave region*, while the common usage of the term "light" is more restrictive, often implying only the visible region. Figure 2.1 defines the wavelength of a wave. Table 2.2 describes the different spectral regions of optical waves characterized by their wavelength (in the units of nm, $1\,nm = 10^{-9}\,m = 10^{-3}\,\mu m$) and their wave number (in cm^{-1}).

Depending on the optical regions, different units are used to characterize the wave. For the visible region, the common practice is to use the nm [or

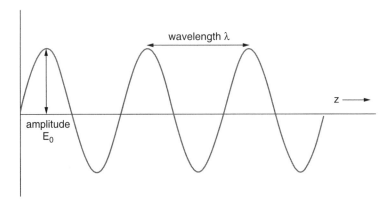

Figure 2.1. Schematic of a wave defining its wavelength.

angstrom $(\text{Å}) = 10^{-1}\,\text{nm}$] unit of wavelength or cm^{-1} unit of wave number. For the near-IR to mid-IR region, one often uses the wavelength in micrometers or microns (μm). From the mid-IR to far-IR region, one uses the wave number in cm^{-1} to characterize a wave.

2.1.2   Propagation of Light as Waves

Most of the interactions between light and molecules of biological interest are electrical in nature. Therefore, the description of a light wave focuses on the nature of the oscillating electric field E, which has both a direction and an amplitude (the value corresponding to maxima and minima of the wave). The direction of the electric field, E, for a plane wave traveling in one direction is always perpendicular both to the direction of propagation and to the oscillating magnetic field, B. However, it can be linearly polarized, when the electric field at each point is in the same direction, as shown in Figure 2.2. When the electric field is distributed equally in a plane perpendicular to the direction of propagation, it is called circularly polarized, as shown in Figure 2.3.

The propagation of light in the z direction with its oscillating electric field $E(z, t)$ is described mathematically as

$$E(z,\ t) = E_0 \cos(\omega t - kz) \tag{2.2}$$

$$k^2 = \frac{\varepsilon\omega^2}{c^2} \tag{2.3}$$

and with E_0 defining the electric field amplitude of the field (Prasad and Williams, 1991). The term ω is the angular frequency of light given as $2\pi\nu$; k, called the *propagation vector*, is defined as

$$k = \frac{2\pi}{\lambda} \tag{2.4}$$

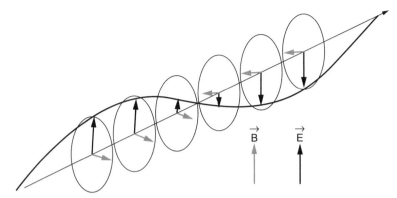

Figure 2.2. Propagation of an *x*-linearly polarized light propagating along the *z*-direction.

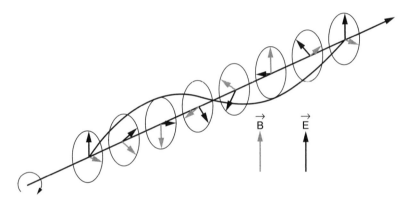

Figure 2.3. Propagation of a circularly polarized light.

It characterizes the phase of the optical wave with respect to a reference point ($z = 0$); thus, kz describes the relative phase shift with respect to the reference point. As an illustration of the phase property, Figure 2.4 represents two waves shifted in phase.

The term ε in equation (2.3) is called the *dielectric constant*, which for optical waves is n^2, with n being the refractive index of the medium. The speed of an optical wave (light) is described by the propagation of waves in a medium. This propagation is characterized by two velocities:

- *Phase velocity*, which describes the travel of a phase front (i.e., displacement of the peak of a wave) of a single wave. The phase velocity is what was defined above as the speed of an electromagnetic wave through a medium.

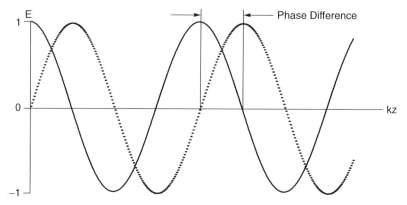

Figure 2.4. Schematics of two waves shifted in phase by an amount labeled in the figure as "phase difference."

- *Group velocity*, which describes the propagation of a wave packet consisting of many waves traveling together.

For a medium with refractive index n, as described above, the phase velocity of a wave is given by

$$v = \frac{c}{n} \tag{2.5}$$

In general, a material as an optical propagation medium shows a dispersion (change) of refractive index as a function of wavelength. The normal dispersion behavior shows an increase of refractive index, n, with a decrease in wavelength. Equation (2.5), therefore, predicts that the phase velocity will increase with an increase in wavelength. In other words, red light will travel faster than blue light. The group velocity of a package of waves behaves similarly. This spread in group velocity for different wavelengths is known as the group velocity dispersion effect. Thus, a short burst of light (such as a laser pulse, discussed in Chapter 5) traveling through a medium such as a fiber broadens due to the group velocity dispersion because the blue end of light pulse spectrum lags in time compared to the red end.

In terms of the wave picture, the energy of an electromagnetic wave is the sum of the electrical and magnetic contributions. The intensity, I, of an electromagnetic wave is the power per unit area carried by the wave and is proportional to the square of the electric field amplitude E_0. In the CGS units, I at an angular frequency ω is given as

$$I(\omega) = (E_0)^2 \frac{cn}{8\pi} \tag{2.6}$$

2.1.3 Coherence of Light

Coherence of light defines the collective wave properties of optical waves produced by a light source. It describes the phase relation among different waves. If a constant phase relation is maintained, the light beam is called *coherent*. If the phase relation is totally random, the light source is labeled as *incoherent*. However, a source may also be partially coherent. A more rigorous and quantitative description of coherence involves the concept of coherence length, the length scale over which a relative phase relation is maintained. The coherence property of a light source determines the divergence properties of a light beam.

The two features defining the complete coherence properties of light are temporal coherence and spatial coherence:

- *Temporal Coherence*. This coherence property is defined by the frequency spread of a wave packet. If all of the waves emanating from a light source are of the same frequency or are in a very narrow range, they possess temporal coherence. The light is then called *monochromatic*. If there is a large spread of frequencies (resulting in color spread), the light is *polychromatic* and does not have temporal coherence.
- *Spatial Coherence*. The second coherence property is defined by the spatial relationship between the phases of different waves emanating from a light source. If a constant phase relation exists and is maintained over the propagation of the wave packet in space, such as in the case of light from a laser source, the light beam is then spatially coherent.

The two phenomena that utilize these coherence properties are interference and diffraction:

- *Interference*. Interference is produced when two light waves are combined. If they are in phase (i.e., the crest and trough of one wave line up with the crest and trough of another wave), as shown in Figure 2.5, they constructively interfere with each other. This leads to an increase in the amplitude and, thus, the brightness of the light. On the other hand, if the two waves are out of phase (i.e., the crests and troughs do not line up), as shown in Figure 2.5, their amplitudes subtract due to destructive interference. This leads to a significant decrease or cancellation of the brightness.
- *Diffraction*. Another important manifestation of the wave nature of light is the phenomenon of diffraction. Diffraction, in general, can be defined as the natural tendency of any wave that is not infinite to spread in space. To understand the reason for such behavior, one often quotes the Huygens principle, which states that any point within the wave may be considered a source of a spherical wave. Thus, the propagation can be

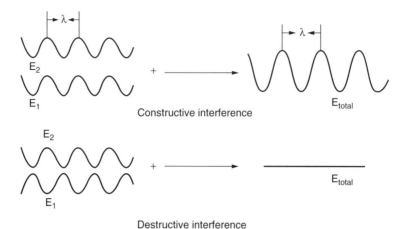

Figure 2.5. Examples of constructive and destructive interference.

seen as the result of summation of all these elementary waves. While diffraction is omnipresent, its practical implications are most obvious when the spatial scale of phenomena is that of the light wavelength. Thus, diffraction gives rise to the bending and spreading of light whenever it encounters sharp obstacles, passes through slits or apertures, or is focused on small spots approximately the size of the light's wavelength.

The net effect of diffraction can be visualized by considering that each point in the opening or slit acts as a source of spherical waves. These spherical waves from all points interfere with each other. At certain angles, constructive interference occurs, resulting in bright spots. At other angles, destructive interference can be found, resulting in dark spots. This pattern of light and dark spots is called the *diffraction pattern*. Figure 2.6 illustrates the interference caused by light passing through two slits, producing a diffraction pattern of bright and dark fringes. According to the Huygens–Fresnel principle, every unobstructed point of a wave front, at a given instant, serves as a source of spherical secondary wavelets (with the same frequency as that of the primary wave). Each slit acts as a source of secondary wavelets; the waves from the two slits interfere constructively or destructively at certain angles, giving rise to modulation of the intensity pattern on the screen (bright and dark fringes).

Diffraction grating in an optical device is often used to separate different wavelengths in a beam of light. It relies on diffraction from thousands of narrow, closely spaced, parallel slits or grooves. The diffracted light from these different slits interacts in such a way that it produces in-phase or constructive interference at certain angles, resulting in light of maximum brightness (intensity) at these angles. These angles are also dependent on the wavelength. The condition for maximum intensity, also known as *Bragg diffraction*, is given as

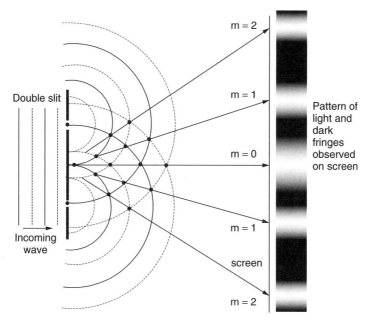

Figure 2.6. Diffraction pattern of light passing through two slits.

$$m\lambda = d\sin\theta, \qquad m = 0, \pm 1, \pm 2, \pm 3, \text{ etc.}$$

where m is the order of diffraction and θ is the angle from the center straight line. The condition $m = 0$ simply represents the central bright spot. The first-order diffraction, $m = 1$, corresponds simply to $\lambda = d\sin\theta$.

Diffraction is also responsible for the spreading of Gaussian beams, such as those usually obtained from laser sources, and for the limitation of the smallest spot size obtainable by focusing such beams (diffraction-limited spot size). Diffraction on periodic structures, slits, reflective surfaces, and two- or three-dimensional patterns of varying absorptive and refractive properties of a medium, together with the accompanying interference of the resulting wave fronts, produces a multitude of phenomena that are in general wavelength-dependent. Therefore, diffraction gratings can be used to separate light of different colors. Photonic crystal structures—a type of three-dimensional (3-D) periodic dielectric structure with the periodicities on the scale of a wavelength—can be used to selectively trap light of chosen characteristics. Photonic crystal structures are discussed in Chapters 9 and 16.

2.1.4 Light as Photon Particles

The wave picture previously described does not adequately explain the way in which light energy is absorbed or scattered. The interaction of light with particles (such as electrons) of matter involves the exchange of energy as well as momentum. These processes can only be described by assuming that the light also behaves like particles called *photons*. As described in Table 2.1, a photon for a light of a specific frequency ν has a discrete, fixed energy of value hν, where h is a constant (called Planck's constant) having a magnitude of 6.63×10^{-34} J sec. Thus, the energy of an electromagnetic wave is quantized (discrete) and is not continuously variable. The smallest energy of an electromagnetic wave is equal to that of a photon. The total energy E is equal to Nhν, where N is the number of photons, resulting in the equation

$$E = Nh\nu \quad \text{or} \quad N = \text{number of photons} = E/h\nu$$

The quantized energy aspect of a photon is used in the description of absorption, emission, or scattering of light by matter, as discussed in Chapter 4.

Photons as particles also carry momentum (a physical quantity described by the product of the mass and the velocity of the particle). As described in Table 2.1, the momentum, p, of a photon is given as

$$p = h/\lambda = h\nu/c$$

The momentum aspect of a photon comes into play when a photon changes its direction of propagation while scattered by another particle or when it is refracted at the surface of a medium. This change of direction of photon propagation creates a change in momentum and can produce a force to trap a particle. This principle is used for optical trapping of biological cells and forms the basis for the operation of optical tweezers, which is covered in Chapter 14.

2.1.5 Optical Activity and Birefringence

The propagation characteristics of light in certain media are dependent on the polarization of incident light. The two relevant effects, which are not related to each other, are optical activity and birefringence.

Optical Activity. The optical activity often relates to certain types of asymmetric molecular structures such as one containing a carbon atom chemically bonded to four different atoms (or groups). A quantum mechanical description of bonding is detailed in Section 2.2.3 on quantized states of molecules. Here, however, a qualitative description should suffice. An asymmetric carbon atom bonded to four different atoms (or groups) is called a *chiral center*. A chiral center exhibits optical activity such as optical rotation in which the plane of polarization is rotated when a linearly polarized light is passed through a medium containing chiral centers (e.g., chiral carbon atom). Chiral structures are further discussed in Section 2.4. These chiral media also show differences in interactions with left and right circularly polarized light, a phenomenon called *circular dichroism*, which is discussed in Chapter 4.

TABLE 2.3. Different Light Sources and Their Characteristics

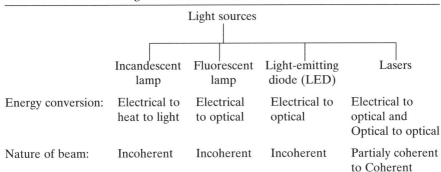

	Incandescent lamp	Fluorescent lamp	Light-emitting diode (LED)	Lasers
Energy conversion:	Electrical to heat to light	Electrical to optical	Electrical to optical	Electrical to optical and Optical to optical
Nature of beam:	Incoherent	Incoherent	Incoherent	Partialy coherent to Coherent

Birefringence. In isotropic media, such as an amorphous material or a liquid, all directions are equivalent. In other words, there is no preferred direction in which molecules are aligned. Such isotropic media exhibit only one refractive index, which is independent of the polarization direction of a linearly polarized light. Therefore, light propagation is independent of the direction in the medium. Anisotropic media are fully ordered or partially ordered materials such as liquid crystals or stretch-oriented polymers in which all directions are not equivalent. An anisotropic medium exhibits different values of refractive index for different linear polarization of incident light. This phenomenon is called *birefringence*. It reflects molecular alignments produced either mechanically (like a stretching of a polymer), by flowing, or by an electric field. Liquid crystals exhibit strong birefringence due to their molecular alignments. Because the velocity of a wave is dependent on its refractive index, light impinging on a birefringent medium is split into normal and extraordinary rays and can propagate along different trajectories at different speeds, due to differences in refractive index.

2.1.6 Different Light Sources

The different light sources available are described in Table 2.3 (Smith, 1989).

2.2 QUANTIZED STATES OF MATTER

2.2.1 Introductory Concepts

The past century produced a major breakthrough in our understanding of the structure of matter. This breakthrough, which was a culmination of the results of many pioneering experiments begun in the early 20th century, comprised the following revolutionary concepts:

TABLE 2.4. Dual Nature of Matter

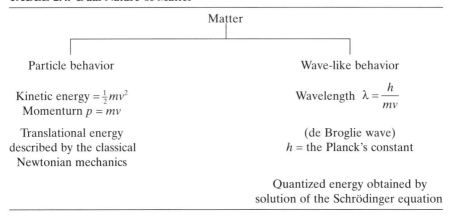

	Matter	
Particle behavior		**Wave-like behavior**
Kinetic energy $= \frac{1}{2}mv^2$		Wavelength $\lambda = \dfrac{h}{mv}$
Momenturn $p = mv$		
Translational energy described by the classical Newtonian mechanics		(de Broglie wave) h = the Planck's constant
		Quantized energy obtained by solution of the Schrödinger equation

- An atom—which was previously considered indivisible—was shown to have a subatomic structure consisting of electrons surrounding a core nucleus.
- The internal energy of matter (an atom, molecule, or molecular assembly) was found to be of discrete values or quantized.
- Matter was shown to possess dual characters of both particles and waves. Table 2.4 describes this dual behavior.
- Heisenberg showed that it was impossible to measure with complete precision both the position and the velocity of an electron (or a particle) at the same instant. This fundamental limitation is known as the *Heisenberg uncertainty principle*. A probabilistic description of the behavior of an electron was therefore necessary.

The probabilistic (statistical) description utilizes the concept of a wave associated with a particle, which leads to quantization of only discrete permissible values of energy states of a matter. The Schrödinger equation is a starting point for obtaining information about the permissible quantized energy states of matter (Atkins and dePaula, 2002; Levine, 2000). It is a second-order differential equation obtained originally from the modification of a wave equation. In its simplest form for a one-dimensional motion, say in x direction, of a particle (whether an electron or a nucleus) the equation is given as

$$\frac{-h^2}{8\pi^2 m}\frac{d^2\psi}{dx^2}+(V-E)\psi = 0 \tag{2.7}$$

In the above equation, h is the Planck's constant and m is the mass of the particle. The wave function $\psi(x)$ is obtained as a mathematical solution of this equation. It is a function of x and relates to the amplitude of the wave. The

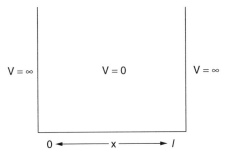

Figure 2.7. Schematics of a particle in a one-dimensional box.

quantity $\psi^2(x)dx$ describes the probability of finding the particle in the length segment between x and $x + dx$. In the above equation, V is the potential energy and E represents the allowed quantized energy (also called the eigenvalue). The energy E is the total energy of a particle, consisting of both the potential energy and the kinetic energy. The allowed discrete values of E obtained from the solution of equation (2.7) represent the energy levels of the particle. The lowest energy level is called the *ground state*.

The second-order differential equation represented in the above equation is solved by defining the potential energy dependence on coordinate x and imposing the boundary conditions applicable to the physical situation of the particle. With these two variables defined, one obtains E and ψ. A similar equation involving three Cartesian coordinates defines the three-dimensional behavior.

A simple model system that illustrates the use of the Schrödinger equation is that of a particle in a one-dimensional box, as depicted in Figure 2.7. The particle is trapped (confined) in a box of length l within which the potential energy is zero. The potential energy rises to infinity at the end of the box and stays at infinity outside the box. This simple model can describe many physical situations such as binding (confinement) of an electron in an atom, the formation of a chemical bond, or the delocalization of electrons (free movement of an electron) over a chain of atoms.

Inside the box, the equation is solved with the following conditions:

$$V(x) = 0$$
$$\psi(x) = 0 \qquad \text{at } x = 0 \text{ and } x = l$$

The solution yields sets of permittable values of E and the corresponding functions for ψ, each defining a given energy state of the particle and labeled by a quantum number n which takes an integral value starting from 1 (Atkins and dePaula, 2002). These values are defined as

$$E_n = \frac{n^2 h^2}{8ml^2} \qquad \psi_n(x) = \left(\frac{2}{l}\right)^{1/2} \sin\left(\frac{n\pi x}{l}\right) \tag{2.8}$$

The lowest value of total energy is $E_1 = h^2/8ml^2$. Therefore, total energy E of a particle can never be zero energy when bound (or confined), even though its potential energy is zero. The discrete energy values are E_1, E_2, and so on, corresponding to quantum numbers $n = 1, 2, 3$, etc. These values represent the various permissible energy levels of a particle trapped in a one-dimensional box. The gap between two successive levels describes the effect of quantization (discreteness). If it were zero, we would have a continuous variation of the energy and there would be no quantization.

The gap ΔE between two successive levels E_n and E_{n+1} can be given as

$$\Delta E = (2n + 1)\frac{h^2}{8ml^2} \qquad (2.9)$$

This equation reveals that the gap between two successive levels decreases as l^2 when the length of the box increases. Translational energies of atoms and molecules, which involve displacement over a large distance compared to the atomic scale, will have very small spacing and can be considered not to be quantized—that is, treatable by classical mechanics. This model also explains that when a bond is formed between two atoms, the length in which the bonding electrons are confined increases (spreads over two atoms). Consequently, the energy E_n is lowered, stabilizing the formation of the bond, because a lower energy configuration is always preferred. Furthermore, the spacing between successive electronic levels also decreases as the electron is spread (delocalized) over more atoms, as in the case of the π electrons in a conjugated structure. This topic of conjugated structures is discussed in Section 2.2.6.

2.2.2 Quantized States of Atoms

The application of Schrödinger's equation to the case of an atom allows one to obtain permissible energy levels of electrons moving around the nucleus of the atom. These are the various quantized electronic energy levels of an atom. For a specific energy level that the electron occupies (in other words, an electron possessing a specific permissible energy value), the probability of finding the electron in the space around the nucleus is described by a wave function that defines an orbital. An orbital can be visualized as the region of space where the probability of finding an electron is high. The Schrödinger equation for a hydrogen atom can be solved mathematically to arrive at an exact solution. However, such a mathematical solution is not possible for a many-electron atom, due to the presence of repulsion between two electrons simultaneously moving around the nucleus. Hence, approximations such as that of self-consistent field theory are introduced. These mathematical descriptions, however, are outside the scope of this book. For such details, refer to books by Atkins and dePaula (2002) or Levine (2000). Here we only provide a flow sheet qualitatively listing the various steps to obtain the quantized energy

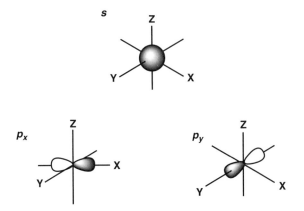

Figure 2.8. Shapes of the s and p atomic orbitals, p_z not shown.

levels and the orbitals for an atom using the Schrödinger equation. These steps are listed in Table 2.5. The orbitals with modified energies, as shown in Figure 2.8, are used to derive electronic distribution (electronic configuration) in many-electron atoms using the following guiding principles:

- Aufbau (a German word meaning "building up") principle, which says the electrons fill in the orbitals of successively increasing energy, starting with the lowest-energy orbital.
- Pauli's principle, which says that each orbital can accommodate a maximum of two electrons, provided that their spins are of opposite signs (i.e., they are paired).
- Hund's rule, which says that if more than one orbital has the same energy (this is called degeneracy, as it exists for the three types of p orbitals), electrons are to be filled singly in each orbital before pairing them up.

Some examples of electronic configurations are He $1s^2$ and Li $1s^2 2s^1$. Here the superscript at the top of an orbital designation represents the number of electrons in the orbital. A single electron in any of these orbitals is characterized by four quantum numbers: n, l, and m_l—derived from the orbital it is in— and m_s from its spin orientation (+1/2 for up and −1/2 for down spin, in its simplest description). The quantum number l represents its orbital angular momentum, with m_l defining the direction of the angular momentum.

For many electrons, the correlation effects (interactions between electrons) produce overall angular momentum quantum numbers L and M_L. Similarly, the coupling of their spins produces an overall quantum number S. Another important manifestation is the spin–orbit coupling, which can be viewed as resulting from the magnetic interaction between the magnetic moment due to the spin of the electron and the magnetic field produced by the electron's

TABLE 2.5. Schematics of Quantum Mechanical Approach for an Atom

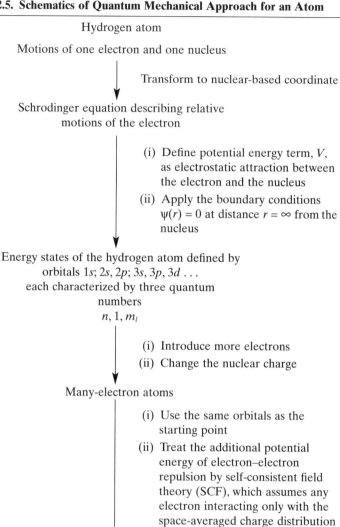

Hydrogen atom

Motions of one electron and one nucleus

Transform to nuclear-based coordinate

Schrodinger equation describing relative
motions of the electron

(i) Define potential energy term, V,
as electrostatic attraction between
the electron and the nucleus

(ii) Apply the boundary conditions
$\psi(r) = 0$ at distance $r = \infty$ from the
nucleus

Energy states of the hydrogen atom defined by
orbitals $1s$; $2s$, $2p$; $3s$, $3p$, $3d$ ...
each characterized by three quantum
numbers
$n, 1, m_l$

(i) Introduce more electrons

(ii) Change the nuclear charge

Many-electron atoms

(i) Use the same orbitals as the
starting point

(ii) Treat the additional potential
energy of electron–electron
repulsion by self-consistent field
theory (SCF), which assumes any
electron interacting only with the
space-averaged charge distribution
due to all other electrons

$1s$; $2s$, $2p$; $3s$, $3p$, $3d$ ... orbitals with modified energies

orbital motion around the nucleus (just like a current in a coil produces a magnetic field). This spin–orbit coupling is dependent on the atomic number (the charge on the nucleus). Therefore, heavier atoms exhibit strong spin–orbit coupling, often called the *heavy atom effect*, which leads to a strong mixing of the spin and the orbital properties. The mixing leads to characterization of overall angular momentum by another quantum number J. The spin–orbit coupling plays an important role in spectroscopy, as discussed in Chapter 4.

The electron orbital and spin correlation effects and the spin–orbit coupling produce shifting and splitting of the atomic energy levels which are characterized by a term symbol (Atkins and dePaula, 2002; Levine, 2000). A term symbol is given as $^{2S+1}\{L\}_J$. Here $2S + 1$, with S as the overall spin quantum number, represents the spin multiplicity. When $S = 0$, $2S + 1 = 1$, it represents a singlet state. The $S = 1$, $2S + 1 = 3$ case represents a triplet state. $\{L\}$ is the appropriate letter form representing the L value, which represents the total orbital angular momentum; for example, 0, 1, 2, 3 values are represented by S, P, D, F. Thus, a term symbol $^1S_0$ for the atom represents an energy level with spin $S = 0$, orbital quantum number $L = 0$, and the overall angular momentum quantum number $J = 0$. These term symbols are often used to designate energy states of an atom and the spectroscopic transitions between them. They are used to designate the transition between two quantized states of an atom or an ion in the production of a laser action, as described in Chapter 5.

2.2.3 Quantized States of Molecules: Partitioning of Molecular Energies

Because a molecule contains more than one nucleus, it represents another level of complexity since one cannot simply choose the origin of electronic displacements (due to electronic motion) with respect to a single nucleus. Motions of other nuclei relative to any nucleus chosen as the reference point have to be considered. The molecular Schrödinger equation, therefore, includes both electronic and nuclear motions, as well as potential energies derived from electron–electron repulsion, electron–nuclear attraction, and nuclear–nuclear repulsion. A wave function, being a solution of the molecular Schrödinger equation, therefore, depends on both the positions of electrons, collectively labeled as r, and the position of nuclei, collectively labeled as R. Since all electrons and nuclei are moving, r and R are continuously changing. No exact mathematical solution can be found for a Schrödinger equation involving such multidimensional coordinates defined by the set of values for r and R. A major breakthrough was produced in 1930 by a simple approximation introduced by Born and Oppenheimer, who postulated that the molecular Schrödinger equation can be partitioned into two parts (Levine, 2000):

- One part describes the fast motions of electrons, which move in a slow varying electrostatic field due to nuclei.
- The other part describes the slow motions of nuclei, which experience an average potential field due to fast moving electrons.

Further decomposition of the nuclear Schrödinger equation is made possible by transformation of the coordinate system to that based inside the molecule (Graybeal, 1988). The total partitioning scheme is shown in Table 2.6. Electronic energy-state spacings are of the largest magnitude, covering the

TABLE 2.6. Partitioning of Molecular Schrödinger Equation

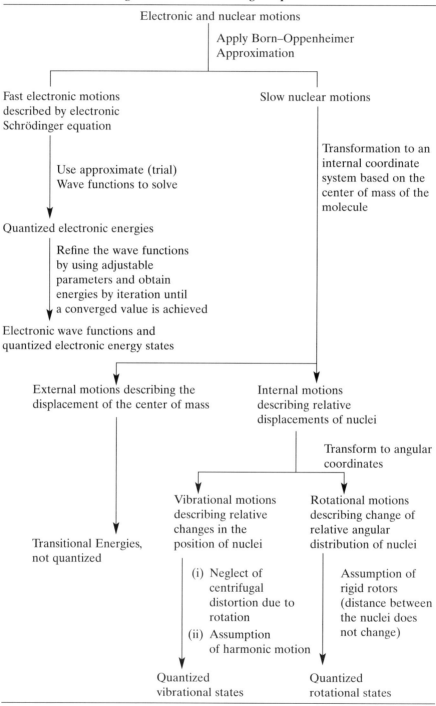

Electronic and nuclear motions

Apply Born–Oppenheimer Approximation

Fast electronic motions described by electronic Schrödinger equation

Slow nuclear motions

Use approximate (trial) Wave functions to solve

Transformation to an internal coordinate system based on the center of mass of the molecule

Quantized electronic energies

Refine the wave functions by using adjustable parameters and obtain energies by iteration until a converged value is achieved

Electronic wave functions and quantized electronic energy states

External motions describing the displacement of the center of mass

Internal motions describing relative displacements of nuclei

Transform to angular coordinates

Vibrational motions describing relative changes in the position of nuclei

Rotational motions describing change of relative angular distribution of nuclei

Transitional Energies, not quantized

(i) Neglect of centrifugal distortion due to rotation

(ii) Assumption of harmonic motion

Assumption of rigid rotors (distance between the nuclei does not change)

Quantized vibrational states

Quantized rotational states

spectral range from deep UV to near IR. Vibrational spacing lies in the mid-IR to far-IR range. The rotational spacing is in the microwave frequency range. Rotational energy states provide important structural information only on small molecules in gaseous phases where intermolecular interactions are weak. In a liquid or solid phase their nature is completely different, becoming hindered rotations (often called vibrations). Therefore, they are of very little practical value for photobiology and thus will not be discussed further here. The following subsections provide further description of quantum mechanical approaches used to elucidate the electronic and vibrational states of a molecule.

2.2.4 Electronic States of a Molecule

The electronic states of a molecule are obtained as a solution of the electronic Schrödinger equation. The electronic wave function ψ_e are functions of the electronic coordinates r and nuclear coordinates R; however, R are treated as variable parameters. In other words, the electronic Schrödinger equation is solved for each set of nuclear coordinates R (assuming them to be clamped at specific values, hence an often-used term: clamped nuclei approximation). Even with this approximation, an exact solution is not derivable because of the various coupled interactions as described in Figure 2.9 using the example of a hydrogen molecule.

The commonly used molecular orbital (MO) method utilizes a similar approach as adopted for the case of atoms. A solution of the electronic Schrödinger equation is obtained first for a one-electron molecule, H_2^+. The one-electron molecular orbitals, analogous to the one-electron atomic orbitals in the case of atoms, are then modified by the self-consistent field method to include electron–electron repulsion. These molecular orbitals are then used with the same guiding principles (e.g., Aufbau principle, Pauli's exclusion principle, and Hund's rule) adopted for many-electron atoms to derive the electronic configuration of many-electron molecules.

The first step in solving the electronic Schrödinger equation for H_2^+ is to use a trial wave function, which is formed by a linear combination of atomic

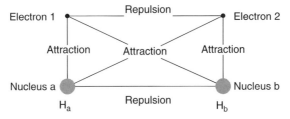

Figure 2.9. The representation of various interactions in a hydrogen molecule involving the bonding of two atoms H_a and H_b. The nuclei a and b are represented by larger circles; the electrons 1 and 2 are represented by small circles.

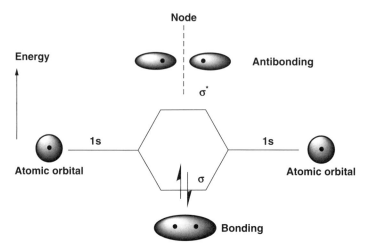

Figure 2.10. Overlap of atomic orbitals to form molecular orbitals in H_2^+.

orbitals. This method is called LCAO-MO (Levine, 2000). For example, for the H_2^+ molecule the lowest-energy atomic orbitals of the constituent hydrogen atoms are $1s$. The lowest set of molecular orbitals 1σ and $2\sigma^*$ (sometimes labeled as $\sigma 1s$ and $\sigma^* 1s$) are formed by two possible linear combinations of $1s$ atomic orbitals of each atom, which describe two different possible modes of overlap of their wave functions. This approach is shown in Figure 2.10.

The plus linear combination describing 1σ involves a constructive overlap (similar to constructive interference of light waves) of the electronic wave functions of the individual atoms leading to an increased electron density between the two nuclei. This increased electron density between the two nuclei acts as a spring to bind them and overcome the nuclear–nuclear repulsion. The energy of the resulting molecular orbital 1σ is lowered (stabilized) compared to that of the individual atomic orbitals $1s_A$ and $1s_B$. This is called a *bonding molecular orbital*. Contrasting this is the minus combination, which leads to a destructive overlap of the wave functions. This results in cancellation of the electron density in the region between the nuclei. The energy of the resulting molecular orbital $2\sigma^*$ is raised compared to that of the constituent atomic orbitals $1s$. The $2\sigma^*$ is, therefore, an antibonding molecular orbital. The star symbol as a superscript on the right-hand side represents an antibonding orbital. The symbol σ represents the overlap of atomic orbitals along the internuclear axis. In general, the mixing of two atomic orbitals centered on two atoms produces two molecular orbitals. This principle can be used to form molecular orbitals of high erenergies derived from the mixing of higher-energy atomic orbitals ($2s$, $2p$; $3s$, $3p$, $3d$, etc.) centered on individual atoms involved in a bond.

In the case of directional orbitals such as p, there are three p orbitals—p_x, p_y, and p_z—directed toward the x, y, and z axes, respectively. If z is taken as

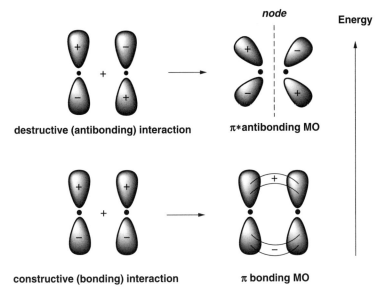

Figure 2.11. Schematics of π and π^* molecular orbital formation by the overlap of two p-type atomic orbitals.

the internuclear axis, then only the p_z orbital can overlap along the internuclear direction and thus form the σ and σ^* molecular orbitals. The p_x and p_y orbitals can then overlap only laterally (in directions perpendicular to the internuclear axis), forming π and π^* molecular orbitals. The formation of the π and π^* orbitals by overlap of two atomic orbitals is schematically shown in Figure 2.11. These molecular orbitals are not as stable as the σ orbitals and are involved in bonding only when a multiple bond is formed between two atoms. The occupation of a given bonding molecular orbital by two electrons in a spin-paired configuration defines the formation of a bond. The occupation of an σ orbital by a pair of electrons defines a single σ bond; the electrons involved are called σ electrons. The occupation of a π orbital by a pair of electrons defines a π bond; the electrons involved are called π electrons. The various σ bonds are shown in Figure 2.12.

Another feature to point out for the bonding and antibonding orbitals is that the energies of these orbitals, obtained as a function of internuclear separation (using the clamped nuclei approximation), behave differently for a σ and σ^* orbital in the H_2^+ molecule. This is shown in Figure 2.13.

The bonding orbital energy E_σ exhibits a minimum corresponding to a bound (stable) state at the internuclear separation R_e, called the equilibrium bond length (Levine, 2000). The amount of the energy lowering, D_e, with respect to the unbound configuration ($R = \infty$), is called the *equilibrium binding energy* or *equilibrium dissociation energy*, which is needed to dissociate the molecule into its constituent atoms. By contrast, the antibonding orbital

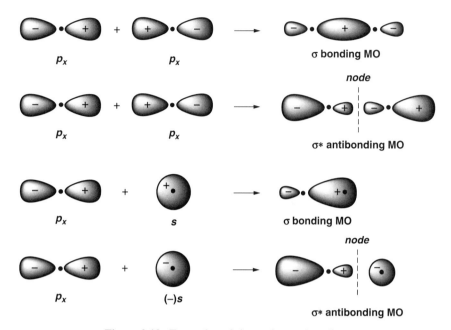

Figure 2.12. Examples of the various σ bonding.

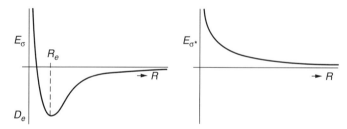

Figure 2.13. The energies of the bonding and the antibonding orbitals in H_2^+ obtained as a function of the internuclear separation R.

energy E_σ^* shows no binding since the energy monotonically increases as the internuclear separation R is decreased. It represents a dissociate state, as opposed to a bound state for the σ orbital.

The electronic configurations of a many-electron diatomic molecule are derived by successively filling electrons in molecular orbitals of increasing energy (the Aufbau principle) while observing Pauli's exclusion principle and Hund's rule. One of the greatest triumphs of the MO theory is its prediction of two unpaired electrons for the lowest-energy (ground-state) configuration of the oxygen molecule, with the overall spin $S = 1$ and thus the spin multiplicity $2S + 1 = 3$ (a triplet state). Because a net spin (nonzero spin) gives rise to paramagnetism, the MO theory was thus successful in explaining the observed paramagnetism of the O_2 molecule.

The behavior described above applies to a homonuclear diatomic molecule for which the same type of atomic orbitals on two binding atoms combine to form molecular orbitals and the coefficient of mixing between the atomic orbitals on each of the binding atoms is the same. This is the case of a true covalent bond where the electronic probability distribution (electron density) on each atom is the same. In the case of a heteronuclear diatomic molecule, such as HF, the following considerations hold true (Levine, 2000):

- The LCAO-MO method now involves mixing of atomic orbitals, which are energetically similar. For example, the $1s$ atomic orbital of H combines with the $2p_z$ orbital on F to form the σ and σ^* molecular orbitals for HF (see Figure 2.12).
- In a more accurate description of bonding, more than one atomic orbital on one atom may be involved in forming a linear combination. The mixing coefficient is determined by the variation principle, which imposes the condition that the energy be minimized with respect to the mixing coefficients (which are treated as adjustable parameters).
- Certain atomic orbitals (forming the inner core of an atom) do not significantly mix in bond formation. As a result, the energies of these atomic orbitals in the molecule are basically the same as that in the unbound atomic state. These are called nonbonding molecular orbitals or n orbitals. In the case of HF, the inner $1s$ orbital of F is a nonbonding orbital.
- The mixing coefficients for atomic orbitals centered on different atoms, when optimized by the variation principle, may not be equal, indicating a higher electron density at a more electronegative atom (such as F in HF). This represents a polar bond where one atom (F) becomes slightly negative (charge δ^-) due to increased electron density at its site and the other atom (H) becomes slightly positive (charge δ^+) due to a decrease in the electron density. The charge separation in a polar bond is represented by a dipole moment $\mu = \delta R_e$, where R_e is the equilibrium bond length and δ represents the charge of opposite sign on each atom.

For a polyatomic molecule, the electronic energy is a function of both the bond lengths and bond angles, which define its geometry. Quantum mechanical methods of geometry optimization are often used to predict the geometry. The quantum mechanical method used to solve the electronic Schrödinger equation involves the integration of a differential equation and thus requires solutions of many integrals. Often these integrals are complex and extremely time-consuming. Therefore, a twofold approach is adopted (Levine, 2000):

- The *ab initio* approach considers all electrons and evaluates all integrals explicitly. This approach requires a great deal of computational time and is used for small or moderately sized molecules. Fortunately, as the speed and efficiencies of computers increase, the size of molecules that can be readily handled by the *ab initio* method will also increase.

- The semiempirical approaches may simplify the Schrödinger equation by considering only the valence electrons in outer orbitals of the bonding atoms and approximate certain integrals by using adjustable parameters to fit certain experimentally observed physical quantities (such as ionization energies required to strip electrons from an atom). This approach is more popular, because large molecules and polymers can be treated with relative ease using even a desktop computer.

In the case of polyatomic molecules, one also takes advantage of the symmetry of a molecule to simplify the LCAO-MO approach by making a linear combination of orbitals, which have the same symmetry characteristics of molecules. These symmetry characteristics are defined by the operations (transformations) of various symmetry elements that lead to indistinguishable configurations. These symmetry elements collectively define the point group symmetry of a molecule. Some of these symmetry elements (Levine, 2000; Atkins and dePaula, 2002) are:

- An axis of rotation C_n, such as a sixfold axis of rotation (C_6) along an axis perpendicular to the plane of a benzene ring and passing through its center
- A plane of symmetry σ, such as the plane of the benzene ring
- A center of inversion i (also called a center of symmetry) such as the center of the benzene ring

As shown in Figure 2.14, inversion of the benzene molecule with respect to its center produces an indistinguishable position. Therefore, it is a centrosymmetric molecule possessing the inversion symmetry. By contrast, the inversion of a chlorobenzene molecule with respect to its center produces a distinguishable configuration (chlorine is now in the down position, as shown in Figure 2.14). Therefore, chlorobenzene is noncentrosymmetric and does not possess inversion symmetry. For molecules possessing inversion symmetry, the

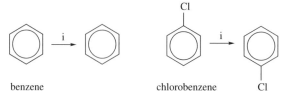

benzene   chlorobenzene   Cl

Figure 2.14. The structures of a benzene molecule and a chlorobenzene molecule and the inversion operation.

molecular orbitals, which do not change sign under inversion, are labeled as g (gerade). Those that change sign under inversion are labeled as u (ungerade).

In a complete description, the molecular orbitals of a molecule are labeled by the representations of the symmetry point group of a molecule. However, a detailed discussion of this symmetry aspect is beyond the scope of this book. For a simple reading, refer to the physical chemistry book by Atkins and dePaula (2002) or any other physical chemistry text.

2.2.5 Bonding in Organic Molecules

Of most interest from a biophotonics perspective are the bonding and energy states involving carbon atoms, which also form the subject of organic chemistry. An unusual feature exhibited by carbon is its ability to form single as well as multiple bonds. Perhaps Nature has chosen carbon to form the basis of life on earth because of this diversity in carbon chemistry. A carbon atom can involve all four orbitals—$2s, 2p_x, 2p_y, 2p_z$—in formation of bonds. When it forms a single bond with another atom (such as C–H bond in CH_4 or a C–C bond in H_3C–CH_3), the energetic closeness of atomic orbitals, $2s, 2p_x, 2p_y, 2p_z$, leads first to their mixing (also known as sp^3 hybridization) to produce four equivalent sp^3-hybridized atomic orbitals directed toward the four corners of the tetrahedron. These are then combined with appropriate atomic orbitals (such as $1s$ on H) using the LCAO-MO method. These four sp^3 orbitals are directed in a tetrahedral geometry defining four single bonds formed by four pairs of σ electrons (one pair of electrons for each bond). Thus, the methane molecule has a tetrahedral geometry with the H–C–H bond angle of 109°28′.

In the case of ethene, which is also called ethylene (Structure 2.1), only one of the two bonds between the two carbon atoms can be of σ type—formed by the overlap of atomic orbitals on the two carbons along the internuclear axis. This is due to the geometric restriction imposed by the directionality of the p orbitals. The second bond between the two-carbon atoms is a π bond, formed by the lateral overlap of another $2p$-type atomic orbital on each atom. Thus, each carbon atom in ethene involves three σ bonds and, therefore, six σ electrons (two with hydrogen and one with another carbon). These σ bonds are described by an sp^2 hybridization scheme in which the $2s$ and two $2p$ atomic orbitals (e.g., $2p_z$ and $2p_x$) on each carbon atom mix to give rise to three sp^2-hybridized atomic orbitals directed in a plane at an angle of 120° from each other. This sp^2 hybridization resulting in three σ bonds also defines the geometry of the molecule. The remaining $2p_y$ orbitals on each carbon atom then laterally overlap perpendicular to the plane of the molecule to form the π

$$\begin{matrix} H & & & H \\ & \diagdown & & \diagup \\ & & C=C & \\ & \diagup & & \diagdown \\ H & & & H \end{matrix}$$

Structure 2.1. Ethene (ethylene)

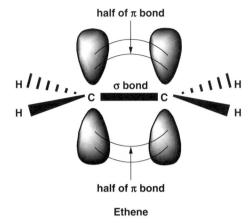

Ethene

Figure 2.15. σ and π bondings in ethene. The σ bonds are formed by sp^2-hybridized atomic orbitals on each carbon.

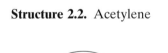

Structure 2.2. Acetylene

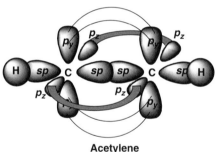

Acetylene

Figure 2.16. σ and π bondings in acetylene. The σ bonds are formed from sp-hybridized atomic orbitals on each carbon.

bond involving two π electrons. The bonding diagram for ethene is shown in Figure 2.15.

In the case of acetylene (Structure 2.2), each carbon atom has two σ bonds and two π bonds. The two σ bonds are now formed involving an sp hybridization of the $2s$ and one $2p$ (say $2p_z$) atomic orbital on each carbon atom. The sp hybrid orbitals are linearly directed and define the direction of the internuclear axis. The overlap of the remaining $2p_x$ and $2p_y$ orbitals from each carbon atom defines the two sets of orthogonal (90° with respect to each other) π orbitals and, thus, the two π bonds containing four π electrons. The σ and π bondings in acetylene are shown in Figure 2.16.

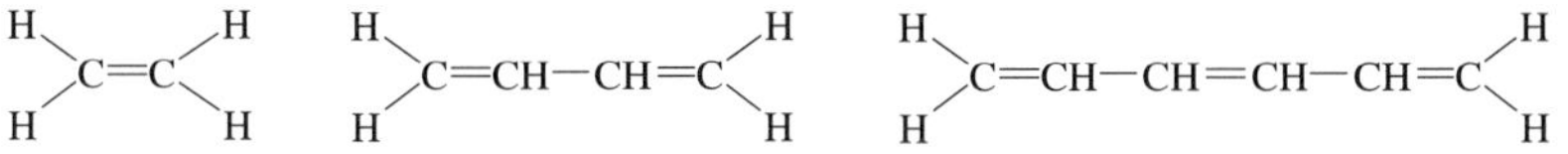

Ethene   Butadiene   Hexatriene

Figure 2.17. Some linear conjugated structures.

2.2.6 Conjugated Organic Molecules

A special class of organic compounds, which involve alternate single and multiple bonds between chains of carbon atoms, are called π conjugated molecules. The length of the carbon chain defines the conjugation length, providing a structural framework over which the π electrons can be spread (delocalized). The picture of a particle in a one-dimensional box, as previously described, can describe the conjugation (delocalization) effect. The length of one-dimensional conjugation, as defined by the chain of carbon atoms involved in alternate single and multiple bonds, determines the length of the box. As conjugation increases, the length of the one-dimensional box increases, leading to the following properties:

- The increase of conjugation leads to lowering of the π electron energy (delocalization energy).
- The energy gap between two successive π orbitals decreases as the conjugation increases. This effect gives rise to the darkening of color of the conjugated structure with the increase of conjugation length (i.e., shift of the absorption band corresponding to an electronic transition between the two levels, from UV to a longer wavelength in the visible region).

A linear conjugated structure series is shown in Figure 2.17. The simplest semi-empirical method to describe π bonding in conjugated structures is the Hückel theory. This theory makes the following assumptions:

- Only π electrons are considered.
- The π molecular orbitals are constructed by the LCAO method involving a $2p$ orbital on each carbon atom.
- Electronic interactions (exchange of electrons) are permitted only between nearest neighbor carbons.
- The π orbitals, lower in energy with respect to the individual $2p$ atomic orbitals, are the bonding π orbitals, while those higher in energy as compared to the $2p$ atomic orbitals are the π^* molecular orbitals.
- The π electron configuration is deduced by placing two electrons with opposite spins in each π orbital.
- The highest occupied molecular orbital is termed HOMO, and the lowest unoccupied molecular orbital is called LUMO.

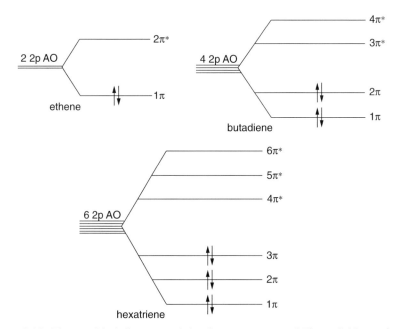

Figure 2.18. The π-orbital diagrams of the three structures of Figure 2.17 are shown.

The π orbital diagrams of the three structures listed in Figure 2.17 are shown in Figure 2.18. In the case of ethene, the two p-type atomic orbitals (abbreviated AO) from the two carbons overlap to form one bonding π and one anti-bonding π * molecular orbital. The pair of electrons is in the lower molecular energy orbital π. Similar descriptions apply to butadiene and hexatriene.

In general, the mixing of N $2p$ orbitals on N carbon atoms in a conjugated structure produces N π orbitals. The gap between the π orbitals decreases as N increases (i.e., as the delocalization length or the length of the one-dimensional box increases). In the limit of very large N, the spacing between successive π levels is very small and forms a closely spaced energy band. Another feature observed from the Hückel theory calculation is that the energy of the π electrons in butadiene is lower than the energy predicted by two isolated ethene type π bonds. This additional energy is called the *delocalization energy*, resulting from the spread of π electrons over all four carbon atoms.

A special case of π electron delocalization is for cyclic structures such as in benzene, which are also referred to as aromatic molecules. Benzene involves a conjugated cyclic structure consisting of three single and three double bonds. However, two resonance structures, describing two structures resulting from the exchange of single and double bonds are possible. The result is that all bonds are equal and have characters between a single and a double bond, which is represented by the commonly used structure for benzene shown in Figure 2.14.

This delocalization behavior provides additional energy (further stabilization by an even greater lowering of energy) compared to the six carbon atoms conjugated in a hexatriene linear structure.

2.2.7 Vibrational States of a Molecule

The vibrational states of a molecule are obtained by solving the vibrational Schrödinger equation of a molecule. A diatomic molecule, which involves only one vibrational degree of freedom (stretching of the bond), provides a simple description. For this solution the electronic energy $E_e(R)$ as a function of the internuclear distance R defines the average potential energy $V(R)$ for the vibrational motion of the nuclei. For a bound state (represented by the left-hand-side electronic curve of Figure 2.13), $E_e(R)$ exhibits a minimum at $R = R_e$. Therefore, $V(R)$ $[E_e(R)]$ can be expanded as a Taylor power series of displacement $x = (R - R_e)$ around R_e, the equilibrium point, as shown below (Graybeal, 1988; Levine, 2000):

$$V(x) = E_e(R) = V(x = 0) + \left(\frac{\partial V}{\partial x}\right)x +$$
$$\text{higher-order terms involving higher powers of } x \qquad (2.10)$$

In the above equation, the derivatives are to be evaluated at the equilibrium point $x = R - R_e = 0$. The first derivative [equation (2.10a)] is at the minimum point $x = 0$:

$$\left(\frac{\partial V}{\partial x}\right)_0 = 0 \qquad (2.10a)$$

If the energy reference point is shifted to $x = 0$ [i.e., $V(x = 0)$ is set equal to zero] and cubic plus higher-order terms, called *anharmonic terms*, are ignored, the potential energy is now

$$V(x) = \frac{1}{2}\left(\frac{\partial^2 V}{\partial x^2}\right)_0 x^2 = \frac{1}{2}kx^2 \qquad (2.11)$$

where equation (2.11a) is called the *force constant*:

$$k = \left(\frac{\partial^2 V}{\partial x^2}\right)_0 \qquad (2.11a)$$

Then the potential energy is quadratic in displacement (a parabola), and this case then defines a harmonic oscillator. The solution of the vibrational Schrödinger equation, using this harmonic approximation of $V(x) = 1/2kx^2$ and the boundary condition that $\psi_{vib}(x) = 0$ at $x = \infty$, yields a set of quantized vibrational energy states labeled by a vibrational quantum number, v. The derived expression for the vibration energy E is

$$E_v = \left(v + \frac{1}{2}\right)h\nu \tag{2.12}$$

with $v = 0, 1, 2$, etc.
ν = vibrational frequency $\sqrt{k/\mu}$
μ = the reduced mass = $m_a m_b/(m_a + m_b)$

where m_a and m_b are the masses of the two nuclei forming the bond.

The following features of the vibrational energy are evident from the above relationship:

- The lowest vibrational energy is $1/2\, h\nu$, called the *zero-point energy*, which a molecule must possess (the vibrational energy cannot be zero even at the lowest temperature of 0 K).
- The spacing between two successive vibrational levels is constant and equal to $h\nu$.
- The vibrational frequency, ν, is inversely proportional to the square root of the reduced mass; the heavier the nuclei, the lower the vibrational frequency. It also explains the isotope effect in that the isotopic molecules such as the D_2 (deuterium) molecule will have a lower vibrational frequency than the H_2 molecule.
- The vibrational frequency, ν, is directly proportional to the square root of the force constant k and thus the strength of the bond (as previously shown, k is given by the second derivative of the electronic energy E_e as a function of the internuclear separation).

The higher-order anharmonic terms [cubic or more in equation 2.10] become important for the higher vibrational states (larger quantum number v). The manifestation of anharmonic interactions in a molecule leads to a continuous decrease of energy levels between successive vibrational levels as the quantum number v increases.

For a polyatomic molecule involving N atoms, the numbers of vibrational degrees of freedoms (number of different vibrational displacements) are

- $3N - 5$ for a linear molecule (the exclusion 5 is due to three translational and two rotational degrees of freedoms for a linear molecule)
- $3N - 6$ for a nonlinear molecule (in this case there are three translational and three rotational degrees of freedom)

These vibrational modes are coupled, leading to complex patterns of simultaneous displacements of all nuclei. These patterns are governed by the symmetry of the molecule and are called normal modes of vibrations. For a linear

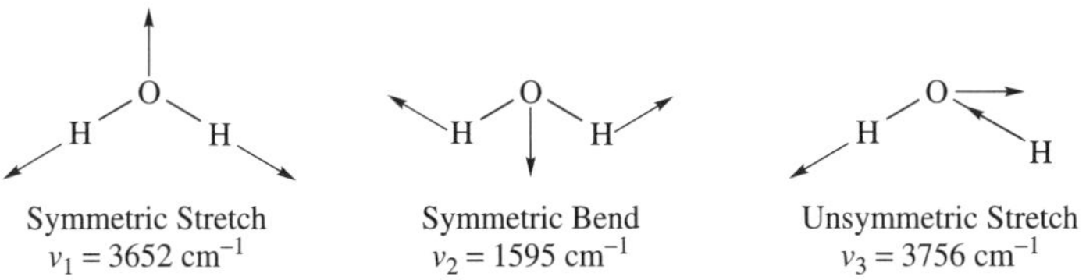

Figure 2.19. Normal modes of vibration of a water molecule.

triatomic molecule, CO_2, $N = 3$ and thus the number of vibrational modes is 4. For a nonlinear molecule such as water (H_2O), the number of vibrational modes is 3. These normal modes of vibrations of water molecules along with their vibrational frequencies are shown in Figure 2.19. However, a description that assigns a specific vibration to the specific bond or angle where it is primarily localized is often found useful as a fingerprint for chemical identification. For example, a C–H stretching vibrational frequency referring to the stretching of the C–H bond is found in the region 2850–2960 cm^{-1}.

2.3 INTERMOLECULAR EFFECTS

When molecules are in a condensed phase such as a liquid or solid or as a specie intercalated in a DNA double helix structure (discussed in Chapter 3), they can interact with each other through a number of different types of interactions. Examples of these interactions are: (i) weak van der Waals interactions (even occurring among neutral molecules); (ii) intermolecular charge transfer interactions whereby one type of molecule (electron donor) transfers an electron, when in excited state, to another type of molecule (electron acceptor); (iii) electrostatic interactions between charged molecular groups; and (iv) specific chemical association such as hydrogen bonding or even chemical bonding (such as that of various monomeric units to form a polymer). These interactions are discussed in detail in Chapter 3.

These interactions produce a modification of the quantized states of individual molecular units (Prasad, 1997). First, a molecule experiences a static potential field due to all other surrounding molecules, which produces a shift of its energy levels. Next, a dynamic resonance interaction between molecules leads to excitation exchange (energy transfer) of the excitation from one molecule to another. It is like the case of coupled pendulums in which oscillation (excitation) of one pendulum is transferred to another. This excitation interaction is also described by the mixing of their excited energy states. If the molecules are identical, the mixing of their excited energy states, which are degenerate (same value), leads to splitting in a manner similar to the one described by the Hückel theory. For example, mixing of a specific excited energy state of identical molecules A and B produces a splitting, Δ, leading to two new levels E_+ with a plus (symmetric) combination and E_- with a minus

Monomer   J-aggregate

Figure 2.20. J-aggregate of a dye.

(antisymmetric) combination of wave functions of A and B, respectively. The magnitude of the splitting, Δ, depends on the strength of the interaction between A and B. The new excited energy states are delocalized (spread) over both molecules A and B (just like the delocalization of the π electrons in the case of the Hückel theory). In the case of N identical molecules interacting together, the excited energy states split into N levels, forming a band for a large value of N (like in the case of a conjugated structure with a very long chain length). This description of an energy band formation applies to both the electronic and vibrational energy excitations of molecules. The band is called an *exciton band*.

Exciton interaction produces a profound effect on the optical properties of fluorescent dyes used for fluorescent tagging in bioimaging and biosensing. Some dyes, when aggregated, form a J-aggregate with new red-shifted excitonic states (Kobayashi, 1996). The J-aggregates represent a structure in which dye molecules align in a certain orientation, as shown in Figure 2.20.

Dyes like fluorescein show concentration quenching derived from dimer and higher aggregates formation (Lakowitcz, 1999). As the fluorescence quenching occurs between identical molecules, it is also called *self-quenching*.

Zhuang et al. (2000) have shown that this type of concentration quenching (self-quenching) can be used to study protein folding at the single-molecule level by attaching multiple dyes to a protein. Folding brings the dyes in close proximity to cause self-quenching, while unfolding moves them apart to reduce self-quenching. Interaction of energy levels between two different molecules produces the shift of their energy levels as well as a unidirectional energy transfer from the higher excited level of one molecule (energy donor) to a lower energy level of another molecule (energy acceptor). This type of electronic energy transfer, called *Forster energy transfer*, forms the basis for fluorescence resonance energy transfer (FRET) bioimaging discussed in Chapter 7. The Forster energy transfer is discussed in Chapter 4.

Another type of interaction which occurs between an electron-rich molecule (electron donor) and an electron-deficient molecule (electron acceptor) through an excited-state charge transfer produces new quantized electronic levels called *charge-transfer states*. The charge-transfer states are new excited states in which an electron is partially (or largely) transferred from the electron donor to the electron acceptor. The absorption from the ground state to the charge-transfer state often makes otherwise colorless electron-donating and electron-accepting molecules acquire colors, due to the absorption being in the visible spectral range.

Another major manifestation of placing a molecule in an ensemble of molecules is that its rotational and translational motions are hindered (spatially restricted). As a result, these motions become to-and-fro vibrations, called *phonons* or *lattice vibrations*. The lattice vibrations or phonons are of frequencies in the range of 0–200 cm^{-1}. The phonons are very sensitive to intermolecular arrangements of the molecules in the solid form, so they can be used as a fingerprint of a given lattice structure. The phonons, as observed by Raman spectroscopy (discussed in Chapter 4), have been shown to be very useful for characterizing different crystalline forms, called *polymorphic forms*, of a drug (Bellows et al., 1977; Resetarits et al., 1979; Bolton and Prasad, 1981). Such information is very useful for drug formulation since it has been shown that the bioavailability of a drug (dissolution rate and subsequent action) depends on its polymorphic form (Haleblian and McCrone, 1969; Haleblian, 1975).

2.4   THREE-DIMENSIONAL STRUCTURES AND STEREOISOMERS

Molecules and biological structures exhibit three-dimensional shapes, which consist of well-defined spatial arrangements of atoms. These shapes at the local level are represented by the three-dimensional disposition of bonds, characterized by bond lengths and bond angles.

Interactions between nonbonded atoms or a group of atoms through space also influence the extended structures of a large molecule. These nonbonded interactions between the atoms play a very important role in determining the complex three-dimensional structure, also known as *conformation*, of a large biological molecule such as a protein and DNA, which are called *macromolecules* or *biopolymers*. These biopolymers involve thousands of small molecules (called *monomers*), which are chemically bonded. This subject is discussed in detail in Chapter 3.

The three-dimensional structures and conformations of biomolecules determine their biological functions. Hence, a determination of the three-dimensional structures of biopolymers such as proteins forms a major thrust of modern structural biology. The methods used to determine these structures are x-ray crystallography (Stout and Jensen, 1989) and NMR (nuclear magnetic resonance) (Cavanaugh, et al., 1996). It is not possible to discuss any of these methods within the scope of this book. Another approach to determining the structure is to use computational biology, which involves theoretical methods of geometry optimization (Leach, 2001).

To represent the three-dimensional structures, a number of atomic models are used. The three models frequently used (Lehninger, 1970) are:

- Space-filling models, which represent the size and configuration of an atom determined by its bonding properties and its size (given by what is known as the *van der Waals radius*). The model atoms are by convention color-coded as follows:

Hydrogen: white   Nitrogen: blue   Oxygen: red

Carbon: black   Phosphorus: yellow

Sulfur: yellow (different in shape from phosphorus)

An example of the space-filling model is provided in Figure 2.21 for alanine, an amino acid building block of proteins.

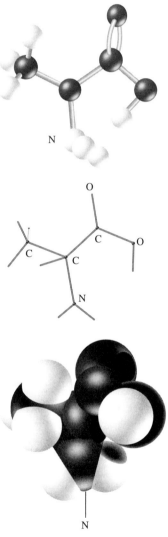

Figure 2.21. Alanine is represented by a ball-and-stick model (*top*), a skeletal model (*middle*) and a space-filling model (*bottom*) (Reproduced with permission from Lehninger, 1970).

- Ball-and-stick models, which depict atoms as a sphere of much smaller radius than that in the space-filling model, so that it is easier to see the bonds explicitly represented by sticks between the balls. Again, Figure 2.21 shows a ball-and-stick model for alanine.
- Skeletal models, which are the simplest of the three. In these models, often only the molecular framework representing bonds by lines are shown. In an even simpler form, atoms are not shown explicitly—their positions are implied by the junctions and the ends of bonds. Again, this representation for alanine is provided in Figure 2.21.

The three-dimensional shapes and conformations of molecules also lead to more than one spatial arrangement, which are not identical. These different spatial arrangements for molecules with the same chemical formula are called *stereoisomers*. Stereoisomers are different from structural isomers. Structural isomers represent differences in bonding of atoms for molecules of the same chemical formula, while stereoisomers have the same chemical bonding order but differ in spatial arrangements.

The two types of stereoisomers are:

- *Geometrical Isomers*. These isomers result from the fact that a double bond does not allow the rotation of atoms or groups around it as can occur around a single bond. Thus a molecule such as 1,2-dichloroethene can have two different structures as depicted in Figure 2.22. When the two identical atoms or groups (such as Cl in the above example) are on the same side of the double bond, the molecule is called a *cis* isomer. When the identical atoms or groups are on the opposite side of the double bond, the molecule is called a *trans* isomer.

- *Enantiomers or Optical Isomers*. When a 3-D arrangement for a molecular structure provides two different spatial arrangements which are mirror images of each other and thus not superimposable (like the left- and right-hand palms), the two isomers are called *enantiomers* or *optical isomers*. The latter name is used because these isomers exhibit optical activity (rotation of plane of polarization) as discussed above in Section 2.1.5. An example is the case of an asymmetric carbon atom, called a *chiral center*, which is bonded to four different atoms (or groups) as shown in Figure 2.23. A molecule containing a chiral center is called *chiral*. The two enantiomers are in current terminology labeled as *R* and *S* forms.

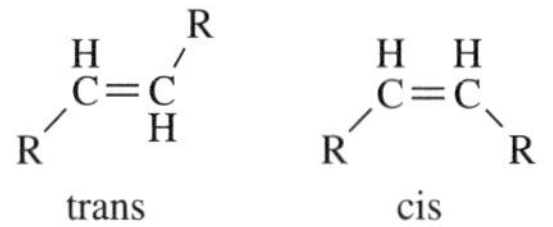

Figure 2.22. *Trans* and *cis* isomers of 1,2-dichloroethene, R = Cl.

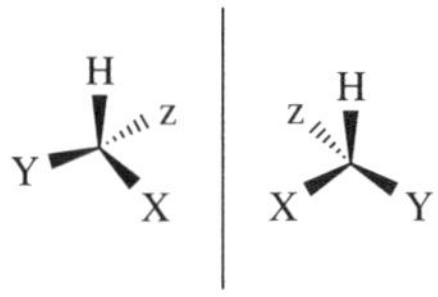

Figure 2.23. Diastereoisomers: isomeric molecules being mirror images of one another that cannot be superimposed by rotation.

The optical activities of the enantiomers are labeled on the basis of the direction they rotate the plane of polarization of a linearly polarized light. If an enantiomer rotates the place of polarization clockwise it is called *dextrorotatory* (*d*) and symbolized (+). If an enantiomer rotates the plane of polarization counterclockwise it is called *levorotatory* (*l*) and is symbolized (−). Not all *R* forms rotate the plane of polarization in the same direction. Thus, one chiral compound may have an *R* (+) form, while a different chiral compound may possess *R* (−) form. The labeling of *R* is based on spatial arrangement determined by structural methods (such as x-ray diffraction), while the direction of optical rotation is an experimental observation.

Biological systems show strict stereospecifity. Thus, common sugars in nature exist only in the *d* form. Enzymes of the body also show stereospecifity by recognizing only one enantiomer. A drug must possess a specific enantiomeric form to be effective.

In biological systems, discussed in Chapter 3, there is another source of chirality, which is not due to the presence of a chiral carbon atom. The chirality is derived from a helical structure, which can be left-handed or right-handed. These left- and right-handed helices, like the mirror image enantiomers discussed above, are not superimposable. These helical chiral structures also exhibit optical rotation.

HIGHLIGHTS OF THE CHAPTER

- Both light and matter simultaneously exhibit dual characters as waves and particles.
- Light as particles consists of photons of discrete energy values, a condition called *quantization of energy*.
- As photons, light exchanges energy and momentum with matter. Energy exchange forms the basis for spectroscopy, while momentum exchange provides physical effects—for example, optical force for trapping of matter.
- As a wave, light behaves as an electromagnetic wave, which consists of electric and magnetic oscillations perpendicular to each other as well as perpendicular to the direction of propagation.

- As a wave, light is characterized by a length, called the *wavelength*, which is the distance between two successive peaks of a wave.
- As waves, light exhibits the phenomena of interference between electromagnetic waves and diffraction at an aperture or a slit.
- The speed of light in a medium is reduced compared to its propagation in a vacuum by a factor called the *refractive index*.
- The refractive index of a medium typically decreases with increasing wavelength, as in going from the blue end to the red end of the visible range.
- A wave packet of light is called *coherent* if all the waves superimpose with their peaks and troughs on top of each other. This is the case for laser light.
- Polarization of light refers to an orientation of the oscillation of electric field of light.
- An optically active medium rotates the plane of polarization of light, while a birefringent medium exhibits different refractive index and, thus, different propagation speed for different polarization of incident light.
- Quantum conditions for the energy states of a matter are derived from the solution of the Schrödinger equation.
- The Heisenberg uncertainty principle places the restriction that the simultaneous knowledge of the exact position of a particle (e.g., an electron) and its velocity cannot be known, necessitating a probabilistic description of the spatial distribution of an electron's position.
- This probabilistic description is provided by the square of a wave function, which is also the solution of Schrödinger equation.
- The energy levels of electrons in an atom are quantized; that is, they can only have certain discrete values.
- Each electronic energy level of an atom is represented by an atomic orbital, which is the region of space where the probability of finding an electron is high.
- A molecule exhibits four types of energies: electronic, vibrational, rotational, and translational; of these, only electronic, vibrational, and rotational levels exhibit quantization effects (possess discrete values).
- The electronic states of a molecule are obtained by using the molecular orbital (MO) approach, which involves the overlap of atomic orbitals of the atoms forming a bond.
- A constructive overlap of atomic orbitals, just like a constructive interference between waves, forms a bonding MO; a destructive overlap produces antibonding MO.
- Overlap of atomic orbitals along the intermolecular axis produces a σ bond, while a lateral overlap produces a π bond.

- Carbon atoms can form multiple bonds, such as double and triple bonds. Of the multiple bonds between two carbon atoms, the first is a σ bond; the others are π bonds.
- Conjugated organic structures contain alternating single and multiple bonds between a chain of carbon atoms.
- The behavior of molecular orbitals and other energy states of a molecule is determined by the symmetry elements of the molecule, which together define its symmetry point group.
- The vibrational energy states of a molecule are described in the harmonic oscillator model. A molecule possesses a minimum energy called *zero-point energy*.
- A molecule exhibits a number of vibrational displacement patterns called *normal modes*, but often the vibrations are described as being associated with the displacement of a bond or deformation of an angle.
- Intermolecular interactions among molecules profoundly affect the electronic, vibrational, and rotational energy levels.
- The three-dimensional arrangement of bonded atoms in a molecule determines its shape and also gives rise to stereoisomers that have the same chemical formula and bonding but different spatial arrangements.
- Geometrical isomers are generally stereoisomers of a molecule containing a double bond. The *cis* isomer has two identical atoms (or groups) on the same side of the double bonds; the *trans* isomer has them on the opposite sides.
- Optical isomers or enantiomers have three-dimensional structures that are mirror images of each other.

REFERENCES

Atkins, P., and dePaula, J., *Physical Chemistry*, 7th edition, W.H. Freeman, New York, 2002.

Bellows, J. C., Chen, F. P., and Prasad, P. N., Determination of Drug Polymorphs by Laser Raman Spectroscopy I. Ampicillin and Griseofulvin, *Drug Dev. Ind. Pharm.* **3**, 451–458 (1977).

Bolton, B. A. and Prasad, P. N., Laser Raman Spectroscopic Investigation of Pharmaceutical Solids: Griseofulvin and Its Solvates, *J. Pharm. Sci.,* **70**, 789–793 (1981).

Born, M., and Wolf, E., *Principles of Optics*, Pergamon Press, Oxford, 1965.

Cavanaugh, J.; Palmer, A., III, Fairbrother, W.; and Skelton, N., *Protein NMR Spectroscopy: Principles and Practice*, Academic Press, New York, 1996.

Feynman, R. P., Leighton, R. B., and Sands, M., *The Feynman Lectures on Physics*, Vol. 1, Addison-Wesley, Reading, MA, 1963.

Graybeal, J. D., *Molecular Spectroscopy*, McGraw-Hill, New York 1988.

Haleblian, K., and McCrone, W., Pharmaceutical Applications of Polymorphism, *J. Pharm. Sci.* **58**, 911–929 (1969).

Haleblian, K., Characterization of Habits and Crystalline Modification of Solids and Their Pharmaceutical Applications, *J. Pharm. Sci.* **64**, 1269–1288 (1975).

Kobayashi, J., *J-Aggregates*, World Scientific Publishing Co., Japan 1996.

Lakowitcz, J. R., *Principles of Fluorescence Spectroscopy*, 2nd edition, Plenum, New York, 1999.

Leach, A., *Molecular Modelling: Principles and Applications*, 2nd edition, Prentice-Hall, Upper Saddle River, NJ, (2001).

Lehninger, A. L., *Biochemistry*, Worth Publishers, New York, 1970.

Levine, I. N., *Quantum Chemistry,* 5th edition, Prentice-Hall, Upper Saddle River, NJ, 2000.

Prasad, P. N.,"Excitation Dynamics in Organic Molecules, Solids, Fullerenes, and Polymers, in B. DiBartolo, ed., *Spectroscopy and Dynamics of Collective Excitations in Solids,* NATO ASI Series, Plenum, New York, 1997, pp. 203–225.

Prasad, P. N., and Williams, D. J., *Introduction to Nonlinear Optical Effects in Molecules and Polymers*, Wiley-Interscience, New York, 1991.

Resetarits, D. E., Cheng, K. C., Bolton, B. A., Prasad, P. N., Shefler, E., and Bates, T. R., Dissolution Behavior of 17β-Estradiol (E_2) from Povidone Coprecipitates: Comparison with Microcrystalline and Macrocrystalline E_2, *Int. J. Pharmaceutics*, **2**, 113–123 (1979).

Smith, K. C., ed., *The Science of Photobiology*, 2nd edition, Plenum, New York 1989.

Stout, G., and Jensen, L., *X-Ray Structure Determination: A Practical Guide*, John Wiley & Sons, New York, 1989.

Zhuang, X., Ha, T., Kim, H. D., Cartner, T., Lebeit, S., and Chi, S., Fluorescence Quenching: A Tool for Single-Molecule Protein-Folding Study, *Natl. Acad. Sci.* **19**, 14, 241–244 (2000).

CHAPTER 3

Basics of Biology

This chapter provides basics of biology and introduces the necessary terminology and concepts of biology used in this book. The chapter is written primarily for those unfamiliar with biological concepts or those wishing to refresh
their background in this subject. The chapter will also serve as a source of
vocabulary of relevant biological terms.

The focus of this chapter is on building a molecular understanding of biological structures and their relation to biological functions. A main focus of
modern biology and the new frontiers of *genomics*, *proteomics*, and *bioinformatics* are derived from this understanding and from profiling of diseases at
the molecular level (Chapters 10 and 11). Such an understanding can lead to
new and effective drug treatments, which are customized for the patient and
are based on molecular profiling using an individual's genetic makeup.

Light–matter interaction, which is the basis for optically probing structure
and function at cellular and tissue levels (see *bioimaging* in Chapters 7 and 8)
as well for the light-activated photodynamic therapy of cancer (Chapter 12)
and other diseases; benefits from a molecular understanding of cellular and
tissue structures and functions. The topics of *biosensing* (Chapter 9), a hotly
pursued area in view of possible threats of bioterrorism and constantly emerging new microbial infections, bioimaging, and multiple analyte detection using
microarray technology (Chapter 10), rely heavily on molecular recognition of
biological species.

This chapter starts with the description of a cell. It describes the various
structural components of the cell and their functions. An important part of a
living organism is the diversity of cells that are present in various organs to
produce different functions. Some important types of cells that are relevant to
this book are introduced.

Next, the chapter provides a description of molecular building blocks and
their assembling to form the major macromolecular components of a cell.
These include nucleic acids, proteins, carbohydrates, and lipids. A description
of important cellular processes then follows.

Introduction to Biophotonics, by Paras N. Prasad
ISBN: 0-471-28770-9   Copyright © 2003 John Wiley & Sons, Inc.

The next higher level of hierarchy in a biological system is a tissue. The organization of a tissue in terms of assembling of cells and utilizing extracellular components is described. A major focus of *photobiology* (Chapter 6) is interaction of light with tissues, where the concepts of this section will help.

Finally, the chapter concludes with an introduction to tumors and cancers.

Further references on cellular and tissue structures and functions are books by Albert et al. (1994) and by Lodish et al. (2000). For biochemical aspects that deal with chemical building blocks, biosynthesis, enzyme catalysis, and cell energy production, suggested references are books by Horton et al. (2002), Stryer (1995), Lohniger (1970), and Voet et al. (2002). General chemical principles are covered in a book by Solomon (1987).

3.1 INTRODUCTORY CONCEPTS

All living creatures are made up of cells. They exist in a wide variety of forms, from single cell in free-living organisms to those in complex biological organisms. Despite the great diversity exhibited by living systems, all biological systems, amazingly, are composed of the same types of chemical molecules and utilize similar principles in replication, metabolism, and, in higher organisms, the ability to organize at the cell levels. This section is intended to provide a basic knowledge of the fundamentals involved in understanding the structure and function of a living cell and its interaction with its environment. A brief description of the fundamentals of cellular and tissue structure is provided. Even at the most elemental cellular level, microorganisms exhibit a large range of length scale, from viruses measuring 20–200 nm to a eukaryotic cell measuring 10–100 μm. Table 3.1 shows the hierarchical structures in the biological evolution of life.

In Table 3.1, viruses are "packages" of nucleic acids, which are not capable of self-replication. While Table 3.1 lists only animal cells as an example of the eukaryotic cells, other types of cells falling in this category are yeasts/fungi, which are also self-replicating single-cell organisms. Cells of plants are also classifed as eukaryotic, but they contain a cell wall structure consisting of cellulose. This cell wall is absent in animal cells. An important feature of the eukaryotic cell is its ability to differentiate and produce a variety of cells, each carrying out a specialized function. The complex assembly of these differentiated cells leads to higher organizations, which eventually form a higher organism, as shown in Table 3.2.

Thus, living organisms formed from cells are highly complex and organized and perform a variety of functions (Stryer, 1995; Lodish et al., 2000):

- They extract and transform energy from their environment (food chain).
- They build and maintain their intricate structures from simple raw materials.
- They carry on mechanical work using muscles.

- They use a highly organized self-replication process to reproduce cells identical in mass, shape, and internal structure.
- They use an intricate set of communications.
- They use internal defenses to fight disease and carry on self-repair of damage due to injury.

TABLE 3.1. Hierarchical Structure of Biological Systems

Small molecules:
Amino acids
Nucleic acids
Water
Lipids
Ions

↓

Polymerization of nucleic acids to form DNA and RNA, and polymerization of amino acids to form protein

Virus	Prokaryotic cells: Bacteria
• Size scale: 20–200 nm	• Size scale: ~1–10 µm
• Structure: Single- or double-stranded RNA or DNA	• Structure: Single-cell organism consisting of single closed compartment that lacks a defined nucleus
• Function: Infectious but not self-replicating	• Function: Free-living and self-replicating

Eukaryotic cells: Animal/plant cells
- Size scale: ~10–100 µm
- Structure: Complex structure surrounded by a lipid membrane, contains an organized nuclear structure
- Function: Self-replicating and able to assemble to form tissues

The following sections describe the structures and functions of cells and tissues that are of particular interest to *biophotonics*. The descriptions of cells and tissues follow a top-to-bottom approach, ending at the bottom with the chemical makeup of cells and tissues. This also illustrates the miracle of Nature: how living organisms conduct live functions using lifeless molecules.

3.2   CELLULAR STRUCTURE

Biological systems are essentially an assembly of molecules where water, amino acids, carbohydrates (sugar), fatty acids, and ions account for 75–80%

TABLE 3.2. Hierarchical Buildup of a Living Organism

Cells

      ↓   Cell differentiation and association

Tissues

      ↓   Organization to perform a function

Organs

      ↓   Integration of various functions

Organism

of the matter in cells. The remainder of the cell mass is accounted for by macromolecules, also called polymers (or biopolymers in the present case), which include peptides/proteins (formed from amino acids), polysaccharides (formed from sugars), DNA (dioxyribonucleic acid, formed from nucleotide bases and dioxyribose sugar), RNA (ribonucleic acid, formed from nucleotide bases and ribose sugar), and phospholipids (formed from fatty acids). These macromolecular polymers organize to form cells. To contain these molecules, a semipermeable membrane (phospholipid bilayer) surrounds them to form a cell. Within this biological universe, two types of organized cells exist, as shown in Table 3.1. Prokaryotic cells (bacteria) are cells with little internal structure and no defined nucleus. Eukaryotic cells have a significantly more complex internal architecture including a defined, membrane-bound nucleus. The smallest organized particle is a virus. The smallest self-replicating cells are bacteria. Eukaryotic cells, for the most part, organize to form complex living organisms. From a single pluripotent cell (a cell with the capacity to differentiate into several cell types) arises tissues and organs, and finally a complex living organism as shown in Table 3.2.

It is speculated that living organisms evolved from the prebiotic conditions in existence during the first one billion years. Although still speculative, it is hypothesized that simple organic molecules (those containing carbon) were formed during the violent electrical discharges in a heated atmosphere containing methane, carbon dioxide, ammonia, and hydrogen. These molecules formed the primordial soup from which primitive proteins and nucleotides were born. From this soup arose the first self-replicating membrane-bound organism.

The ability to self-replicate endows an organism with the ability to evolve. According to the Darwinian principle, organisms vary randomly and only the fittest survive. In living systems, genes (discussed below) define the cell constituents, structures, and cellular activities. Alteration in structure and organization nurtures the evolutionary changes necessary for survival of an organism. Recent progress in determining the nucleotide sequence for a variety of organisms has revealed that subtle, not drastic, changes are responsible for the evolutionary process.

The structure of a cell—specifically, eukaryotic cells—can be described in terms of the various subcellular compartments and the constituent chemical species they contain. The main structural components of a cell are:

- Plasma membrane, which defines the outer boundary of a cell. This is present in all cells.
- Cell wall, which exists in the prokaryotic cells as well as in the eukaryotic cells of plants but not animals.
- Cytoplasm, which represents everything within a cell, except the nucleus.
- Cytosol, which is the fluid of the cytoplasm.
- Organelle, which is the name used for a subcellular compartment in a cell where a specific cellular function takes place.
- Nucleus, which contains the chromosomes (genetic information).

Figure 3.1 compares the schematic representation of a eukaryotic cell versus a prokaryotic cell. It is readily seen that the prokaryotic cell, or bacteria, has a much less complex internal structure compared to the eukaryotic cell. In addition, a complex outer wall structure exists in most bacteria and is composed of unique outer membrane and inner membrane structures between which are sandwiched a rigid unique polysaccharide cell wall. This wall consists of a macromolecule known as peptidoglycan that enables an organism to survive in changing environments.

Organelles are like little organs of a cell that perform various cellular functions, just like organs perform various tasks in a living system. Organelles are intracellular (or subcellular) structures: specifically, nucleus, mitochondria, Golgi apparatus, endoplasmic reticulum, cytoskeleton, lysosomes, and peroxisomes. In the case of plant cells, other organelles are plastids, chloroplasts, vacuole, and cell wall (already listed above). The following describes some structural aspects of various cellular components and functions they perform (Audesirk et al., 2001).

Plasma Membrane. This forms a semipermeable outer boundary of both prokaryotic and eukaryotic cells. This outer membrane, about 4–5 nm thick, is a continuous sheet of a double layer (bilayer) of long-chain molecules called *phospholipids*. A phospholipid molecule has a long tail of alkyl chain, which is hydrophobic (repels water), and a hydrophilic head (likes water) which carries a charge (and is thus ionic). Phospholipid molecules spontaneously orient (or self-organize) to form a bilayer in which the hydrophobic tails are pointed inwards (shying away from the outer aqueous environment). The hydrophilic, ionic head groups are in the exterior and are thus in contact with

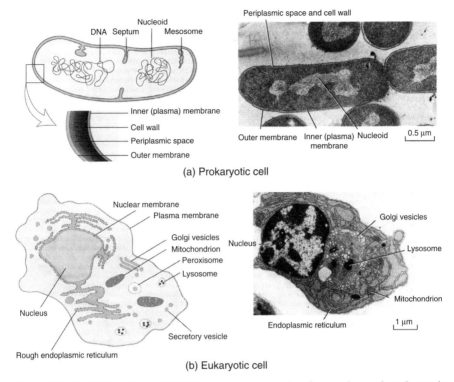

Figure 3.1. *Left*: Drawings; *right*: Electron micrographs. Comparison of prokaryotic and eukaryotic cells. (Reproduced with permission from Lodish et al., 2000).

the surrounding aqueous environment. This structure is shown in detail in Figure 3.2. The membrane derives its rigidity by inclusion of cholesterol molecules, which are interdispersed in the phospholipid bilayer. Also embedded are membrane proteins (receptors, pores, and enzymes) that are important for a number of cell activities including communication between the intracellular and extracellular environments. The plasma membrane controls the transport of food, water, nutrients, and ions such as Na^+, K^+, and Ca^{2+} (through so-called ion channels) to and from the cell as well as signals (cell signaling) necessary for proper cell function.

Cytoplasm. As indicated above, cytoplasm represents everything enclosed by the plasma membrane, with the exclusion of the nucleus. It is present in all cells where metabolic reactions occur. It consists mainly of a viscous fluid medium that includes salts, sugars, lipids, vitamins, nucleotides, amino acids, RNA, and proteins which contain the protein filaments, actin microfilaments, microtubules, and intermediate filaments. These filaments function in animal and plant cells to provide structural stability and contribute to cell movement. Many of the functions for cell growth, metabolism, and replication are carried out within the cytoplasm. The cytoplasm performs the functions of energy pro-

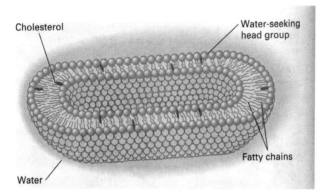

Cholesterol

Water-seeking
head group

Fatty chains

Water

Figure 3.2. Schematics of the phospholipid membrane bilayer structure. (Reproduced with permission from Lodish et al., 2000.)

duction through metabolic reactions, biosynthetic processes, and photosynthesis in plants. The cytoplasm is also the storage place of energy within the cell. *Cytosol*, a subset of cytoplasm, refers only to the protein-rich fluid environment, excluding the organelles.

Cytoskeleton. The cytoskeleton structure, located just under the membrane, is a network of fibers composed of proteins, called *protein filaments*. This structure is connected to other organelles. In animal cells, it is often organized from an area near the nucleus. These arrays of protein filaments perform a variety of functions:

- Establish the cell shape
- Provide mechanical strength to the cell
- Perform muscle contraction
- Control changes in cell shape and thus produce locomotion
- Provide chromosome separation in mitosis and meiosis (these processes are discussed below)
- Facilitate intracellular transport of organelles

Nucleus. The nucleus is often called the control center of the cell. It is the largest organelle in the cell, usually spherical with a diameter of 4–10 μm, and is separated from the cytoplasm by an envelope consisting of an inner and an outer membrane. All eukaryotic cells have a nucleus. The nucleus contains DNA distributed among structures called *chromosomes*, which determine the genetic makeup of the organism. The chromosomal DNA is packaged into chromatin fibers by association with an equal mass of histone proteins. The nucleus contains openings (100 nm across) in its envelope called *nuclear pores*, which allow the nuclear contents to communicate with the cytosol.

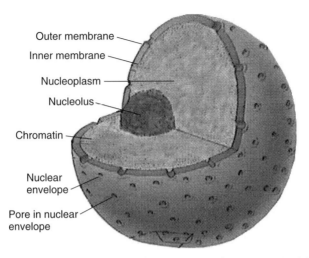

Outer membrane
Inner membrane
Nucleoplasm
Nucleolus
Chromatin
Nuclear envelope
Pore in nuclear envelope

Figure 3.3. Schematics of the structure of the nucleus. (Reproduced with permission from http://wing-keung.tripod.com/cellbiology.htm.)

Figure 3.3 shows a schematic of a nucleus. The inside of the nucleus also contains another organelle called a *nucleolus*, which is a crescent-shaped structure that produces ribosomes by forming RNA and packaging it with ribosomal protein. The nucleus is the site of replication of DNA and transcription into RNA. In a eukaryotic cell, the nucleus and the ribosomes work together to synthesize proteins. These processes will be discussed in a later section.

Mitochondria. Mitochondria are large organelles, globular in shape (almost like fat sausages), which are 0.5–1.5 µm wide and 3–10 µm long. They occupy about 20% of the cytoplasmic volume. They contain an outer and an inner membrane, which differ in lipid composition and in enzymatic activity. The inner membrane, which surrounds the matrix base, has many infoldings, called *cristae*, which provide a large surface area for attachment of enzymes involved in respiration. The matrix space enclosed by the inner membrane is rich in enzymes and contains the mitochondrial DNA. Mitochondria serve as the engine of a cell. They are self-replicating energy factories that harness energy found in chemical bonds through a process known as *respiration*, where oxygen is consumed in the production of this energy. This energy is then stored in phosphate bonds. In plants, the counterpart of mitochondria is the chloroplast, which utilizes a different mechanism, photosynthesis, to harness energy for the synthesis of high-energy phosphate bonds.

Endoplasmic Reticulum. The endoplasmic reticulum consists of flattened sheets, sacs, and tubes of membranes that extend throughout the cytoplasm of eukaryotic cells and enclose a large intracellular space called *lumen*. There is a continuum of the lumen between membranes of the nuclear envelope. The rough endoplastic reticulum (rough ER) is close to the nucleus, and is the site of attachment of the ribosomes. Ribosomes are small and dense structures, 20 nm in diameter, that are present in great numbers in the cell, mostly attached to the surface of rough ER, but can float free in the cytoplasm. They are manufactured in the nucleolus of the nucleus on a DNA template and are then transported to the cytoplasm. They consist of two subunits of RNA (a large, 50S, and a small, 30S) that are complexed with a set of proteins. Ribosomes are the sites of protein synthesis. The process of protein synthesis using a messenger RNA template is described below. The rough ER transitions into a smooth endoplastic reticulum (smooth ER), which is generally more tubular and lacks attached ribosomes. The smooth ER is the primary site of synthesis of lipids and sugars and contains degradative enzymes, which detoxify many organic molecules.

Golgi Apparatus. This organelle is named after Camillo Golgi, who described it. It consists of stacked, flattened membrane sacs or vesicles, which are like shipping and receiving departments because they are involved in modifying, sorting, and packaging proteins for secretion or delivery to other organelles or for secretion outside of the cell. There are numerous membrane-bound vesicles (<50 nm) around the Golgi apparatus, which are thought to carry materials between the Golgi apparatus and different compartments of the cell.

Lysosomes. These are bags (technical term: vesicles) of hydrolytic enzymes that are 0.2–0.5 μm in diameter and are single-membrane bound. They have an acidic interior and contain about 40 hydrolytic enzymes involved in intracellular digestions.

Peroxisomes. These are membrane-bound vesicles containing oxidative enzymes that generate and destroy hydrogen peroxide. They are 0.2–0.5 μm in diameter.

Chloroplast. This cell organelle exists only in plants. It contains pigments, called *chlorophylls*, which harvest light energy from the sun. The chloroplast is the site of photosynthesis, where light energy from the sun is converted into chemical energy to be utilized by the plant cell (synthesis of ATP).

3.3   VARIOUS TYPES OF CELLS

Cells come in many shapes, sizes, and compositions. The human body is made up of over 200 different types of cells, some of which are living cells. The human body also consists of nonliving matter such as hair, fingernails, and hard parts of bone and teeth, which are also made of cells. These cell variations are produced by cell differentiation. Also, different types of cells assemble together to form multicellular tissues or organisms. Presented here is a small selection representing the most common types of cells in animal organisms.

Epithelial Cells. Epithelial cells form sheets, called *epithelia*, which line the inner and outer surfaces of the body. Some of the specialized types of cells are (i) *absorptive cells*, which have numerous hair-like microvilli projecting from their surface to increase the absorption area, (ii) *ciliated cells*, which move substances such as mucus over the epithelial sheet, (iii) *secretory cells*, which form exocrine glands that secrete tears, mucus, and gastric juices, (iv) *endocrine glands*, which secrete hormones into the blood, and (v) *mucosal cells*, which protect tissues from invasive microorganisms, dirt, and debris.

Blood Cells. These cells are contained in blood. Blood, in fact, is a heterogeneous fluid consisting of a number of different types of cells. These cells comprise about 45% of the blood's volume and are suspended in a blood plasma, which is a colloidal (small, suspendable particle) suspension of proteins in an electrolyte solution containing mainly NaCl. The three different types of blood cells are (i) *erythrocytes* (commonly known as red blood cells; often abbreviated as RBC), (ii) *leucocytes* (commonly known as white blood cells), and (iii) *thrombocytes* (also known as platelets). Erythrocytes or red blood cells are very small cells, 7–9 μm in diameter, with a biconcave, discotic shape. They usually have no nucleus. One cubic centimeter of blood contains about 5 billion erythrocytes, the actual number depending on a number of factors such as age, gender, and health. They contain an oxygen-binding protein called *hemoglobin* and thus perform the important function of transporting O_2 and CO_2.

Leucocytes or white blood cells provide protection against infection. They exist in a ratio of one white blood cell for about every 1000 red blood cells. They are usually larger than red blood cells. There are a number of different types of leucocytes, including (i) *lymphocytes*, which are responsible for immune responses (the two kinds of lymphocytes are T cells, which are responsible for cell-mediated immunity, and B cells, which produce antibodies), and (ii) *macrophages* and *neutrophils*, which move to sites of infection where they ingest bacteria and debris. Platelets are organelles devoid of a nucleus and 2–5 μm in diameter (smaller than red blood cells). They produce specific substances for blood coagulation. Blood cells are further discussed in Chapter 11 on flow cytometry.

Muscle Cells. These specialized cells form muscle tissues such as skeletal muscles to move joints, cardiac muscles to produce heartbeat, and smooth muscle tissues found around the internal organs and large blood vessels. Muscle cells produce mechanical force by their contraction and relaxation.

Nerve Cells or Neurons. Neurons are cells specializing in communication. The brain and spinal cord, for example, are composed of a network of neurons that extend out from the spinal cord into the body.

Sensory Cells. These cells form the sensory organs such as (a) hair cells of the inner ear, which act as detectors of sound, and (b) retina, where rod cells specialize in responding to light. The rod cells contain a photosensitive region consisting of light-sensitive pigments (chromophores) called *rhodopsin.*

Germ Cells. Germ cells are haploids (cells containing one member or a copy of each pair of chromosome). The two types of germ cells specialized for sexual fusion, also called *gametes*, are (i) a larger, nonmotile (nonmoving) cell called the *egg* (or ovum) from a female and (ii) a small, motile cell referred to as *sperm* (or spermatozoon) from a male. A sperm fuses with an egg to form a new diploid organism (containing both chromosomes). Bacteria are another example of haploid cells.

Stem Cells. Another type of cell that has received considerable attention during recent years is the stem cell. Stem cells can be thought of as blank cells that have yet to become specialized (differentiated), giving them the characteristics of a particular type of cell such as the ones described above. Stem cells thus have the ability to become any type of cell to form any type of tissue (bone, muscle, nerve, etc.). The three different types of stem cells are (i) embryonic stem cells, which come from embryos, (ii) embryonic germ cells, which come from testes, and (iii) adult stem cells, which come from bone marrow.

Embryonic stem cells are classified as pluripotent because they can become any type of cell. Adult stem cells, on the other hand, are multipotent in that they are already somewhat specialized. The pluripotent type (which are in the early stage of specialization after several cell divisions) are more useful than the adult stem cells. However, recent research suggests that multipotent adult stem cells can have pluripotent capability. Stem cells can provide a solution in regard to curing diseases caused by cell failure and repairing tissues that do not repair by themselves by allowing one to produce appropriate cells and grow needed tissues. Some of the diseases for which stem cell research is projected to benefit are heart damage, spinal cord injuries, Parkinson's disease, leukemia, and diabetes.

3.4 CHEMICAL BUILDING BLOCKS

This section describes the chemical makeup of the principal constituents of a cell: (i) nucleic acids in the form of DNA and RNA, (ii) proteins, (iii) saccharides (sugar derivatives), and (iv) lipids. (Albert et al., 1994; Stryer, 1995).

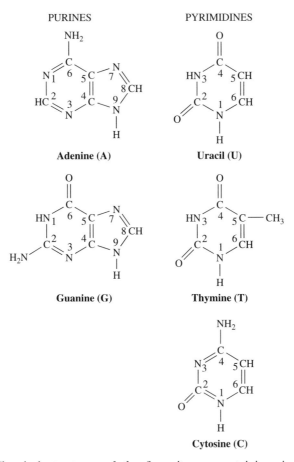

Figure 3.4. Chemical structures of the five nitrogen-containing ring compounds: purines and primidine bases (A, G, U, T, and C) in nucleic acids.

Nucleic Acids. Nucleic acids exist in a cell in two forms: (i) deoxyribonucleic acids (DNA) and (ii) ribonucleic acids (RNA). Both forms consist of three chemical building blocks: (i) nitrogen-containing ring compounds that are either purine or pyrimidine bases, (ii) a sugar (either deoxyribose for DNA or ribose for RNA), and (iii) a phosphate. The four bases constituting DNA are adenine (A), guanine (G), thymine (T), and cytosine (C). In the case of RNA, the base thymine (T) is replaced by another base, uracil (U). The structures of these bases are shown in Figure 3.4. The constituent sugars are ribose for RNA and 2-deoxyribose for DNA. Their chemical structures are shown in Figure 3.5. The phosphate group is the tetrahedral PO_4 group.

The following hierarchy (3.1) of chemical coupling represents the nucleic acid formation. It also provides the nomenclature used to describe the nucleic acid chemistry (Stryer, 1995; Voet et al., 2002):

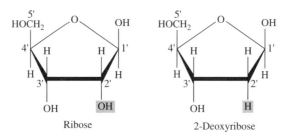

Figure 3.5. Chemical structures of the five-membered pentose sugar molecules: ribose and 2-deoxyribose. (Reproduced with permission from Lodish et al., 2000.)

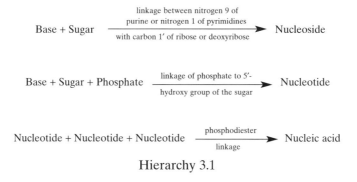

Hierarchy 3.1

The polymerization of the nucleotides to form nucleic acids occurs in a directed order. The convention is that the synthesis occurs from 5′ to 3′. In this manner, the phosphate group located on the 5′ carbon is added to the hydroxyl group on the 3′ carbon of the growing chain. And, by convention, the polynucleotide sequences are also read 5′ to 3′. For example, AUG represents 5′-AUG-3′. The sequence of nucleotides in the DNA defines the genetic makeup of the organism. All nucleotides involve a phosphate group linked by a phosphoester bond to a five-membered sugar molecule, pentose, which in turn is bonded to an organic base. As an example, the adenosine 5′-monophosphate nucleotide consists of the base adenine (A) as shown in Figure 3.6.

The DNA chemical structure involves the four nucleotides A, G, T, C as a continually varying sequence where the nucleotide bases are connected by a phosphodiester bond involving condensation of the hydroxyl (OH) group on one, with the acid, —OH, group on the phosphate of the other as shown in Structure 3.1. The sugar unit is a five-membered ring, deoxyribose. Here, B_1 and B_2 are the two-nucleotide bases. The OH group on the other sugar unit provides a site for bonding with a phosphate group on another nucleotide to repeat the process, ultimately yielding the polymeric nucleic acid of DNA.

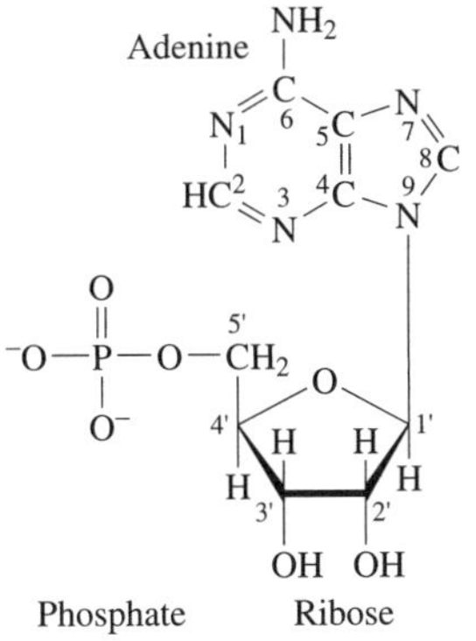

Adenosine
5'-monophosphate
(AMP)

Figure 3.6. Chemical structure of a nucleotide, adenosine 5'-monophosphate. (Reproduced with permission from Lodish et al., 2000.)

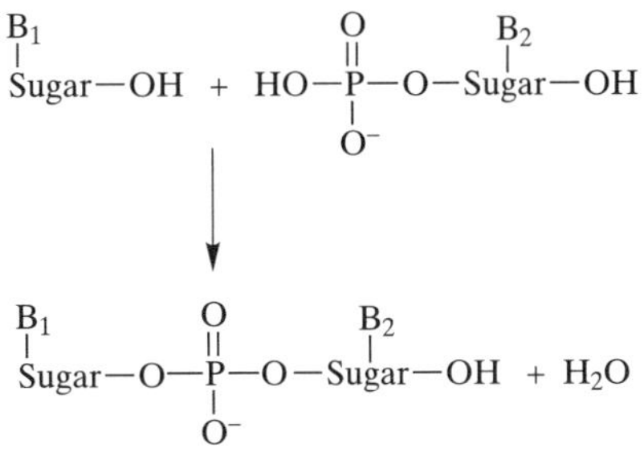

Structure 3.1

The chemical structure of RNA is similar to DNA, but the sugar component of RNA has an additional hydroxyl group at the 2' position (the sugar is ribose) and the thymine (T) in DNA is replaced by uracil (U) in RNA. DNA is contained in the nucleus of a cell and is distributed in supercoiled structures called *chromosomes*. The number of chromosomes is the same in all types of cells of a specific organism, but varies from one organism to another.

Proteins. Proteins are formed during a polymerization process that links amino acids. Only 20 different amino acids (listed in Table 3.3) form a vast array of proteins capable of highly diverse tasks. The functional nature of each of these amino acids resides in the side-chain group (R group). The amino acids link together by a peptide bond formed by condensation of the amino

TABLE 3.3. Twenty Amino Acids Constituting Most Proteins

Formula	Name (Abbreviation)[a]	Formula	Name (Abbreviation)[a]
H₂N−CH−C−OH with O (double bond) and H below	Glycine (Gly, G)	H₂N−CH−C−OH with O (double bond), CH₂, C=O, OH	Asparagine (Asn, N)
H₂N−CH−C−OH with O (double bond), CH₃	Alanine (Ala, A)	H₂N−CH−C−OH with O (double bond), CH₂, CH₂, C=O, OH	Glutamic acid (Glu, E)
H₂N−CH−C−OH with O (double bond), CH−CH₃, CH₃	Valine (Val, V)	H₂N−CH−C−OH with O (double bond), CH₂, CH₂, C=O, NH₂	Glutamine (Gln, Q)
H₂N−CH−C−OH with O (double bond), CH₂, CH−CH₃, CH₃	Leucine (Leu, L)	H₂N−CH−C−OH with O (double bond), CH₂, SH	Cysteine (Cys, C)
H₂N−CH−C−OH with O (double bond), CH−CH₃, CH₂, CH₃	Isoleucine (IIe, I)	H₂N−CH−C−OH with O (double bond), CH₂, CH₂, S, CH₃	Methionine (Met, M)
H₂N−CH−C−OH with O (double bond), CH₂, OH	Serine (Ser, S)	H₂N−CH−C−OH with O (double bond), CH₂, benzene ring with OH	Tyrosine (Tyr, Y)
H₂N−CH−C−OH with O (double bond), CH−OH, CH₃	Threonine (Thr, T)	H₂N−CH−C−OH with O (double bond), CH₂, benzene ring	Phenylalanine (Phe, F)

TABLE 3.3. (Continued)

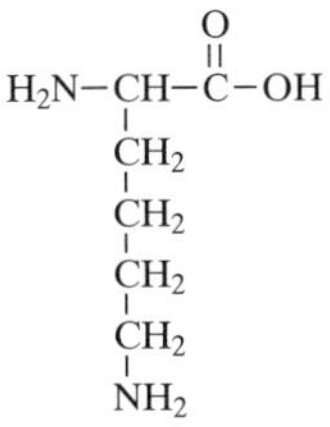

H₂N−CH−C−OH with CH₂, CH₂, CH₂, CH₂, NH₂ chain	Lysine (Lys, K)

Lysine (Lys, K)

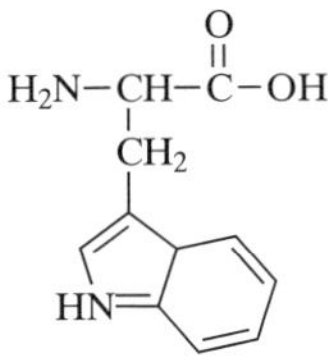

Tryptophan (Trp, W)

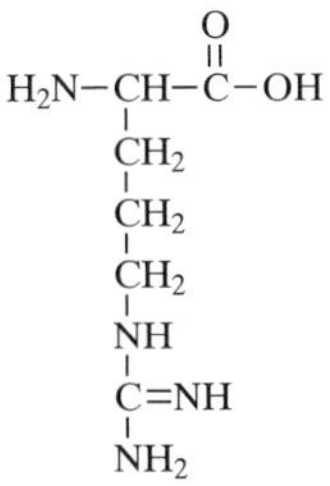

Arginine (Arg, R)

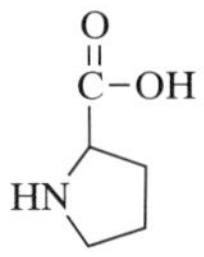

Proline (Pro, P)

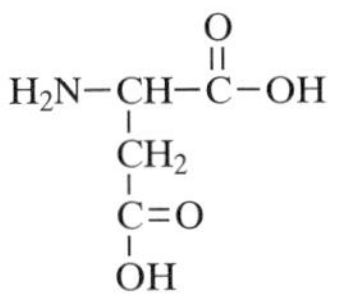

Aspartic acid (Asp, D)

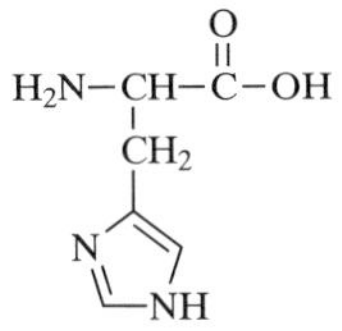

Histidine (His, H)

ᵃThe abbreviations include both three-letter and single-letter designations.

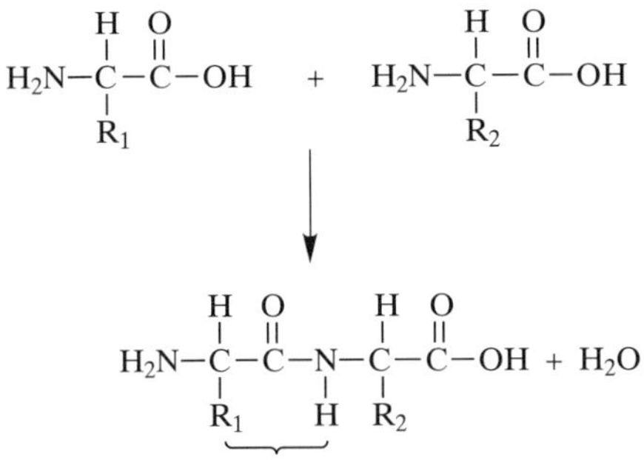

$$H_2N-\underset{R_1}{\overset{H}{\underset{|}{C}}}-\overset{O}{\overset{||}{C}}-OH \quad + \quad H_2N-\underset{R_2}{\overset{H}{\underset{|}{C}}}-\overset{O}{\overset{||}{C}}-OH$$

$$\downarrow$$

$$H_2N-\underset{R_1}{\overset{H}{\underset{|}{C}}}-\overset{O}{\overset{||}{C}}-\underset{H}{\overset{}{\underset{|}{N}}}-\underset{R_2}{\overset{H}{\underset{|}{C}}}-\overset{O}{\overset{||}{C}}-OH \ + \ H_2O$$

Peptide
bond

Structure 3.2

group of one amino acid with the carboxylic group of another, as shown in Structure 3.2. This process continues on both ends (involving the NH₂ and COOH groups) to produce the polymeric chemical structures of a protein. R₁ and R₂ are side groups (or chains) that may simply be an alkyl, acidic, basic, or aromatic group. Proteins are ubiquitous throughout the cell.

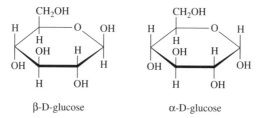

β-D-glucose   α-D-glucose

Figure 3.7. Chemical structure of glucose.

Sugars. Sugars are the primary source of food molecules for a cell. They provide the needed energy for various cellular functions. The simplest sugars are called monosaccharides and are of general chemical formula $(CH_2O)_n$. Glucose is a six-membered ring and is one of the simplest sugars used as a source of energy for cells. Its structure is shown in Figure 3.7. The five-membered ring sugars, ribose and deoxyribose, are utilized by cells only in the formation of RNA and DNA, respectively. Glucose has a number of stereoisomers, galactose and mannose, which differ in the orientation of their hydroxy groups. Also, natural glucose only exists in one optically active *D*-form, which can have two conformations, α and β. Complex sugars form when multiple monosaccharide units are linked together. Two monosaccharides can link together to form a disaccharide. An oligosaccharide involves the linkage of more than two monosaccharides. A polysaccharide is formed by the linkage of thousands of monosaccharide units, also called *residues*. Examples of polysaccharides are glycogens in animal cells and starches in plants, which are composed solely of glucose residues. Complex polysaccharides form important extracellular structures and are also found covalently linked to proteins in the form of glycoproteins, as well as to lipids in the form of glycolipids.

Lipids. Lipids refer to a group of compounds in living systems that are soluble in nonpolar solvents like hexane (a hydrocarbon, C_6H_{12}). The various kinds of lipids found in living systems are represented in Figure 3.8. Most lipids are derived from fatty acids with a general formula of R-COOH. It is a carboxylic acid in which the R group is a long alkyl chain. The most nonpolar of the lipids is triacylglycerol, classified as a fat, which has no polar group. It is stored in the body as a fat and serves as an energy reservoir that is metabolized when needed. Triacylglycerol is formed when one molecule of glycerol reacts with three fatty acid molecules. Phosphoglycerides, the structural components of the membrane, also known as phospholipids are formed from a combination of two fatty acids, one glycerol, one phosphoric acid, and one other hydrophilic group. These groups include choline (phosphatidylcoline), ethanolamine (phosphatidylethanolamine), serine (phosphatidylserine), and inositol (phosphatidylinositol).

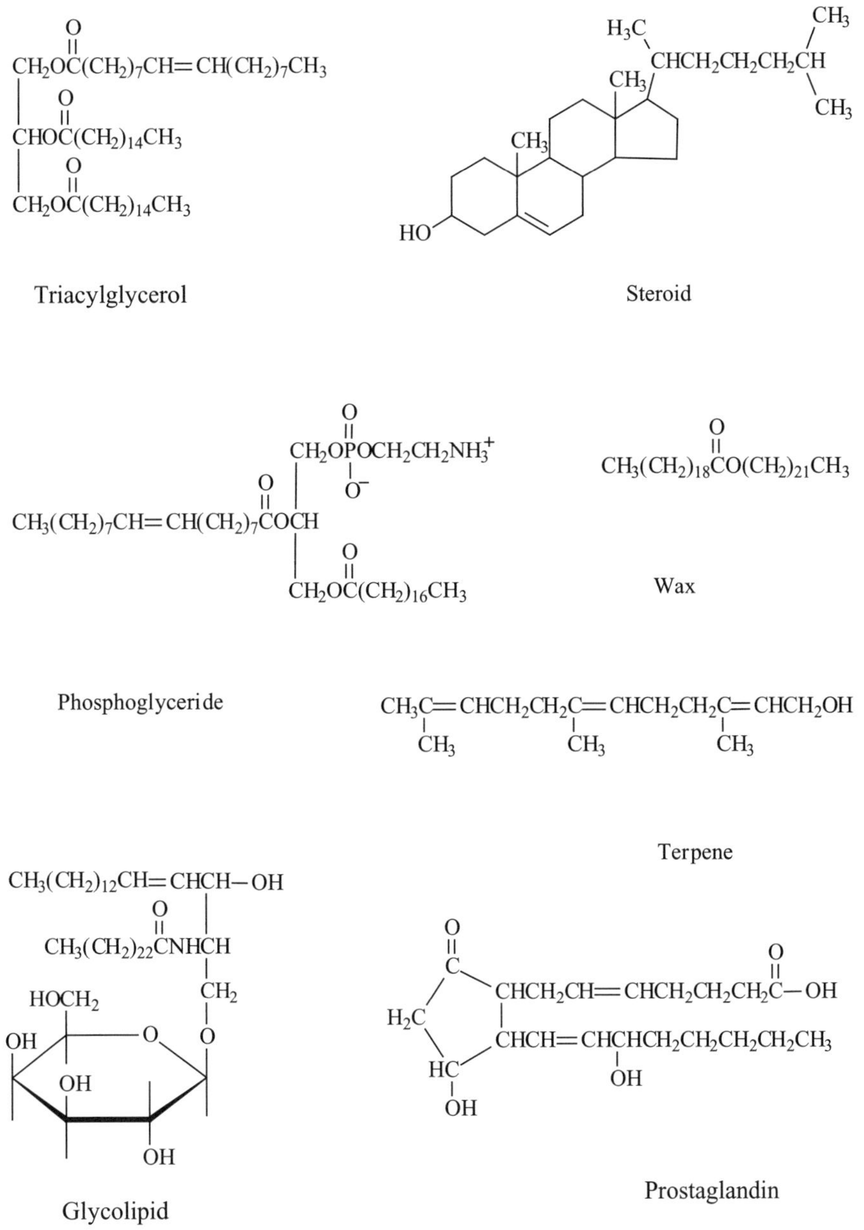

$$CH_2OC(CH_2)_7CH=CH(CH_2)_7CH_3$$

$$CHOC(CH_2)_{14}CH_3$$

$$CH_2OC(CH_2)_{14}CH_3$$

Triacylglycerol

Steroid

$$CH_3(CH_2)_7CH=CH(CH_2)_7COCH$$

$$CH_2OPOCH_2CH_2NH_3^+$$

$$CH_2OC(CH_2)_{16}CH_3$$

Phosphoglyceride

$$CH_3(CH_2)_{18}CO(CH_2)_{21}CH_3$$

Wax

$$CH_3C=CHCH_2CH_2C=CHCH_2CH_2C=CHCH_2OH$$
$$\quad CH_3 \qquad CH_3 \qquad CH_3$$

Terpene

$$CH_3(CH_2)_{12}CH=CHCH-OH$$

$$CH_3(CH_2)_{22}CNHCH$$

Glycolipid

Prostaglandin

Figure 3.8. Various lipids found in living systems.

Steroids found in living systems are hydroxy-substituted ring hydrocarbons. The most abundant steroid in living systems is cholesterol, whose structure is shown in Figure 3.8. Prostaglandins are lipids found in seminal fluids, as in prostate glands, as well as in other animal tissues. Glycolipids, also called sugar lipids, are predominantly found in membranes of brain tissues; however, selected types are universal in all cell membranes. Lipid-soluble vitamins form a minor subset of hydrophobic molecules, which are important in nutrition. The precursor molecules for most of these vitamins are terpenes, which are multiples of 2-methyl-1,3-butadiene or isoprene. Wax, another group of lipids present in nature, serves as a protective coating in leaves, skin, feathers, fur, and the outer skeleton of some insects.

3.5 INTERACTIONS DETERMINING THREE-DIMENSIONAL STRUCTURES OF BIOPOLYMERS

Biopolymers actually are three-dimensional structures. Their spatial distributions (3-D structures) are most frequently determined by noncovalent interactions which do not involve any chemical bond. These interactions make a long flexible chain (such as protein) fold, and two complementary chains (strands) of DNA pair up to form the double-helix structure. These noncovalent interactions are explained below (Albert et al., 1994).

Hydrogen Bond. This bond involves a weak electrostatic interaction between the hydrogen atom bonded to an electronegative atom (a highly polar bond) in one molecule and another electron-rich atom (consisting of a non-bonded electron pair) on another molecule. An example is the hydrogen bond in H_2O, shown in Figure 3.9. The groups capable of forming the hydrogen bonds are $-NH_2$, $-C=O$, and $-OH$ in proteins.

It is this hydrogen bonding between the base pairs of two chains of nucleic acids that gives rise to the double-stranded Watson–Crick model of DNA (Watson and Crick, 1953). The base pairs, which can form hydrogen bonding, are quite specific. For example, the T nucleotide on one chain (strand) can hydrogen bond only with the A nucleotide on the second chain (strand). Similarly the G nucleotide can hydrogen bond only with the C nucleotide. Often, multiple hydrogen bonds are the key factors in determining the architecture of large biomolecules in aqueous solutions.

Ionic Interactions. These are interactions between ions or ionic groups of opposite charges. Also, ionic interaction in the form of ion–dipole can occur

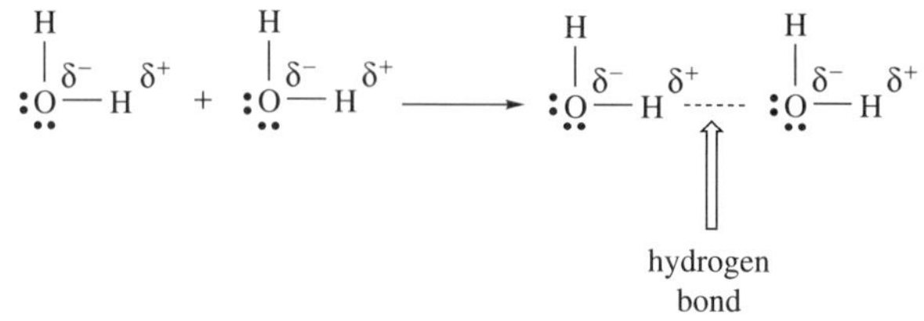

Figure 3.9. The hydrogen bond is represented by a dashed line.

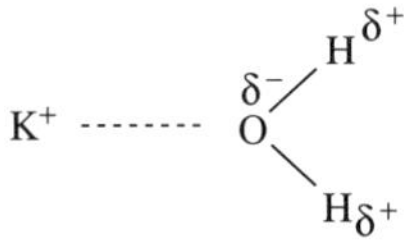

Figure 3.10. Solvation of an ion by water.

in solvation of an ion by water, as depicted in Figure 3.10. This hydration process promotes the solubility of a specific ionic specie in water.

Van der Waals Interactions. These are short-range, nonspecific interactions between two chemical species. They are nonspecific interactions resulting from the momentary random fluctuations in the distribution of the electrons of any atom, giving rise to a transient, unequal electric dipole. The attraction of the unequal dipoles of two noncovalently bonded atoms gives rise to the van der Waals interaction. These forces are responsible for the cohesiveness of nonpolar liquids and solids that cannot form hydrogen bonds.

Hydrophobic Interactions. These interactions involve the nonpolar segment (segments containing only C—C and C—H bonds) of a molecule. An example is the interaction of nonpolar tail parts between the phospholipids, which forms the outer membrane bilayer structure. Another example is the clustering of nonpolar molecules or groups in water, produced by these hydrophobic interactions, to form vesicles.

Ionic interactions, hydrogen bonding, van der Waals interactions, and hydrophobic interactions are all considered noncovalent bonds or interactions in biological systems since no chemical bonds, in the true sense, are formed. Although these interactions individually may be weaker than a covalent chemical bond, collectively they play a critical role in maintaining the three-dimensional structures of DNA, proteins, and biomembranes. Noncovalent interactions of more than one type often determine how large biopolymers (proteins) fold or unfold. Another example of noncovalent bonding is the stacking of the nitrogenous bases in the core of the Watson–Crick double helix or of the aromatic/aliphatic side chains of proteins.

Four levels of structure describe the shapes of bioploymers such as proteins and DNA. These levels are described below, using protein as an example (Stryer, 1995).

Primary Structures. In the case of proteins, primary structure refers to the sequence of the constituent amino acids that are linked through the peptide bond (hence, also called polypeptides). It is the nucleotide sequence of the gene that determines the primary structure of DNA. The primary sequence of proteins is represented in Figure 3.11.

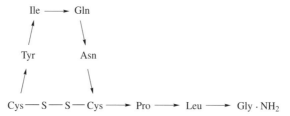

Figure 3.11. Primary sequence of a protein. (Reproduced with permission from Wade, 1999.)

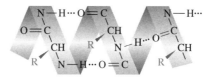

Figure 3.12. Geometric arrangement showing secondary structure of the α form of a protein. (Reproduced with permission from Wade, 1999.)

Secondary Structures. Secondary structure describes the local organization of various segments of the polymer chain. The local organization can take various spatial arrangements, also called conformation. The diverse conformations found in proteins are primarily determined by the sequence of the R groups of its amino constitutents. In the case of proteins, hydrogen bonding between certain amino acid groups leads to folding of the polymer backbone in two geometric arrangements: (i) as an α helix with a rod-like spiral structure and (ii) as a β sheet with a planar structure composed of alignments of two or more strands, which are relatively short and fully extended segments. Secondary structure of an α form is represented in Figure 3.12.

Tertiary Structures. Tertiary structures represent the overall conformation of a polypeptide chain and thus the three-dimensional arrangements of the entire amino acid polymer. An example is shown in Figure 3.13. The tertiary structures are determined by hydrophilic (ionic) and hydrophobic interactions and, in some proteins, by disulfide bonds. For monomeric proteins, which consist of a single polypeptide chain, the three-dimensional organization is determined by the primary, secondary, and tertiary structures. The complex tertiary structure of proteins is determined through its amino acid composition and sequence. In some amino acids, such as cysteine, tertiary structures enhance structural stability. The β-SH groups of cysteine are involved in inter- and intrachain disulfide linkages as represented in Figure 3.14. Also, U-shaped four-residue segments, called *turns*, are stabilized by hydrogen bonds between their arms. They are located at the surfaces of proteins.

Figure 3.13. The folded tertiary structure of a protein. (Reproduced with permission from Wade, 1999.)

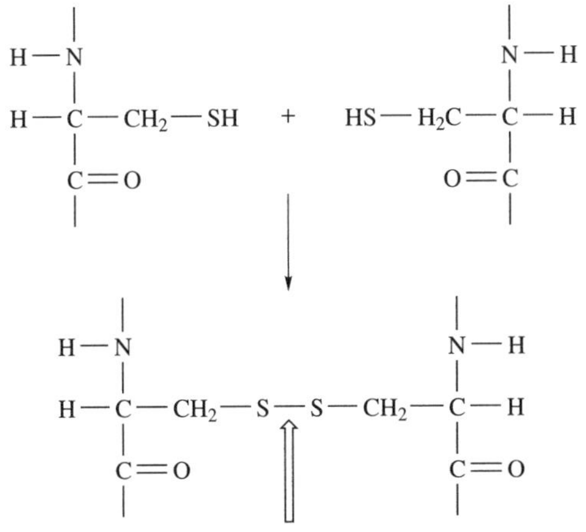

Figure 3.14. Disulfide bond formation in cysteine.

Quaternary Structures. Quaternary structures comprise the three-dimensional organization of a multimeric protein that consists of more than one polypeptide chain. The quaternary structure describes the number and relative positions of individual polypeptide chains, also called subunits.

Similarly, the primary, secondary, and tertiary structures associated with DNA determine its three-dimensional helical structure deduced in 1953 by James Watson and Francis Crick. The two polynucleotide strands, each possessing a complementary helical structure, are bound together by hydrogen bonding, which involves a specific base pairing between an A base on one strand and a T base on another (A•T base pairs) or a G base on one and a C base on another (G•C base pairs). The two polynucleotide strands can form either a right-handed helix (B form) or a left-handed helix (Z form). The

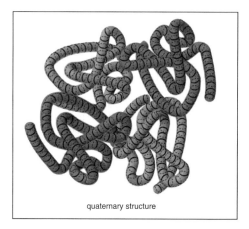

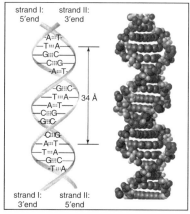

Figure 3.15. *Left*: The quaternary structure of a protein consisting of four polypeptide chains (shown in four different colors). *Right*: The double-stranded structure of DNA in two representations: the base pairing through hydrogen bonding, shown on the left, and the space-filling model, shown on the right. (Reproduced with permission from Wade, 1999). (See color figure.)

supercoiled, condensed form of DNA involves a complex interaction of tertiary structures of DNA and spherical proteins called histones. RNA is single-stranded or a complex combination of a single- and double-stranded domain. The quarternary structure of a protein and the double-helix structure of DNA are shown in Figure 3.15.

3.6 OTHER IMPORTANT CELLULAR COMPONENTS

Other important cellular constituents are AMP (adenosine monophosphate), ADP (adenosine diphosphate), and ATP (adenosine triphosphate).

During a chemical process called *oxidative/phosphorylation*, ATP synthesis in the mitochondria provides the primary source of chemical energy to drive many cellular processes. The structure of ATP is shown in Figure 3.16. It involves the formation of phosphoanhydride bonds to store energy that is released in a reaction with water (hydrolysis) to produce adenosine diphosphate (ADP), which contains two phosphate groups, or adenosine monophosphate (AMP), which contains only one phosphate group. Within the mitochondria, phosphate is added back to AMP and ADP to form ATP (stored chemical energy).

Nicotinamide adenine dinucleotide acts as a coenzyme in enzymatically catalyzed reactions that involve electron/proton transfer. The structure of NADH is shown in Figure 3.17. As discussed in Section 3.4, ribose is a five-membered ring sugar and adenosine is a nucleotide. A related compound is $NADP^+$ and its protonated form NADPH (nicotinamide adenine dinucleotide phosphate) in which a phosphate group replaces an H atom in the adenosine group.

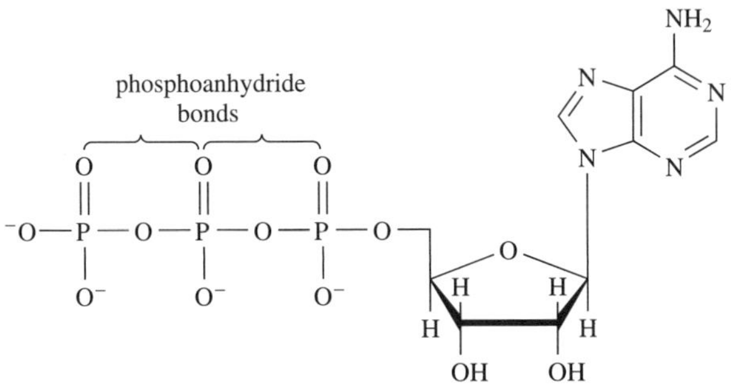

Figure 3.16. The structure of ATP.

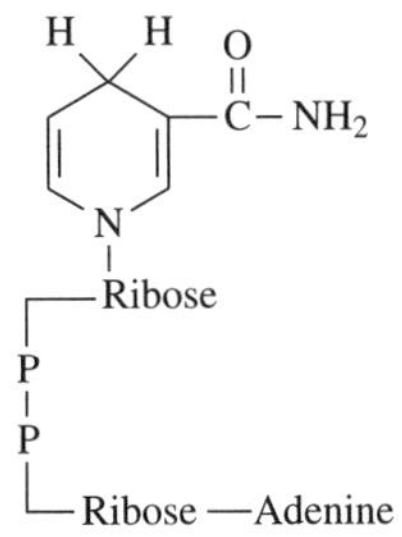

Figure 3.17. The structure of NADH.

3.7 CELLULAR PROCESSES

A living cell is a highly dynamic system in all its functions, from replication, to operation, to communication between cells. The processes involve both physical and chemical changes. Furthermore, the chemical changes can be permanent (such as protein synthesis, DNA replication) or cyclic (such as conversion of ATP into ADP and back). The chemical changes occurring are highly complex, often catalyzed (accelerated) by enzymes (reaction-specific proteins) and coenzymes (small molecules such NADH). The various dynamic processes occurring in a cell are illustrated in Table 3.4. Some classes of cellular processes fall in more than one category. For example, most enzyme-catalyzed reactions are irreversible; however, a few are reversible.

On the basis of their biological function, the cellular processes can also be divided as shown in Table 3.5. The cellular processes listed in Table 3.5 are further discussed here.

Cell Replication. The complex cellular metabolic process is necessary not only for the cell to survive but, even more important, to replicate. The primary driving force in life is the ability to replicate for the continuing survival of the species. In eukaryotic and prokaryotic cells, replication produces identical

TABLE 3.4. Cellular Processes Classified on the Basis of Their Chemical and Physical Nature

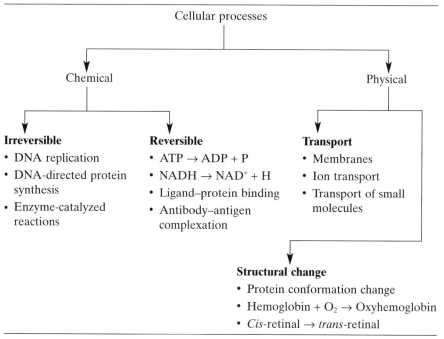

Cellular processes

Chemical — Physical

Irreversible
- DNA replication
- DNA-directed protein synthesis
- Enzyme-catalyzed reactions

Reversible
- $ATP \rightarrow ADP + P$
- $NADH \rightarrow NAD^+ + H$
- Ligand–protein binding
- Antibody–antigen complexation

Transport
- Membranes
- Ion transport
- Transport of small molecules

Structural change
- Protein conformation change
- Hemoglobin + $O_2 \rightarrow$ Oxyhemoglobin
- *Cis*-retinal → *trans*-retinal

daughter cells. Eukaryotic cells replicate through an orderly series of events constituting the cell cycle (Lodish et al., 2000). During this cycle, the chromosomes are replicated and one copy segregates into each daughter cell. Strict regulation of the cell cycle is required for normal development of the organism. This regulation is found in the nucleus of the cell coupled with the replication of the nuclear material. Loss of this control ultimately leads to cancer. In eukaryotic cells, the process of formation of the daughter cells is mitosis.

Most animal cells are diploids, because they contain two copies of each chromosome. Mitosis is the process in which a diploid parent cell duplicates its DNA, condenses it into chromosomes, and then splits to form two new cells. Thus, a parent cell produces two genetically equivalent daughter cells, each receiving two copies of each chromosome. Mitosis is the most common type of cell division.

Meiosis is another type of cell division process by which a cell duplicates its DNA, condenses it into chromosomes, but then splits in successive steps to form four new cells, each containing only one copy of each chromosome—that is, half the number of chromosomes of the parent cell. Therefore, only one step duplicates DNA; the second step does not produce DNA. Meiosis produces sex cells, also know as *gametes*. The four genetically nonequivalent

TABLE 3.5. Various Cellular Processes in Biological Systems

- **Cell replication:** Process to produce or replenish cells
  Mitosis: Process whereby the nucleus divides into two genetically equivalent
    daughter cells
  Meiosis: Specialized process whereby the two successive rounds of nuclear and
    cellular division with a single round of DNA replication yields four genetically
    nonequivalent haploid cells (germ cells)
- **Cell biosynthesis:** Production of cellular macromolecules
  Transcription: Replication of single strand of DNA to complementary RNA
  Translation: Ribosome-mediated production of proteins from messenger RNA
- **Cell energy production:** Conversion of energy forms
  Glycolysis: Conversion of chemical energy of carbohydrates to ATP
  ATP synthesis: Process of producing ATP from ADP from either:
    Chemical bond energy: Mitochondria (respiration)
    Light energy: Chloroplasts (photosynthesis)
- **Cell signaling:** Use of biomolecules initiated specific processes in adjacent or
  distant cells
  Endocrine: Hormonal signaling of distant cells
  Paracrine: Cell to adjacent cell signal
  Autocrine: Same cell stimulation signal
- **Cell Death:** Control of multicellular development
  Apoptosis: Programmed cell death through a well-defined sequence of
    morphological changes
  Cell killing: Cell death caused by ancillary cells of the immune system
- **Cell Transformation:** Genetic conversion of a normal cell into a cell having
  cancer-like properties (oncogenesis)

cells formed containing only one copy of each chromosome are also called *haploid cells* or *germ cells*.

Cell replication by either mitosis or meiosis involves a number of stages which constitute what is called the *cell cycle* (shown in Figure 3.18). A cell cycle is composed of two major phases: (i) the M phase, where the division of a parent cell into two daughter cells for a mitosis process takes place, and (ii) an interphase, which defines processes occurring within the cell before its division. The interphase itself is composed of three phases: the G1, S, and G2 phase. The main processes of these phases and their average time durations are also indicated in Figure 3.18. The actual time durations depend on the type of cell. The average time of 20 hours for the cell cycle is only representative of the time a dividing cell may take and should not be confused with the time an organ may take to double the cell population by division. The G1 phase, which is the first gap (G for gap) phase, represents the period during which proteins, lipids, and carbohydrates are synthesized to prepare a cell for DNA replication in the subsequent S phase. The S phase, where S stands for synthesis, represents the period during which the DNA is replicated (new copies of DNA are synthesized) in the nucleus, by enzyme (DNA polymerase)-

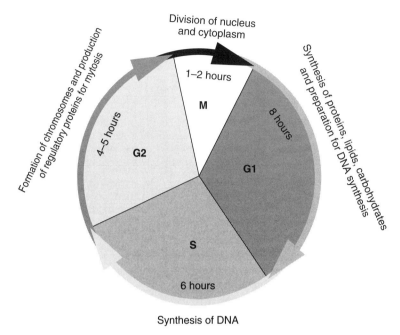

Figure 3.18. Schematics of DNA replication in the sphere of the cycle.

catalyzed reaction. This process is the key step in cell replication where the genetic code is replicated. The G2 phase is the second gap phase that prepares conditions for cell division to occur in the M phase. In the M phase, division of a nucleus followed by the division of cytoplasm produces division of a cell. The step of the division of cytoplasm, and thus the physical separation of the new daughter cells, is called *cytokinesis*.

The process of duplication of DNA in the S phase of the cell cycle is described by a replication fork scheme, as illustrated in Figure 3.19. In this process a portion of the double helix is unwound by an enzyme called *helicase*. Then another enzyme, DNA polymerase, binds to one strand of the DNA and begins moving along it, reading in the 3′ to 5′ direction to assemble (synthesis goes 5′ to 3′) a "leading strand" of nucleotide using this strand as the template. However, this process requires a free 3′ OH group, which does not exist on the chain when the two separate. To begin synthesis, an RNA primer molecule synthesized on the site provides the free 3′ OH group to begin the DNA synthesis. A second type of DNA polymerase binds to the other template strand; however, it can only synthesize discontinuous segments of polynucleotides, which are called *Okazaki fragments*. These fragments are stitched together by another enzyme, DNA ligase, into what is called the "lagging strand." When the replication process is complete, each strand of the original DNA duplex remains intact since it served as the template for the synthesis of a complementary strand. For this reason, this mode of replication is

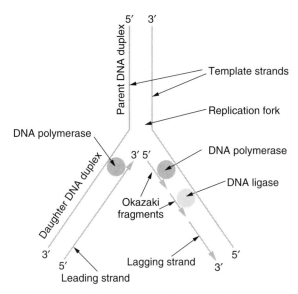

Figure 3.19. Schematics for DNA replication.

termed "semiconservative" because one half of each new DNA is old, and the other half is new (Figure 3.19).

Cell Biosynthesis. Every process in a living cell involves cellular macro-molecules, or proteins. Thus, an important function of the cell is the biosynthesis of proteins utilizing the constituent amino acids. The process of biosynthesis of proteins involves two steps: transcription and translation. In protein synthesis, three kinds of RNA cooperatively perform to lead to protein synthesis (Lodish et al., 2000; Stryer, 1995). These are:

- Messenger RNA (abbreviated as mRNA), which carries the code (genetic information) copied on it in the form of a series of three-base code called "words" or codons. Each word is associated with a specific amino acid constituent of the protein.
- Transfer RNA (abbreviated as tRNA), which deciphers the code in mRNA and binds to a specific amino acid, carrying it to the growing end of the peptide chain.
- Ribosomal RNA (abbreviated as rRNA), which complexes with a set of proteins and RNA called *ribosomes*. These complexes, moving along an mRNA molecule, catalyze the assembling of amino acids into protein chains in the process of translation.

Transcription is the process of copying the DNA template to produce mRNA, thereby transcribing the gene code from the DNA onto a strand of

RNA. It takes place in the nucleus where the DNA unwinds in small segments, allowing access to an enzyme called RNA polymerase to the single-stranded DNA. The RNA polymerase moves along the whole length of a gene (a length of DNA), producing an unzipping action and making an accurate copy of the gene in the form of mRNA. The transcription process is highly regulated and controlled at several levels.

Translation is the process by which the template encoded in the mRNA strand leads to the synthesis of proteins in a specific sequence of amino acids, thereby translating the code. The mRNA passes out of the nucleus into the cytoplasm, where it attaches to ribosomes. The ribosomes "knit" together the amino acids in a sequence dictated by mRNA. As dictated by the genetic code, specific tRNAs carrying the appropriate amino acid transfer the new amino acid to the growing polypeptide chain. The completed protein chain is then threaded into the lumen of the endoplasmic reticulum, where it folds up into its 3-D shape. Subsequently, the protein is transported to the region of the cell where it is needed. It is further processed as it passes through the Golgi apparatus. The process of translation is a high-volume activity, because several ribosomes can be traveling along a single mRNA, producing the same protein.

Cellular Energy Production. Metabolic and reproductive processes integrate the macromolecules and processes to produce energy for synthetic events and replication of cells (Voet et al., 2002; Stryer, 1995). Almost all synthetic reactions require input of energy into the reaction. There are two principal forms of energy in the cell: kinetic and potential. Kinetic energy in the cell is in the form of heat, and to do work it must flow from areas of higher to lower temperature. Differences in temperature often occur in different regions of the cell. However, most cells cannot use these heat differentials as a source of energy. In warm-blooded animals, kinetic energy is chiefly used for maintaining constant organismic temperature. However, biological systems have developed an efficient mechanism to harness kinetic energy in photons of light. Within specific organelles of plant cells (and certain prokaryotic cells), the process of photosynthesis converts light energy into potential energy, which is stored in the form of chemical bonds (i.e., glucose and fats).

There are several forms of potential energy that are biologically significant. Central to biology is the energy stored in chemical bonds. Indeed, all of the reactions described in this chapter involve the making or breaking of chemical bonds. The best example of energy production from chemical bonds is *glycolysis*, where glucose is degraded so that its stored energy can be used to do synthetic work. The second important form of potential energy is referred to as energy in a *concentration gradient*. The concentration gradients between the interior and exterior of a cell or across a compartment in a cell can exceed 500-fold (e.g., protons in a lyzosome). This gradient can be utilized to move ions and waste products across membranes. Finally, the potential energy stored as electric potential (separation of charge) is a source of energy for the cell. In almost all cells, this gradient is ~200,000 V/cm across the plasma membrane.

The interconvertibility of energy provides the cell with mechanisms to store and utilize energy for cellular processes when needed. Living cells extract energy from their environment (food) and convert it into stored energy. Because most biological systems are held at a constant temperature and pressure, it is possible to predict the direction of a chemical reaction by measuring the potential energy called Gibb's free energy, or G. The change in free energy is represented as $\Delta G = G_{products} - G_{reactants}$.

The natural direction of a reaction (spontaneous process) is that for which there is a decrease in the free energy (ΔG is negative). Two factors determine the ΔG in a system: the change in bond energy between the reactants and products (enthalpy, or H) and the change in randomness of the system (entropy, or S). Where T = temperature, ΔG is defined as $\Delta G = \Delta H - T\Delta S$.

In biological systems, most reactions lead to an increase in order and thus a decrease in entropy. An example of this is the formation of bilayers by lipids or proteins from amino acids. Both bilayers and proteins have lower entropy because they restrict the movement of the lipids and amino acids, respectively. Only a few osmotic processes decrease the entropy of the biological system. Cells must expend energy to overcome this osmotic pressure when ionic gradients are needed for cell metabolism. To this end, active transport systems are used to "pump" ions in or out of the cell in order to generate a gradient state. The energy used to perform this transport is chemical energy, usually supplied by adenosine triphosphate (ATP). The process involves two high-energy phosphoanhydride bonds linking the three phosphate groups. The conversion of adenosine diphosphate (ADP) to ATP requires energy (ΔG is positive), which is then defined as the phosphate chemical bond. It occurs in the mitochondria through coupled processes known as glycolysis, electron transport, and oxidative phosphorylation. In this system, a powerful proton gradient is generated across the mitochondrial membrane through the transfer of electrons via the electron transport or respiratory chain. The primary source of energy is that which is stored in the glucose molecule. In glycolysis, glucose is sequentially broken down to yield ATP, NADH (a reduced form of nicotinamide adenine dinucleotide), and pyruvate. The pyruvate produced in this process is utilized by the mitochondria in the Krebs or Citric Acid cycle to generate more NADH and FADH (reduced flavin adenine dinucleotide). Electrons are moved down the chain via coenzymes and eventually are transferred to O_2 to produce water. The electron translocation down this chain is coupled to proton translocation across the mitochondrial membrane. On the inner membrane of the mitochondria (these organelles have two membranes; the electron transport chain is located on the outer membrane) is located the protein machinery for the translocation of ADP and free phosphate and the ATP synthase complex. This complex utilizes the proton gradient generated by the electron transport chain to synthesize ATP.

As previously described, production of ATP utilizes aerobic respiration processes, which use oxygen as the terminal electron acceptor. This energy production system is universally used in mammalian systems to convert glucose

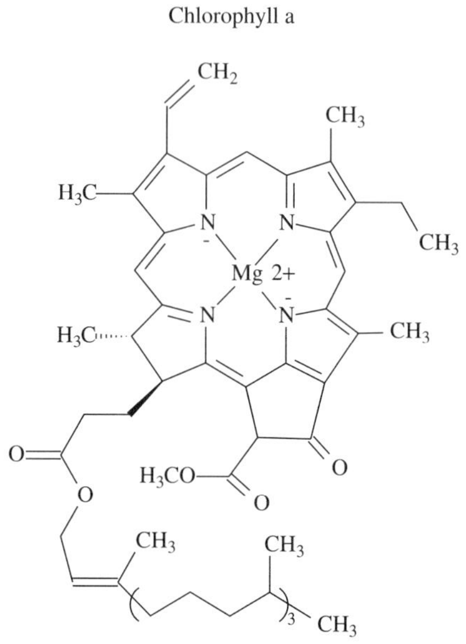

Chlorophyll a

Figure 3.20. Structure of chlorophyll *a*.

into energy. However, plants have evolved to utilize light to produce ATP, which then is used to produce glucose/sucrose needed for plant growth. This is accomplished through a specialized organelle, similar to the mitochondria, known as the *chloroplast* (see Section 6.4). In the chloroplast, light energy is trapped in the thylakoid membrane by the principal pigment (also called chromophore): chlorophyll *a* (Figure 3.20). These plant chloroplasts have two distinct photosystems. As in the mitochondria, a complex mechanism of electron movement and proton gradient formation is created within the chloroplast to produce NADPH and ATP from harvested light energy. Chlorophyll is a highly conjugated system in which electrons are delocalized among three of the four central rings and the atoms interconnecting them in the molecule. In chlorophyll, an Mg^{2+} ion is the center of the porphyrin ring and an additional five-membered ring is present. Its structure is similar to that of heme found in molecules such as hemoglobin and cytochromes where the central ion is Fe^{3+}.

Most intracellular and intercellular processes are energetically unfavorable; that is, they require energy and, thus, in terms of the free energy change, ΔG is positive. Some examples are synthesis of proteins from amino acids, synthesis of nucleic acids from nucleotides, and the contraction of muscles. Energy is released by the hydrolysis (reaction with water) of ATP to produce ADP and a phosphate, which in turn provides the needed energy for other cellular functions. ATP thus "stores" energy by converting ADP to ATP using processes such as metabolism of glucose to CO_2 and H_2O.

Cell Signaling. Cell signaling refers to a two-step process. In the first step, a signaling molecule (e.g., a ligand) is bound by a receptor at a target cell. In the second step, the receptor is activated. Membrane-bound receptors respond to a large variety of extracellular signals, which range from light and odor to hormones, growth factors, and cytokinesis. The receptors are proteins, which can undergo a chemical change or a conformational change to switch from a dormant, inactive state to an active state. The receptors can transmit signals from the surface of the cell to the interior and finally to the nucleus, resulting in up- or down-regulation of specific genes. In such a case, the signaling ligand molecule binds to the extracellular domain of the membrane—inserted receptor and subsequently triggers, in a cooperative manner, a change in the domain inside the cell.

Signaling introduced by a growth factor leads to a global response, such as growth, proliferation, and differentiation of cells. The different types of cell stimulation produced by cell signaling due to growth factors and hormones are:

- *Autocrine*: Refers to the resulting cell stimulation when the receptor resides on the same cell where the ligand is expressed (produced).
- *Paracrine*: Describes cell stimulation when the growth factor diffuses from the cell to neighboring cells in the same organ.
- *Endocrine*: Describes cell stimulation when a factor is transported through the bloodstream from the place of synthesis of the signaling molecule to other tissues (distant cells) equipped with receptors that recognize the factor. This is the case with hormones.

Cell Death. Cell death is a necessary biological response for the control of multicellular development. Cell death occurs by two principal mechanisms: apoptosis and cell killing by injurious agents, which are listed in Table 3.5. These processes are further discussed here.

Apoptosis. This process is often called *programmed cell death* (PCD) because the destiny of a cell is to die by itself. It is also referred to as "suicide by a cell." A programmed cell death is very orderly and necessary to destroy cells that represent a threat to the organism, such as cells infected with a virus, cells of the immune system, and cells with DNA damage. There are two mechanisms that cause apoptosis to occur. One is triggered by internal signals from within the cell such as an internal damage to a protein in the outer membranes of the mitochondria. The other is triggered by external signals, also called *death activators*, which bind to receptors at the cell surface. During apoptosis, cells undergo shrinkage, their mitochondria break down, releasing cytochrome C, then finally the chromatin in the nucleus degrades and eventually breaks into small, membrane-wrapped fragments.

Cell Killing. This process refers to cell death by injury. The term *necrosis* refers to a cell's response to overwhelming injury and is often called "accidental cell death." The injury may be caused by mechanical damage or by exposure to toxic chemicals. Death by injury may involve a set of changes such as swelling of cells and their organelles like mitochondria, with subsequent leakage of cellular contents, leading to inflammation of surrounding tissues.

Cell Transformation. This process involves the permanent, inheritable alteration of a cell resulting from the uptake and incorporation of foreign DNA. Normal cells can undergo this "transformation" as a result of exposure to viral or other cancer-causing agents such as mutagens. The result is the transformation of the normal cell into a cell having cancer-like properties (oncogenesis).

Cellular Processes of Current Interest. Some of the cellular processes that are receiving considerable interest today are:

- *Genomics*: Study of genetic code and its regulation.
- *Proteomics*: Study of protein function as determined by genetic code, cellular mechanism of drug–cell interactions, membrane dynamics, and transport across membrane.
- *Cloning*: Insertion of foreign DNA into cells or duplication of an organism through genetic manipulation.

3.8 PROTEIN CLASSIFICATION AND FUNCTION

Properties of Proteins. Each protein has a normal three-dimensional structure (shape) called its *native conformation* (Stryer, 1995; Lodish et al., 2000). In its native (natural) form, a protein is also sometimes referred to as the *wild type*. Genetically varied proteins, prepared artificially, are called *mutant or variant proteins*. The three-dimensional arrangement of a protein is required to perform a biological role specific to the protein. The destruction or change of this native configuration is called *denaturation*, which leads to loss of the protein's ability to carry out its intended biological function. The process of denaturation, leading to a change of the native conformation, involves breaking of the noncovalent bonds described earlier, whereby the secondary and tertiary structures are disrupted. The resulting conformations are like random coils that clump together (a process called *coagulation*) and precipitate as an insoluble aggregate, and are thus unable to carry on the intended biological activity. Denaturation can be caused by heat, alcohol, and other organic solvents, acids, and bases (large changes in the pH), certain metal ions, and various oxidizing and reducing agents. Denaturation can also be produced by mechanical disruption. Denaturation, such as that produced by heat, is often irreversible. However, in some cases, renaturation of the protein can occur, resulting in the return to the original structure/confirmation and function.

Other important properties of a protein are derived from its large size. Proteins, because of their large size, form colloids rather than solutions in water. Proteins are too large to pass through the openings of the cell membranes and, therefore, they contribute to maintaining the osmotic pressure (pressure due to concentration difference) of body fluid. Proteins can contain both hydrophilic and hydrophobic domains (amphipatic). Charged proteins are in response to the total net charge of the ionizable polar groups at the specific pH of the medium. Protein degradation occurs in a process called *hydrolysis*, which involves breaking of peptide bonds (chemically or enzymatically), by addition of water molecules, to produce free amino acids.

Types of Proteins. Proteins are very diverse in their structure and function and thus can be classified into various groups. Based on structure and shape, proteins can be divided into the following classes:

- **Fibrous proteins**, which are the main components of supporting and connective tissues such as skin, bone, and teeth. An example is a collagen, which is the most abundant protein in the body; it is a triple helix formed by three extended chains arranged in parallel.
- **Globular proteins**, which consist of polypeptides tightly folded into the shape of a ball. Most globular proteins are soluble in water. Examples are albumin and gamma globulins of the blood. Hemoglobin is another important protein belonging to this class. However, it is also an example of a protein group, often classified separately as *conjugated proteins*, which carry a conjugated group for their function. In the case of hemoglobin, this conjugated group is a heterocyclic ring called *heme*, which binds and releases molecular oxygen. The quaternary structure of protein is tightly folded and compact, with the heme group contained in the center pocket.

On the basis of their function, proteins can be classified in the following categories:

- **Enzymes**, which act as catalysts for a specific biological reaction.
- **Structural proteins**, such as collagen, which form major connective tissue and bone.
- **Contractile proteins**, such as actin and myosin, which are found in muscles and allow for stretching or contraction.
- **Transport proteins**, like hemoglobin, which carry small molecules like oxygen through the bloodstream. Other proteins transport lipids and iron.
- **Hormones**, which consist of proteins and peptide molecules. These are secreted from the endocrine glands to regulate chemical processes. An example is insulin, which controls the use of glucose.

- **Storage proteins**, which act as reservoirs for essential chemical substances. An example is ferritin, which stores iron for making hemoglobin.
- **Protective proteins**, which provide protection to various cells and tissues. Antibodies are globular proteins that provide protection against a foreign protein called antigens. Others are fibrinogen and thrombin, which are involved in blood clotting. Interferons are small proteins that provide protection against viral infection.

Protein Function. Proteins provide a wide variety of functions. Many functions involve binding with specific molecules called *ligands* or *substrates* that produce catalytic chemical reactions as well as enable them to work as switches and machines. Some of these functions are described here.

- **Enzymatic Catalysis**. Chemical reactions occurring in a cell are catalyzed by proteins called *enzymes* (Voet et al., 2002; Stryer, 1995). A specific enzyme (E) binds reversibly with a substrate, S, which may be a small molecule (e.g., glucose) or a polymer and catalyzes the conversion of the substrate to a product (such as the conversion of glucose to H_2O and CO_2). The mechanism is represented as

$$E + S \rightleftharpoons ES \rightarrow E + P \qquad (3.1)$$

- **Immune Protection**. Another important biological process describes the immunoresponse of a cell as the antibody–antigen binding. Plasma cells (found in bone marrow, lymphatics, and blood) produce antibodies in response to invasion by infectious agents (such as bacteria or a virus). The antigens are the agents inducing the formation of antibodies. A specific antibody binds to a specific antigen. The antibodies are Y-shaped molecules, and the specific binding is like a lock-and-key combination.
- **Transport Across Cell Membranes**. Another function of protein is the transport of small molecules to a specific organ by binding with them. Here a specific site (a chromophore) unit such as the heme group in the protein, hemoglobin, binds reversibly with O_2 to carry it to various sites. There is also a considerable traffic of ions and small molecules into and out of a cell. Gases and small hydrophobic molecules can readily diffuse across the phospholipid bilayer, but ions, sugars, and amino acids sometimes are transported by a group of integral membrane proteins. These involve channels, transport proteins, and ATP-powered ion pumps. Transmembrane proteins are also involved in signal transduction, allowing a cell to communicate with its "environment." This specific set of membrane proteins recognizes specific biologic signals and, by a complex mechanism involving conformation change, sends a signal to the inside of the cell.

3.9 ORGANIZATION OF CELLS INTO TISSUES

A tissue is a multicellular bioassembly in which cells specialized to perform a particular task contact tightly and interact specifically with each other. The functions of many types of cells within tissues are coordinated, which collectively allows an organism to perform a very diverse set of functions such as its ability to move, metabolize, reproduce, and conduct other essential functions. The various constituents forming tissue are shown in Table 3.6.

Figure 3.21 presents a schematic view of molecules and components that bind cells to cells and also bind cells to the extracellular matrix.

What follows is a brief description of the intercellular components.

Cell-Adhesion Molecules (CAM). These are cell-surface-bound proteins that mediate adhesion between cells of the same type (homophilic adhesion) as well as between cells of different types (heterophilic adhesion). Most CAMs are uniformly distributed within plasma membranes that contact other cells. The five principal classes of CAMs are: cadherins, immunoglobulins (Ig), selectin, mucins, and integrins. Figure 3.22 shows the major families of CAMs.

TABLE 3.6. The Various Constituents of a Tissue

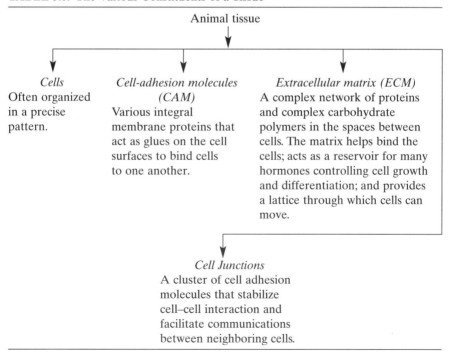

Animal tissue

Cells	Cell-adhesion molecules (CAM)	Extracellular matrix (ECM)
Often organized in a precise pattern.	Various integral membrane proteins that act as glues on the cell surfaces to bind cells to one another.	A complex network of proteins and complex carbohydrate polymers in the spaces between cells. The matrix helps bind the cells; acts as a reservoir for many hormones controlling cell growth and differentiation; and provides a lattice through which cells can move.

Cell Junctions
A cluster of cell adhesion molecules that stabilize cell–cell interaction and facilitate communications between neighboring cells.

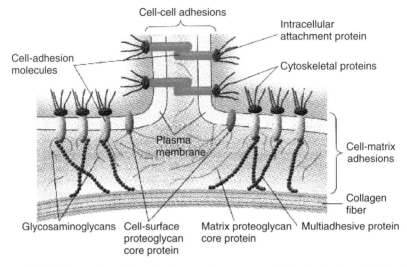

Figure 3.21. Schematics of molecules and components involved in cell adhesion. (Reproduced with permission from Lodish et al., 2000.)

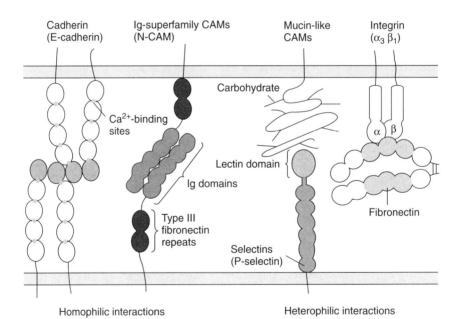

Figure 3.22. Major families of cell-adhesion molecules. Integral membrane proteins are built of multiple domains cadherin, and the immunoglobin (Ig) superfamily of CAMs mediate homophilic cell–cell adhesion. (Reproduced with permission from Lodish, et al., 2000.)

Extracellular Matrix (ECM). Cells in tissues are generally in contact with a complex network of secreted extracellular materials called the *extracellular matrix*. This matrix performs the function of holding cells and tissues together and provides an organized lattice within which cells can migrate and interact with one another. The extracellular matrix consists of a variety of polysaccharides and proteins that are secreted locally and assemble into an organized meshwork. Connective tissues, which largely consist of the extracellular matrix, form the architectural framework of an organism. An amazing diversity of tissue forms (skin, bone, spinal cord, etc.) is derived from a variation in the relative amount of the different types of matrix macromolecules and the manner in which they organize in the extracellular matrix. Connective tissues consist of cells sparsely distributed in the extracellular matrix, which is rich in fibrous polymers such as collagen. The cells are attached to the components of the extracellular matrix. In contrast, epithelial tissues consist of cells that are tightly bound together into sheets called *epithelia*. In this case, most of the tissue's volume is occupied by cells, and the extracellular matrix content is relatively small.

The extracellular matrix consists of three major proteins: (i) highly viscous proteoglycans providing cushions for cells, (ii) insoluble collagen fibers, which provide strength and resilience, and (iii) multiadhesive matrix proteins, which are soluble and bind to receptors on the cell surface. Collagen is the single most abundant protein in all living species. Although there are at least 16 types of collagen, 80–90% belong to three types classified as Types I, II, or III, depending on the type of tissue they are found in. These collagen molecules are packed together to form long, thin fibrils. Basal lamina, as tough matrix, forms a supporting layer underlying cell sheets and prevents cells from ripping apart.

Cell Junctions. Cell junctions occur at many points of cell–cell and cell–matrix contact in all tissues. There are four major classes of junctions:

- *Tight junctions*, which connect epithelial cells that line the intestine and prevent the passage of fluids, through the cell layers
- *Gap junctions*, which are distributed along the lateral surfaces of adjacent cells and allow the cells to exchange small molecules for metabolic coupling among adjacent cells
- *Cell–cell junctions*, which perform the primary function of holding cells into a tissue
- *Cell–matrix junctions*, which also perform the primary function of holding cells into a tissue

3.10 TYPES OF TISSUES AND THEIR FUNCTIONS

There are more than 200 distinguishable kinds of differentiated cells that organize to form a variety of tissues in the human body. This section describes some examples of the different types of tissues.

Epithelial Tissues. Epithelial tissues form the surface of the skin and line all the cavities, tubes, and free surfaces of the body. They are made of closely packed epithelial cells arranged in flat sheets. They function as the boundaries between cells and a cavity or space. They perform the function of protecting the underlying tissues, as in the case of skin. In the case of intestines, the columnar epithelium secretes digestive enzymes and absorbs products from the intestine. The specialized junctions between the cells enable these sheets to form barriers to the movement of water, solutes, and cells from one compartment of the body to another.

Muscle Tissues. The three kinds of muscle tissues are

- Skeletal muscles, which are made of long fibers that contract to provide the locomotion force
- Smooth muscle lines of the intestines, blood vessels, and so on.
- The cardiac muscle of the heart

Connective Tissues. The cells of connective tissues are embedded in the extracellular materials. Examples of supporting connective tissues are cartilage and bone. Examples of binding connective tissues are tendons and ligaments. Another type is fibrous connective tissues, which are distributed throughout the body and serve as a packing and binding material for most of the organs.

Nerve Tissues. These tissues are composed primarily of nerve cells (neurons). They specialize in the conduction of nerve impulses.

Tissues provide coordinated functions of the constituent cells. The functions of the tissues, however, are not just those provided by the constituent cells, but are also derived from intercellular communications and from the extracellular matrix (ECM) components. ECM acts as a reservoir for many hormones that control cell growth and differentiation. In addition, cells can move through ECM during the early stages of differentiation. ECM also communicates with the extracellular pathways, directing a cell to carry out specific functions. Through gap junction, two adjacent cells are metabolically coupled, because small molecules can pass from one cell to another.

3.11 TUMORS AND CANCERS

Understanding the molecular basis of cancer growth is of great significance to society. This understanding can perhaps lead to the prevention, early detection, and cure of cancer. *Tumor* is a general term used to describe aberrations in normal cellular behavior. Tumors differ from their normal counterparts in growth control, morphology, cell-to-cell interactions, membrane properties, cytoskeletal structure, protein secretion, and gene expression. Tumors can either be benign or malignant (neoplasms). Neoplasms multiply rapidly even in the absence of any growth-promoting factors, which are required for pro-liferation of normal cells. Furthermore, neoplastic cells are resistant to signals that program normal cell death (apoptosis). These cells also produce elevated levels of certain cell-surface receptors and produce specific enzymes. One process by which normal cells can transition to tumor cells is known as *trans-formation*. In this process, any foreign genetic material that causes cancer and has been incorporated into the chromosome is activated by factors in a process that is not yet clearly understood. These "oncogenes" interfere with normal regulatory processes, resulting in loss of replicative control in the cell.

Tumors that remain localized are called *benign*. In the case of a benign tumor, the tumor cells resemble and function like the normal cells. The benign tumors pose little threat to life and can be removed without affecting any normal functions of a tissue. Tumors in which the cells exhibit rapidly growing features and possess a high nucleus-to-cytoplasm ratio are called *malignant* tumors, *neoplasms*, or *cancers*. They often do not remain localized and they may invade the surrounding tissues. Neoplastic cells spread through the body's circulatory system and establish secondary areas of growth. This behavior is called *metastasis*. Metastatic cells also break their contacts with other cells, thus creating a degeneration of the tissue function.

HIGHLIGHTS OF THE CHAPTER

- All living creatures are made up of cells.
- Biological systems are classified as viruses (non-self-replicating), prokary-otes (no well-defined nucleus), and eukaryotes (well-organized cell and nucleus).
- Living systems are aggregates of water, amino acids, carbohydrates, fatty acids, ions, and other macromolecules called biopolymers (e.g., DNA, RNA, and proteins).
- Cells undergo differentiation to form different types of cells and as-semble to form tissues, which organize to form organs with different functions. Various organs integrate to form an organism.
- A plasma membrane consists of a structure lipid bilayer that forms a semipermeable boundary around the cell. The membrane controls the transport of food, water, ions, and signals.
- The interior contents of a cell, enclosed by the membrane, consists of various organelles (compartments) such as the cytoplasm, cytoskeleton, nucleus, mitochondria, endosplasmic reticulum (ER), Golgi apparatus, lysosomes, and peroxisomes. Plant cells have an additional organelle, called the *chloroplast*.
- Cytoplasm consists of the entire contents of a cell (except the nucleus) enclosed by a membrane. Cytoplasm forms the center of activities such as cell growth, metabolism, and replication.

- The cytoskeleton structure consists of protein filaments and is responsible for cell shape, the mechanical strength of a cell, muscular contraction, and locomotion, and it facilitates intracellular transport of organelles.
- The central control house of the cell is the nucleus, which contains DNA distributed among the chromosomes. It communicates with the cytoplasm by means of nuclear pores.
- The mitochondria, the "engine" of the cell, is responsible for the energy production in the cell.
- The endoplasmic reticulum (ER) consists of flattened sheets, sacs, and tubes of membranes and are of two types: smooth and rough. Ribosomes are present in rough ER and are responsible for protein synthesis. Smooth ER is where the synthesis of lipids and sugars occurs.
- The Golgi apparatus is responsible for monitoring the movement of proteins in and out of cells.
- There are more than 200 different types of cells present in the human body. The common types are epithelial cells, which form the lining on the inner surfaces of the body; blood cells present in blood; nerve cells, which are responsible for communication within the body; sensory cells, which detect sound and light; germ cells, which are responsible for cell reproduction; and stem cells, which are cells waiting to be assigned functions.
- The basic building blocks of cells are nucleic acids (DNA and RNA), proteins, saccharides (sugars), and lipids. DNA consists of four bases: adenine, guanine, thymine, and cytosine. In RNA, the thymine is replaced by uracil. The sequence of these bases in the DNA represents the genetic makeup of the organism.
- Proteins are formed by the polymerization of essentially only 20 different types of amino acids.
- The three-dimensional spatial distribution in biopolymers, which determine their functions, are governed by the noncovalent interactions such as hydrogen bonds, ionic bonds, van der Waals bonds, and hydrophobic bonds.
- Proteins are diverse in structures, such as fibrous and globular. They are responsible for functions such as enzyme catalysis, immune protection, and transport across cell membranes.
- Important cellular processes are: (i) cell replication to produce DNA and, subsequently, new cells; (ii) cell biosynthesis by a process of transcription to produce RNA and a process of translation to synthesize protein; (iii) cell energy production in a usable form; (iv) cell signaling to initiate processes in adjacent and distant cells; (v) cell death by programmed cell death and death due to injury; and (vi) cell transformation that produces a tumor.

- The constituents of a tissue are: (i) various cells; (ii) cell-adhesion molecules which bind cells to one another; (iii) extracellular matrix filling the space between cells; and (iv) the cell junctions.
- The extracellular matrix helps to bind the cell, acts as a reservoir for hormones controlling cell growth and differentiation, and provides a lattice through which cells can move.
- Cell junctions stabilize cell–cell infrastructure and facilitate communication between neighboring cells.
- Tumors consist of cells with structural changes produced by transformation that show aberration in the normal behavior.
- Cancer is a special type of tumor that exhibits much more rapid multiplication (proliferation) of cells compared to normal cells.

REFERENCES

Albert, B., Bray, D., Lewis, J., Raff, M., Roberts, K., and Watson, J. D., *Molecular Biology of the Cell*, 3rd edition, Garland, New York, 1994.

Audesirk, T., Audesirk, G., and Byers, B. E., *Biology: Life on Earth*, 6th edition, Prentice-Hall, Upper Saddle River, NJ, 2001.

Horton, H. R., Moran, L. A., Ochs, R. S., Rawn, J. D., and Scrimgeour, K. G., *Principles of Biochemistry*, 3rd edition, Prentice-Hall, Upper Saddle River, NJ, 2002.

Lohniger, A. L., *Biochemistry*, Worth Publishers, New York, 1970.

Lodish, H., Bark, A., Zipersky, S. L., Matsudaira, P., Baltimore, D., and Darnell, J., *Molecular Cell Biology*, 4th edition, W. H. Freeman, New York, 2000.

Solomon, S., *General Organic & Biological Chemistry*, McGraw-Hill, New York, 1987.

Stryer, L., *Biochemistry*, 4th edition, W. H. Freeman, New York, 1995.

Voet, D., Voet, J. G., Pratt, C. W., *Fundamentals of Biochemistry*, John Wiley & Sons, New York, 2002.

Wade, L. G., Jr., *Organic Chemistry, 4th edition*, Prentice-Hall, Upper Saddle River, NJ, 1999.

Watson, J. D., and Crick, F. H. C., *Molecular Structure of Nucleic Acid. A Structure of Deoxyribose Nucleic Acid*, *Nature*, **171**, 737–738 (1953).

CHAPTER 4

Fundamentals of Light–Matter Interactions

Biophotonics involves interaction of light with biological matter. Therefore, an understanding of light–matter interactions provides the fundamental basis for biophotonics. This chapter, written for multidisciplinary readership with varied backgrounds, provides knowledge of the necessary tools of optical interactions utilized in biophotonics applications which are covered in Chapters 7–16. The emphasis again is on introducing concepts and terminologies without getting into complex theoretical details.

The interaction of light at the molecular level, producing absorption, spontaneous emission, stimulated emission, and Raman scattering, is described. Then interaction at the bulk level, producing absorption, refraction, reflection, and scattering during the propagation of light through a bulk sample, is introduced. The various photophysical and photochemical processes produced in the excited state that is generated by light absorption are discussed.

A major branch of interaction between light and matter is spectroscopy, which involves the study of a transition between quantized levels. As discussed in Chapter 2, the quantized levels of biological interests are electronic and vibrational. The various spectroscopic approaches are then introduced and discussed in relation to their utilities in biological investigation.

Another major area of light–matter interaction is light emission, which can be either (a) intrinsic due to a biomaterial or (b) extrinsic due to an added molecule. This emission is utilized in a number of applications such as bioimaging (Chapters 7 and 8), biosensors (Chapter 9), microarray technology (Chapter 10), and flow cytometry (Chapter 11). The concepts of fluorescence emission and its associated properties are introduced here, providing the necessary backgrounds for these subsequent chapters.

Many biological molecules are chiral, a type of stereoisomers defined in Chapter 2. An active area of spectroscopy is the differences in interaction of a chiral molecule with left and right circularly polarized light. This difference in interactions as probed by electronic, vibrational, and Raman spectroscopy

Introduction to Biophotonics, by Paras N. Prasad
ISBN: 0-471-28770-9   Copyright © 2003 John Wiley & Sons, Inc.

and utilized to investigate conformation and dynamics of biopolymers is discussed. Another technique presented is fluorescence correlation spectroscopy, which is useful for the study of diffusion and association of biopolymers.

For further reading on the topics covered here, some general references suggested here are:

Atkins and dePaula (2002): General introduction to light-matter interaction and spectroscopy

Sauer (1995): Broad coverage of spectroscopic techniques to biochemistry

Lakowicz (1999): Coverage of various aspects of fluorescence spectroscopy

Chalmers and Griffiths (2002): Vibrational spectroscopy and its application to biology, pharmaceutics, and agriculture

Griffiths and deHaseth (1986): Principles and applications of Fourier transform infrared spectroscopy

4.1 INTERACTIONS BETWEEN LIGHT AND A MOLECULE

4.1.1 Nature of Interactions

As described in Section 2.1, light is an electromagnetic radiation consisting of oscillating electric and magnetic fields. Biological systems are molecular media. For such a medium the interaction with light can be described by the electronic polarization of a molecule subjected to an electric field. This approach is also referred to as the *electric dipole* (or simply *dipole*) *approximation*.

The linear field response, which is defined by linear dependence of the dipole moment on the electric field, gives the total molecular dipole as

$$\mu_T = -er = \mu + \alpha\varepsilon(v') \tag{4.1}$$

In the above equation, μ_T is the total electronic dipole moment vector given by the product of the electronic charge e and its position r. The term μ is the permanent dipole term in the absence of any field, and the term $\alpha\varepsilon(v')$ is the electric-field-induced dipole moment, μ_{in}, describing the polarization of the electronic cloud of a molecule in the field. In the case of polarization due to the oscillating electric field $\varepsilon(v')$ of light, the induced polarization is characterized by the dynamic polarizability term α, which is a second rank tensor that relates the directions of two vectors, the electric field ε and, as in this case, the resulting dipole, μ_{in}.

The dipolar interaction V between the molecule and a radiation field $\varepsilon(v)$ can be described as

$$V = \varepsilon(v)\mu + \varepsilon(v)\alpha\varepsilon(v') \tag{4.2}$$

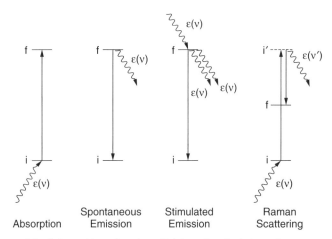

Figure 4.1. Schematics of various light–molecule interaction processes.

The first term in equation (4.2) describes interaction with a photon field of frequency ν leading to the phenomenon of absorption and emission of a photon by the molecule. The second term represents inelastic scattering, Raman scattering, where a photon of frequency ν is scattered inelastically (with a change in energy) by a molecule creating a photon of a different frequency ν′ and exchanging the energy difference with the molecule. The energy diagrams in Figure 4.1 describe these processes. The absorption process describes the transition from a quantized lower energy initial level, i, to a higher energy level, f, with the energy gap between them matching the photon energy. For electronic absorption, generally the initial electronic level i is the ground state (the lowest electronic level). If the initial level is an excited level, the resulting absorption is called an *excited state absorption*. The spontaneous emission process describes the return of the molecule from the excited state, f, to its lower energy state, i, by emission of a photon of energy corresponding to the energy gap between the two levels. The stimulated emission is a process of emission triggered by an incident photon of an energy corresponding to the energy gap between i and f. In the absence of an incident photon of same energy, there can be no stimulated emission, but only spontaneous emission.

The Raman scattering describes a process that is a single-step scattering of a photon of energy $h\nu$, being scattered into another photon of energy $h\nu′$, the difference $h(\nu − \nu′)$ corresponding to the energy gap $\Delta E = E_f − E_i$. In the schematics shown in Figure 4.1, the scattered photon of energy $h\nu′$ is of lower energy than the incident photon ($h\nu$), depositing the energy difference $h(\nu − \nu′)$ in the molecule to produce an excited state f. This process is called *Stokes Raman scattering*, which is normally studied in Raman spectroscopy. The case where $h\nu′$ is higher than $h\nu$ represents anti-Stokes Raman scattering. Very often Raman scattering is described by the photon $\varepsilon(\nu)$ taking the molecule to a virtual intermediate level, $i′$ (as shown in Figure 4.1), from which the molecule emits a photon $\varepsilon(v')$ to end up in the final state f. This level, i', is generally not a real level (no energy level exists at this energy value). If i' is a real level, then the scattering process is considerably enhanced and the process is called *resonance Raman scattering*.

As we shall see later, the absorption and emission processes are exhibited by both the electronic and vibrational states of a molecule. They also are exhibited by the quantized electronic states of atoms. However, the Raman scattering processes of significance involve vibrational states of a molecule or a molecular aggregate.

4.1.2 Einstein's Model of Absorption and Emission

The Einstein model is often used to describe the absorption and emission processes (Atkins and dePaula, 2002). In this model, the absorption process from a lower energy state i to a higher energy state f is described by a transition rate W^{abs} which is proportional both to the number of molecules, N_i, present in state i and to the density of photons ρ. Hence

$$W^{abs} = B_{if} N_i \rho \tag{4.2a}$$

where B the proportionality constant is called the *Einstein's coefficient* and the subscripts i and f simply designate that the coefficient is for states i and f.

The stimulated emission, which also requires a photon to trigger it, is also given by a similar expression for its rate, W_{st}^{emi}:

$$W_{st}^{emi} = B_{if} N_f \rho \tag{4.2b}$$

This rate is proportional both to the number N_f in the excited state f from where emission originates and to the density of photons present. The proportionality constant is the same coefficient B_{if}.

The spontaneous emission rate, however, is only proportional to the number, N_f, of molecules in the excited state f because this process does not require triggering by another photon. Hence,

$$W_{sp}^{emi} = A_{if} N_f \tag{4.2c}$$

where A_{if} is called the *Einstein's coefficient of spontaneous emission*.

The total emission rate is then given by

$$W_{st}^{emi} + W_{sp}^{emi} = N_f (A_{if} + B_{if}\rho) \tag{4.2d}$$

The net absorption of a photon is given as

$$W_{net}^{abs} = B_{if} N_i \rho - N_f (A_{if} + B_{if}\rho) \tag{4.3}$$

In the presence of stimulated emission, this process dominates over the spontaneous emission, which can then be ignored. Hence, the absorption rate of equation (4.3) becomes

$$W_{net}^{abs} = B_{if}N_i\rho - B_{if}N_f\rho = (N_i - N_f)B_{if}\rho \qquad (4.4)$$

A net absorption process takes place when this rate is positive: in other words, when $N_i > N_f$. This situation, when the lower energy state, i, has more molecules than the higher energy state, f, is called *normal population condition*.

In the case where $N_f > N_i$ the net absorption rate of equation (4.4) will have a negative sign, implying that a net stimulated emission rather than a net absorption will occur under these conditions. This net stimulated emission rate is given as

$$W_{net}^{emi} = (N_f - N_i)B_{if}\rho \qquad (4.5)$$

The situation $N_f > N_i$ for the stimulated emission where more molecules are in the higher energy (excited) state than in the ground state is called the *population inversion condition*. This population inversion is one of the conditions to achieve laser action as discussed in Chapter 5.

The quantum mechanical description of these processes provides a formal theoretical foundation for them. Quantum mechanical formulation of a transition from state i to state f is described by a quantity called the *transition dipole moment*, μ_{if}, which connects states i and f through charge/electron redistribution (hence, dipole interaction). This transition dipole moment, μ_{if}, is evaluated as an integral using standard quantum mechanical procedures described in Levine (2000).

Quantum mechanics also provides expressions relating the coefficients of stimulated absorption and emissions with the transition dipole moments as

$$B_{if} = 4\pi^2 \frac{|\mu_{if}|^2}{6\varepsilon_0 h^2} \qquad (4.6)$$

$$A_{if} = \left(\frac{8\pi h\nu^3}{c^3}\right)B_{if} \qquad (4.7)$$

The term ε_o is the dielectric constant of the medium. Therefore, the strength of a transition from state i to state f is proportional to the square of the transition dipole moment. Similarly, the coefficient of spontaneous emission, A_{if}, being related to B_{if} can be calculated from μ_{if}. Here, c = speed of light, and ν is the frequency of light.

Although, in order to quantify the strength of a transition, one needs to evaluate the transition dipole moment, one can often get qualitative information about it whether a transition is dipole-allowed ($\mu_{if} \neq 0$) or dipole-forbidden ($\mu_{if} = 0$), based on symmetry consideration of orbitals i and f (whether atomic or molecular). For example, for a molecular system with an inversion symmetry i (discussed in Chapter 2), the Laporte rule provides the following guidance:

(i) A transition from a g state (overall wave function symmetric, g, under inversion, as described in Chapter 2) to a u state or a u state to a g state is dipole-allowed ($\mu_{if} \neq 0$).

(ii) A transition from a g state to another g state or from a u state to another u state is dipole-forbidden ($\mu_{if} = 0$).

While (ii) of this simple rule always holds, (i) is not strictly true. One needs to consider the overall representations of the energy states i and f under the point group symmetry of the molecule to determine if a transition between i and f is allowed.

4.2 INTERACTION OF LIGHT WITH A BULK MATTER

Interaction of light with a bulk matter such as a molecular aggregate with a size scale comparable to or larger than the wavelength produces new manifestations such as reflection, refraction, and scattering, in addition to the absorption process. These manifestations also play an important role in understanding the interaction of a biological bulk specimen such as a tissue with light. The interaction of light with biological tissues is discussed in detail in Chapter 6, for which the materials presented here form the basis.

The bulk property is derived from an average sum of the corresponding molecular properties. Again, in the linear response theory (only linear term in electric field ε), one considers the bulk polarization P of a bulk medium induced by the external electric field ε (such as that due to light of frequency v) and given as (Prasad and Williams, 1991)

$$P(v) = \chi^{(1)}(v) \cdot \varepsilon(v) \tag{4.8}$$

P for bulk is analogous to ($\mu_T - \mu$) of equation (4.1), describing the dipole moment per volume induced by an electric field ε. The proportionality constant $\chi^{(1)}$, called *linear optical susceptibility*, is a second rank tensor that relates the vector P with another vector ε.

As described in Chapter 2, Section 2.1.2, the optical response of a medium with respect to propagation of light through it is described by a refractive index, n, which determines the phase as well as the velocity of propagation. It is related to the linear susceptibility $\chi^{(1)}$ by the relation

$$n^2(v) = 1 + 4\pi\chi^{(1)}(v) \tag{4.9}$$

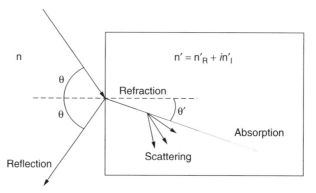

Figure 4.2. Propagation of a light ray from one medium (air) to a medium of interest (a biological tissue).

The refractive index n at frequencies corresponding to the gap between the two states of a medium (near a resonance) becomes a complex quantity given as

$$n = n_R + in_I \qquad (4.10)$$

The real part, n_R, determines the refraction and scattering, while the imaginary part, n_I, describes the absorption of light in a medium.

Figure 4.2 represents the various processes when light enters from one medium (such as air) into another bulk medium of interest. The reflection from an interface between the two bulk media (air and tissue, for example) and refraction (change of angle of propagation when entering from one medium to another) are governed by principles called *Fresnel's law* (Feynman et al., 1963) and their relative strengths are determined by the relative values of their refractive indices.

In propagation from air ($n \approx 1$) to a tissue of refractive index n', the reflectance, R (ratio of the reflected to the incidence intensities of light), is given as (in the case of normal incidence)

$$R = \left(\frac{n'_R - 1}{n'_R + 1} \right)^2 \qquad (4.11)$$

Equation (4.11) is useful in calculating reflections from various tissues.

In general, the reflectance is dependent also on the polarization of light and the angle of incidence. The two polarizations often used are s, where the polarization of light is perpendicular to the plane of incidence, and p, where the polarization is parallel to the plane of incidence.

The refraction behavior is given by Snell's law:

$$n \sin \theta = n' \sin \theta' \qquad (4.12)$$

The scattering behavior is more complex. It will be discussed in the section on light–tissue interactions in Chapter 6.

4.3 FATE OF EXCITED STATE

This section discusses the processes that can take place following an excitation created by light absorption, which takes a molecule to an excited state. These processes can be radiative, where a photon is emitted (emission) to bring the molecule back to the ground state. They can be nonradiative, where the excited-state energy is dissipated as a heat or in producing a chemical reaction (photochemistry). The return to the ground state may also involve a combination of both. The nonradiative processes producing heat involve crossing from one electronic level to another of lower energy, with the excess energy converting to vibrational energy by an interaction called *electronic–vibrational state coupling*. Subsequently, the excess vibrational energy is converted to heat by coupling to translation (this process is called *vibrational relaxation*). These processes are schematically represented in Table 4.1. In this table the star sign, such as in A*, signifies that A is in the excited state. The processes of

TABLE 4.1. Schematic Representation of Processes Involved in Electronic Excitation

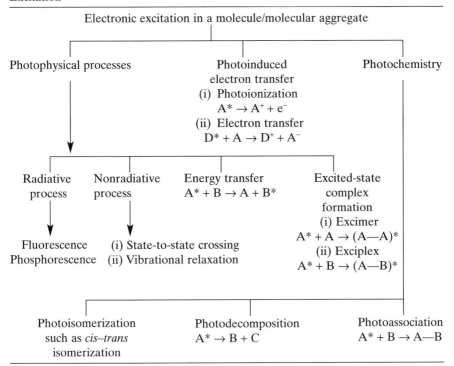

energy transfer and excited-state complex formation occur only when more than one molecule are interacting. Therefore, for such processes a minimum size molecular aggregate is a dimer (A_2) or a bimolecular (AB) unit. An excimer [an excited-state dimer, (A–A)*] or an exciplex [an excited-state complex, (A–B)*] may return to the ground state radiatively (by emitting light) or nonradiatively. An example of excimer formation is provided by an aromatic dye, pyrene, which shows a broad structureless fluorescence peaked at ~500 nm, well shifted to the red from the emission (at ~390 nm) of the single pyrene molecule. In biological fluid media, the excimer formation is diffusion-controlled. Therefore, excimer emission (such as from the pyrene dye) has been used to study diffusion coefficient (a quantity defining the diffusion rate) in membranes.

Exciplexes are excited-state complexes formed between two different molecules (or molecular units), A and B, when one of them (e.g., A) is excited (designated as A* in Table 4.1). Exciplexes can form between aromatic molecules, such as naphthalene and dimethylaniline. Just like in the case of an excimer, the resulting emission from the exciplex (A–B*) is red-shifted, compared to that from A*. Of biological interest has been the exciplex formation between a metalloporphyrin and a nucleic acid or an oligonucliotide, which can provide structural information on the microenvironment of the metalloporphyrin (Mojzes et al., 1993; Kruglik et al., 2001). Exciplex formation has been studied for double-stranded polynucleotides and natural DNA having regular double-helix structures.

The photochemical processes, listed under photochemistry in Table 4.1, are of considerable significance to biology, because they occur in biological materials with important consequences. These processes in biological materials are discussed in detail in Chapter 6, "Photobiology," with specific examples provided there.

The state-to-state crossing and the various possible radiative and nonradiative processes in an organic structure are often represented by the so-called Jablonski diagram shown in Figure 4.3. In this diagram, the radiative processes are represented by a straight arrow, whereas nonradiative processes (also sometimes referred to as *radiationless transition*) are represented by a wiggly arrow.

The ground state of most molecules (organics in particular) involves paired electrons; therefore, their total spin $S = 0$ and the spin multiplicity $2S + 1 = 1$. These are singlet states and, in the order of increasing energy from the ground-state, singlets are labeled S_0, S_1, S_2, and so on. An exception is the common form of O_2, where the ground-state is a triplet with the spin $S = 1$ and the spin multiplicity $2S + 1 = 3$. Therefore, the ground state of oxygen is T_0. This case is not represented in Figure 4.3, which only depicts the case of molecules with a singlet ground state, S_0. For molecules whose ground states are S_0, the excitation of an electron from a paired electron pair to an excited state can produce either a state where the two electrons are still paired (like S_1) or where the two electrons are unpaired (a triplet, T state). The excited

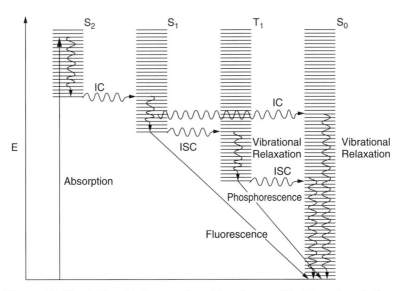

Figure 4.3. The Jablonski diagram describing the possible fates of excitation.

triplet configurations are labeled as T_1, T_2, and so on, in order of increasing energies.

Quantum mechanical considerations show that for the excitation to the same orbital state, the energy of the excited triplet state (say T_1 state) is lower than that of its corresponding singlet state (S_1 in this case). In Figure 4.3 the possibilities for the fate of an excitation to a higher singlet S_2 manifold are described. The horizontal closely spaced lines represent the vibrational levels. Suppose the excitation is to an electronic level, S_2. A nonradiative crossing from the S_2 state to S_1 is generally the dominant mechanism. Only very few molecules (e.g., azulene), show emission (radiative decay) from S_2. This crossing between two electronic states of the same spin multiplicity (such as from S_2 to S_1) is called *internal conversion* (IC). This IC process is then followed by a rapid vibrational relaxation where the excess vibrational energy is dissipated into heat, the molecule now ending up at the lowest, zero-point vibration level ($v = 0$, see Chapter 2 on vibration) of the S_1 electronic state. From here, it can return to the ground electronic state S_0 by emitting a photon (radiatively). This emission from a state (S_1) to another state (S_0) of same spin multiplicity is called *fluorescence* and is spin-allowed (observes the rule of no change of spin value). It, therefore, has a short lifetime of emission, generally in the nanoseconds (10^{-9}-sec) range. Alternatively, the excitation may cross from S_1 to T_1 by another nonradiative process called *intersystem crossing* (ISC) between two states of different spin. This crossing (change) of spin violates the rule of no change of electron spin during a change of electronic state and is thus called a *spin-forbidden transition*. This spin violation (or occurrence of a spin-forbidden transition) is promoted by spin–orbit coupling, described in

Chapter 2, which relaxes the spin property by mixing with an orbital character. Followed by a rapid vibrational relaxation, the excitation ends in the zero-point vibrational level of the T_1 state. A radiative process of emission from here leading to S_0 is spin-forbidden and is called *phosphorescence*. Again, the spin-violation occurs because of spin–orbit coupling (Chapter 2). This is a weaker emission process and, therefore, has a long lifetime. Some of the phosphorescence lifetimes are in seconds. Many photochemical processes originate from this type of long-lived triplet state. Heavy metals, molecular oxygen (having a triplet ground state), paramagnetic molecules, and heavy atoms such as iodine increase the intersystem crossing rate, thus reducing the fluorescence and enhancing the process taking place from the excited triplet state.

Finally, there can also be a nonradiative intersystem crossing from T_1 to S_0.

4.4 VARIOUS TYPES OF SPECTROSCOPY

Spectroscopy deals with characterization and applications of transition between two quantized states of an atom, a molecule, or an aggregate. A description of the nature of interactions and various spectroscopic transitions has already been presented in Section 4.1.

The various spectroscopic transitions and methods useful for biophotonics are described in Table 4.2. Electronic transitions are not efficiently excited by a Raman process. A vibrational excitation generally decays by a nonradiative process and, therefore, exhibits no fluorescence; exceptions are small molecules such as CO_2.

A spectrum is a plot of the output intensity of light exiting a medium as a function of its frequency (or wavelength). For absorption, a broad band light source generally is used, and its transmission (and hence, attenuation or absorption) is obtained as a function of frequency or wavelength. For an emission or a Raman process, the medium is excited at a specific wavelength (called *excitation wavelength*), and the emitted or scattered radiation intensity is monitored as a function of wavelength.

Two types of spectrometers are used for obtaining the spectral information on the intensity distribution as a function of wavelength:

- *Conventional Spectrometers*. In this case a dispersive element such as a prism or a diffraction grating separates light with different frequencies into different spatial directions. The intensities of spatially dispersed radiation of different wavelengths may be obtained by using a multielement array detector where each array detects radiation of a narrow spectral range centered at a specific frequency. This type of spectrometer allowing simultaneous detection of lights of all wavelengths is also referred to as a *spectrograph*. Alternatively, one may scan the angle of a diffraction grating (or a prism) so that at a given angle only a narrowly defined wave-

TABLE 4.2. The Various Spectroscopies Useful for Biophotonics

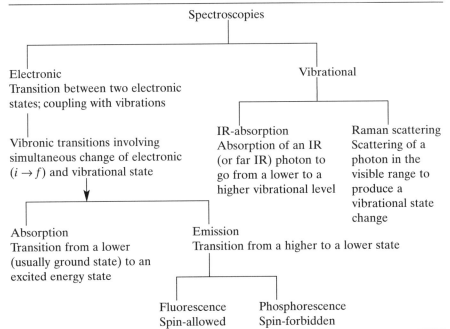

length region passes through as narrow aperture (a slit) to impinge on the photodetector.

- *Fourier Transform Spectrometers.* Most modern IR-absorption spectrometers employ this technique and hence are often called FT–IR spectrometers (FT is the abbreviation for Fourier transform). Here, instead of using a dispersive element, such as a diffraction grating or a prism to disperse the different frequencies, the information on intensity distribution as a function of frequency is obtained by using an optical device called a *Michelson interferometer.* In a Michelson interferometer, the beam from a broad-band light source (infrared light source in the case of FT–IR) is split into two beams by a beam splitter, with one beam going to a fixed mirror and the other incident on a movable mirror. After reflection from the two mirrors, the beams recombine at the beam splitter and then pass through the sample. This arrangement is shown in Figure 4.4. Depending on the relative positions of the two mirrors, the beams can constructively or destructively interfere for various frequencies. The plot of the interference intensity, called the *interferogram,* as a function of the position of the movable mirror, is related to the intensity of light as a function of frequency by a mathematical relation called *Fourier transform.* Thus, by performing a Fourier transform with the help of a computer, one can obtain

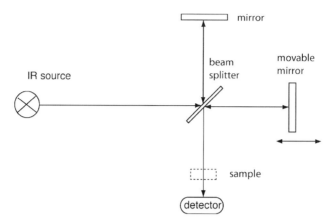

Figure 4.4. Schematics of Michaelson interferometer.

the plot of intensity versus. the frequency of light, which gives the spectrum. For more details of this technique, the reader is referred to the book by Griffiths and deHaseth (1986). A major advantage of the Fourier transform method is that one can monitor the entire spectrum continuously with a good sensitivity. Recently, FT–Raman spectrometers have also become commercially available (Chase and Robert, 1994; Lasema, 1996). In this case, a near-IR monochromatic source, such as the beam at 1064 nm from a CW Nd: YAG laser, is used to generate the Raman spectra. This laser is described in Chapter 5. The same Fourier transform technique using a Michelson interferometer is used for FT–Raman spectroscopy.

Spectral transition from a quantized state i (initial) to another quantized state f (final) does not occur at one (monochromatic) frequency ν (or wavelength λ). There is a spread of frequency ν of transition, called the *linewidth*, which is quantified by the term $\Delta\nu$ and is often defined as the width (spread of frequency) of a spectroscopic transition at half of the maximum value (also called full width at half maximum, or FWHM). The width corresponds to the broadening of a spectroscopic transition, also known as *line broadening*. There are two mechanisms of line broadening:

1. Inhomogeneous broadening caused by a statistical distribution of the same type of molecule or biopolymer among energetically inequivalent environment. For example, the biopolymer molecules may be distributed at various sites with different local structures, and thus interactions, producing a distribution of their site energies. This type of statistical broadening gives rise to a Gaussian distribution of the intensity, also called a *Gaussian lineshape*.

2. Homogeneous broadening due to the limited lifetime of the states involved in the transition. Here the frequency (energy) spread is caused by the condition imposed by the Heisenberg's uncertainty principle that

a finite lifetime (uncertainty in time) produces a corresponding uncertainty in energy (frequency), the products of their uncertainty being equal to the Planck's constant.

4.5 ELECTRONIC ABSORPTION SPECTROSCOPY

Electronic absorption is often used for a quantitative analysis of a sample (Tinoco et al., 1978). The basic absorption process uses a linear absorption of light from a conventional lamp (e.g., a Xe lamp), which provides a continuous distribution of the electromagnetic radiation from UV to near IR. The spectrometer used for this purpose is often called a *UV-visible spectrometer*, and it measures linear electronic absorption. This linear absorption is defined by the Beer–Lambert's law, according to which the attenuation of an incident beam of intensity I_o at frequency v is described by an exponential decay whereby the output intensity I is given as

$$I(v) = I_0 e^{-k(v)bc} = I_0(v)10^{-\varepsilon(v)bc} \tag{4.13}$$

The more frequently used coefficient $\varepsilon(v)$ expressed in L (liters) mol^{-1} cm^{-1}, rather than the coefficient k, is called the molar extinction coefficient at frequency v; c is the molar concentration (mol/L). This ε should not be confused with ε used in 4.1.1 and 4.2. This concentration c is not to be confused with the term c used earlier to represent the speed of light in vacuum. The term b (in cm) is the optical path length defined by the length of the absorbing medium through which the light travels. This situation is illustrated by Figure 4.5. In terms of the photon picture, a linear absorption process involves the absorption of a single linear photon by a molecule to excite an electron from a lower (ground) level to an excited level.

Other terms used to describe absorption or attenuation are

Absorbance: $A(v) = \log_{10}\left(\dfrac{I_0(v)}{I(v)}\right) = \varepsilon(v)bc$

(∴   A more familiar, simpler equation $A = \varepsilon(v)bc$)   (4.14)

Transmittance: $T(v) = \dfrac{I(v)}{I_0(v)}$

Optical density (OD) $= \log_{10}(1/T)$

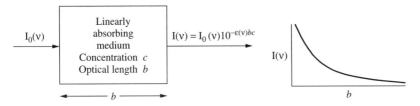

Figure 4.5. A linear absorption process.

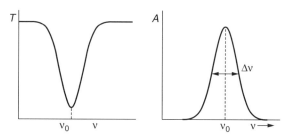

Figure 4.6. Absorption spectra in two different representations, obtainable on most commercial spectrometers.

The terms transmittance and optical density consider intensity losses due to both absorption and scattering when light travels through a medium. If the dominant contribution is due to absorption, then $OD = A$.

A typical absorption spectrum is exhibited as a plot of either T versus v (or λ) or A versus v. Typical absorption curves are shown in Figure 4.6. In the case of transmission, a continuous output from a lamp (or any broadband light source) shows a dip at the absorption frequency, v. In the case of absorption, a peak appears at the absorption frequency v_0. Δv, the full width at half maximum (FWHM), defines the width (frequency spread) of an absorption band (transition) at half of the maximum absorbance for the band.

The absorption spectra can be used to identify a molecular unit called a *chromophore* where an electron being excited is primarily localized. The transition in a chromophore produces absorption at a specific frequency (or frequency range). This absorption frequency can be dependent on the environment of the chromophore. Therefore, from the shift of the absorption band one can also probe the interactions in which the chromophore or the chromophore containing bioassemblies may be involved. Quantitatively, knowing the molar extinction coefficient $\varepsilon(v)$ at frequency v for an identified chromophore, one can obtain the number density (concentration) of the chromophore.

In the case where a bioassembly (biological medium) may contain many absorbing chromophores of known molar excitation coefficients, the absorbance A is measured at a number of frequencies to obtain the concentrations of various chromophores.

Types of Electronic Transitions. The various electronic transitions encountered in a bioassembly are listed here (Atkins and de Paula, 2002):

(a) σ–σ^* *Transitions.* They involve the promotion of an electron from a bonding σ orbital to an antibonding σ^* orbital. The required energy for this transition is large. For example, methane which only consists of C–H σ bonds exhibits a σ–σ* transition at 125 nm. These transitions, being of high energy (vacuum u.v.), are not suitable as spectroscopic probes of biomolecules.

(b) *d–d Transitions*. These transitions are encountered in an organometal-lic biomolecule involving a transition metal complex with organic ligands. Examples are hemoglobin involving Fe or a porphyrin involv-ing Mn or Zn. The *d–d* transitions involve the excitation of an electron from one *d* orbital of the transition metal atom to another *d* level, the splitting between the *d* orbitals being determined by the surrounding ligands. The rare-earth complexes, also used for probing and imaging biological structures, involve transitions of *f* electrons in rare-earth ions (e.g., Eu^{3+}). The molar extinction coefficient ε for these transitions are low.

(c) *π–π* Transitions*. Associated with double bonds (e.g., C=C or C=O) or a conjugated structural unit, they involve the promotion of an electron from a bonding π orbital to an antibonding π* orbital. This absorption is often also represented as a ππ* transition. An important example of this type of transition is provided by the absorption in the 11-*cis*-retinal chromophore in eye which forms the basis of the photochemical mech-anism of vision. These transitions are relatively strong, with the molar extinction coefficients ε being between 1000 to 10,000 L mol^{-1} cm^{-1}.

(d) *n–π* Transition*. It involves the excitation of an electron from a non-bonding orbital to an empty π* orbital. An example is the excitation of an electron of the electron pair in the outer nonbonding orbital of oxygen in a >C=O group to the π* MO of the C=O double bond. This absorption is also represented as an nπ* transition. These are weak transitions (symmetry forbidden) with molar extinction coefficients in the range of 10 to 100 L mol^{-1} cm^{-1}.

(e) *Charge Transfer Transition*. This transition, giving rise to a charge transfer band, involves the excitation of an electron from the highest occupied orbital centered on one atom (or a group) to the lowest un-occupied orbital centered on another atom or a group. In the case of an organometallic molecule involving a transition element, a *dπ** tran-sition promoting a *d* electron of the metal to an empty π* orbital of the ligand is called a metal-to-ligand charge-transfer transition (MLCT). The reverse π*d* or *nd* transition indicating a photoinduced charge trans-fer from the ligand to the metal is called the ligand-to-metal charge-transfer transition (LMCT).

Another type of charge transfer transition involves an asymmetric mole-cule containing both an electron donor and an electron acceptor. These transitions also occur where two neighboring molecules are involved, one of them of donor type, the other of acceptor type. Here a charge-transfer band describes the transition to an excited state where there is an additional charge transfer from the electron donor to the electron acceptor producing an increase in the permanent dipole moment (i.e., the excited state is more ionic). Sometimes, a reverse process also occurs by which the dipole moment is reduced in the excited state. The molar extinction coefficients of the charge-transfer transitions can be quite large (greater than $10,000 \, L \, mol^{-1} \, cm^{-1}$).

A molecular specie, particularly a biomolecule, can exhibit a complex absorption spectra consisting of many absorption bands due to the various types of spectroscopic transitions discussed above.

A class of biological molecules, called *macrocycles*, which consist of a large-size π-electron-rich fused ring with a considerable delocalization of the π electrons, exhibits a number of π–π* transitions, because there are many π and π* orbitals (discussed in Chapter 2). An important example is a porphyrin such as the heme group in hemoglobin. The absorption spectra of porphyrins exhibit an intense π–π* transition in blue region at 400 nm which is called the "Soret band." In addition, there are a series of weaker π–π* transitions in the region, 450–650 nm which are called "Q bands." Chlorophyll *a*, discussed in Chapters 3 and 6, is a porphyrin derivative. Its absorption spectrum is shown in Figure 4.7. The absorption at 650 nm is responsible for the green color of chlorophyll, as the absorption in the red (650 nm) produces a green transmitted or scattered light (complementary color). Another examples is HPPH, a

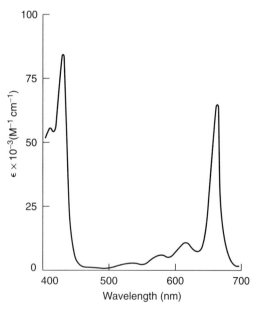

Figure 4.7. Absorption spectrum of chlorophyll *a* in ether. (Reproduced with permission from Goodwin and Mercer, 1972.)

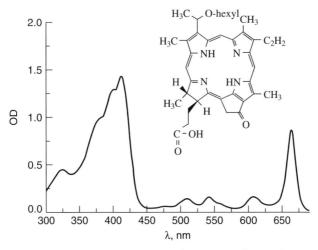

Figure 4.8. Absorption spectrum of HPPH (2-divinyl-2-(1-hexyloxyethyl)pyrophe-ophorbide), a drug for photodynamic therapy. Water solution, $C = 22\,\mu M$.

porphrin based drug which is used in photodynamic therapy discussed in Chapter 12. The absorption spectrum of HPPH is shown in Figure 4.8.

4.6 ELECTRONIC LUMINESCENCE SPECTROSCOPY

Luminescence spectroscopy deals with emission associated with a transition from an excited electronic state to a lower state (generally the ground state) (Lakowicz, 1999; Lakowicz, 1991–2000). Biological molecules at room temperature exhibit fluorescence. Phosphorescence from a triplet excited state to the singlet ground state is rarely observed at room temperature. One-photon absorption produces a fluorescence band that is red-shifted (to a lower energy). This shift between the peak of the absorption band and that of the fluorescence band is called *Stokes shift*. The amount of Stokes shift is a measure of the relaxation process occurring in the excited state, populated by absorption. The difference in the energy of the absorbed photon and that of the emitted photon corresponds to the energy loss due to nonradiative processes. The Stokes shift may arise from environmental effect as well as from a change in the geometry of the emitting excited state. Figure 4.9 shows the absorption and the emission spectra of fluorescein, a commonly used dye.

Although fluorescence measurements are more sophisticated than an absorption (transmission) experiment, they provide a wealth of the information about the structure, interaction, and dynamics in a bioassembly. Also, fluorescence imaging is the dominant optical bioimaging technique for biophotonics.

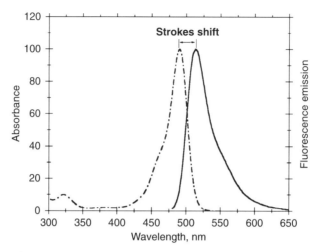

Figure 4.9. Absorption and fluorescence spectra of fluorescein in buffer (pH 9.0).

The fluorescence spectroscopy includes the study of the following features to probe the interaction and dynamics:

- Fluorescence spectra
- Fluorescence excitation spectra
- Fluorescence lifetime
- Fluorescence quantum efficiency
- Fluorescence depolarization

The fluorescence spectrum is obtained by exciting the molecules in a medium using a conventional lamp (a xenon lamp or a mercury xenon lamp). For excitation, a wavelength range corresponding to the absorption band is selected by a broad-band cutoff filter that only allows light at frequencies higher than that of emission. The fluorescence spectrum comprised of the fluorescence intensity as a function of frequency is obtained in a fluorescence spectrometer which includes a dispersive element (grating). Lasers are often used as a convenient and powerful source for one-photon excited fluorescence in which case it is called *laser-induced fluorescence* (LIF).

The fluorescence excitation spectra (sometimes simply referred to as *excitation spectra*) give information on the absorption (excitation) to the state that produces maximum fluorescence. Here the total fluorescence or fluorescence at the maximum frequency is monitored and the excitation frequency of a lamp or a tunable laser source is scanned to obtain the excitation spectrum. A maximum in the excitation spectrum corresponds to the frequency of a photon, where absorption produces maximum fluorescence.

Fluorescence lifetime represents the decay of fluorescence intensity. A simple fluorescence decay is exponential (first-order kinetics) involving a rate constant k which describes the decay of the fluorescence intensity I as $I = I_o e^{-kt}$ where I_o is the fluorescence intensity at the start of fluorescence (at $t = 0$). This behavior is called a *single exponential decay*. The rate constant k has two contributions, a radiative decay constant k_r characterized by a radiative lifetime τ_r and a nonradiative decay constant k_{nr}, characterized by a nonradiative lifetime τ_{nr}. Thus:

$$k = k_r + k_{nr} = \frac{1}{\tau_r} + \frac{1}{\tau_{nr}} = \frac{1}{\tau}$$

(4.15)

$$I = I_0 e^{-t/\tau}$$

From experimental measurements and fit of the decay to a single exponential, one obtains the overall fluorescence lifetime τ.

The radiative lifetime τ_r is inversely proportional to the strength of the transition dipole moment. It can be shown that it is related to the maximum extinction coefficient, $\varepsilon_{max}(v)$, of the absorption to the emitting state as follows:

$$\tau_r = \frac{10^{-4}}{\varepsilon_{max}(v)} \quad \text{sec}$$

(4.16)

In this equation, $\varepsilon_{max}(v)$ is in the unit of $L\,mol^{-1}\,cm^{-1}$.

Two methods of measurement of fluorescence lifetimes are:

(a) *Time Domain Measurement.* Here a short pulse, generally from a pulse laser source, excites the fluorescence, and decay of fluorescence is measured. The fluorescence lifetimes are generally in the range of nanoseconds to hundreds of picoseconds. For nanosecond decay, one utilizes a fast scope or a boxcar technique, whereas for lifetimes in hundreds of picoseconds, one utilizes a streak camera.

(b) *Phase Modulation Measurement.* This method utilizes a modulated excitation source (a lamp or a mode-locked laser, the latter of which is discussed in Chapter 5) and is based on the principle that a finite fluorescence lifetime causes the fluorescence waveform to be phase-shifted by an amount φ with respect to the waveform of the exciting light. This phase shift φ is related to the lifetime by the following equation:

$$\tan \varphi = \omega\tau$$

(4.17)

where ω is the modulation frequency (rate of modulation of exciting light). Therefore, from a measurement of the phase shift using a phase-sensitive detector (a lock-in amplifier), one can obtain the fluorescence lifetime τ. Several companies now sell instruments for phase-modulation lifetime measurements.

A rapidly growing field in photobiology is time-resolved fluorescence spectroscopy. Here the entire fluorescence spectrum is obtained as a function of time to monitor a spectral change induced by any dynamic change in the local configuration of the fluorescent unit called *fluorophore* or *fluorochrome*.

A nonexponential decay or a multiexponential decay (fit into a weighted sum of a number of exponentials) represents more complicated decay kinetics of the excited states. Some of the processes are (i) decay of the excited states through a number of channels (to different lower states), (ii) bimolecular decay involving interaction between two molecules, (iii) diffusion-controlled decay, and (iv) Förster energy transfer from an excitation donor molecular unit (the molecule absorbing the photon) to an excitation acceptor (the molecule which accepts the excitation and then may emit). Förster energy transfer is efficient when the emission spectrum of the donor molecule overlaps with the absorption spectrum of the acceptor molecule. With a significant overlap, the energy transfer is also called a *resonance energy transfer* and the fluorescence from the acceptor molecule is also called *fluorescence resonance energy transfer* (FRET). FRET has also found useful application for bioimaging, as discussed in Chapters 7 and 8 on bioimaging.

The rate of energy-transfer under a dipole–dipole transfer mechanism is inversely proportional to the sixth power of their separation. This dependence of energy transfer has been used to determine distance of separation between the excitation donor and acceptor sites and their mobilities.

The fluorescence quantum efficiency (also called *quantum yield*) Φ is defined as

$$\Phi = \frac{\tau_{nr}}{\tau_{nr} + \tau_r} = \frac{\kappa_r}{\kappa_r + \kappa_{nr}} \tag{4.18}$$

The quantum yield is a quantitative measure of the ratio of the number of photons emitted to the number of photons absorbed. In the absence of any nonradiative decay, the quantum yield Φ equals 1; that is, the excited state decays only by a radiative (fluorescence) process. This is the case producing the most efficient fluorescence; therefore, ideal fluorophores to be used as fluorescent probes should have a quantum yield as close as possible to 1. Fluorescence efficiency (quantum yield) serves as an excellent probe for the environment surrounding a fluorophore in a bioassembly.

Fluorescence depolarization is a measure of the loss of polarization of fluorescence by a number of dynamic effects such as rotation of the fluorophore. The polarization P of fluorescence is defined as

$$P = \frac{(I_\| - I_\perp)}{(I_\| + I_\perp)} \tag{4.19}$$

Another quantity also representing polarization of fluorescence is called *fluorescence emission* anisotropy, defined as $r = (I_\| - I_\perp)/(I_\| + 2I_\perp)$. Here $I_\|$ and $I_\perp$

are the fluorescence intensities polarized parallel and perpendicular to the polarization of excitation light.

The polarization ratio is determined by the relative orientation of the transition dipole moment (a vector) connecting the emitting excited state to the ground state (also called *emission dipole*) and the transition dipole moment connecting the ground state to the absorbing excited state (also called *absorption dipole*). For a randomly oriented rigid medium (molecular not being able to change the orientation) averaging over all possible molecular orientation yields $P = +1/2$ for the case when absorption (excitation) and emission dipoles are parallel, and $P = -1/3$ for the case when they are perpendicular to each other. A significant reduction in the magnitude of P indicates fluorescence depolarization. Therefore, a study of P or r for a fluorophore attached to a biopolymer or a biomembrane can provide information about the rotational mobility of its microenvironment. The P and r measurements are also used to measure rotational diffusion of molecules in biological systems such as membranes and cytosols (the biological systems are described in Chapter 3).

4.7   VIBRATIONAL SPECTROSCOPY

Vibrational spectroscopy comprises IR spectroscopy and Raman spectroscopy (Chalmers and Griffiths, 2002). In IR spectroscopy, the absorption of an IR (or far IR) photon produces a change in vibrational levels. The selection rule for a vibrational transition using a harmonic oscillator model discussed above is $\Delta v = 1$ for any vibrational mode. Overtone ($\Delta v > 1$) absorption is possible, but it is much weaker. The overtone absorption in water is, however, important in some wavelength ranges (i.e., $\sim 1.9\,\mu_m$.) The IR absorption spectrum consists of a series of $\Delta v = 1$ vibrational transitions for different vibrational modes of a molecule. For most vibrations, it involves an absorption from a $v = 0$ (zero-point vibrational level) to a $v = 1$ level. However, some low-frequency vibrations can be thermally populated, leading to absorption starting from $v \neq 0$. These are called *hot bands*. For a truly harmonic vibrational mode, all vibrational spacings between adjacent levels are equal and, therefore, all $\Delta v = 1$ transition will be at the same IR frequency. However, anharmonic interactions make the spacings change; therefore, different $\Delta v = 1$ transitions occur at different IR frequencies. The anharmonic effect has been discussed in Chapter 2.

The strength of an IR transition for a vibrational mode (described by a normal coordinate that consists of a vibrational displacement pattern) is determined by the dipole moment (μ) derivative $d\mu/dQ_k$, where Q_k is the normal coordinate for vibrational mode k. The normal mode of vibration (displacement) producing the largest change in the dipole moment exhibits the strongest vibrational transition in IR (most intense band in the IR spectra). For a molecule with an inversion symmetry, only u-type vibrational modes are excited by IR absorption.

In Raman spectroscopy, a photon of frequency in the visible spectral range (generally of an argon-ion laser line of wavelength 488 nm or 514.5 nm) is scattered to a shifted frequency light, the difference in energy being the vibrational energy (Farrano and Nakamoto, 1994). If the frequency of the scattered photon is lower than that of the incident photon, a vibrational transition from a lower level to a higher level is induced. This process is called *Stokes Raman scattering*, and the corresponding peaks in the Raman spectra are called *Stokes Raman lines*. If the scattered photon is of higher frequency than the frequency of the incident photon, a transition from a thermally populated higher vibrational level to a lower level is induced, giving rise to what is known as *anti-Stokes lines*.

As discussed in Section 4.1.1, Raman scattering is derived from the polarizability term, α. The strength of the Raman transition for a vibrational mode (normal mode) with normal coordinate Q_k is determined by the polarizability derivative $d\alpha/dQ_K$. The larger the derivative (the larger the change in the polarizability due to the vibrational displacement of the normal mode), the stronger is the Raman transition due to this normal mode of vibration.

The efficiency of Raman scattering is weak: Typically, one out of 10^5 photons is scattered to produce a frequency-shifted photon. For this reason, one uses a laser source (a source of high photon density), which provides a monochromatic excitation laser at frequency v_o. The intensity of scattered photon frequency is plotted as a function of frequency shift $(v_o - v)$. The different peaks correspond to Raman excitations of various vibrations of frequency $v_R = (v_o - v)$. Thus, one obtains a vibrational spectrum.

For a centrosymmetric molecule with an inversion symmetry, only *g*-type vibrational modes are excited by Raman scattering. Therefore, for a centrosymmetric molecule, there is a mutual exclusion between the vibrational modes that are Raman active (*g*-type) and those that are IR-active (*u*-type).

The $\Delta v = 1$ transition gives rise to what are called *fundamental bands*, while $\Delta v > 1$ are called *vibrational overtones*. Also, Raman scattering can excite a combination of two vibrational modes. This type of coupled transitions gives rise to what are called *combination bands*.

IR spectroscopy (more routinely now in the form of FT–IR spectroscopy) and Raman spectroscopy are used as complementary techniques to provide information on various vibrational transitions or vibrational bands. These vibrational bands provide a detailed fingerprint of different bonds, functional groups, and conformations of molecules, biopolymers, and even microorganisms. Even though vibrational transitions are considerably weaker than the electronic transitions, they are much richer in structures (a large number of vibrational modes and corresponding bands, well resolved in the spectra) compared to room-temperature electronic spectra (whether absorption or fluorescence), which are relatively featureless. Therefore, vibrational spectroscopy has found wide application in structural characterization of biological materials and in probing interaction dynamics (Table 4.3) (Stuart, 1997; Thomas, 1999). As examples of illustrations of vibrational spectra of molecules of

TABLE 4.3. Representative Vibrational Frequencies of Some Bonds

Hydroxyl (OH)	3610–3640 cm^{-1}
Amines (NH)	3300–3500 cm^{-1}
Aromatic rings (CH)	3000–3100 cm^{-1}
Alkenes (CH)	3020–3080 cm^{-1}
Alkanes (CH)	2850–2960 cm^{-1}
Triple bonds (C≡C)	2500–1900 cm^{-1}
Double bonds (C=C)	1900–1500 cm^{-1}

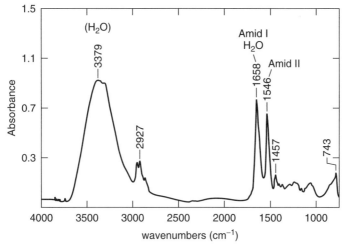

Figure 4.10. Typical IR absorption spectrum of hydrated protein film, in this case intrinsic cell membrane protein bacteriorhodopsin. (Reproduced with permission from Elsevier Science; Colthup et al., 1990.)

biological interested, presented here are the IR spectra of a protein, bacteriorhodopsin, and the Raman spectra of insulin (Figures 4.10 and 4.11).

As discussed above, Raman spectroscopy is not as sensitive as IR spectroscopy because of the relative inefficiency of Raman scattering. Furthermore, if the sample exhibits any intrinsic fluorescence (also called *autofluorescence*, which is discussed in Chapter 6), the fluorescence signal is many orders of magnitude higher than that of Raman. Therefore, the fluorescence background can overwhelm the Raman bands, limiting the utility of Raman spectroscopy. However, Raman spectroscopy offers a number of distinct advantages over the IR spectroscopy for vibrational analysis and probing; some of them are as follows:

- Ability to obtain vibrational spectra in an aqueous medium, because water shows very weak Raman scattering. On the other hand, IR absorp-

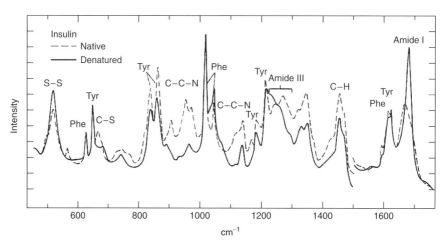

Figure 4.11. Raman spectra of native and denatured insulin in the solid state. (Reproduced with permission from Elsevier Science; Yu et al., 1972.)

tion by water is very strong, which overwhelms absorption by other cellular constituents.

- Ability to use samples in their natural form, (liquid, solid, gels). No special sample preparation is needed.
- Ability to focus the visible wavelength laser excitation source to a micron size spot. This allows one to obtain Raman spectra of microsize dimensions such as a single cell.
- Ability to selectively probe a specific chemical segment or subcellular component. This goal is achieved by resonantly enhancing the Raman scattering from the desired chemical unit by using an excitation frequency close to its absorption band.

The last point is illustrated by the selective enhancement of the Raman bands due to the β-carotene structural unit in the photosynthetic protein (Ghanotakis et al., 1989), by using the 488-nm excitation wavelength, which is close to the absorption band of β-carotene. The resonance-enhanced Raman spectrum is compared with the ordinary Raman spectrum in Figure 4.12.

A growing field is ultraviolet resonance Raman spectroscopy, whereby the excitation provided around 230–250 nm resonantly enhances the bands due to the aromatic residues of proteins. The UV resonance Raman spectroscopy has been used to derive specific interactions concerning the noncovalent interactions (such as hydrogen bonding) of important aromatic residues, tyrosine and tryptophan (Chi and Asher, 1998).

Another method of enhancement of Raman transitions is provided by surface-enhanced Raman spectroscopy (Lasema, 1996). Surface-enhanced Raman spectroscopy refers to the method where the intensity of the Raman

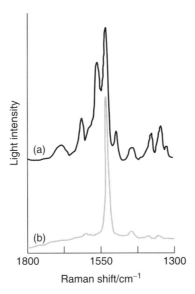

Figure 4.12. (a) The Raman spectra of photosynthetic protein obtained using 407-nm excitation and exhibiting Raman bands of both chlorophyll *a* and β-carotene; (b) resonantly enhanced Raman spectra, obtained with 488-nm excitation, showing the bands of β-carotene. (Reproduced with permission from Ghanotakis et al., 1989.)

vibrational transitions of a molecule is enhanced by deposition of molecules on the surface of a microscopically rough metal, metal colloids, and metal nanoparticles. The enhancement is by several orders of magnitude. The metals providing the largest enhancement are silver and gold and have been used to increase the Raman spectroscopic sensitivity for the study of molecules and biopolymers such as proteins.

4.8   SPECTROSCOPY UTILIZING OPTICAL ACTIVITY OF CHIRAL MEDIA

As discussed in Chapter 2, optically active chiral media consist of structures such as those containing an asymmetric carbon (chiral center) or a helical structure (such as in protein and DNA). These media interact differently with right- and left-circularly polarized light. One manifestation, already discussed in Chapter 2, is rotation of plane of polarization of a linearly polarized light as it propagates through a chiral medium. This effect results from a difference in the phase velocities due to different refractive indices for right- and left-circularly polarized light. As a linearly polarized light can be considered as composed of right- and left-circularly polarized light with equal amplitudes, this difference in phase velocity creates a phase difference between the right- and left-circularly polarized components, which amounts to rotation of the plane of polarization. The variation of optical rotation as a function of wavelength is called *optical rotary dispersion* (ORD).

Spectroscopic response of a chiral medium also shows differences in interaction with left- and right-circularly polarized light. The three spectroscopic effects discussed here are circular dichroism (CD), vibrational circular dichroism (VCD), and Raman optical activity (ROA). These effects provide valuable spectroscopic probes for structure, interactions, and functions of chiral biological matter. These three spectroscopic methods utilizing chirality are discussed here.

Circular Dichroism (CD). Circular dichroism refers to the difference in the absorption of left- and right-circularly polarized light to create an electronic transition (Berova et al., 2000). In other words, the extinction coefficient ε (or the absorbance *A*) of equation (4.13) for an electronic transition (electronic band) are different for the left and right circular polarizations. This is to be expected from optical principles. As shown in equation (4.10), the real part of the refractive index gives rise to phase information (propagation, refraction, etc.). A change in the real part of refractive index, from left- to right-circularly polarized light, determines optical rotary strength and its dispersion. The imaginary part of the refractive index represents absorption. The corresponding change in the imaginary part of the refractive index determines circular dichroism. The change in the real part of the refractive index produces a corresponding change in the imaginary part of the refractive index because the two are related by a well-known equation called *Kronig–Kramers transformation*. Thus, ORD, which gives optical rotation as a function of wavelength, and CD are related.

The CD spectra are typically measured as the difference in the absorbance of a molecule for the left- and right-circularly polarized light for varying wavelength λ (or wavenumber ν) as

$$\Delta A(\lambda) = A_{\mathrm{L}}(\lambda) - A_{\mathrm{R}}(\lambda) \tag{4.19a}$$

Here A_L is the absorbance for the left-circularly polarized light and A_R is the absorbance for the right-circularly polarized light. The CD spectra are also frequently expressed as the difference in the molar extinction coefficient defined as

$$\Delta\varepsilon(\lambda) = \varepsilon_{\mathrm{L}}(\lambda) - \varepsilon_{\mathrm{R}}(\lambda) \tag{4.19b}$$

Based on this definition, the unit used for representing circular dichroism is molar circular dichroism, also called delta epsilon in $L\,mol^{-1}\,cm^{-1}$. Another unit used for CD is mean residue ellipticity in degree $cm^2\,dmol^{-1}$. The ellipticity unit is derived from the conceptual visualization that two equal amplitudes of opposite circular polarization (left and right) form a linearly polarized light. If the material exhibits circular dichroism—that is, left- and-right circularly polarized lights are absorbed to a different extent—then their amplitudes (contributions) will not be the same, resulting in an elliptical behavior of the distribution of electric field. The ellipticity is quantified as the tangent of the ratio of the two elliptical axes (perpendicular contributions) called minor and major.

Circular dichroism spectroscopy has been used for a number of applications in structural biology. Some of them are as follows:

- Determination if a protein is folded and thus of its secondary and tertiary structure
- Comparison of structures of proteins obtained from different sources or structures of different mutants of the same protein
- Study of conformational stability of a protein under various environmental perturbations (temperature, pH, buffer composition, addition of stabilizers and excipients)
- Determination of the effect of protein–protein interactions on the protein conformation

The application of circular dichroism to biology is illustrated here with the example of identification of secondary and tertiary structures of proteins. The far-UV spectral region (190–250 nm) is representative of the peptide bond. Thus, the CD spectra in this region arise if the peptide bond is in a regular, folded environment, thus providing information on the secondary structure of a protein. Figure 4.13 compares the CD spectra of different conformations: alpha helix, beta sheet, and random coil structures of poly-lysine.

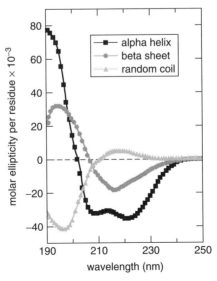

Figure 4.13. CD spectra of the three conformations of poly-lysine. (Reproduced with permission from http://www.ap-lab.com/circular_dichroism.htm.)

The appropriate fraction of each structure present in any protein can be obtained by fitting its far-UV CD spectrum as a weighted sum of reference spectra for each structure.

The CD spectra of proteins in the near-UV spectral region of 250–350 nm are characteristics of aromatic amino acids and disulfide bonds. The CD signal in this region, therefore, is a sensitive probe of the overall tertiary structure of a protein. For example, the presence of a significant CD signal in the near-UV region indicates that the protein is folded into a well-defined structure. Thus, the near-UV CD spectra can be used as a sensitive probe for any change in the tertiary structure due to protein–protein interactions or any external perturbation such as changes in the solvent.

Vibrational Circular Dichroism (VCD). Just like the electronic transitions discussed above giving rise to CD, vibrational transitions also exhibit optical activity in their response to left- versus right-circularly polarized light (Berova et al., 2000). The general term for optical activity of vibrational transitions is vibrational optical activity (VOA). The vibrational circular dichroism (VCD) is one type of VOA which specifically refers to the difference in vibrational spectrum of a molecule for left- versus right-circularly polarized light as obtained by using the IR spectroscopic technique discussed above (Nafie et al., 2002). It is, therefore, an extension of CD spectroscopy from UV–visible (electronic transitions) to near-IR and IR (vibrational transitions).

Dramatic progress in the instrumentation has lead now to the availability of a VOA spectrometer from a number of commercial sources, which makes it possible for a nonspecialist in this field to use VCD for a variety of applications. For example, a dedicated FT–IR spectrometer is available from Biomen-Bio Tools of Quebec, Canada. This instrument is constructed on one optical platform that includes the interferometer (for FT–IR) and all the VCD optical components. The advantage of VCD over the UV–visible CD (due to electronic absorption) is derived from the rich structural sensitivity of IR spectroscopy due to a large number of vibrational transitions representing the various vibrations (vibrational modes) of a molecule. An important application of VCD is the determination of absolute three-dimensional configuration of a biomolecule. This determination utilizes a comparison of the experimentally measured VCD spectrum to that theoretically calculated. If there is a good correlation of the bands and their signs, then the theoretical absolute configuration corresponds to that of the unknown sample. The theoretical VCD spectra can be calculated using accurate *ab initio* methods, introduced briefly in Chapter 2 (Nafie et al., 2002). Figure 4.14 provides an example of the correlation between the experimental and the calculated IR spectra and the corresponding VCD spectra of (+) *trans*-pinene (Nafie et al., 2002). The molecule α-pinene is routinely used as a standard for VCD because it has a strong VCD signal., it can be sampled in high concentration as a neat (pure) liquid and has a rigid stereoconfiguration due to its fused ring structure. According to the convention, a positive VCD band corresponds with the case

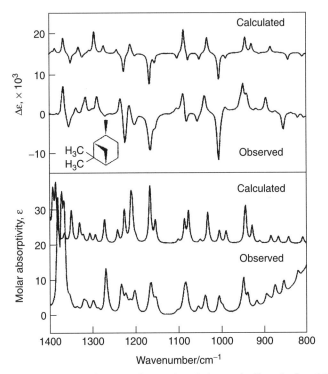

Figure 4.14. Comparison of the experimental and theoretically calculated IR spectra and the corresponding VCD spectra of (+) *trans*-pinene under conditions of neat liquid using a sample pathlength of 55 μm. (Reproduced with permission from Nafie et al., 2002.)

where the left-circularly polarized light is absorbed more than the right-circularly polarized light.

Proteins exhibit strong VCD in the amide vibrational band regions of the IR spectrum (Keiderling, 1994). VCD can be used to study the secondary structures of peptides and proteins, as well as the conformations of nucleic acids and sugars. The chirality in nucleic acids is derived from the sugar-phosphate backbone. The VCD technique is particularly sensitive to the base stacking regions between 1750 and 1550 cm^{-1} (Wang and Keiderling, 1992).

VCD is also emerging as an important tool for pharmaceutical research (Dukor and Nafie, 2000). VCD can be used to determine the optical purity of manufactured drugs and to characterize a biologically active enantiomeric form of a particular drug.

Raman Optical Activity (ROA). The definition of Raman optical activity (ROA) is more complex than that of VCD, because a Raman transition, as discussed above, involves the polarization characteristics of both the incident beam and the frequency shifted Raman scattered beam. The original form of

ROA, sometimes also called as incident circular polarization (ICP)ROA is defined as the difference in the intensity of Raman scattering, measured using right- and left-circularly polarized incident light (Barron et al., 2002; Nafie and Freedman, 2001). Hence, the convention used for (ICP)ROA is opposite (right minus left) to that used for VCD where the difference is measured between left-and right-circularly polarized light (left minus right for VCD). (ICP)ROA is commonly used for studying ROA spectra. Other forms of ROA involve changing the polarization state of the scattered radiation between the right-circularly polarized and the left-circularly polarized states. Another variable in the ROA is the scattering angle; nearly all current ROA measurements are carried out using a backscattering geometry which minimizes the interference due to high background scattering from the solvent and any residual fluorescence.

Like VCD, ROA probes the effect of chirality on vibrational transitions. As discussed above, IR spectra and Raman spectra can provide complementary information on the vibrational bands and thus can be used as fingerprints for a molecular structure and its conformation. Similarly, VCD and ROA are complementary techniques to study vibrational optical activity and use it to determine secondary and tertiary structure of biopolymers, study protein folding, elucidate the conformation of nucleic acids and sugars, and determine the optical purity of a pharmaceutical compound. However, due to the lower sensitivity of the available Raman techniques compared to that of FT–IR spectroscopy, ROA has not been used as widely as VCD. Thus, no commercial ROA instrument is available at the time of writing of this book.

Despite its lower sensitivity compared to VCD, ROA provides some distinct advantages. These advantages are derived from those of Raman over IR as discussed above. A primary one is the use of H_2O and D_2O as excellent solvents for Raman studies of biopolymers. An important application of ROA has been in fold determination, which is of special importance in post-genome structural biology. ROA can be used to discriminate between extended helix (as in the coat proteins of filamentous bacteriophases), the globin fold (as in the serum albumins), and the helix bundle (as in tobacco mosaic virus). As a sufficiently large set of protein reference spectra becomes available, ROA may become a routine technique for reliable determination of protein fold. ROA can also be useful for the study of non-native protein states such as molten globules and native states containing mobile regions.

4.9 FLUORESCENCE CORRELATION SPECTROSCOPY (FCS)

The contents of this section are derived from the following website: **www.probes.com/handbook/boxes/1571.html**. This website provides a very lucid description of the method of fluorescence correlation spectroscopy. Another suggested reference is a review by Thompson (1991).

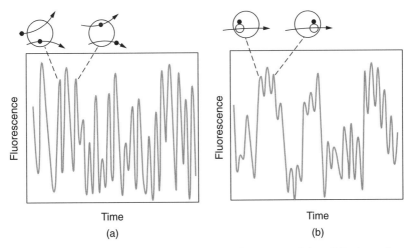

Figure 4.15. (a) Fluorescence intensity fluctuations caused by diffusion of small molecules; (b) fluorescence intensity fluctuations caused by diffusion of less mobile biopolymers. (Reproduced with permission from the highlighted website.)

In fluorescence correlation spectroscopy, often abbreviated as FCS, spontaneous fluorescence intensity fluctuations in a microscopic volume consisting of only a small number of molecules is monitored as a function of time. The volume typically sampled is about 10^{-15} L (or a femtoliter) compared to that of 0.1–1.0 mL or even larger, typically sampled by conventional fluorescence spectroscopy. The fluorescence intensity fluctuations measured by FCS relates to dynamical processes occurring in the interrogation volume. These dynamical processes can be due to changes in the number of fluorescing molecules due to their diffusion in and out of the microscopic volume sampled. They can also represent a change in the fluorescence quantum yield due to processes occurring in the interrogation volume.

Fluctuations caused by diffusion of molecules depend on their size. Rapidly diffusing small molecules produce rapid intensity fluctuations as shown in Figure 4.15a. In contrast, large molecules and biopolymers such as proteins or protein bound ligands exhibit slowly fluctuating patterns of bursts of fluorescence, as shown in Figure 4.15b. Quantitatively, the fluorescence intensity fluctuation is characterized by a function $G(\tau)$, called the *autocorrelation function*, which correlates the fluctuation $\delta F(t)$ in fluorescence intensity at time t with that $[\delta F(t + \tau)]$ at time $(t + \tau)$, where τ is a variable time interval., averaged over all data points in the time series. Thus, $G(\tau)$ is defined as

$$G(\tau) = \frac{<\delta F(t) \cdot \delta F(t+\tau)>}{<F(t)>^2} \tag{4.19c}$$

The brackets in this expression represent the average over all data points at different times t.

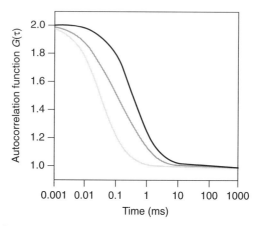

Figure 4.16. Simulated FCS autocorrelation functions representing a free ligand and a corresponding bound ligand. The intermediate curve represents a mixture. (Reproduced with permission from the above highlighted website.)

A typical autocorrelation function $G(\tau)$ plotted as a function of time interval τ is represented in Figure 4.16 for a free ligand (low molecular weight) which can diffuse faster and a bound ligand which diffuse slower. The initial amplitude of the autocorrelation function is inversely proportional to the number of molecules in the sampled volume. The decay of the autocorrelation function is fast for a free ligand and relatively slow for a bound ligand. Thus, the decay behavior of $G(\tau)$ provides information on the diffusion rates of the fluorescing species.

FCS is an excellent probe for monitoring biomolecular association and dissociation processes. With the recent progress of increase of detection sensitivity for fluorescence to the limit of single molecule detection, FCS has emerged as a valuable tool to investigate a variety of biological processes such as protein–protein interactions, binding equilibria for drugs, and clustering of membrane bound receptors. Another extension of FCS is dual color cross-correlation, which measures the cross-correlation of the time-dependent fluorescence intensities of two different dyes fluorescing at different wavelengths (Schwiller et al., 1997). This method has the advantage that cross-correlated fluorescence is only generated by molecules or biopolymers fluorescently labelled (chemically attached) with both dyes, allowing quantitation of interacting dyes.

HIGHLIGHTS OF THE CHAPTER

- Light–matter interactions involve four types of energy exchange between them: (i) absorption of a photon, (ii) spontaneous emission of a photon, (iii) stimulated emission of a photon, and (iv) Raman scattering.

- The absorption of a photon leads to a transition (jump) from a quantized lower energy state of a molecule (or atom), often called the *ground state*, to its higher state, often called an *excited state*.

- The spontaneous emission of a photon brings back the molecule (or atom) from its excited state to a lower energy state.

- Stimulated emission of a photon occurs when there is a population inversion—that is, more molecules (or atoms) are in the excited state than in the ground state. The stimulated emission requires triggering by (thus the presence of) another photon of same frequency.

- Raman scattering produces photons of energy (frequency) different from the energy (frequency) of the incident photon; the difference in energy is either deposited in a molecule to create an excited state, usually a vibrationally excited state (in a Stokes process), or taken away from a vibrationally excited molecule (in an anti-Stokes process).

- The processes of absorption, spontaneous emission, and stimulated emission are phenomenologically described by Einstein's model.

- Quantum mechanically, the strength of a transition between two energy states is characterized by the magnitude of a quantity called the transition dipole moment, which connects the two states through charge redistribution.

- The Lapote rule provides a qualitative guide if a transition between two states is dipole allowed (i.e., the transition dipole moment has nonzero value) and will occur with great probability (strength) or whether it is dipole forbidden (i.e., the transition dipole moment has a zero value) and may only be weakly manifested because of other interactions.

- Interaction of light with bulk matter is described in terms of the process of reflection, refraction, scattering, and absorption and is determined by the refractive index properties of the bulk and the surrounding medium.

- An excited energy state may dispose of the excess (excitation) energy by a photophysical process, by a photoinduced electron transfer process, or by performing a photochemical process.

- The various photophysical processes are (i) radiative, in which the excess energy is emitted as a photon (spontaneous and stimulated emission); (ii) nonradiative, in which the excess energy is dissipated as heat; (iii) energy transfer in which the excess energy is transferred to another neighboring molecule, and (iv) excited state complex formation (association) between neighboring molecules.

- The various radiative and nonradiative processes in an organic molecule are often described by the so-called Jablonski diagram.

- A nonradiative transition between two electronic states of same spin being spin-allowed (spin conserving) is called the process of *internal conversion*, and that between two states of different spin occurs by a spin-forbidden (and thus weak) process of *intersystem crossing*.

- The emission from a higher electronic state to a lower electronic state of same spin (spin conserving) is called *fluorescence* and is much stronger than that of the process of phosphorescence, which is between two states of different spin.
- Spectroscopy is a branch of light interactions which deals with the study of dependence of light absorption or emission on the wavelength of light; the plot of the strength of the transition as a function of wavelength is called the *spectrum*.
- Electronic absorption spectroscopy dealing with absorption between two electronic states is quantitatively expressed by Beer–Lambert's law under ordinary light intensity consideration. Here only one photon absorption per molecule occurs at a time (hence, linear absorption).
- Electronic fluorescence spectroscopy allows for a multiparameter analysis using its emission spectra (emission as a function of wavelength), excitation spectra (strength of emission as a function of excitation wavelength), lifetime of emission (decay of emission), and the polarization characteristics of the emitted light.
- Vibrational spectroscopy gives information on the vibrational frequencies (energies) associated with different chemical bonds and associated bond angles. These vibrational frequencies are used as a detailed chemical fingerprint for various bonds, bond angles, and chemical units and, thus, for identification of molecules.
- The two types of vibrational spectroscopy are (a) IR, which involves absorption of an IR photon to create vibrational transition, and (b) Raman, where a Raman scattering process generates a vibrational excitation.
- IR is more sensitive than Raman and is often used to get detailed structural information on organic molecules in solid or liquid or nonaqueous forms.
- For biological samples, often in an aqueous environment, Raman spectroscopy is more useful because water produces only weak Raman scattering but has very strong IR transitions.
- Resonance Raman spectroscopy offers the prospect of selectively exciting the vibrations of a particular molecular unit by choosing the incident wavelength at which this unit absorbs.
- Circular dichroism, referring to the difference in the electronic absorption of left- and right-circularly polarized light in a chiral structure, is very useful for determining the secondary and tertiary structures, and interactions of a biopolymer.
- Vibrational circular dichroism refers to the difference in IR vibrational spectra of a chiral molecule for left- versus right-circularly polarized light. It is useful for determining absolute three-dimensional configuration of a biopolymer.

- Raman optical activity is the difference in the Raman scattering of a chiral molecule using the right- and left-circularly polarized incident light.
- Fluorescence correlation spectroscopy measures the correlation between fluorescence intensities at two different times, from a microscopic volume containing only a small number of molecules. It provides information on fluorescent intensity fluctuation due to processes such as molecular diffusion and protein–ligand association.

REFERENCES

Atkins, P., and dePaula, J., *Physical Chemistry*, 7th edition, W.H. Freeman, New York, 2002.

Barron, L. D., Hecht, L., Blanch, E. W., and Bell, A. F., Solution, Structure and Dynamics of Biomolecules from Raman Optical Activity, *Prog. Biophys. Mol. Biol.* **73**, 1–49 (2000).

Berova, N., Nakanishi, K., and Woody, R. W., eds., *Circular Dichroism: Principles and Applications*, 2nd edition, Wiley-VCH, New York, 2000.

Chalmers, J. M.. and Griffiths, P. R., *Handbook of Vibrational Spectroscopy*, Vol. 5, *Applications in Life, Pharmaceutical and Natural Sciences*, Wiley Milan, Italy, 2002.

Chase, D. B., and Robert, J. F., eds., *Fourier Transform Raman Spectroscopy: From Concept to Experiment*, Academic Press, San Diego, 1994.

Chi, Z., and Asher, S. A., UV Raman Determination of the Environment and Solvent Exposure of Tyr and Trp Residues, *J. Phys. Chem. B*. **102**, 9595–9602 (1998).

Colthup, N. B., Daly, L. H., and Wiberly, S. E., *Introduction to Infrared and Raman Spectroscopy*, Academic Press, Boston, 1990.

Dukor, R. K., and Nafie, L. A., Vibrational Optical Activity of Pharmaceuticals and Biomolecule, in R. A. Meyers, ed., *Encyclopedia of Analytical Chemistry*, John Wiley & Sons, Chichester, 2000, pp. 662–676.

Farrano, J. R., and Nakamoto, K., *Introductory Raman Spectroscopy*, Academic Press, San Diego, 1994.

Feynman, R. P., Leighton, R. B., and Sands, M., *The Feynman Lectures on Physics*, Vol. 1, Addison-Wesley, Reading, MA, 1963.

Ghanotakis, D. F., dePaula, J. C., Demetriou, D. M., Bowlby, N. R., Petersen, J., Babcock, G. T., and Yocum, C. F., Isolation and Characterization of the 47kDa Protein and the D1, D2, Cytochrome B-559 Complex, *Biochim. Biophys. Acta* **974**, 44–53 (1989).

Goodwin, T. W., and Mercer, E. I., *Introduction to Plant Biochemistry*, Pergamon Press, New York, 1972.

Griffiths, P. R., and deHaseth, J. A., *Fourier Transform Infrared Spectrometry*, John Wiley & Sons, New York, 1986.

Keiderling, T. A., Vibrational Circular Dichroism Spectroscopy of Peptides and Proteins, in K. Nakanishi, N. D. Berova, and R. W. Woody, eds., *Circular Dichroism Principles*, Wiley-VCH, New York, 1994, pp. 497–521.

Kruglik, S. G., Mojzes, P., Mizutani, Y., Mizutani, Y., Kitagawa, T., and Turpin, P.-Y., Time-Resolved Resonance Raman Study of the Exciplex Formed Between Excited Cu-Porphyrin and DNA, *J. Phys. Chem. B* **105**, 5018–5031 (2001).

Lakowicz, J. R., *Principals of Flouresence Spectometry*, Kluwer/Plenum, New York, 1999.

Lakowicz, J. R., ed., *Topics in Fluorescence Spectroscopy,* Vol. 1 (1991), Vol. 2 (1991), Vol. 3 (1992), Vol. 4 (1994), Vol. 5 (1997), Vol. 6 (2000), Plenum, New York.

Lasema, J. J., ed., *Modern Techniques in Raman Spectroscopy*, John Wiley & Sons, New York, 1996.

Levine, I. N., *Quantum Chemistry*, 5th edition, Prentice-Hall, Upper Saddle River, NJ, 2000.

Mojzes, P., Chinsky, L., and Turpin, P.-Y., Interaction of Electronically Excited Copper (II) Porphyrin with Oligonucleotides and Polynucleotides—Exciplex Building Process by Photoinitiated Axial Ligation of Porphyrin to Thymine and Uracil Residues, *J. Phys. Chem. B* **97**, 4841–4847 (1993).

Nafie, L. A., Dukor, R. K., and Freedman, T. B., Vibrational Circular Dichroism, in J. M. Chalmers and P. R. Griffiths, eds., *Handbook of Vibrational Spectroscopy*, Vol. 1, John Wiley & Sons, Chichester, 2002, pp. 731–744.

Nafie, L. A., and Freedman, T. B., Biological and Pharmaceutical Applications of Vibrational Optical Activity, in H. U. Gremlich and B. Yan, eds., *Infrared and Raman Spectroscopy of Biological Materials*, Marcel Dekker, New York, 2001, pp. 15–54.

Prasad, P. N., and Williams, D. J., *Introduction to Non-Linear Optical Effects in Molecules and Polymers*, John Wiley & Sons, New York, 1991.

Sauer, K., Biochemical Spectroscopy, *Methods in Enzymology*, Vol. 246, Academic Press, San Diego, 1995.

Schwiller, P., Meyer-Almes, F. J., Rigler, R., Dual-Color Fluoescence Cross-Correlation Spectroscopy for Multicomponent Differential Analysis in Solution, *Biophys. J.* **72**, 1878–1886 (1997).

Stuart, B., *Biological Applications of Infrared Spectroscopy*, John Wiley & Sons, New York, 1997.

Thomas, G. J., Jr., Raman Spectroscopy of Protein and Nucleic Assemblies, *Annu. Rev. Biophys. Biomol. Struct.* **28**, 1–27 (1999).

Thompson, N. L., Fluorescence Correlation Spectroscopy, in J. R. Lakowicz, ed., *Topics in Florescence Spectroscopy*, Vol. 1, Kluwer/Plenum, New York, 1991.

Tinoco, I., Jr., Sauer, K., and Wang, J. C., *Physical Chemistry, Principles and Applications in Biological Sciences*, Prentice-Hall, Englewood Cliffs, NJ, 1978.

Wang, L., and Keiderling, T. A., Vibrational Circular Dichroism Studies of the A-to-B Conformational Transition in DNA, *Biochemistry* **31**, 10265–10271 (1992).

Yu, N. T., Liu, C. S., and O'Shea, D. C., Lasar Raman Spectroscopy and the Confirmation of Insulin and Proinsulin, *J. Mol. Biol.* **70**, 117–132 (1972).

CHAPTER 5

Principles of Lasers, Current Laser Technology, and Nonlinear Optics

Lasers are devices that produce highly directional, monochromatic, and intense beams of light. They are the most commonly used light source for biophotonics. Every subsequent chapter in this book involves the use of lasers. The usage of lasers falls into two categories. The first utilizes lasers as a convenient and highly concentrated source of photons. The second utilizes the highly coherent nature of the light beam. Because of the wide usage of lasers for a diverse range of applications, many monographs and textbooks can be found on this topic. This chapter provides a brief introduction to the topic of lasers and the nonlinear optical effects that are manifested by interaction of matter with the intense and coherent beam of laser light. This chapter is written at a basic level so that even those who are being exposed to these subjects for the first time can develop some general understanding of principles of lasers, become familiar with the various terminologies used in laser technology, and thus appreciate the features offered by lasers for biophotonics.

This chapter describes the principle of laser action, relying on a simple diagrammatic description. The various steps involved and components used in laser operation are briefly explained. The conditions to produce a continuous laser operation and to produce a pulse laser operation (periodic bursts of intense light) are described. Various types of lasers are described according to a number of different classification schemes.

An advantage of using lasers for biophotonics is the use of time resolution provided by pulse lasers. Techniques of Q-switching to produce nanosecond (10^{-9}-sec) pulses and mode-locking to produce picoseconds to femtoseconds pulses are qualitatively described. Some important lasers used for biophotonics are described, followed by a discussion of the current laser technology.

An important concept in biophotonics deals with the level and amount of light exposure needed for a given application. This is the topic of radiometry, which is also covered in this chapter.

Introduction to Biophotonics, by Paras N. Prasad
ISBN: 0-471-28770-9   Copyright © 2003 John Wiley & Sons, Inc.

Nonlinear optical processes such as infrared-light-to-visible-light conversion and two-photon excited fluorescence, which occur under the illumination of an intense beam of laser light, are also covered in this chapter. These nonlinear optical processes have found important applications in bioimaging and are also showing promise for light-activated therapy and tissue engineering. These applications are discussed in separate chapters.

Pulse lasers offer the prospect of studying various biophysical and biochemical processes in real time. Time-resolved techniques to probe these processes are described. Finally, the chapter is concluded with a discussion of the laser safety issue.

For further reading, suggested texts are:

Svelto (1998): Basics of lasers and laser technology

Menzel (1995): Applications of laser to spectroscopy

Prasad and Williams (1991): Nonlinear optical processes in organic systems

Boyd (1992): General coverage of nonlinear optical processes

Two magazines that cover development in laser technologies and applications are (i) *Laser Focus World* (PennWell Publication; website: www.optoelectronics-world.com) and (ii) *Photonics Spectra* (Laurin Publishing Co. Inc.; website: www.photonics.com).

5.1 PRINCIPLES OF LASERS

5.1.1 Lasers: A New Light Source

The term "laser" is an acronym for *l*ight *a*mplification by *s*timulated *e*mission of *r*adiation. However, it has now expanded into the term "lasing," a verb representing laser action. It is my opinion that the invention of lasers ranks among the top 10 most significant inventions of the 20th century. Since the first demonstration of laser action in 1960, lasers have enriched all aspects of life: from grocery store scanners that read the price of an item from its barcode label, to CDs and DVDs for home entertainment, to colorful laser light displays in nightclubs, to laser surgery, to laser cancer treatment (photodynamic therapy).

The history of laser development is just as colorful. Townes and Schallow theoretically proposed the concept of laser design and action in 1958 and hold the patent on it. However, the first laser was not developed until 1960, when Maiman demonstrated the ruby laser.

For many biophotonics applications, a laser is simply an intense light source that is

- Monochromatic (one color or wavelength)
- A highly directional beam with low divergence

- Capable of being focused into a very small spot
- Capable of producing short bursts of light of very high intensity (in less than a trillionth of a second)

Therefore, many applications utilize a laser source for convenience and effectiveness, even though a conventional lamp can be used instead. Lasers are uniquely capable of producing a coherent light beam and ultra-short pulses (in 10^{-9}–10^{-15} sec, i.e., nanoseconds to femtoseconds). Many applications are made possible by using these unique properties of a laser beam.

Recent developments in laser technology have produced highly compact and energy-efficient solid state lasers which can readily be integrated with efficient delivery systems (such as optical fibers) to tissues or used endoscopically with a minimally invasive approach into a specific site of a live object. These developments have further increased the scope of light-based optical probing and light-based treatment. The properties of lasers also make them uniquely suited to creating nonlinear optical effects, which are discussed in Section 5.4. The ease with which these nonlinear optical effects can be produced has given renewed impetus to biophotonics, creating new approaches for optical imaging, sensing, and diagnostics, as well as new modalities for laser treatment and therapy.

5.1.2 Principles of Laser Action

The basic principle of the laser, as the name "light amplification by stimulated emission of radiation" indicates, is based on stimulated emission from a higher level f to a lower level i (not necessarily the ground state) (Svelto, 1998). As discussed in Chapter 4, this will require that a population inversion is created so that the number of atoms or molecules in level f is higher than that in level i (i.e., $N_f > N_i$). Most lasers utilize electronic levels for laser actions. A widely used carbon dioxide laser is an example of exceptions which utilize vibrational levels (f and i are vibrational levels with different quantum numbers, v). This population inversion is created either by electrical excitation of level f or by optical excitation (absorption) to a higher level f' from which nonradiative relaxation stores the energy into state f to reach population inversion. The population inversion has also been achieved in chemical lasers by utilizing energy released in a chemical reaction (Basov et al., 1990). However, chemical lasers have not found applications in biophotonics and therefore will not be discussed here. Table 5.1 describes the various steps involved in laser action.

A simple diagram of laser design is shown in Figure 5.1. The components of a laser are:

- An active medium, also called a gain medium, in which an atom or a molecule can be excited by a suitable pumping mechanism to create population inversion so that the spontaneously emitted photons at some site in the medium stimulates emission at other sites as it travels through it.

TABLE 5.1. Various Steps in Laser Operation

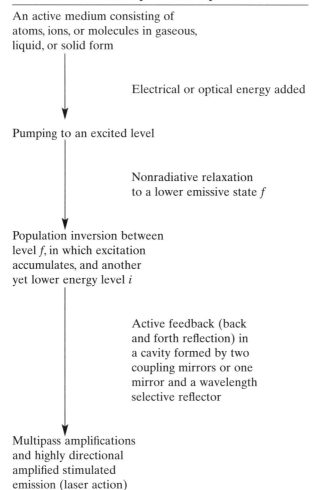

An active medium consisting of atoms, ions, or molecules in gaseous, liquid, or solid form

Electrical or optical energy added

Pumping to an excited level

Nonradiative relaxation to a lower emissive state f

Population inversion between level f, in which excitation accumulates, and another yet lower energy level i

Active feedback (back and forth reflection) in a cavity formed by two coupling mirrors or one mirror and a wavelength selective reflector

Multipass amplifications and highly directional amplified stimulated emission (laser action)

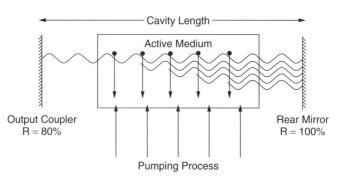

Figure 5.1. The schematics of a laser cavity. R represents percentage reflection.

- An energy pump source, which can be an electrical discharge chamber, an electrical power supply, a lamp, or even another laser.
- Two reflectors, also called rear mirror and output coupler, to reflect the light in phase (determined by the length of the cavity) so that the light will be further amplified by the active medium in each round-trip (multipass amplification). The output is partially transmitted through a partially transmissive output coupler from where the output exits as a laser beam (e.g., $R = 80\%$ as shown in Figure 5.1).

Both the process of stimulated emission and cavity feedback impart coherence to the laser beam. Only the waves reflected in phase and in the direction of the incident waves contribute to multipass amplification and thus build up intensity. Emission coming at an angle (like from the sides) does not reflect to amplify. This process provides directionality and concentration of beam in a narrow width, making the laser highly directional with low divergence and greater coherence. For an incident beam and reflected beam to be in phase in a cavity, the following cavity resonance conditions must be met (Svelto, 1998), as shown in equation (5.1):

$$n\frac{\lambda}{2} = l \tag{5.1}$$

where λ = the wavelength of emission; l = the length of the cavity, and n is an integral number.

In the case where the fluorescence from the active medium is broadened inhomogeneously, different emitting centers in the medium may emit at different wavelengths that form a continuous band describing the fluorescence lineshape. However, the stimulated emission curve is generally much narrower because extremes of the emission profile cannot lase due to lack of sufficient population needed to create threshold population inversion. The range of wavelengths over which sufficient stimulated emission and lasing action can be achieved defines what is called the *gain curve*.

Within the gain curve of an active medium, equation (5.1) can be satisfied for many wavelengths with different integral numbers n. These are called the *longitudinal cavity modes*. Wavelength selection can be introduced by replacing the rear mirror in Figure 5.1 with a spectral reflector (grating or prism) that permits the return of only a certain narrow wavelength of light (monochromatic beam) to be multipass-amplified. However, these optical elements do not provide the resolution necessary to select a single longitudinal mode of the cavity (narrowest bandwidth possible). One introduces other optical elements such as a Fabry–Perot etalon or Liot filter in the cavity to isolate a single longitudinal mode. Lasers that provide the laser beam output in a single longitudinal mode and at the same time are stable so that the optical output

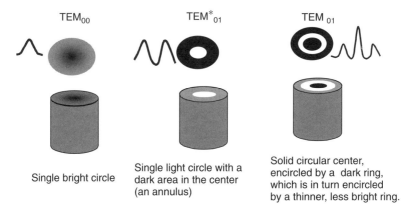

Figure 5.2. Transverse electromagnetic modes and the corresponding shapes of the laser beam.

does not undergo mode-hopping (jumping from one mode to another) or mode-beating (mode interference) are called *single-frequency lasers*.

Another feature of a laser is the spatial profile of its beam in the transverse direction (the plane perpendicular to its propagation direction). These profiles are called *transverse modes* and are represented in the form TEM_{mn}, where m and n are small integers that describe the intensity distribution in the transverse x and y directions (the plane perpendicular to the beam). Some mode structures with the beam shape (intensity distribution) are shown in Figure 5.2. The beam with a TEM_{00} transverse mode characteristic is called a Gaussian beam because the intensity distribution from the center (the brightest point) to the edge of the beam falls off as a Gaussian function given by equation (5.2):

$$I(r) = I(0) \ \exp\left(\frac{-2r^2}{\omega_0^2}\right) \qquad (5.2)$$

where $I(r)$ is the intensity of the beam as a function of distance r in any direction from the center. The parameter ω_0 is called the *beam radius* or the *spot size*, which in the case of a TEM_{00} mode is the distance from the center where by the intensity has dropped by a factor of $1/e^2$.

The TEM_{00} beam has the minimum possible beam divergence and can be focused to a "diffraction limited" size, which is the minimum attainable beam spot possible. If a beam is not Gaussian (TEM_{00}), the minimum spot it can be focused to is not diffraction limited. For this reason, it is preferable to use a Gaussian beam where a tight focus or a spatially uniform intensity distribution is needed (e.g., in microscopy).

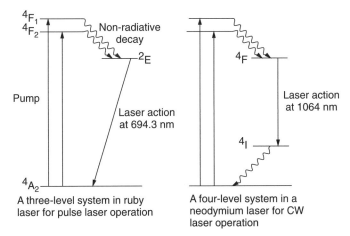

Figure 5.3. Schematics of a three-level system for pulse operation and a four-level system for CW as well as pulse operation.

Depending on the energy levels scheme, the population inversion, a key step in lasing, can be achieved only for a short time or can be maintained continuously. They respectively produce a pulse laser or a continuous wave (termed CW) laser. The two possible schemes, as shown in Figure 5.3, determine pulse or CW operations. In Figure 5.3, the energy levels of the atoms (ions) are represented by their term symbols as defined in Chapter 2. In the three-level system, such as in ruby, the excitation is to a set of two closely spaced levels $^4F_1$ and $^4F_2$ from where the energy relaxes by nonradiative decay to a lower level ^2E where excitation builds up to create a population inversion between it and the ground state. In the case of the neodymium laser, a continuous population inversion is maintained between two intermediate levels ^4F and ^4I.

5.1.3 Classification of Lasers

Lasers can be classified into different categories utilizing different considerations. Some of these are shown in Tables 5.2, 5.3, and 5.4.

Pulse Lasers. Pulse lasers are characterized by high gain achieved due to a large population inversion for a short time. Pulsed operation can be realized by using a pulse excitation (electrical or optical). This approach is called *gain modulation*. High peak power pulse operation is achieved by controlling cavity feedback, such as in Q-switching, described below. A main advantage of lasers is their ability to provide short-duration pulses of very high intensity. The pulse durations, as shown in Table 5.4, can range from milliseconds to several femtoseconds. However, to achieve different ranges of pulse widths one requires different techniques. Some of these pulse operations are described below:

TABLE 5.2. Classification on the Basis of the Pumping Process

Electrically pumped: direct conversion from electrical to laser energy
   Diode laser
   Argon ion laser
   He–Ne laser
   Excimer laser
   CO_2 laser
Optically pumped by lamps
   Dye laser
   Nd:YAG laser
   Er:YAG laser
   Alexandrite laser
Optically pumped by another laser
   Dye laser
   Ti:sapphire laser
Diode laser pumped solid-state lasers
   Nd:YAG laser
   Nd:vanadate laser
   Er:YAG laser

TABLE 5.3. Classification on the Basis of Laser Medium

Gas lasers
   CO_2 laser
   Argon ion laser
   Krypton ion laser
   Copper vapor laser
   Excimer laser
   Nitrogen laser
   He–Ne laser
Liquid lasers
   Dye laser
Solid-state lasers
   Diode laser
   Nd:YAG laser
   Er:YAG laser
   Ti:sapphire laser
   Holmium:YAG laser
   Alexandrite laser

TABLE 5.4. Classification on the Basis of Temporal Feature

CW (continuous wave)
Modulated, chopped, gated (milliseconds–microseconds)
Pulsed (10^{-6} seconds:microseconds)
Superpulsed (10^{-3} seconds:milliseconds, repetitive train pulses)
Q-switched (10^{-9} seconds:nanoseconds)
Mode-locked (10^{-12} seconds:picoseconds); (10^{-15} seconds:femtoseconds)

Normal Pulse Operations (Free-Running). In the case of an electrically pumped system, the electronic driver produces modulated, chopped, and gated beams by providing an appropriate pump pulse. The same technique is used to produce a superpulsed laser. In the case of lamp-pumped lasers (such as a Nd:YAG laser), a flash lamp with a controlled duration of the flash, achieved by appropriately designing the electronic circuit charging the flash lamp (inductance and capacitance of the circuit), is used to achieve pulse durations in the range of 100 μsec to several milliseconds. In such a case, the laser is called a *free-running pulsed system*.

Q-Switched Operations. For Q-switched operations, one controls the quality, Q, of the laser cavity by using an optical element (called Q-switching element) that can be switched from very low optical transmission to very high optical transmission. Initially, the Q of the cavity is kept at a low value where the feedback of the cavity does not work. In other words, a shutter is introduced into the cavity. Under such a condition, the energy is stored inside the active medium and the population in the excited laser level further builds up. Then the Q of the cavity is suddenly switched to a high transmission by an appropriate action on the Q-switching element. Thus, the shutter is suddenly opened. The cavity now becomes transmissive and a large amount of energy (a very high intensity pulse) emerges from the cavity, rapidly depleting the stored energy in the excited lasing state of the medium.

A variety of Q-switching elements have been used to control the Q of a laser cavity. They include rotating optical rear reflector or output coupling mirror, electro-optic and acousto-optic Q-switches, and passive saturable absorber. The electro-optic Q-switch utilizes a device called a *Pockels cell*, which changes the plane of polarization of light by the application of an electrical field, an effect known as the *electro-optic effect*. This Pockels cell acts as a fast shutter when used in combination with a polarizer in such a way that with the application of a specific voltage, the Pockels cell acts as a shutter by change of the polarization of the light. When the voltage is removed, the Pockels cell shutter is opened. Electro-optic Q-switching allows one to achieve 5- to 20-nsec pulses.

An acousto-optic Q-switch utilizes a shutter action produced by the acousto-optic effect, which involves diffraction of light by an acoustic (ultrasonic) wave. The acousto-optic Q-switch consists of an acoustic-optic material to which is bonded a piezoelectric transducer. When the piezoelectric transducer is driven by a radiofrequency oscillator, it launches an acoustic wave in the acousto-optic materials which then diffracts light away from the cavity (hence, acting as a shutter). The acousto-optic Q-switching generally is slow, producing pulses of ~100 nsec. This method is often used for repetitive Q-switching of a continuous-wave laser such as a CW Nd:YAG laser. All these methods are externally controllable, and thus the function provided by them is called active Q-switching. In contrast, a passive Q-switching process utilizes a saturable absorber (dye or crystal), the absorption in which decreases as the intensity of beam increases, to produce high transmission at high buildup of energy. The current trend is to use electro-optic Q-switches, whereby the Q of a switch is controlled by an electrical pulse to achieve stable, high-quality nanosecond pulse operation.

Mode-Locked Operations. In this operation, the different longitudinal modes of a cavity, which are randomly generated in time inside the laser cavity, are locked together in phase to produce a train of extremely short pulses (from picoseconds to femtoseconds). In this method, the radiation parameters (amplitude or phase) are modulated inside the cavity by an optical process at the frequency

$$\frac{c}{2l};$$ (5.2a)

the difference in frequency between the adjacent longitudinal modes of a cavity of length l (c is the speed of light). It also corresponds to the round trip time of a wave in the cavity. The modulation in some active media (such as in a Ti:sapphire laser) occurs spontaneously within the cavity, under certain conditions. This is called self-mode-locking. In most cases, one utilizes an acousto-optic modulator, operating at a resonant frequency of

$$\frac{c}{2l}$$ (5.2b)

to provide gating of pulses so that an incoming pulse and another coming after completing a round trip of the cavity are in-phase at any point in the cavity. Therefore, they add their intensities (constructive interference). One of the requirements here is that the cavity length l has to be matched to the acoustic frequency of the acoustic-optic mode-locker by relation

$$v = \frac{c}{2l}$$ (5.2c)

where v = the acoustic frequency of the mode-locker, generally about 100 MHz. The minimum duration of pulses achievable by mode-locking is dictated by the line width of the optical transition because of the Heisenberg uncertainty principle, which couples the time uncertainty (pulse width) with energy uncertainty (the line width ΔE or Δv). The pulse width τ is given as

$$\frac{1}{\Delta v}$$ (5.2d)

For an Nd:glass laser, these mode-locked pulses are in the subpicoseconds range. For a Ti:sapphire laser, which is becoming more popular and has a broader linewidth, the mode-locked pulses are several femtoseconds.

Cavity-Dumped Operations. This method is similar to Q-switching and is used to achieve a periodic control that switches the direction of beam propagation relative to cavity axes by acousto- or electro-optic modulation. It is generally used to obtain higher peak power pulses from a system that ordinarily operates in a CW mode or which is mode-locked at a high repetition rate but producing low peak power pulses. The pulses emerging from a CW cavity-dumped laser system (such as an Nd:YAG) can be at a repetition rate in range of several megahertz.

5.1.4 Some Important Lasers for Biophotonics

Some selected examples of lasers useful for biophotonics are described in Tables 5.5–5.11:

5.2 CURRENT LASER TECHNOLOGIES

The current emphasis in laser technologies has moved to solid-state lasers, which are compact, energy efficient, and reliable with long operational lifetimes. The basic foundation of the solid-state laser technologies is provided by diode lasers. Diode lasers are highly efficient (>20% conversion of the electrical energy into light emission) and cover a broad range of wavelengths, but are continuously tunable only over a narrow wavelength range, allowing selection of one appropriate monochromatic wavelength at a time. A major driving force has been their applications in optical telecommunications, for which diode lasers of wavelengths ~980 nm, ~1300 nm, and ~1550 nm are needed. The 980- and 1300-nm diode lasers can be of significant value in biophotonics for bioimaging because of their greater penetration in biological tissues (a topic discussed in Chapter 6).

Diode lasers of ~800-nm wavelength are particularly useful for optical pumping of Nd:vanadate and Nd:YAG lasers. These diode-pumped YAG and vanadate lasers are replacing Ar ion lasers for many applications because their

TABLE 5.5. CO$_2$ Laser

Wavelength:	10,600 nm or 10.6 μm (most commonly used, mid-infrared)
Laser-Medium:	CO$_2$ gas (mixed with nitrogen and helium)
Excitation:	Electrical
Applications:	Numerous, one of the main workhorses
	20 W CW: Applications in vaporization of tissues, cosmetic applications
	Pulsed, 500 W peak power, 20–40 W average power: adjusted to appropriate level, it can be used for both vaporization and cutting of tissues without charring
	Skin resurfacing (pulsed, superpulsed)

TABLE 5.6. Diode Lasers (Semiconductor Lasers)

Wavelength:	From 400 nm to 1900 nm (dependent on the laser medium)
Laser medium:	GaN ~400 nm
	5 mW, 20 mW
	AlGaAs ~800 nm (near IR)
	5 mW, 50 mW, 4 W
	InGaAs ~670 nm (red)
	5 mW, 40 mW, 400 mW
	~635 nm (bright red)
	5 mW
Excitation:	Electrical
Applications:	Aiming beam
	Low-level laser therapy: pain relief, wound healing
	Ophthalmology
Diode bars containing many emitters:	~800 nm, 1–20 W
Applications:	Cutting of tissues
	Optical pumping of Nd:YAG laser (more efficient and compact)
	Heat treatment of tissues

TABLE 5.7. Nd:YAG Laser

Wavelength:	1064 nm or 1.064 μm (most common)
Laser medium:	Nd ions dispersed in a crystal of yttrium–aluminum–garnet (YAG)
Excitation:	Lamp pumped: old technology
	Diode laser pumped: new technology (very durable, compact and energy efficient but expensive)
Applications:	Numerous, one of the main workhorses
	100-W CW for surgery (e.g., prostate)
	Coagulation and vaporization of a bladder tumor

TABLE 5.8. KTP Laser

Frequency-Doubled Nd:YAG Laser Using a Nonlinear Crystal, KTP (the principle described in Section 5.4.2)

Wavelength:	$\dfrac{1064\,\text{nm}}{2} = 532\,\text{nm (green)}$
Applications:	CW KTP: Ophthalmic
	Q-switched KTP (nanosecond pulses): Tattoo removal, Port wine stain removal

TABLE 5.9. Dye Lasers

Wavelength:	Continuously tunable from 400 to 800 nm with the change of dye.
Laser medium:	Fluorescent dyes dissolved in solvents
Excitation:	Pulsed dye laser: Flash lamp, KTP laser, excimer laser
	CW dye laser: Ar-ion laser
Applications:	Photodynamic therapy
	Dermatology
	Ophthalmology
	Vascular disorders, selective target damages of blood vessels or pigment cells

TABLE 5.10. Argon-Ion Laser

Wavelength:	488 nm/514.5 nm
Laser medium:	Argon gas at about 1 torr of pressure
Excitation:	Electrical
Output:	CW mode, power range typically in the range of 100 mW to 20 W
Applications:	Most important medical use is in ophthalmology (retinal detachment)
	Light source for bioimaging
	Raman spectroscopy

TABLE 5.11. Ti:Sapphire Laser

Wavelength:	Tunable between 690 and 1000 nm
Laser medium:	Ti^{3+} ions in a sapphire solid host (Al_2O_3)
Excitation:	Optical, using an argon laser or a frequency-doubled Nd:YAG laser
Output:	CW, nanoseconds and femtoseconds pulses, power typically in hundreds of milliwatts range
Applications:	Multiphoton microscopy
	Multiphoton photodynamic therapy
	Tissue contouring, ablation

frequencies can be doubled (using nonlinear optical techniques described in Section 5.4) to produce green radiation at 532 nm.

Diode lasers at 635 nm are replacing the helium–neon gas lasers for applications for beam aiming and for low-level light therapy.

Recently, diode lasers operating near 400 nm with CW powers up to 20 mW have been introduced in the laser market. They provide a convenient excitation source for many dyes for usage in fluorescence microscopy. Another very recent addition is that of a solid-state laser from Coherent, Inc., which provides a CW output at ~488 nm, the lasing line of an argon-ion laser. Currently, the maximum output provided by this laser ranges in power from 10 mW to 200 mW. This power level is sufficient to replace the argon-ion laser for many applications such as in bioimaging, flow cytometry, and Raman spectroscopy.

This laser involves an optically pumped semiconductor laser that is frequency-doubled inside the laser cavity (intracavity frequency doubling) to convert the lasing wavelength of 976 nm to 488 nm. The nonlinear optical process of frequency doubling is described in Section 5.4.2.

Another area of significant current interest is the femtosecond laser technology for nonlinear optical techniques used in imaging and in laser treatment. The Ti:sapphire laser has captured most of the femtosecond laser market. Current technological developments are focusing on (i) the use of diode-pumped and frequency-doubled neodymium lasers for optical pumping of the Ti:sapphire laser, (ii) techniques to produce amplified pulses, and (iii) methods for generating extremely short pulses (<<100 fsec; 100 fsec are typical values for the current mode-locked Ti:sapphire lasers commercially sold).

Another direction for current laser technology is the development of high-power optical parametric oscillators (OPO) and optical parametric amplifiers (OPA). These devices utilize the nonlinear optical processes of optical parametric generation discussed in Section 5.4.2. The OPOs and the OPAs use nonlinear crystals to generate a widely tunable wavelength range and are candidates to replace dye lasers which for a long time have been the only tunable laser source (Schäfer, 1990).

Free electron lasers (FELs) form another area of laser technology development (Brau, 1990). These lasers work on a different physical principle in which the kinetics energy of electrons in an electron beam is directly converted into light. The structure of an FEL includes a series of magnetic wigglers providing an alternating field to the electron beam. The trajectory of an electron is included in an optical cavity to produce a laser action. An FEL provides the advantage of very broad tunability, covering from far-IR to UV and, potentially, high power. Therefore, the FEL can be a very useful laser light source for many biophotonics applications. However, the very large size and the very high cost of the current FEL lasers severely limit their applications at the present time.

5.3 QUANTITATIVE DESCRIPTION OF LIGHT: RADIOMETRY

In interaction of light with matter, a number of different quantitative descriptions are used to specify the level and the amount of light exposure. This forms the topic of radiometry as defined in Table 5.12.

For a pulse laser source the average power, the peak power, and the energy per pulse are related. The average power, defined as the energy per second averaged over pulses, is related to the energy as

$$P_{ave} = \text{energy per pulse} \times \text{number of pulses per second}$$

For example, the average power of a laser producing 100 mJ at 10-Hz repetition rate is given as

TABLE 5.12. Radiometry: Quantitative Description of Light

Light energy:	joules (J), mJ (10^{-3} J)
power output	
(energy/time):	watts (W) (1 joule/sec)
Peak power:	W, KW (10^3 W)

(in the case of a pulse laser; it is the power of the laser pulse at its peak position)

power density:	W/cm^2
(Irradiance)	Power per unit area
	Power/area of the laser beam spot
Fluence:	Energy density (joules/cm^2)
(Flux)	Energy per unit area
	Energy/area of the laser beam

$$P_{ave} = 0.1 \text{ J} \times 10 \text{ sec}^{-1} = 1 \text{ J sec}^{-1} = 1.0 \text{ W}$$

The peak power is calculated from the energy per pulse assuming a specific form of the pulse shape (in time). For example, assuming a rectangular shape (also called top-hat pulse), the peak power, P_{peak}, is given as

$$P_{peak} = \frac{\text{energy per pulse}}{\text{pulsewidth (pulse duration)}} \tag{5.3}$$

For example, the peak power of a laser pulse of 1-mJ energy and 10 nsec pulse duration is

$$P_{peak} = \frac{1 \times 10^{-3} \text{ J}}{10 \times 10^{-9} \text{ sec}} = 10^5 \text{ W} \tag{5.3a}$$

5.4 NONLINEAR OPTICAL PROCESSES WITH INTENSE LASER BEAM

5.4.1 Mechanism of Nonlinear Optical Processes

The description of interaction of light with matter as presented in Section 4.1 is applicable only under linear response when the polarization of a molecule (or a medium) is linearly proportional to the applied electric field (whether a dc field or an oscillating electric field of light, represented here by $\vec{E}$ but in Section 4.1 by ε). Under the illumination with an intense light source such as a laser beam where the associated electric field is strong, one has to consider the polarization (distortion) of an electron of a molecule and the corresponding medium by using a power series expansion in the applied field strength $\vec{E}$. For a molecule, equation (4.1)—which describes the induced dipole moment—now becomes

$$(\mu_T - \mu) = \alpha \cdot \vec{E} + \beta : \vec{E} \cdot \vec{E} + \gamma : \vec{E} \cdot \vec{E} \cdot \vec{E} \tag{5.4}$$

where β and γ describe the higher-order polarizations and are called, respectively, the first hyperpolarizability (second-order nonlinear optical term) and the second hyperpolarizability (the third-order nonlinear optical term). The corresponding bulk polarization P, as defined by equation (4.10), can be expanded in a similar power series:

$$P = \chi^{(1)} \cdot \vec{E} + \chi^{(2)} : \vec{E} \cdot \vec{E} + \chi^{(3)} : \vec{E} \cdot \vec{E} \cdot \vec{E} \tag{5.5}$$

For a bulk form (such as liquids, molecular solids, or organic glasses) consisting of weakly interacting molecules, the bulk polarization P is derived from the distortion of electronic clouds in constituent molecules. The bulk nonlinear optical susceptibilities $\chi^{(2)}$ and $\chi^{(3)}$ are, therefore, obtained from the corresponding molecular nonlinear optical coefficients β and γ by using a sum of the molecular coefficients over all molecule sites. The sum considers their orientational distribution (different orientations at different sites). Furthermore, a local field correction factor F is introduced to take into account that the fields experienced at any site are modified from the applied fields ($\vec{E}_1, \vec{E}_2$, etc.) due to interaction with the surrounding molecules. For example, $\chi^{(2)}$ and β, both being third-rank tensors, (relating three vectors: P and $\vec{E}$, $\vec{E}$) are related as

$$\chi^{(2)} = F \Sigma \beta^n (\sigma, \varphi) \tag{5.6}$$

In equation (5.6), β^n represents the β value for a molecule at site n with angular orientation σ, φ. For an ordered molecular aggregate with identical molecules at constant spacing and same orientation, we have

$$\chi^{(2)} = FN\beta \tag{5.7}$$

In equation (5.7), N = number density of molecules. Nonlinear optical interactions in a molecular medium, described by the bulk nonlinear optical susceptibilities, now become primarily molecular properties, described by the values of β and γ. Various quantum mechanical approaches are available to calculate these coefficients β and γ and, thus, derive an understanding of which chemical structures will produce a large β or a large γ. Any description of these approaches is beyond the scope of this book; for information on this topic see Prasad and Williams (1991).

In general, charge distribution due to π electrons are readily deformable. Therefore, conjugated π-electron structures (Chapter 2) give rise to large optical nonlinearities. In comparison, contributions due to σ electrons are considerably smaller.

The following sections describe some selected nonlinear optical effects that are of significance to biophotonics.

5.4.2 Frequency Conversion by a Second-Order Nonlinear Optical Process

Using trigonometric relations, one can see that the second-order term $\chi^{(2)}$ involving two electric vectors $\vec{E}(v_1)$, $\vec{E}(v_2)$ due to two light beams of frequencies v_1 and v_2 will produce terms with $v_1 + v_2$ and $v_1 - v_2$. These terms lead to mixing of two light beams to produce a light beam of new frequencies $v_1 + v_2$ and $v_1 - v_2$ or vice versa (see Figures 5.4–5.7).

The SHG process, actually, is a special case of sum frequency mixing where the frequencies of the photons from the two incident beams are equal ($v_1 = v_2$). In a $\chi^{(2)}$ medium all these processes can occur. However, by choosing the appropriate conditions, one can selectively enhance one process. Some of these conditions are (i) selection of appropriate input beam (or beams) and (ii) fulfillment of the phase-matching condition according to which a new output component (a sum frequency or a parametric process) builds up (is significant) only when the phase velocities [determined by the refractive indices as given by equation (2.5)] of the input and output waves are the same (in phase).

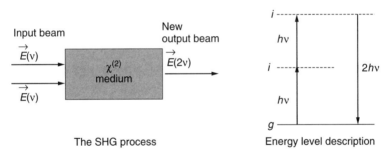

The SHG process   Energy level description

Figure 5.4. Second harmonic generation (SHG, also called frequency doubling); SHG process (*left*); energy level description (*right*).

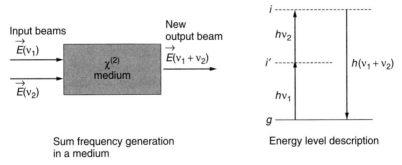

Sum frequency generation in a medium   Energy level description

Figure 5.5. Sum frequency generation in a medium; mixing process in a medium (*left*); energy level description (*right*).

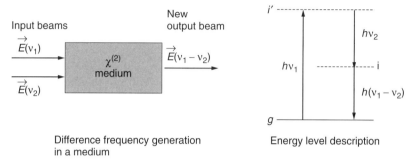

Figure 5.6. Difference frequency generation: the mixing process in a $\chi^{(2)}$ medium (*left*); energy level description (*right*).

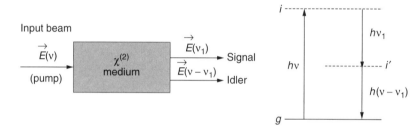

Figure 5.7. Optical parametric generation describing the splitting of a photon of frequency v to two photons of frequency and v_1 and $v - v_1$; energy level description (right).

The energy level description shows that the new photons of shifted frequencies are not generated by absorption, followed by an emission of a photon. They are generated in a one-step process (the medium acts as an energy transformer). In other words, states i and i' connecting the photons to the ground state g are generally not real levels. Therefore, no absorptions to levels i and i' take place; it is just a conceptual visualization of the process involving virtual levels i and i'. If i and or i' are real states of a molecule, these processes are considerably resonance-enhanced. One way to distinguish these processes from an emission such as fluorescence is to look at the line width (frequency distribution) of the output. For the $\chi^{(2)}$ nonlinear optical process, the new frequency, say $2v$ for SHG, is sharply peaked at $2v$ and is determined by the frequency distribution of the input beam at v. In contrast, fluorescence is considerably broad, involving a homogeneous or an inhomogeneous broadening mechanism of the medium.

5.4.3   Symmetry Requirement for a Second-Order Process

For a medium to exhibit the above frequency conversion processes mediated by $\chi^{(2)}$, the medium must have $\chi^{(2)} \neq 0$. This condition requires that at molecular level the nonlinear coefficient β (see relation between $\chi^{(2)}$ and β [equations (5.6) and (5.7)] must not be zero. Furthermore, the orientationally averaged sum of β^n at all sites which gives $\chi^{(2)}$ in the bulk form [equation (5.6)] should not be zero. These two conditions lead to the following symmetry requirements for $\chi^{(2)} \neq 0$:

- The molecules are noncentrosymmetric (do not possess an inversion symmetry). For such a structure, β—being an odd rank (3rd rank) tensor—is not zero.
- The molecules in the bulk form are arranged in a noncentrosymmetric structure. Only then the orientationally averaged sum values of all β^n, an odd rank tensor, is not zero.

A molecular design often used to make molecules with large β values (say for second harmonic generation) is

$$A–\pi \text{ conjugation–D}$$

where a molecular unit involving π conjugation is connected to an electron donor group, D (such as —NH_2), at one end and to an electron accept group, A (such as —NO_2), at the other end. A classic example is *p*-nitroaniline, which involves the NO_2 and NH_2 groups substituted at the para (*p*) positions of the benzene ring, as shown in Structure 5.1:

The second condition, requiring a noncentrosymmetric bulk form, permits a second-order nonlinear process such as SHG only for the following bulk forms:

- A noncentrosymmetric crystal
- An interface between two media (an interface is always asymmetric)
- An artificially aligned molecular orientation under the influence of an electric field (such structures are called *electrically poled*)

Thus, a solution or a glass, in its natural form being random, does not exhibit any second-order effect. For biological systems, important second-order effects are associated with the interface and with electrical field poling. Surface second harmonic generation from a biological membrane provides a powerful method for second-harmonic imaging to selectively probe interactions and dynamics involving membranes. The electric-field-induced second harmonic generation provides an excellent probe for membrane potential. These applications are discussed in Chapter 7 on bioimaging.

$$O_2N \underbrace{} NH_2$$

Structure 5.1

5.4.4   Frequency Conversion by a Third-Order Nonlinear Optical Process

The third-order term $\chi^{(3)}$ also leads to new optical frequency generation by mixing of three waves of frequencies v_1, v_2, and v_3. The two examples presented here, which are of importance for bioimaging are third-harmonic generation, abbreviated as THG, and coherent anti-Stokes Raman scattering, abbreviated as CARS.

The third harmonic generation is the third-order nonlinear optical analogue of second-harmonic generation. This process is schematically represented in Figure 5.8.

In this process, the incident photons are of frequency v (wavelength λ). In a $\chi^{(3)}$ medium, three incident photons combine together to generate a new photon of frequency $3v$ (wavelength $\lambda/3$). Thus, an incident fundamental light of wavelength at 1064 nm (from an Nd:YAG laser) will produce a third-harmonic beam at ~355 nm in the UV. Again, like in the case of second-harmonic generation, it is a coherent process and does not involve absorption of three photons of frequency v. The frequency conversion occurs through virtual states, as shown by the right-hand-side energy diagram of Figure 5.8.

Coherent anti-Stokes Raman scattering is also a $\chi^{(3)}$ process, but unlike third-harmonic generation, it is a resonant process. In other words, it involves a real level of the molecule (Figure 5.9). In this process, two input light beams of frequencies v_1 and v_2 generate a new output beam at frequency $(2v_1 - v_2)$, provided that the $\chi^{(3)}$ medium consists of molecules with a vibrational frequency v_R observed in Raman spectra such that $v_1 - v_2 = v_R$. In other words,

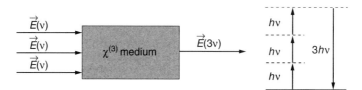

Figure 5.8. Third harmonic generation (abbreviated as THG) in a $\chi^{(3)}$ medium. The THG process (*left*); energy level description (*right*).

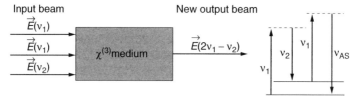

Figure 5.9. Coherent anti-Stokes Raman scattering, abbreviated as CARS. The CARS process (*left*); energy level description (*right*).

this frequency mixing process is a resonant process involving a vibration of the molecule at frequency $v_R = v_1 - v_2$. Thus, by monitoring the new output CARS signal at frequency $v_{AS} = 2v_1 - v_2$ as a function of $v_1 - v_2$ where v_1 or v_2 is fixed and v_2 or v_1 is varied, one can obtain the Raman spectra of the molecule.

Unlike the second-order nonlinear optical processes, which can take place only in a noncentrosymmetrically oriented medium, the third-order processes can take place in any medium, random or ordered. Hence, they can be observed in liquids, amorphous solid media, or crystalline media. But because it is a higher-order process, third-harmonic generation is a less efficient process than second-harmonic generation, for a medium where both can occur. In other words, it will take a more intense optical pulse (higher electric field $\vec{E}(v)$ to generate third-harmonic) than what will be required for second-harmonic generation.

5.4.5   Multiphoton Absorption

Multiphoton absorption is a nonlinear optical absorption process in which more than one photon, either of the same energy (frequency) or of different energies, are absorbed to reach a real excited state. Two-photon absorption is a third-order nonlinear optical process, described by the third-order nonlinear optical susceptibility $\chi^{(3)}$. Thus, this process has no symmetry restriction and can occur in any medium, symmetric or noncentrosymmetric. Extended conjugation (delocalization of π electrons) gives rise to a large molecular third-order coefficient, γ.

In order to see a two-photon process, one can cast the nonlinear optical effect as a manifestation on its refractive index. In Chapter 4, Section 4.2, it was discussed that the refractive index n is related to the susceptibility χ [equation (4.11)]. Under the influence of nonlinear optical interactions, the bulk susceptibility becomes dependent on the electric field strength $\vec{E}$ (through $\chi^{(2)}$, $\chi^{(3)}$ etc.). The refractive index n^2 related to χ through equation (4.11) of Chapter 4 now becomes dependent on $\vec{E}^2$ or I of the light beam and can be written as

$$n(I) = n_0 + n_2 I + n_3 I^2 + \cdots \tag{5.8}$$

Here n_0 is the linear refractive index (independent of the intensity of light) describing the linear optical processes, such as refraction and absorption, and relates to $\chi^{(1)}$ as discussed in Section 4.2. The terms n_2 and n_3 describe the nonlinear refractive indices and relate to the odd term nonlinear susceptibilities $\chi^{(3)}, \chi^{(5)}$, and so on. The real part of n_2 describes the nonlinear refractive index contribution, linearly dependent on the intensity of light. It relates to the real part of $\chi^{(3)}$ and, thus, the real part of the molecular coefficient, γ. Just like for linear absorption, the imaginary part of n_2, relating to the imaginary part of

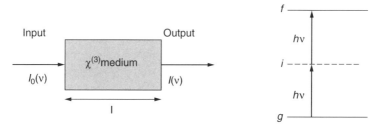

Figure 5.10. (*Left*) Two-photon absorption for photons of the same frequency, ν (only one laser beam of frequency ν). (*Right*) The energy-level description.

$\chi^{(3)}$ and, therefore, the imaginary part of γ, describes the absorption process, which is linearly dependent on the intensity I of light. This process is called *two-photon absorption* and is depicted in Figure 5.10.

The absorption is described by a two-photon absorption coefficient α_2 (unfortunately, the same letter α as used for the linear polarizability of the molecule). The above description applies when both photons absorbed simultaneously are of the same frequency ν. One can also have a two-photon absorption involving simultaneous absorption of photons of two different frequencies ν_1 and ν_2. The energy level diagram shows that a simultaneous two-photon absorption can be viewed as an absorption of one photon leading to a virtual intermediate level i from where the absorption of another photon leads to a final level f.

A simultaneous absorption of two photons, sometimes also called *direct two-photon absorption*, is different from a two-photon absorption involving a real level i which occurs in the case of many rare-earth ions. The latter process is a sequential two-photon absorption in which each sequential step is a linear absorption to a real level.

The direct two-photon absorption process was predicted in 1930, but for a long time it was of academic interest. With the availability of suitable light sources (short-pulse lasers) and materials with large two-photon absorption cross section (Bhawalkar et al., 1996), this process is now receiving tremendous attention for both technological applications and biophotonics.

For biophotonics, of particular interest is a two-photon process that produces an up-converted emission, as shown in Figure 5.11. The direct absorption of two photons of energy $h\nu_a$ each, creates an excitation of some excited level S_i (a singlet state) from which excitation cascades to the state S_1. The emission and thus the return to the ground state S_0 produces a photon of higher energy, $h\nu_e$—hence the term *up-converted emission*. The absorbed photons may be in the near-IR, but fluorescence produced is in the visible.

Since a two-absorption process is linearly dependent on the intensity, the population of the generated excited state is quadratically dependent on

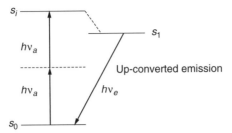

Figure 5.11. Two-photon pumped up-conversion emission.

the intensity. For this reason, a high peak power and highly focused laser beam produces the most efficient two-photon absorption as is also the case with all nonlinear optical processes. One, therefore, uses a pulse laser source for two-photon absorption. A convenient source is a femtoseconds Ti:sapphire laser, which provides high peak power with a very low average power because of the ultra-short duration of the pulse.

For a centrosymmetric molecule, there is a symmetry selection rule for two-photon absorption. The two-photon absorption cross section involves a product of two transition dipole moments, $\langle g|\mu|i\rangle\langle i|\mu|f\rangle$, with the first transition dipole connecting the ground state to an intermediate (virtual) level i and the second dipole connecting i to the final (real) level f. Applying the Laporte rule as discussed in Section 4.1.2, i connecting to g has to be a u state. Hence, the two-photon active f state has to have a g symmetry. Therefore, while a one-photon absorption leads to a u-type excited state from the ground g state, a two-photon excitation leads to a g-type excited state, not accessible by one-photon absorption.

Another point to be noted is that the two-photon excited up-converted process, being a third-order nonlinear optical process, is different from a second-harmonic generation process (or sum-frequency generation process) which also produces an up-converted photon of higher energy. In the case of second-harmonic generation, the up-converted photon is exactly of doubled frequency and the emitted light is coherent (directional). In the case of a two-photon excited up-conversion, the emitted photons are less than double the frequency with a broad and incoherent emission (nondirectional).

A three-photon absorption is a fifth-order nonlinear optical process, described by $\chi^{(5)}$. In equation (5.8), the imaginary part of the nonlinear refractive index, n_3, describes three-photon absorption that involves simultaneous absorption of three photons. This absorption process is quadratically dependent on the intensity I of light. The resulting up-converted fluorescence shows an I^3 dependence on light intensity. Efficient three-photon-induced up-converted emission has recently been reported (He et al., 2002). Three-photon-induced fluorescence has also been used for fluorescence imaging (Hell et al., 1995; Maiti et al., 1997).

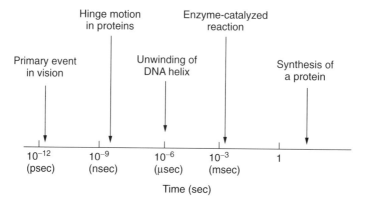

Figure 5.12. Typical time scales for some biological processes. (Reproduced with permission from Stryer, 1995.)

5.5 TIME-RESOLVED STUDIES

A major advantage provided by a laser source is its ability to produce pulses of very short duration, leading to the ability to monitor a biochemical or biophysical process in real time (Svelto et al., 1996). As shown in Figure 5.12, the various biological processes span a wide range of time scales, from hours to picoseconds (Stryer, 1995). For example, enzyme-catalyzed reactions can take milliseconds; the unwinding of a DNA double helix occurs in a microsecond. The hinge motion in proteins involving rotation of a domain can occur in nanoseconds. Even faster is the primary event in the process of vision, where the initial photoprocess of conformational change occurs in less than a few picoseconds after light absorption. Similarly, the primary step of photoinduced electron transfer in photosynthesis occurs in picoseconds. The use of appropriate lasers and laser techniques provides the capability to cover all these time scales.

Time-resolved studies using lasers, where an event is monitored as a function of time, can be used for the following investigations:

- *Decay of Emission to Obtain Lifetime*. Here an initial laser pulse of appropriate duration can be used to excite the emission. The emission decay can be monitored in real time.
- *Real-Time Monitoring of Fluorescence Resonance Energy Transfer (FRET)*. Here one selects the wavelength of the pulse laser to excite only the energy donor molecule and monitors the emission from the energy acceptor molecule as a function of time.
- *Transient Spectroscopy*. In this case, laser pulses are used to obtain absorption, emission, or Raman spectra of the excited states of an absorbing molecule or a reactant, photo-generated intermediates, and any eventual photoproducts. The current methods of generating a wide range of optical frequencies using nonlinear optical techniques (such as frequency

mixing and optical parametric generation) allows one to obtain transient spectra from UV to far IR. Another nonlinear technique frequently used is that of a white-light continuum generation using ultra-short pulses (femtoseconds to picoseconds). A white-light continuum (light of wavelengths covering the entire visible range) is generated by a pulse exhibiting fast intensity change (hence, ultra-short pulse) when it is focused on a medium such as water.

- *Pump-Probe Studies.* It is another variation of transient spectroscopy in which a strong pulse is often used to create photoexcited species. A probe pulse, which is time-delayed with respect to the pump pulse by a predetermined amount (which can be continuously varied), monitors the changes in the spectral characteristics of the photoexcited molecules as it goes through conformational changes, forms intermediates or eventually generates a photoproduct.

- *Laser-Induced Temperature Jump Studies.* Laser pulses can be used to create a jump in the temperature of a biological medium to initiate a thermally induced change such as unfolding of a protein. This process can be followed by transient spectroscopy using another probe pulse.

The pump-probe method, using laser pulses, is the most versatile technique for time-resolved studies of biological processes. For longer time scales (nanoseconds or longer), two separate lasers can simply be used to provide pump and probe pulses. Electronic delay in firing of each laser can be adjusted to delay the probe pulse from one laser with respect to the pump pulse from another laser.

For time resolutions in picoseconds and femtoseconds, one often uses the same laser source to derive the pump and the probe pulses. A schematic for this arrangement is shown in Figure 5.13. An appropriate pump wavelength can be derived from the fundamental wavelength of a laser source by frequency shifting using an appropriate nonlinear optical technique (frequency doubling, optical parametric generation, etc.) Similarly, the probe pulse can be selected in wavelength by using an appropriate nonlinear optical technique. Alternatively, the probe pulse can be a broad-band white-light continuum, described above. The time delay between the pump beam and the probe beam is achieved by passing the probe beam through an optical delay line (longer optical path length) that includes two reflecting prisms (or corner cubes) which are mounted on a motorized stage to vary the optical path. Thus, if the optical path of the probe pulse is 3 mm longer than that traveled by the pump pulse (path length delay $\Delta l = 3$ mm), the corresponding time delay Δt is $\Delta l/c \approx 10$ psec, where c is the speed of light.

As presented in 6.4, transient spectroscopy has provided insight into the nature of intermediates formed during the processes of vision and photosynthesis (Birge, 1981; El-Sayed, 1992). More recently, the emphasis has shifted from transient electronic spectroscopy to transient vibrational spectroscopy

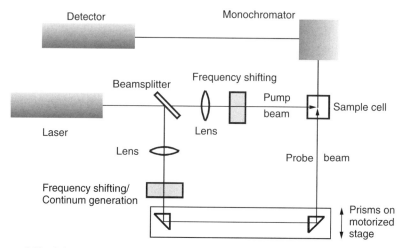

Figure 5.13. Schematic of a pump-probe experimental arrangement using the same ultra-fast laser source to generate both the pump and the time-delayed probe pulses.

which provides more structural details such as bond dissociation and changes in geometry (Hochstrasser, 1998; Hamm et al., 1998; Wang and El-Sayed, 2000). An important development is that of coherent femtosecond infrared laser pulses for the bond-specific approach to probe structure and functions of biological systems (Hamm and Hochstrasser, 2001).

5.6 LASER SAFETY

A laser as a high-intensity light source providing concentrated photon energy in a small beam size also poses potential safety hazards. Therefore, it requires a laser operator to be aware of these hazards and to exercise appropriate caution while using lasers. A good source of information on laser safety is the booklet *Laser Safety Guide* (Marshall and Sliney, 2000), published by the Laser Institute of America. This section only provides a brief discussion of the possible laser hazards and some precautions, particularly eye protection, that are to be taken.

Laser Hazards. When the eye and the skin are exposed to a laser beam over a maximum permissible exposure, called the MPE value, damage can occur. The MPE value depends on the wavelength and temporal characteristics (CW versus pulse) of a laser beam as well as on the exposure time. It also depends on the nature of the tissue being exposed. Thus, MPE values for skin exposure are usually higher than the corresponding values for eye exposure, because the retina is more sensitive than the skin.

As an example of MPE values, using the 488/514-nm beam from an argon-ion laser (often used in ophthalmology; see Chapter 13), the ocular exposure limit is $0.32 \, mJ/cm^2$.

The nature of these hazards is described below:

- *Eye Hazard.* Eye hazard can lead to injury to different structures of the eye (see Section 6.4.1), depending on the wavelength. Laser radiation of wavelength 400 nm to 1.4 µm causes retinal damage. This wavelength, transmitting through the cornea, can be focused by the lens on the retina to a spot 10–20 µm in diameter, thereby increasing the irradiance 100,000 times. Therefore, even specular reflection or scattering from an object in the path of the beam can cause retinal damage.

  UV radiation in the wavelength range of 180–315 nm is absorbed in the cornea and can do corneal damage. The near-UV radiation in the range 315–400 nm is absorbed in the lens and may produce certain forms of cataracts.

  The IR radiation such as that from a CO_2 laser (wavelength 10.6 µm) is strongly absorbed by water and can damage any biological tissue, as they contain water.

- *Skin Hazard.* Skin hazard becomes more serious while using UV radiation as well as high-power lasers. Light in the wavelength range of 230–380 nm can cause sunburn, skin aging, and skin cancer. The most significant effect in the infrared range is skin burn.

- *High-Power Laser Hazards.* The most serious hazard associated with high-power lasers is due to high electrical voltage circuits. Accidents associated with this can be electric shock and electrocution. Fire hazards also exist with high electrical power. Stray light in the presence of flammable materials may also create fire hazards.

Laser Hazard Classification. Many organizations worldwide have adopted the classification schemes employed on the basis of standards proposed by the American National Standards Institute (ANSI).

The ANSI standard has four hazard classifications based on the laser output power or energy. The higher the classification number, the greater the risk. Under this scheme, Class 1 lasers (like laser pointers) do not pose any significant hazard under normal operating conditions.

Laser Safety. In view of the potential hazards described here, it is imperative that safety procedures be strictly observed while operating a laser or utilizing a laser beam. Most of these precautions are just common sense. A detailed discussion of laser safety has been provided by Sliney and Wolbarsht (1980). Here, a few selected points are highlighted:

- All high-power electrical connections and cords should be well-insulated and kept far away from any source of water leakage.

- Laser beams should be blocked when not in use.
- All stray light and uncontrollable reflections should be eliminated.
- Commercially available eye safety goggles, appropriate for the laser wavelength in use, should be worn at all times while the laser beam is in use.
- All flammable substances should be distant from the laser beam path or the electrical power unit.
- With IR laser beams (such as the one from a Nd:YAG laser), a commercially available IR viewer or an IR card should be used to identify the beam path for alignment purposes.

HIGHLIGHTS OF THE CHAPTER

- A laser, which is an acronym for *l*ight *a*mplification by *s*timulated *e*mission of *r*adiation, is an intense source of highly directional, coherent, and monochromatic source of light with low beam divergence, capable of being focused into a very small spot.
- The two important steps to generating laser action are (i) excitation to create a population inversion between two energy levels of molecules in an active medium and (ii) optical feedback between two reflectors that form an optical cavity where stimulated emission and amplification occur.
- If the active medium involves three energy levels, also known as a *three-level structure*, it produces only a pulse laser action.
- If the active medium involves four energy levels, also known as a *four-level structure*, it can produce a pulse as well as a continuous laser action.
- Different ranges of the laser pulse width and pulse repetition rates can be achieved by a variety of optical gating or switching techniques such as (a) Q-switching to produce nanoseconds pulses and (b) mode-locking to produce picoseconds or femtoseconds pulses.
- Lasers can be classified on the basis of the pumping (excitation) process (electrical, optical, etc.), the lasing medium (gas, liquid, or solid), and the temporal features (continuous wave, also abbreviated as CW, Q-switched pulses, mode-locked pulses).
- Radiometry quantitates the level and amount of light exposure that a material is subjected to effect light–matter interactions.
- When the intensity of light is high, such as from a laser, new types of processes called *nonlinear optical processes* take place. They are dependent on the electric field (hence, intensity) of light in higher order.
- Two important types of nonlinear optical processes are higher-harmonic generation and intensity-dependent multiphoton absorption.

- Second-order nonlinear optical processes depend quadratically on the intensity, an important example being second-harmonic generation. Here, an input beam of frequency ν generates a new beam of frequency 2ν, but no energy is absorbed or emitted by the medium (nonresonant).

- Third-order nonlinear optical processes depend cubically on the intensity, an important example being third-harmonic generation. Here, an input beam of frequency ν generates a new input beam of 3ν, again through a nonresonant interaction (no absorption or emission from the medium).

- For second-order effect to be manifested in a medium, it must be non-centrosymmetric (not an amorphous solid or a liquid). No such symmetry restrictions apply for a third-order effect.

- Examples of multiphoton processes are two-photon and three-photon absorption in which two photons or three photons, respectively, are simultaneously absorbed to reach an excited state of a molecule.

- An up-converted emission at a higher energy (higher frequency or shorter wavelength) is generated when a molecule is excited by multiphoton absorption of lower energy (longer wavelength) photons.

- Coherent anti-Stokes Raman scattering, abbreviated as CARS, is another nonlinear optical frequency conversion process. Here, two beams of frequencies ν_1 and ν_2 generate a new output beam of frequency $2\nu_1 - \nu_2$ under the condition that the difference $\nu_1 - \nu_2$ corresponds to the frequency of a Raman active vibrational mode of the molecule.

- Pulse lasers are capable of monitoring the formation of intermediates or products in real time. Biophysical and biochemical processes using time-resolved studies, such as transient spectroscopy or pump-probe experiments, can be studied this way.

- Laser safety guidelines must be followed while operating lasers, because laser light poses hazards to eyes and skin.

REFERENCES

Basov, N. G., Bashkin, A. S., Igoshin, V. I., Oraevsky, A. N., and Shcheglov, V. A., *Chemical Lasers*, Springer-Verlag, Berlin, 1990.

Bhawalkar, J. D., He, G. S., and Prasad, P. N., Nonlinear Multiphoton Processes in Organic and Polymeric Materials, *Rep. Prog. Phys.* **59**, 1041–1070 (1996).

Birge, R. R., "Photophysics of Light Transduction in Rhodopsin and Bacteriorhodopsin, *Annu. Rev. Biophys. Bioeng.* **10**, 315–354 (1981).

Boyd, R. W., *Nonlinear Optics*, Academic Press, New York, 1992.

Brau, C. A., *Free-Electron Lasers*, Academic Press, Orlando, FL, 1990.

El-Sayed, M. A., On the Molecular Mechanisms of the Solar to Electric Energy Conversion by the Other Photosynthetic System in Nature, Bacteriorhodopsin, *Acc. Chem. Res.* **25**, 279–286 (1992).

Hamm, P., and Hochstrasser, R. M., Structure and Dynamics of Proteins and Peptides: Femtosecond Two-Dimensional Infrared Spectroscopy, in M. D. Fauer, ed., *Ultrafast Infrared and Raman Spectroscopy*, Marcel Dekker, New York, 2001, pp. 273–348.

Hamm, P., Lim, M. H., and Hochstrasser, R. M., Ultrafast Dynamics of Amide-I Vibrations, Biophys. J. **74**, A332–A332 (1998).

He, G. S., Markowicz, P. P., Lin, T.-C., and Prasad, P. N., Observation of Stimulated Emission By Direct Three-Photon Excitation, Nature, **415**, 767–770 (2002).

Hell, S. W., Bahlmann, K., Schrader, M., Soini, A., Malak, H., Gryczynski, I., and Lakowicz, J. R., Three-Photon Excitation in Fluorescence Microscopy, *J. Biomed. Opt.* **1**, 71–74 (1995).

Hochstrasser, R. M., Ultrafast Spectroscopy of Protein Dynamics, *J. Chem. Ed.* **75**, 559–564 (1998).

Maiti, S., Shear, J. B., Williams, R. M., Zipfel, W. R., and Webb, W. W., Measuring Serotonin Distribution in Live Cells with Three-Photon Excitation, *Science* **275**, 530–532 (1997).

Marshall, W., Sliney, D. H., eds. *Laser Safety Guide*, Laser Institute of America, Orlando, FL, 2000.

Menzel, E. R., *Laser Spectroscopy: Techniques and Applications*, Marcel Dekker, New York, 1995.

Prasad, P. N., and Williams, D. J., *Introduction to Nonlinear Optical Effects in Molecules and Polymers*, John Wiley & Sons, New York, 1991.

Schäfer, F. P., ed., *Dye Lasers*, 3rd edition, Springer-Verlag, New York, 1990.

Sliney, D. H., Wolbarsht, M., *Safety with Lasers and Other Optical Sources, A Comprehensive Handbook*, Plenum Press, New York, 1980.

Stryer, L., Biochemistry, 4th edition, W. H. Freeman, New York, 1995.

Svelto, O., *Principles of Lasers*, 4th edition, Plenum Press, New York, 1998.

Svelto, O., DeSilvestri, S., and Denardo, G., ed., *Ultrafast Processes in Spectroscopy*, Plenum Press, New York, 1996.

Wang, J., and El-Sayed, M. A., The Effect of Protein Conformation Change from α_{II} to α_I on the Bacteriorhodopsin Photocycle, *Biophys. J.* **78**, 2031–2036 (2000).

CHAPTER 6

Photobiology

This chapter covers the topic of photobiology, which deals with the interaction of light with biological matter. Thus, it is a natural extension of Chapter 4, which discussed the interaction of light with matter in general. The topic of photobiology, hence, forms the core of biophotonics, which utilizes interactions of light with biological specimens. This chapter utilizes a number of concepts, optical processes, and techniques already covered in Chapters 2–5.

This chapter discusses the interactions of various molecular, cellular, and tissue components with light. The various light-induced radiative and nonradiative processes are described, along with a discussion of the various photochemical processes. Photochemistry in cells and tissues can also be initiated by externally added exogenous substances, often called *photosensitizers*, which form the basis for photodynamic therapy, a topic Covered in Chapter 12.

The various types of scattering processes occurring in a tissue are covered. These processes, together with light absorption, determine the penetration of light of a given wavelength into a particular type of tissue. Methods of measurement of optical reflection, absorption, and scattering properties of a tissue are introduced. Some important manifestations of nonradiative processes in a tissue, used for a number of biophotonics applications such as laser tissue engineering (covered in Chapter 13) and laser microdissection (covered in Chapter 14), are thermal, photoablation, plasma-induced ablation, and photodisruption. These processes are defined.

Photoprocesses occurring in biopolymers play a major role in biological functions. Examples are the processes of vision and of photosynthesis. These processes are also covered.

An emerging area of biophotonics is *in vivo* imaging and spectroscopy for optical diagnosis. This topic is covered, along with the various methods of light delivery for *in vivo* photoexcitation. Another exciting *in vivo* biophotonics area is that of optical biopsy to detect the early stages of cancer. This topic is covered as well.

Finally, the chapter concludes with the coverage of single molecule detection. Understanding of structure and functions at the single biomolecule and

Introduction to Biophotonics, by Paras N. Prasad
ISBN: 0-471-28770-9   Copyright © 2003 John Wiley & Sons, Inc.

bioassembly levels is a major thrust of molecular and structural biology. The use of novel optical techniques allows one to probe processes at the single molecule level.

For supplementary reading on the contents of this chapter, suggested books are:

Grossweiner and Smith (1989): A general reference on photobiology

Niemz (1996): A comprehensive coverage of laser–tissue interactions

6.1 PHOTOBIOLOGY—AT THE CORE OF BIOPHOTONICS

Photobiology deals with the interaction of light with living organisms, from cellular specimens, to sectional tissues, to *in vivo* live specimens. Therefore, it deals with interactions of light with matter ranging in size scale from ~100 nm (viruses) to macro-objects (live organisms). It is an area that still offers many challenges in understanding the nature of light-induced processes. Chapter 4 dealt with the interaction of light with matter. The concepts developed in that chapter are of direct relevance to the topics discussed here, providing the basic foundation for them. However, the interaction of light and biological media, whether individual cells or tissues, is much more complex, often introducing a chain of events. The interactions can induce physical, thermal, mechanical, and chemical effects, as well as a combination of them. These interactions form the basis for the use of light for optical diagnostics and light-induced therapy as well as for medical procedures, which are discussed in subsequent chapters. These light-induced processes are mostly initiated by linear absorption of light. However, under intense field using a short laser pulse, one can induce nonlinear optical processes. For example, one can observe second-harmonic generation (SHG) from the surface of a cell membrane. Also, two-photon absorption can be induced in many chromophores. Here we discuss only the linear optical absorption. Table 6.1 lists various absorbing biological components.

6.2 INTERACTION OF LIGHT WITH CELLS

Biological cells span the size scale from submicron dimensions to over 20 μm. Therefore, they can be smaller than the wavelength of light or much larger. Interaction with light can lead to both scattering and absorption. Of particular interest in this regard is the Rayleigh scattering where even the subcellular components, organelles, can be a scattering center. Rayleigh scattering is dependent on three parameters: (i) the size of the scattering centers (cells or organelles), (ii) the refractive index mismatch (difference) between a scattering center and the surrounding medium, and (iii) the wavelength of light. The Rayleigh scattering is inversely proportional to the fourth power of wavelength. Therefore, a blue light (shorter wavelength) will be more scattered than a red light. Therefore, on the basis of scattering alone as the optical loss mech-

TABLE 6.1. Various Molecular, Cellular, and Tissue Components that Interact with Light

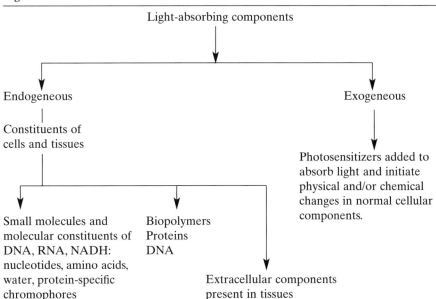

anism (attenuation of light transmission), the longer the wavelength of light, the deeper it would penetrate in a biological specimen. However, an upper wavelength limit of transmission is set up by absorptive losses in IR due to water absorption and absorption by the —CH and the —OH vibrational overtone bands. Bulk scattering becomes more pronounced in tissues and will be discussed in the next section.

In this section, light absorption by various components of the cell and the effects caused by light absorption will be discussed. Primary photoinduced cellular effects are produced by light absorption to induce transition between two electronic states (electronic or coupled electronic–vibrational [vibronic] transitions). Purely vibrational transitions (such as IR and Raman) are of significance only in structural identification and in conformational analysis.

In this section we discuss first the absorption by the various constituent molecules and biopolymers. Subsequently we discuss the various photochemical and photophysical processes induced by light absorption. Then we discuss the effects produced from light absorption by an exogenous chromophore added to the cell.

6.2.1 Light Absorption in Cells

Proteins are the most abundant chemical species in biological cells. They also are the most diversified chemical unit in living systems, from smaller-sized enzymes to larger proteins and from colorless to highly colored systems. They also exhibit a diversity of functions, from carrying oxygen to providing a light-induced neurological response for vision.

The basic constituents (building blocks) of proteins are amino acids, which can be aliphatic or aromatic (containing benzene or fused benzene type π-electron structures; see Chapter 2). The aliphatic amino acids absorb the UV light of wavelengths shorter than 240 nm (Grossweimer and Smith, 1989). Colorless aromatic amino acids such as phenylalanine (Phe), tyrosine (Tyr), and tryptophan (Trp) absorb at wavelengths longer than 240 nm, but well below the visible. However, the absorption by a protein is not completely defined by those of the constituent amino acid residues. Protein bonding involving the polypeptide bonds and disulfide linkage also absorb and contribute to the overall absorption of a protein. Furthermore, a protein may contain a chromophore such as the heme group (in hemoglobin) and *cis*-retinal (in case of retinal protein), which provide strong absorption bands. Hemoglobin has absorption peaks around 280 nm, 420 nm, 540 nm, and 580 nm. Melanin, the basic pigment of skin, has a broad absorption, covering the entire visible region, but decreasing in magnitude with the increase of wavelength. These features are evident from Figure 6.1, which exhibits the absorption characteristics of oxyhemoglobin (HbO_2) and melanin.

The constituents of DNA and RNA are the nucleotides that contain carbohydrates and purine and pyrimide bases (A, C, T, G, and U discussed in Chapter 3). The absorption by carbohydrates is below 230 nm; the absorption by the carbohydrate groups generally does not produce any significant photophysical or photochemical effect. The purine and pyrimidine bases absorb light of wavelengths in the range of 230–300 nm. This absorption is mainly responsible for DNA damage. A cellular component, exhibiting absorption in the visible, is

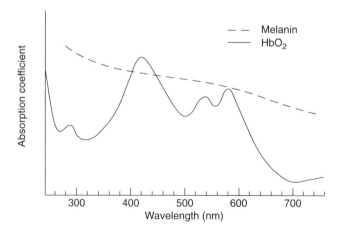

Figure 6.1. The absorption spectra of two important cellular constituents.

NADH, with absorption peaks at ~270 nm and 350 nm. Water does not have any bands from UV to near IR, but starts absorbing weakly above 1.3 μm, with more pronounced peaks at wavelengths ≥2.9 μm and very strong absorption at 10 μm, the wavelength of a CO_2 laser beam. Therefore, most cells exhibit very good transparency between 800 nm (0.8 μm) and 1.3 μm.

6.2.2 Light-Induced Cellular Processes

Cells exhibit a wide variety of photophysical, as well as photochemical, processes followed by light absorption. Some of these processes are shown in Table 6.2. A number of cellular constituents fluoresce when excited directly or excited by energy transfer from another constituent (Wagnieres et al., 1998). This fluorescence is called *endogenous fluorescence* or *autofluorescence*, and the emitting constituent is called an *endogenous fluorophore* (also called *fluorochrome*). As discussed in Chapter 4, fluorescence originates from an excited singlet state and has typical lifetimes in the range of 1–10 nsec. Phosphorescence, which is emission from an excited triplet (usually T_1), is generally not observed from cellular components.

Some of the fluorophores native to cells are NADH, flavins and aromatic amino acid constituents of proteins (e.g., tryptophan, tyrosine, phenylalanine). Various porphyrins and lipopigments such as ceroids and lipofuscins, which are end products of lipid metabolism, also fluoresce.

In addition, some important endogenous fluorophores are present in the extracellular structures of tissues. For example, collagen and elastin, present in the extracellular matrix (ECM), fluoresce as a result of cross-linking between amino acids. The absorption and emission spectra of some endogenous fluorophores are shown in Figure 6.2 (Katz and Alfano, 2000).

An important fluorescing protein that has received considerable attention during recent years for fluorescence-tagging of proteins is the green fluorescent protein (GFP) derived from jellyfish (Pepperkok and Shima, 2000; Hawes et al., 2000). In its native form, it absorbs at 395 nm and 475 nm with emission maximum in green, around 508. Intensive mutagenesis of the primary sequence has produced a wide variety of GFPs with broad spectral and biochemical properties. As shall be discussed in Chapter 8, the GFP and its variants have been utilized as multicolor fluorescent markers to be used as subcellular probes.

The thermal effects induced by light become more pronounced at the tissue level and will be discussed in the next section. Photochemical processes involving a chemical reaction in the excited state of a cellular constituent (or a chemical unit such as thymine in DNA) are varied, as exhibited in Table 6.2. Here are some examples (Grossweiner and Smith, 1989; Kochevar, 1995):

(i) *Photoaddition.* An important photoaddition process responsible for UV-induced molecular lesions in DNA is the photodimerization of thymine as illustrated below:

TABLE 6.2. Various Light-Induced Cellular Processes

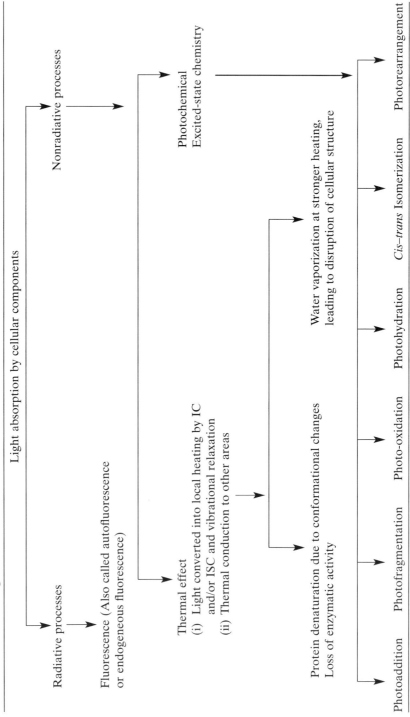

Light absorption by cellular components

Radiative processes

Fluorescence (Also called autofluorescence or endogeneous fluorescence)

Nonradiative processes

Thermal effect
(i) Light converted into local heating by IC and/or ISC and vibrational relaxation
(ii) Thermal conduction to other areas

Protein denaturation due to conformational changes
Loss of enzymatic activity

Water vaporization at stronger heating, leading to disruption of cellular structure

Photochemical
Excited-state chemistry

Photoaddition   Photofragmentation   Photo-oxidation   Photohydration   *Cis–trans* Isomerization   Photorearrangement

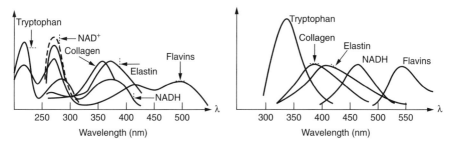

Figure 6.2. The absorption (*left*) and fluorescence (*right*) spectra of important tissue fluorophores. The y axes represent the absorbance (*left*) and fluorescence intensity (*right*) on a relative scale. (Reproduced with permission from Katz and Alfano, 2000.)

$$(6.1)$$

Another important photoaddition is that of cysteine (in protein) to thymine (in DNA), which can lead to photochemical cross-linking of DNA to protein as illustrated below:

$$(6.2)$$

(ii) *Photofragmentation.* In a photofragmentation reaction the original molecule, when photoexcited, decomposes into smaller chemical fragments by the cleavage of a chemical bond. This type of reaction is very common in biomolecules when exposed to short wavelength UV light. An example is the photofragmentation of riboflavin as illustrated below:

$$(6.3)$$

(iii) *Photooxidation.* Here the molecule, when excited, adds an oxygen molecule from the surroundings (a chemical process called oxidation). An example is the photooxidation of cholesterol:

Cholesterol   3β-Hydroxy-5α-hydroperosy-Δ⁶-cholestene

$$\text{(6.4)}$$

(iv) *Photohydration.* This type of reaction is also responsible for creating lesions in DNA. In this reaction, an excited molecule adds a water molecule to produce a new product, as illustrated for uracil.

Uracil   6-Hydroxy-5-hydrouracil

$$\text{(6.5)}$$

(v) *Photoisomerization.* Photoisomerization here specifically refers to the change in geometry or conformation of stereoisomers (discussed in Chapter 2). An important photoisomerization process responsible for retinal vision is that of 11-*cis*-retinal which upon excitation rotates by 180° around a double bond to produce a geometric isomer, the all-*trans*-retinal. This process is shown below:

11-*cis*-Retinal   All-*trans*-retinal

$$\text{(6.6)}$$

(vi) *Photorearrangement.* In this photoinduced process the chemical formula of the molecule does not change, only a rearrangement of bonds occurs as illustrated for 7-dehydrocholesterol in skin which upon UV exposure produces vitamin D_3:

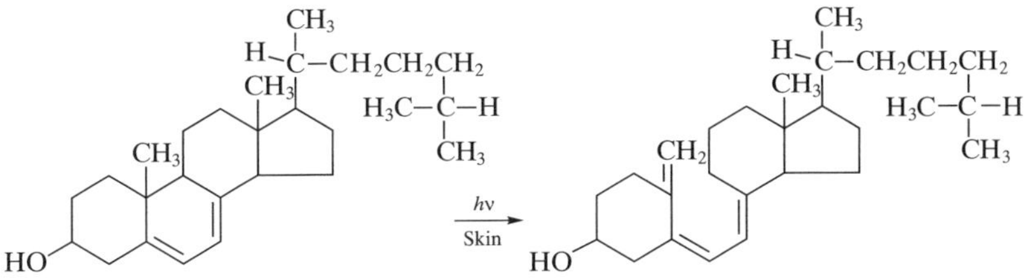

7-Dehydrocholesterol   Vitamin D$_3$ (6.7)

Photochemistry of proteins is derived from those of its amino acid residues. In addition, important photochemical processes in proteins involve the splitting of the disulfide bridges. Furthermore, DNA–protein cross-linking involves the addition of thymine in DNA with cystine in protein. This addition process has been discussed above.

In the case of DNA, UV irradiation can also lead to the breaking of either one or both DNA strand and intra- and intermolecular DNA cross-linking. The photochemistry of RNA is very similar to that of DNA.

6.2.3 Photochemistry Induced by Exogenous Photosensitizers

There are some important photochemical reactions that are produced by light absorption in chemicals introduced in a cell (or a tissue). These are exogenous chromophores that perform the function of photosensitization—that is, sensitizing a photoprocess (Kochevar, 1995; Niemz, 1996). The mechanism for most photosensitization involves photoaddition and photooxidation. These processes have been discussed in the previous section. However, in the present context, these processes are initiated by light absorption by the exogenous photosensitizers.

In the photoaddition reaction, a photosensitizer, when excited by light absorption, covalently bonds to a constituent molecule of the cell. An important example is the photoaddition reaction between the photosensitizer, 8-methoxypsoralen (8-MOP) with a pyridine base in DNA as shown below (Kochevar, 1995).

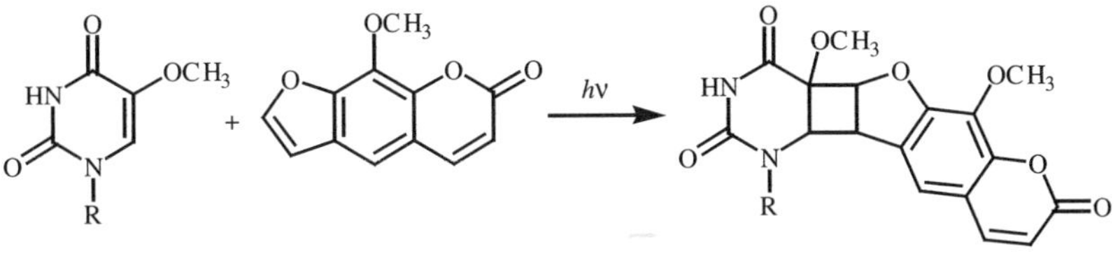

Thymine in DNA   8-Methoxypsoralen   Cycloadduct   (6.8)

This photochemical reaction occurs through the singlet excited state of 8-MOP. Because the singlet-state lifetime is short (in the nanosecond range), 8-MOP must be in a close proximity to a pyrimidine base. Therefore, the photoaddition is more likely to involve those 8-MOP molecules that are intercalated into the double-stranded DNA. This type of photoaddition reaction is supposed to be responsible for the phototoxicity of 8-MOP in human skin (Yang et al., 1989).

Photosensitized oxidation reactions involve a process in which the excited state of a photosensitizer produces a highly reactive oxygen specie such as an excited singlet oxygen ($^1O_2$), a superoxide anion (O_2^-), or a free radical (these are neutral chemical species with an unpaired electron, often represented by a dot as a superscript on the right-hand side) such as a hydroxyl radical ($OH^\bullet$). In fact, a photosensitized oxidation reaction often involves a chain reaction as shown below (Niemz, 1996):

(i) S_0 (photosensitizer) $\xrightarrow{h\nu}$ S_i (photosensitizer) $\longrightarrow T_1$

(ii) T_1(photosensitizer) + T_0 (oxygen) $\longrightarrow S_1$ (oxygen) + S_0 (photosensitizer)

(iii) S_1 (oxygen) + A cellular component $\longrightarrow$ Photooxidation of the cellular component

This photosensitized oxidation process forms the basis for light-activated cancer therapy, called *photodynamic therapy*, which is discussed in Chapter 12.

6.3   INTERACTION OF LIGHT WITH TISSUES

Nature of Optical Interactions.   A tissue is a self-supporting bulk medium. In other words, unlike cells, which have to be supported in a medium (in an aqueous phase as *in vitro* specimen or in a tissue either as an *ex vivo* or an *in vivo* specimen), tissues do not need a medium. Tissues, therefore, behave like any bulk medium in which light propagation produces absorption, scattering, refraction, and reflection as discussed in Chapter 4 (Niemz, 1996). These four processes are shown in Figure 6.3. The reflection of light from a tissue is reflection from its surface. The greater the angle of incidence, the larger the reflection from the surface. Therefore, maximum light will be delivered to the tissue (penetrate the tissue), when it is incident on the tissue at 90° (the light beam is perpendicular to the tissue).

The absorption of light under weak illumination (such as under lamp or a CW laser source) is a linear absorption described by Beer–Lambert's law, discussed in Chapter 4. The absorption is due to various intracellular as well as extracellular constituents of the tissue.

However, the most pronounced effect in a tissue is scattering. A tissue is a highly scattering turbid medium. The turbidity or apparent nontransparency of a tissue is caused by multiple scattering from a very heterogeneous structure consisting of macromolecules, cell organelles, and a pool of water. This

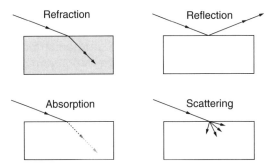

Figure 6.3. The four possible modes of interaction between light and tissue.

TABLE 6.3. The Various Light Scattering Processes in a Tissue

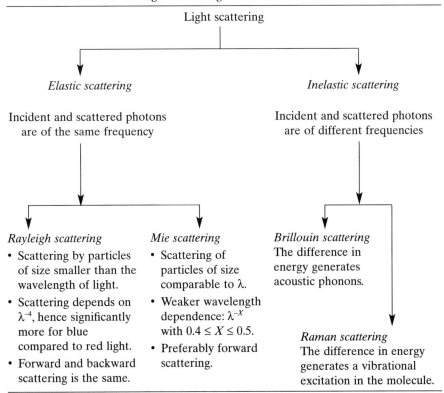

scattering leads to spreading of a collimated beam, resulting in a loss of its initial directionality as well as in defocusing (spread) of the light beam spot. The scattering process in a tissue is more complex and involves several mechanisms (Niemz, 1996). They are represented in Table 6.3.

The inelastic scattering in a biological tissue is weak. Brillouin scattering becomes significant only under the condition of generation of shockwaves discussed below. The acoustic phonons are ultrasonic frequency mechanical waves. Raman scattering in cells produces excitation of molecular vibrations, as discussed in Chapter 4. However, neither Rayleigh scattering nor Mie scattering completely describes the elastic scattering of light by tissue where photons are preferably scattered in the forward direction. The observed scattering shows a weaker wavelength dependence than that predicted by Rayleigh scattering but stronger than wavelength dependence predicted by Mie scattering. Other theories of scattering in a turbid medium (such as a tissue) have been proposed (Niemz, 1996). However, any detailed discussion of this subject is out of the scope of this book.

Raman scattering from tissues can provide valuable information on chemical and structural changes occurring as a result of diseases as well as due to mechanical deformations induced by aging or a prosthetic implant. Morris and co-workers have applied Raman spectroscopy to bone and teeth tissues. They have obtained spectra of both minerals (inorganic components) and proteins (organic components) of these tissues (Carden and Morris, 2000; Carden et al., 2003). In response to mechanical loading/deformation, they reported changes in the vibrational spectra in both the inorganic and organic components regions. Raman spectra taken at the edge of the indents revealed increases in the low-frequency component of the amide III band ($1250\,cm^{-1}$) and high-frequency component of the amide I band ($1665\,cm^{-1}$). These changes were interpreted as indicative of the rupture of collagen cross-links due to shear forces exerted by the indenter passing through the bone. More recently, Morris and co-workers have also applied Raman scattering for studying craniosynostosis (premature fusion of the skull bones at the sutures), which is the second most common birth defect in the face and skull (Tarnowski et al., submitted).

Like absorption, scattering creates a loss of intensity as the light propagates through a tissue. This loss is also described by an exponential function of the same nature as discussed for absorption in Chapter 4. Therefore, the total intensity attenuation in a tissue can be described as

$$I(z) = I_0 e^{-(\alpha + \alpha_s)z} \tag{6.9}$$

In this equation, $I(z)$ is the intensity at a depth z in the tissue; I_0 is the intensity when it enters the tissue, α = absorption coefficient, and α_s = scattering coefficient. Therefore, $\alpha + \alpha_s$ is the total optical loss. Another term used to describe the optical transparency of a tissue is the *optical penetration depth*, δ, which measures the distance z in the tissue after traveling which the intensity $I(z)$ drops to a fraction $1/e$ (= 0.37) of the incident value I_0. The term δ provides a measure of how deep light can penetrate into a tissue, and thus the extent of optical transparency of a tissue. From equation (6.9) one can find that the penetration depth δ is equal to $1/(\alpha + \alpha_s)$. The initial intensity I_0 is reduced to approximately 90% at a depth of 2δ in a tissue. In general, δ

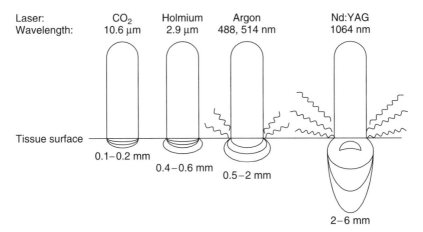

Figure 6.4. Penetration depths for commonly used laser wavelengths.

decreases with the vascularity (blood content) of a tissue. Furthermore, δ is significantly less for blue light than for red light and is the largest in the region 800–1300 nm. Figure 6.4 illustrates the penetration depths in a typical tissue for light of wavelengths of some commonly used lasers.

Measurement of Optical Properties of a Tissue. This subsection describes a method to determine reflection, absorption, and scattering properties of a tissue (Niemz, 1996). In a typical transmission experiment, one measures the transmission of a collimated beam (a laser source being the most convenient source) through a tissue of a finite length (the tissue specimen may be a dissected tissue). This method, in its simplest form, provides a total attenuation coefficient including optical losses from reflection, absorption, and scattering. In order to get information on each of these processes, a more sophisticated experimental arrangement has to be made which also takes into account the angular distribution of the scattered intensity. A commonly used experimental arrangement to simultaneously determine the reflectance, absorption, and scattering is that of double-integrating spheres first applied by Derbyshire et al. (1990) and Rol et al. (1990). The schematic of this experimental arrangement is shown in Figure 6.5.

In this method, two nearly identical spheres are located in front of and behind the tissue sample. These spheres have highly reflective coatings on their inner surface. Therefore, light-reaching detectors D_R and D_T are collected from all angles (hence the term *integrating spheres*). The first sphere integrates all the light that is either reflected or scattered backward from the sample. The light that is transmitted through the sample and scattered forward is detected by the second sphere at two ports. The port with detector D_T integrates all the forward scattering of the transmitted light, while the detector D_C measures the intensity of light in the forward direction of propagation. From these two

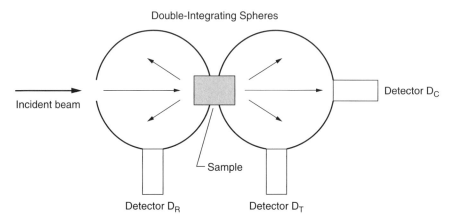

Figure 6.5. Schematics of an experimental arrangement utilizing a double-integrating sphere geometry for simultaneous measurement of reflection, scattering and absorption. The detectors D_C, D_T, and D_R at three different parts measure three quantities respectively: (i) transmitted coherent light that passes through the tissue in the direction of light propagation, (ii) transmitted diffuse intensity, and (iii) reflected diffuse intensity.

measurements one can separate the contributions due to scattering and absorption.

Light-Induced Processes in Tissues. Interaction of light with a tissue produces a variety of processes, some from its cellular and extracellular components and some derived from its bulk properties (Niemz, 1996). These processes are listed in Table 6.4. Each of these manifestations is briefly discussed below.

Autofluorescence. The autofluorescence subject has already been discussed in Section 6.2.2. Autofluorescence arises from the endogenous fluorophores that are present either as a cellular component or in the extracellular matrix. Any given tissue has in general a nonuniform distribution of many fluorophores that may also vary as a function of depth below the tissue surface. Therefore, the fluorescence spectrum measured at tissue surfaces may be different from that within the tissue from different layers. Furthermore, the autofluorescence may be different from a premalignant or malignant tissue compared to a normal tissue, thus providing a method for optically probing and even for early detection of cancer. Also metabolic changes induce changes in autofluorescence. For example, NADH is highly fluorescent, but its deprotonated form NAD^+ is not.

Photochemical Processes. The various photochemical processes in tissue components, initiated by absorption of light have been discussed above in

TABLE 6.4. Light-Induced Various Processes in Tissues

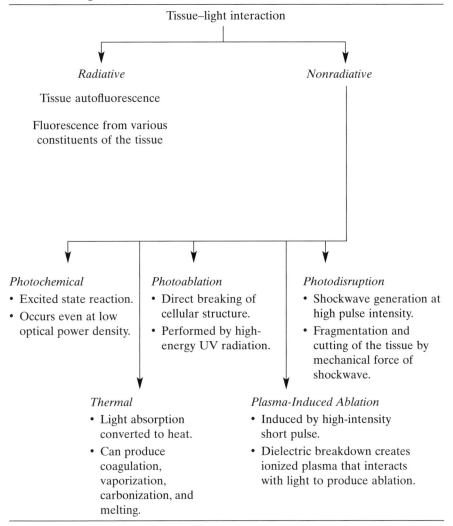

Section 6.2.2. These effects occur even at very low power densities (typically at $1\,W/cm^2$) when the absorption process is a simple one-photon absorption (linear absorption). These processes are dependent on fluence (irradiance) rather than intensity. Even though a conventional lamp source can be used for this purpose, one often uses a laser beam as a convenient light source. Recent interest has focused on nonlinear optical excitations such as multiphoton absorption, particularly two-photon absorption, discussed in Chapter 5, to induce photochemical processes, particularly photosensitzed oxidation in photodynamic therapy. The advantage offered by two-photon absorption is that the same photochemistry can be affected deeper inside the tissue, compared

to that induced by one-photon absorption which remains localized within microns of depth from the surface. One of the most important photochemical processes in tissues, from the biophotonics perspective, is the photosensitized oxidation discussed in Section 6.2.2. This subject will be further discussed in Chapter 12 on photodynamic therapy.

Thermal Effects. The thermal effects result from the conversion of the energy of light, absorbed in tissues, to heat through a combination of nonradiative processes such as internal conversion (IC), intersystem crossing (ISC), and vibrational relaxations. These topics have been discussed in Chapter 4. Thermal effects can be induced both by lamp as well as by CW and pulse laser sources and they are nonspecific; that is, they do not show a wavelength dependence, implying that no specific excited state need to be populated to create these effects. In the case of the use of a monochromatic laser beam, the choice of wavelength (absorption strength) and the duration of laser beam irradiance (pulse width, in the case of a pulse laser) may determine how the thermal effect manifests. The two important parameters are the peak value of the tissue temperature reached and the spread of the heating zone area in the tissue. The heating of an area in a tissue can produce four effects: (i) coagulation, (ii) vaporization, (iii) carbonization, and (iv) melting. For coagulation, the local temperature of a tissue has to reach at least 60°C, where the coagulated tissue becomes necrotic. Both CW (e.g., Nd:YAG) and pulse (e.g., Er:YAG) lasers have been used for different tissues. For a vaporization effect to manifest, the local temperature of a tissue has to reach 100°C, where water starts converting into steam, producing thermal ablation (or photothermal ablation) of the tissue. This ablation is a purely thermomechanical effect produced by the pressure buildup due to steam formation and is thus different from photoablation discussed below. In this process the tissue is torn open by the expansion of steam, leaving behind an ablation crater with lateral tissue damage. In a living tissue, the blood vessels can transport heat away from the ablation site, creating damage at other sites and thus the spread of the lateral damage. If one wishes to reduce the lateral thermal damage from thermal diffusion, one must ablate the tissue with a short pulse laser. Based on the thermal diffusion characteristics of biological tissues, it can be assumed that if the energy is deposited in the tissue in tens of microseconds, the thermal ablation remains primarily localized around the focal spot of the beam and the lateral thermal damage is minimized. However, the use of ultrashort and high-intensity pulses can lead to other complications such as nonlinear optical processes that may produce other effects.

Carbonization occurs when the tissue temperature reaches above 150°C, at which tissue chars, converting its organic constituents into carbon. This process has to be avoided because it is of no benefit and leads to irreparable damage of a tissue.

At sufficiently high power density from a pulse laser (generally in microseconds to nanoseconds), the local temperature of a tissue may reach above its melting point. This type of process can be used for tissue welding and shall be further discussed in the chapter on tissue welding.

Photoablation. This is a process whereby the various cellular and extracellular components are photochemically decomposed by the action of an intense UV laser pulse. The result is the release of photofragmented species from a tissue, causing etching (or ablation). This ablation is localized within the beam spot and is thus very clean. Typical power densities are 10^7–10^{10} W/cm^2. A convenient UV laser source is an excimer laser that provides a number of lasing wavelengths in the range of 193–351 nm. The pulses from these lasers are typically 10–20 nsecs. This method is very useful for tissue contouring (sculpturing), such as in refractive corneal surgery. This topic is also covered later in Chapter 13, entitled Tissue Engineering with Light.

Plasma-Induced Ablation. When exposed to a power density of 10^{11} W/cm^2, the tissue experiences an electric field of 10^7 V/cm associated with the light. This field is considerably larger than the average coulombic attraction between the electrons and the nuclei (a subject covered in Chapter 2) and causes a dielectric breakdown of the tissue to create a very large free electron density (plasma) of ~10^{18} cm^3 in the focal volume of the laser beam in an extremely short period (less than hundreds of picoseconds). This high-density plasma strongly absorbs UV, visible, and IR light, which is called *optical breakdown* and leads to ablation.

Photodisruption. This effect occurs in soft tissues or fluids under high-intensity irradiation that produces plasma formation. At higher plasma energies, shock waves are generated in the tissue which disrupt the tissue structure by a mechanical impact. When a laser beam is focused below the tissue, cavitation occurs in soft tissues and fluids produced by cavitation bubbles that consist of gaseous vapors such as water vapor and CO_2. In contrast to a plasma-induced ablation, which remains spatially localized to the breakdown region, photodisruption involving shockwaves and cavitation effects spreads into adjacent tissues. For nanosecond pulses, the shockwave formation and its effects dominate over plasma-induced ablation. However, for shorter pulses both plasma-induced ablation and photodisruption may occur and it is not easy to distinguish between these two processes. Their application is described in Chapter 13.

Photodisruption has found a number of applications in minimally invasive surgery such as posterior capsulotomy of the lens, often needed after cataract surgery and laser-induced lithotripsy of urinary calculi.

6.4 PHOTOPROCESSES IN BIOPOLYMERS

Photoprocesses in biopolymers are excellent examples of how Nature involves biophotonics more efficiently to perform various biological functions. The amazingly high efficiency of these processes have provided further impetus to the field of biomimicry which focuses on copying the underlying principles utilized by Nature to design and make useful materials. One example is design of multiarmed (dendritic) structures containing antenna molecules to harness solar energy for solar cells (Shortreed et al., 1997; Swallen et al., 1999; Adranov and Frechet, 2000), just as plants harvest solar energy for photosynthesis. The light-harvesting dendrimers are described in Chapter 16, entitled Biomaterials for Photonics. The photoprocesses in biopolymers are of great biological importance and are, therefore, discussed here as a separate section.

The photoprocesses in biopolymers covered in this book relate to three types of systems: (i) rhodopsin in the photoreceptors of a vertebrate eye, (ii) photosynthetic system in the chloroplast of a plant, and (iii) bacteriorhodopsin in the purple membrane of *Halobacterium halobium*. In each case, a light-absorbing chromophore initiates a highly complex series of photoinitiated processes. Some important features of these light initiated processes are as follows:

- They are highly efficient.
- The light-induced primary step leads to a series of complex steps forming a number of intermediates by a variety of physical and chemical processes.
- The subsequent intermediates following the light-induced primary process are formed in the dark; hence the steps involving them are often called *dark reactions*.
- The photoinduced process is cyclic, regenerating the original biopolymeric structure. Thus the entire cycle of going through the intermediate back to the original species is often called the *photocycle*.

In this section, only the process of vision by eye and the photosynthesis by plant are covered. The topic of bacteriorhodopsin is covered in Chapter 16, because bacteriorhodopsin has been shown to be useful for holographic data storage.

6.4.1 The Human Eye and Vision

Structure. At a very simple level, the visual system is comprised of the eye (Figure 6.6) and a long chain of neural connections. The pattern of excitation within the retina is processed by a neuronal machinery to create sensory perception in our brain. The light from an object enters the eye through the clear and transparent cornea. The pupil, an opening of the iris, in the front of the eye regulates the amount of light allowed to enter the eye. Much like the diaphragm in a camera, the iris controls the size of the pupil opening, to adjust to the light intensity level. The lens of the eye lies directly behind the cornea, and focuses the light rays on the receptor cells of

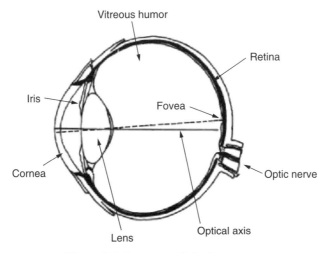

Figure 6.6. Structure of the human eye.

the retina. As in a camera, the lens of the eye creates an inverted image of the outside world on the retina. The retina is the innermost layer, which contains the specialized cells called the *light-sensitive receptors*. Perception of brightness is derived from the amplitude (intensity) of a light source, whereas the perception of a color is derived from the wavelength of light source. A human eye can detect only a portion of the light spectrum, ranging from approximately 380 (violet) to 760 (red) nms.

The human retina is organized into three primary layers: the outer nuclear cell layer, the inner nuclear cell layer, and the ganglion cell layer. The first one is the photoreceptive layer and is made up of rods and cones. The human retina consists of approximately 100–120 million rods and 7–8 million cones. The rods are extremely sensitive to light and provide achromatic vision (only shade of gray, no color) at lower (scotopic) levels of illumination. The cones are less sensitive than the rods, but provide color vision at higher (photopic) levels of illumination. The fovea covers only a small 1.5-mm area where photoreceptor cells are directly exposed to light. The bipolar cells in the second layer of the retina provide communication between the first and third layers. The ganglion cells constitute the third layer of the retina and convey the visual information, as encoded by the retina, via the optic nerve to the brain. Horizontal cells in the outer plexiform layer provide lateral interconnections between receptors.

The rods contain the visual pigment, *rhodopsin*, and are sensitive to blue-green light with peak sensitivity at the 500-nm wavelength.

In rhodopsin, 11-*cis*-retinal is covalently bonded to the protein opsin by the so-called *Schiff base linkage*, which involves acid-catalyzed (presence of proton) coupling of the aldehyde ($-\overset{\overset{\displaystyle O}{\|}}{C}-H$) group of retinal to the amine group of the protein. This coupling is represented below:

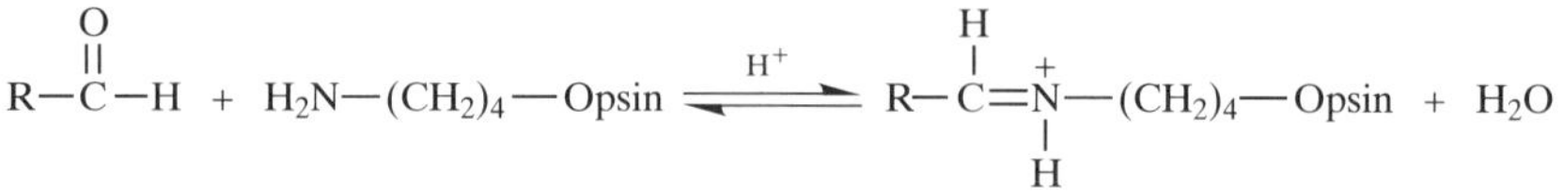

$$\underset{\text{11-}cis\text{-Retinal}}{R-\overset{\overset{\text{O}}{\|}}{C}-H} + \underset{\text{Protein}}{H_2N-(CH_2)_4-\text{Opsin}} \underset{}{\overset{H^+}{\rightleftharpoons}} \underset{\substack{\text{Rhodopsin} \\ \text{(protonated Schiff base)}}}{R-\overset{\overset{\text{H}}{|}}{C}=\overset{+}{\underset{\underset{\text{H}}{|}}{N}}-(CH_2)_4-\text{Opsin}} + H_2O \qquad (6.10)$$

Physical and Chemical Processes. Light reaching the eye is converted from stimulus to sensation to perception via three steps: reception, transduction, and coding. Reception is the process where photons are absorbed by rhodopsin. Transduction involves the utilization of the absorbed photon energy to cause a chemical reaction that initiates processes to generate nerve impulses. Coding of the light is a result of complex interactions between retinal neurons within the various layers, producing impulses that are later decoded in the visual cortex resulting in the perceived visual image.

Rhodopsin is a protein in the disk membrane of the rod photoreceptor cells in the retina. It provides only the light-sensitive step in vision. The 11-*cis*-retinal chromophore lies in a pocket of the protein and is isomerized to *all-trans*-retinal when light is absorbed. This isomerization is shown in scheme (6.6) earlier in this chapter and is represented again here in Figure 6.7. The isomerization of retinal leads to a change of the conformation of rhodopsin, which triggers a cascade of reactions.

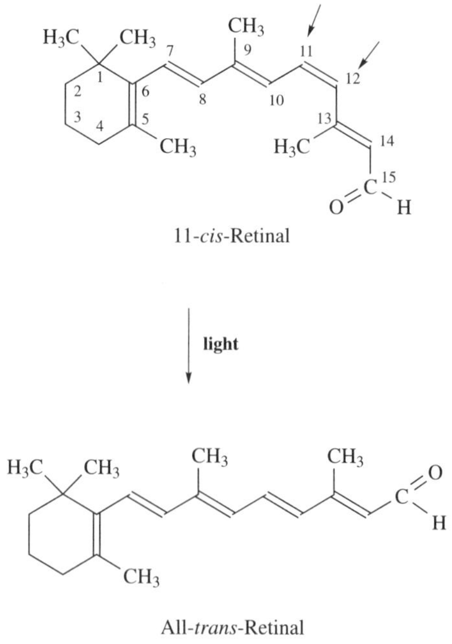

11-*cis*-Retinal

light

All-*trans*-Retinal

Figure 6.7. Retinal isomerization under light exposure.

Retinal consists of a system of alternating single and double bonds. In the dark, the hydrogen atoms attached to the #11 and #12 carbon atoms of retinal (pointing arrows in Figure 6.7) point in the same direction, producing a kink in the molecule. This configuration is designated *cis*. When light is absorbed by retinal, the molecule straightens out forming the all-*trans* isomer. This physical change in retinal triggers the following chain of events culminating in a change in the pattern of electrical impulses sent back along the optic nerve to the brain (http://users.rcn.com/jkimball.ma.ultranet/BiologyPages/V/Vision.html):

1. Formation of all-*trans*-retinal activates its opsin.
2. Activated rhodopsin, in turn, activates many molecules of a protein complex called *transducin*. (Transducin is one of many types of G proteins, rhodopsin is the prototypical member of a large family of G-protein-coupled receptors, GPCRs.)
3. Transducin activates an enzyme that breaks down cyclic GMP. (GMP is the guanine-containing cousin of AMP, discussed in Chapter 3.)
4. The drop in the cyclic GMP concentration closes Na^+ and Ca^{2+} channels in the plasma membrane of the rod. Because these positive ions can no longer enter (even though Ca^{2+} can still leave through a separate transporter pathway), the interior of the cell becomes more negative (hyperpolarized), increasing its membrane potential from -40 to as much as $-60\,mV$.
5. The electrical signal through a series of complex processes is finally transmitted by the optical nerve.

Reprocessing of all-*trans*-retinal to 11-*cis*-retinaldehyde occurs in the supporting retinal pigment epithelial cells. The 11-*cis*-retinaldehyde is then transported and incorporated into the photoreceptor membranes where it becomes available again for light transduction. Photoreceptor disk membranes have a finite lifetime of about 10 days. They are constantly shed and reprocessed, and their components are reincorporated by advanced recycling mechanisms outside of the photoreceptors.

The range of light energy we experience in the course of a day is vast. The light of the noonday sun can be as much as 10^8 times more intense than starlight. Our visual system copes with this huge range of brightness by adapting to the conditions of illumination, which range over 7–9 orders of magnitude. Adaptation is achieved through the coordination of mechanical, photochemical, and neural processes in the visual system. It involves coordinated action of the pupil, the rod, and the cone systems, producing bleaching and regeneration of receptor photopigments, modulation of the steps in phototransduction, and changes in neural processing.

Photoinduced Intermediates. Rhodopsin forms a number of intermediates upon light absorption, exhibiting a series of color changes because the different intermediates have different absorption maxima. Through these interme-diates, both the retinal molecule and the protein (opsin) continue to change their conformations that produce changes in their absorption maxima. Time-resolved spectroscopy has played an important role in elucidating the mechanism of photocycle of rhodopsin and in the identification of various intermediates. Table 6.5 lists the reported intermediates in the photoinitiated cycle of rhodopsin (their λ_{max}^{abs} are given in the parenthesis).

TABLE 6.5. The Various Intermediates Formed After Light Absorption by Rhodopsin

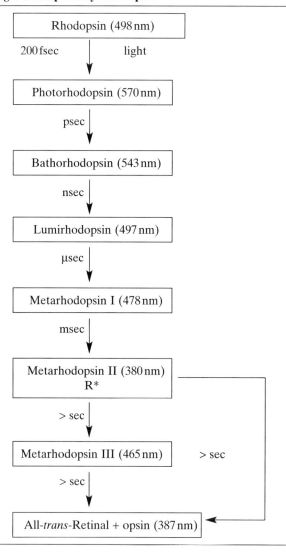

Light absorption by 11-*cis*-retinal invokes a singlet-to-singlet $\pi\pi^*$ transition, leading to a rapid photoisomerization within 200 fsec to form photorhodopsin which is thermally hot and conformationally distorted (Kim et al., 2001; Pan et al., 2002). It then relaxes in picoseconds to form bathorhodopsin. Time-resolved spectroscopy described in Chapter 5 (Section 5.5) has played a valuable role in the identification of various short-lived intermediates. More recently, time-resolved Raman spectroscopy has been used to obtain more detailed information on the chemical structure of the intermediates (Pan et al., 2002).

In these experiments, a pump pulse initiates the photocycle of rhodopsin, and a probe pulse near the optical absorption of a given intermediate form is used to obtain the resonance Raman spectra. The time-resolved resonance Raman spectra have also established the existence of another intermediate called *blue-shifted intermediate* (BSI) between bathorhodopsin and lumirhodopsin (Pan et al., 2002). Figure 6.8 shows a representative time-resolved Raman spectra of rhodopsin and its intermediates.

Metarhodopsin initiates an enzymatic cascade starting with the dissociation of opsin from all-*trans*-retinal, subsequently generating an electrical impulse. The Schiff base linkage becomes deprotonated during the transformation of metarhodopsin I to II. Metarhodopsin II, sometimes also referred to as activated opsin or photoexcited rhodopsin intermediate, still contains opsin covalently bonded, but now to all-*trans*-retinal. The dissociation of activated opsin, releasing opsin and all-*trans*-retinal, triggers a set of enzymatic reactions, eventually generating the electrical impulse. In the dark, all-*trans*-retinal is converted back to 11-*cis*-retinal by a series of reactions catalyzed by enzymes in the membranes. The 11-*cis*-retinal then recombines with opsin, to generate back rhodopsin.

6.4.2 Photosynthesis

Photosynthesis is another example of Nature's utilization of biophotonics in a very clever way to harness light and utilize photon energy for conducting various cellular processes to sustain life. Photosynthesis occurs in plants, algae, and a variety of bacteria; these organisms utilize sunlight to power various cellular processes. A good general coverage of the topic of photosynthesis is provided by Ksenzhek and Volkov (1998).

A true appreciation for this marvelous photosynthetic machinery of Nature can readily be seen by considering the following features:

- Nature utilizes a large number of light-absorbing chromophores that act as antenna to collect photon energy and funnel it efficiently to a reaction center. The antenna system thus allows an organism to greatly increase light absorption without having to build a reaction center for each absorbing chromophore.

- The excitation energy of the light absorbing antenna molecules is transferred to the reaction center in tens of picoseconds. This time period is

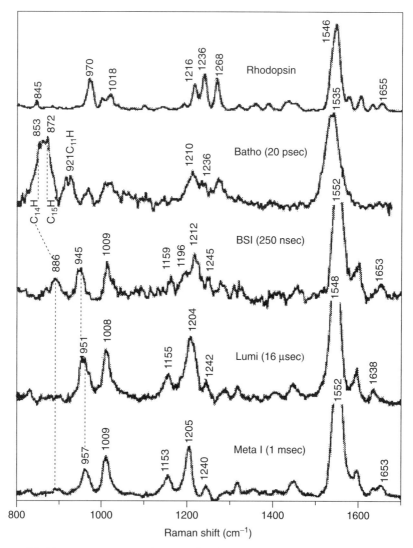

Figure 6.8. Room-temperature time-resolved resonance Raman spectra of rhodopsin and its intermediates. The rhodopsin spectrum is obtained using excitation at 458 nm. (Reproduced with permission from Pan et al., 2002.)

significantly shorter than the excited-state lifetime (1- to 5-nsec range) of isolated antenna molecules, thus allowing almost all the absorbed photon energy to reach the reaction center where it is used to perform appropriate photochemistry.

• A number of different types of antenna molecules (chlorophylls, carotenoids, etc.) are used such that their absorption spectra cover the solar spectrum. This way the antenna system can efficiently harvest (absorb) photons of wavelengths covering the solar spectral range.

- The process of photosynthesis is also self-regulatory. An example is Nature's use of quenchers (carotenoids) to dispose of the absorbed excess photon energy before photooxidation damage can occur as a result of strong absorption (excess photon harvesting) during periods of high irradiance such as midday.

All oxygen-evolving (oxygenic) photosynthetic organisms such as plants, cyanobacteia, and algae have two reaction centers. Anoxygenic photosynthetic organisms have only one reaction center complex.

The organelle chloroplast, described in Chapter 3, is bound by two membranes that do not contain the light-absorbing chromophore (also referred to as pigments in this context). The outer membrane is somewhat permeable, while the inner membrane provides the permeability barrier. However, unlike mitochondria of the animal cells, the chloroplast also contains a third membrane called the *thylakoid membrane* that forms flattened vesicles, thylakoids. It is the thylakoid membrane that contains the light-absorbing pigments. Hence the thylakoid membrane is the site of photosynthesis.

Some features of the photosynthetic unit are described below.

Light-Absorbing Photosystems. In oxygenic photosynthetic organisms such as plants, photons are absorbed by two large membrane protein complexes called Photosystem I, abbreviated as PS-I, and Photosystem II, abbreviated as PS-II. The main absorbing pigments are chlorophyll and carotenoid. The absorbing chromophore molecules transfer energy to specialized chlorophyll molecules that are reaction centers for electron transfer processes. These specialized chlorophylls in PS-I are referred to as P_{700} as their absorption maxima are 700 nm. Similarly, the specialized chlorophylls in PS-II are referred to as P_{680}, because of their wavelength of maximum absorption being at 680 nm.

The atomic level structural information on PS-I and PS-II as well as the molecular details of the energy levels and the detailed nature of excitation transfer resulting ultimately in electron-transfer reactions is still not resolved. It is thus an area currently being hotly pursued. An approach to getting atomic resolution structural information is by high-resolution crystal structures of the photosystems isolated from systems in Nature (e.g., bacteria). An example is a recent high-resolution crystal structure of PS-I of cyanobacterium *Synechococus elongatus* (Jordan et al., 2000).

For detailed mapping of the energy transfer and electron-transfer processes, a valuable method has been time-resolved spectroscopy (Connelly et al., 1997; Agarwal et al., 2000; van Amerongen and van Grondelle, 2001; Kennis et al., 2001; Melkozernov, 2001).

Each photosynthetic unit includes a spatial arrangement of membrane protein in complex with some 200–400 molecules of chlorophyll and other pigments such as carotenoids and phycobilins. The PS-I system involves mainly chlorophyll *a*, while PS-II includes various chlorophylls (e.g., chlorophyll *a* and chlorophyll *b*). The thylakoid membrane contains about 1 million photosynthetic units. The number of chloroplasts in a plant cell varies over a wide range depending on its nature (color green reflects the presence of chloroplasts containing chlorophylls). An average number is about 40.

In biological terms, the leaves of a plant interact with the environment, harvest light energy, utilize carbon dioxide to form sugar, and convert water into oxygen. The reactions involved in photosynthesis are electrochemical, involving transfer of electrons and protons. An electrochemical reaction in which a molecule gains electrons is called *reduction*; the opposite process of loss of electrons is called *oxidation*. The two reactions are coupled and called *oxidation–reduction* or *redox processes*. In oxygen-producing organisms, water is the prime source of electrons used to reduce $NADP^+$ to NADPH, which in turn is used to reduce carbon dioxide into hexose sugars.

The reaction center, P_{680}, of PS-II is excited by energy transfer from the light-gathering antenna molecules and transfers the electron to the primary acceptor pheophytin (abbreviated as Ph). An electron transfer chain (ETC) initiates various enzyme-catalyzed reactions in PS-II. In photosynthetic organisms involving both PS-II and PS-I, the terminal point of ETC of PS-II is the reaction center P_{700} in PS-I. The coupled PS-II and PS-I systems are shown in Figure 6.9. The electron transfer chain in PS-II uses the energy of the electrons to make ATP. The PS-I system produces NADPH. The protons generated in conjunction with the oxidation of water are used by a separate complex, the ATP synthase, to generate ATP.

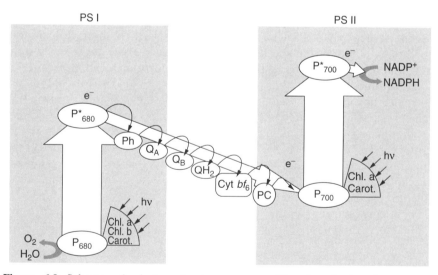

Figure 6.9. Scheme of photosynthesis processes. Ph, pheophytin; Q_A and Q_B, quinones; QH_2, hydrogenated (reduced) quinones; Cyt b_6 and Cyt f, cytochromes; pc, plastocyanine.

Light-Harvesting Antennas. The array of light-absorbing pigments are called *antenna molecules* because they absorb light analogous to how a radio antenna absorbs radio waves. The light-harvesting antenna molecules are mainly chlorophylls that absorb in the red and the blue spectral regions, reflecting in the green. This feature imparts a green color to plant leaves. There are approximately 200–400 molecules of antenna chlorophyll per photosynthetic unit. Photosynthetic units PS-I and PS-II contain mainly chlorophyll, chlorophyll *a*. In addition to chlorophylls, other antenna molecules are carotenoids (π-conjugated system discussed in Chapter 2), which absorb in the blue and green regions, and phycocynin, which absorbs in the green and yellow region. The color of a photosynthetic microorganism (such as a bacteria) is determined by the relative amounts of these antenna molecules. Nature utilizes a combination of these antenna molecules to extend the range of wavelength of light that can be absorbed for photosynthesis (harvesting of light over a broad spectral range). The antenna molecules do not produce, by themselves, the electron transfer process, which takes place at the reaction center.

The light-harvesting antenna molecules are packed together, interacting with the transmembrane proteins to form what is known as *light-harvesting complexes* (LHCs). Multiples of LHCs are associated with each reaction center. Some photosynthetic bacteria contain two types of LHCs: LHCI and LHCII. An area of current investigation is an understanding of the structures and functions of the basic unit of LHCII which associates with the PSII reaction center.

The transmembrane proteins in LHCs perform the role of maintaining the pigment molecules in an orientation and position, optimum for light absorption and energy transfer. The funneling of excitation energy of the antenna molecules to the reaction center occurs by multistep energy transfer. The nature of the energy transfer pathways from the absorbing antenna molecules to the reaction center has been a subject of extensive studies (van Amerongen and van Grondelle, 2001). These antenna assemblies are an example of molecular aggregates discussed in Chapter 2, Section 2.5. The energy transfer between energetically identical chromophores proceeds through an exciton mechanism, described in Chapter 2. The energy transfer between energetically different chromophores, as well as that from the antenna molecules to the reaction center, proceeds by the Fluorescence resonance energy transfer (FRET) mechanism. This process involves a transfer of excitation from a higher-energy donor molecule to a lower-excitation-energy acceptor molecule. Time-resolved spectroscopic measurements yield highly efficient and extremely fast first-step resonance energy transfer time of 200 fs. The overall energy transfer rate exhibits a multiple exponential form derived from a broad distribution of transfer times that extends to many picoseconds.

The Chemical Process. The net chemical process by the combined action of PS-I and PS-II in a plant is represented below:

$$6CO_2 + 6H_2O \longrightarrow \underset{\text{Glucose}}{(CH_2O)_6} + 6O_2 \qquad (6.11)$$

The reaction involves a number of electrochemical steps initiated by electron transfer at the reaction center. These electron transfer redox processes produce energy-rich ATP and NADPH. The ATP and NADPH are used in a series of light-independent reactions called the *Calvin cycle*, by which the plants convert carbon dioxide eventually to glucose or other carbohydrates. Since the focus of this book is not on biochemical aspects, readers interested in this subject are referred to the books by Stryer (1995) and Lodish et al. (2000). It will suffice here to say that in PS-I, $NADP^+$ is converted to NADPH, involving an electron transfer process. A coupled movement of protons across the membrane powers the synthesis of chemical energy in storing molecule ATP (see Chapter 3). PS-II utilizes the electron transfer process to convert water to oxygen (O_2). These processes are shown in Figure 6.9, which also shows the involvement of various species (abbreviated) along the electron transfer chain. Again, the detailed nature of these species can be found in Stryer (1995) and Lodish et al. (2000).

6.5   *IN VIVO* PHOTOEXCITATION

A very active area of research currently is *in vivo* optical excitation for spectroscopic analysis, bioimaging, laser surgery, and light-activated therapy. The *in vivo* spectroscopic studies are described in the next subsection, while imaging and light-activated therapies and tissue engineering are discussed in later chapters. This section describes the various methods of bringing light, particularly laser light, for photoexcitation to a specific section of a tissue or an internal organ of a live biological specimen or a human subject. The principal methods used are listed in Table 6.6.

6.5.1   Free-Space Propagation

Here the light is delivered directly to the photoexcitation site by propagating it through free space. In the case of a lamp, the light may be used to propa-

TABLE 6.6. Methods of Light Delivery for *In Vivo* Photoexcitation

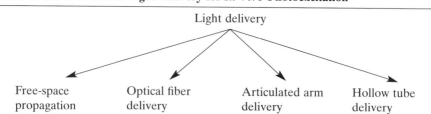

gate as a collimated beam and then focused/defocused to various spot sizes at the desired point. In the case of a laser beam, the light output generally is already collimated with a very low divergence. The method of free-space propagation is used for UV–visible light photoexcitation of external tissues/organs such as skin, eye, and so on. With the availability of solid-state diode lasers, which now can cover from blue to near IR and are very small in size, one can even use them as hand-held units for *in vivo* photoexcitation.

6.5.2   Optical Fiber Delivery System

Optical fibers are long, thin, flexible cylindrical elements of diameters in the range of several microns to several hundred microns that are capable of trapping light and then transmitting it down its length. The optical fibers are used for light delivery in the wavelength range of 200 nm (UV) to 1600 nm (IR) and are made of glass or fused silica. They trap light entering at one end by total internal reflection from the interface between the fiber edge and an outer coating material, called *cladding* (generally a plastic). These processes are shown in Figure 6.10.

In order for light to be trapped within the fiber core by total internal reflection at the interface of the core and the cladding media, the refractive index n_2 of the core (glass or quart) has to be higher than n_1 of the cladding medium. The main advantage offered by the optical fiber transmission is that it is small in size to be used within an endoscopic or catheter delivery system to reach an internal organ. Very often, to increase the power delivery, a bundle of fibers is used. These fiber bundles, called *light guides*, can transmit higher optical powers than single fibers and have greater mechanical strength. Medical

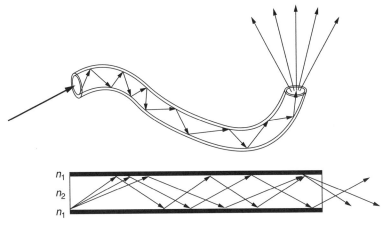

Figure 6.10. Light propagation in an optical fiber. The bottom diagram shows light trapping by total internal reflection from the interface between the fiber and the cladding medium.

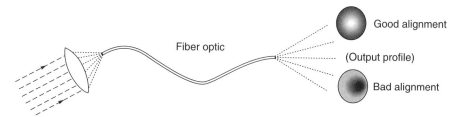

Figure 6.11. The geometry of a fiber optic permits flexible delivery of laser energy to tissue. Note the difference of the laser energy upon exit from the fiber.

endoscopy provides a good example of a major biological application of fiber bundles.

As shown in Figure 6.11, the fiber bundle light guides use lenses to couple light into the bundle and also to collimate or focus the output light because the output beam exiting from the fiber is fairly divergent. The alignment of the entering beam to the fiber is fairly crucial because it determines the nature of the exiting beam. This behavior is also illustrated in Figure 6.11.

Optical fiber technology also offers options to use a dual core fiber in which two optical fibers can be used, one for photoexcitation and the other to collect photo-response (either reflected light or fluorescence). Also, a bifurcated optical cable or "Y guide" can be used in which the incident light is transmitted down one arm of the Y and exits out of the common leg. The reflected scattered light is also collected by the common leg but measured at the other arm of the Y. An optical device, called an *oximeter*, uses this principle to measure the oxygen content of blood *in vivo* by detecting the diffusely scattered light from hemoglobin at several wavelengths (Reynolds et al., 1976).

Depending on the size (core diameter) and the refractive index of the core and the cladding media, an individual optical fiber can support different types of mode propagation which define the electric field distribution of the electromagnetic radiation within the fiber. Typically, fibers of diameter 2–4 μm can support a single mode of propagation and are called *single-mode fibers*. Larger-diameter fibers are multimode fibers that are capable of propagating many modes simultaneously. Single-mode fibers are desirable for many applications, such as nonlinear optical excitation, which require a well-defined mode. There are fibers that preserve the polarization of the coupled polarized light. They are called *polarization preserving fibers*.

The propagation characteristics of an optical fiber are defined by the optical loss that the light suffers by traveling in its length. The optical loss, α_T, in a fiber is generally defined in the units of dB/Km as

$$\alpha_T = \frac{-10}{L} \log\left[\frac{P_T}{P_0}\right] \qquad (6.12)$$

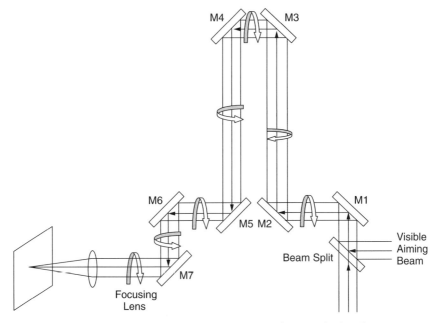

Figure 6.12. Reflecting mirrors arrangement in an articulated arm.

where P_0 is the initial power launched into the fiber and P_T is the power exiting at length L in kilometers. The fiber losses in the visible range are generally 1–10 dB/km.

6.5.3 Articulated Arm Delivery

Articulated arm delivery utilizes an arrangement involving a series of hollow tubes and mirrors. An articulated arm delivery system is shown in Figure 6.12. The light beam (a laser source) is transmitted through each tube and is then reflected into the next tube by an appropriate angled mirror. At the exit port, there is a small hand piece that can be hand held to manipulate the beam.

This mode of beam delivery is used for mid-infrared range, such as 10.6 μm radiation from a CO_2 laser. The articulated arms are bulky, awkward, and expensive, requiring more maintenance than optical fibers. However, at the present time, there are no suitable optical fibers for the CO_2 laser beam delivery. The silica, glass, or quartz fibers strongly absorb at the 10.6 μm CO_2 laser wavelength. There are specialty optical fibers made of chalcogenides that transmit at 10.6 μm, but they are not widely used because optical losses are still high. Furthermore, these fibers tend to be brittle.

In the case of using an articulated arm to propagate a CO_2 laser beam that is in the infrared (invisible), a co-propagating visible beam (usually a red beam from a He–Ne laser or a diode laser) is used as an aiming beam as shown in Figure 6.13.

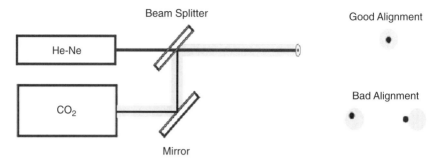

Figure 6.13. Articulated arm laser beam delivery with an aiming beam from an He–Ne laser.

6.5.4 Hollow Tube Waveguides

Another light delivery system involves a hollow tube made of a metal, ceramic, or plastic (Cossman et al., 1995). The inside wall of the tube is coated with a high reflector. The laser light is propagated down the tube by reflection from the inner wall. The advantage of using this type of plastic waveguide over an articulated arm is its semiflexibility. The plastic waveguides generally have an inner diameter of ~1 mm and an outer diameter of ~2 mm.

An important recent development in this area is the use of a photonic crystal waveguide. The photonics crystals are discussed in Chapters 9 and 16. An example of a photonic crystal is a multilayered medium that reflects (does not allow the propagation through it) light of a given wavelength range. Using an appropriately designed multilayered plastic hollow waveguide, one can make light within a given wavelength range to reflect from the inner walls with very high reflectivity.

6.6 *IN VIVO* SPECTROSCOPY

In vivo spectroscopy has emerged as a powerful technique for biomedical research covering a broad spectrum, from study of cellular and tissue structures, to biological functions, to early detection of cancer. The spectroscopic methods used have involved study of electronic absorption spectra by analysis of back-scattered light, Raman scattering, and fluorescence. Fluorescence spectroscopy has been the most widely used technique because of its sensitivity and specificity (Wagnieres et al., 1998; Brancoleon et al., 2001). For fluorescence studies, both endogenous fluorescence (autofluorescence) and exogenous fluorescence have been used.

Earlier reports focused on the use of optical absorption properties of tissues for clinical applications (Wilson and Jacques, 1990). An example is blood oximetry, which is widely used clinically to monitor continuously blood oxygenation with the help of an optical fiber probe as described in the previous section. In this method the diffuse reflectance also collected by the fiber is analyzed based on the differences in the absorption bands of oxy- and deoxyhemoglobins. Diffuse reflectance from the skin can be used to monitor changes induced, for example, by the UV radiation. Endoscopic reflectance spectroscopy from mucosa of the gastrointestinal tract has been used to determine blood content and oxygenation (Leung et al., 1987).

6.7 OPTICAL BIOPSY

A major focus of *in vivo* spectroscopy has been to use it for early detection of cancer. Optical biopsy refers to detection of the cancerous state of a tissue using optical methods. This is an exciting area, offering the potential to use noninvasive or minimally invasive *in vivo* optical spectroscopic methods to identify a cancer at its various early stages and monitor its progression. One can envision that one day noninvasive optical spectroscopic methods would find routine usage in doctors' offices, clinics, and operating rooms for diagnosis of diseases, monitoring its progression and determining the efficacy of a medical treatment or procedure.

The basic principle utilized for the method of optical biopsy is that the emission and scattered light are strongly influenced by the composition and the cellular structures of tissues. The progression of a disease or cancer causes a change in the composition and the cellular structure of the affected tissues, producing a change in emission and scattering of light. Thus, the optical biopsy can be represented by the schematics of Figure 6.14 (Katz and Alfano, 1996).

The primary optical methods used in the past have been fluorescence and Raman spectroscopic techniques. The changes in tissue from a normal state to a cancerous state have been shown to alter the fluorescence and the Raman spectra. These methods have successfully differentiated normal tissues from those with breast, gynecological, colon, and prostate cancers.

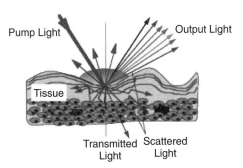

Figure 6.14. Schematics of various optical interactions with a tissue used for optical biopsy. (Reproduced with permission from Katz and Alfano, 1996.)

The benefits provided by optical biopsy are (Katz and Alfano, 1996) as follows:

- Optical biopsy is noninvasive or minimally invasive, utilizing endoscopic or needle based probes. Hence, removal of a tissue specimen is not required.
- The method provides rapid measurements. Hence, real-time measurements can be made.
- High spatial resolution offered by the optical methods provides the ability to detect small tumors.
- Optical biopsy provides the ability to detect precancerous conditions. This feature is derived from the presence of distinguishable spectral characteristics associated with molecular changes that manifest even before a cancer actually can be detected.

For fluorescence detection, endogenous fluorescence (autofluorescence) from a tissue is preferred over exogenous fluorescence (fluorescence from an added dye). The endogenous fluorescence is derived from a number of fluorophores that are constituents of a tissue or a cell. Examples of endogenous fluorophores are tryptophan, elastin, collagen, NADH, and flavin. The absorption and the fluorescence spectra of these molecules have been covered in Section 6.2 (Figure 6.2).

Alfano and co-workers were the first to use the fluorescence property of an animal tissue to study cancer (Alfano et al., 1984). Since then the area of optical biopsy has received a great deal of attention as is evidenced by a number of symposia held on optical biopsy (Alfano, 2002). Fluorescence measurements have used both laser excitation (e.g., 488 nm from an argon ion laser) and excitation (particularly in the UV region) from a xenon lamp. The optical biopsy studies have been extended both to *ex vivo* and *in vivo* human tissues. Furthermore, *in vivo* studies have utilized both endoscopic and needle-based probes to a variety of cancers, some of which are listed below (Alfano, private communications):

Endoscopic-Based Probes

- Stomach
- Colon
- Intestines
- Lungs
- Gynecological tract

Needle-Based Probes

- Breast
- Prostate
- Kidney

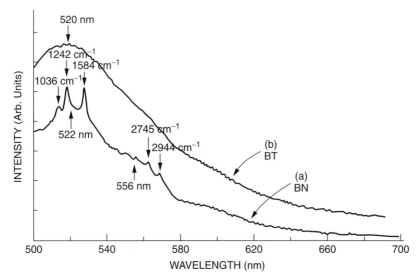

Figure 6.15. Fluorescence spectra of the normal breast (BN) and the tumor breast tissue (BT) excited at 488 nm. (Reproduced with permission from Alfano et al., 1989.)

In the endoscopic approach, large-diameter optical fibers coupled to an endoscope are used to excite and collect the emission. In needle-based probes, small-diameter optical fibers are mounted in stereotactic needles to provide high spatial resolution for *in vivo* examination of a breast or prostate tissue.

Alfano et al. (1989) used this method to show differences between the fluorescence and Raman spectra of normal and malignant tumor breast tissue. The observed fluorescence spectra of the normal breast tissue (BN) and that of a tumor breast tissue (BT), observed with excitation at 588 nm from an argon ion laser, is shown in Figure 6.15. The fluorescence is associated with flavins. The normal breast tissue also shows sharp Raman vibrational peaks in the region of $1000–1650\,cm^{-1}$ associated with hemeproteins, lipids, hemoglobin, and porphysims. Frank et al. (1995) also reported the use of Raman spectroscopy to distinguish normal and diseased human breast tissues. The advantage of vibrational spectroscopy is that it provides detailed information relating to molecular structure and composition and, thus, can be used as a detailed signature associated with abnormality.

Another study is the use of *in vivo* autofluorescence spectroscopy of human bronchial tissue for early detection of lung cancer (Zellweger et al., 2001). An optical fiber bundle was adapted to fit the biopsy channel of a standard flexible bronchoscope. Clear differences in the autofluorescence spectra were observed for the healthy, inflammatory, and early-cancerous lung tissue when excited at 405 nm *in vivo*.

The *in vivo* tissue autofluorescence also was used to distinguish normal skin tissue from nonmelanoma skin cancer (NMSC) (Brancoleon et al., 2001). They reported that in both basal cell carcinomas and squamous cell carcinomas, the autofluorescence (endogenous fluorescence) at 380 nm due to tryptophan residues and excited at 295 nm was more intense in tumors than in the normal tissue, probably due to epidermal thickening and/or hyperproliferation. In contrast, the fluorescence intensity at ~480 nm produced by excitation at 350 nm and associated with cross-links of collagen and elastin in the dermis was found to be lower in tumors that in the surrounding normal tissue. The authors suggested this change to be due to degradation or erosions of the connective tissues due to enzymes released by the tumor.

Recently, a fiber-optic diagnostic analyzer with a trade name of Optical Biopsy System has been introduced by Spectra Science, Inc. of Minneapolis, Minnesota (FDA approved). It is used as an adjunct to lower gastrointestinal (GI) endoscopy for evaluation of polyps less than 1 cm in diameter and can aid a physician to decide whether they should be removed. In this device, light is transmitted through a long fiber inserted in a colonoscope and is directed to a polyp. The autofluorescence from the polyp is collected back through the optical fiber.

Raman spectroscopy, particularly near-infrared FT-Raman (see Chapter 4), has been applied to *in vitro* studies of human breast, gynecological, and arterial tissues. The advantage of Raman spectroscopy is that it provides detailed information (with sharp vibrational transitions) relating to molecular structure and composition and, thus, can be used as a detailed signature associated with abnormality. The use of infrared excitation for Raman is advantageous because it reduces the probability of interference (background) from autofluorescence and provides deeper penetration in a tissue.

Alfano et al. (1991) used 1064 nm from a Nd:YAG laser to acquire the Raman spectra of a normal breast tissue, as well as of a benign and malignant tumor. These spectra in the region of 700–1900 cm^{-1} are shown in Figure 6.16. These different types of tissues exhibit differences in relative intensities and number of vibrational transitions.

Frank et al. (1995) also reported the usefulness of Raman spectroscopy to distinguish normal and diseased human breast tissues. More recently, Vo-Dinh et al. (2002) have used surface-enhanced Raman spectroscopy (Chapter 4) to detect cancer.

Other spectroscopic techniques used for optical biopsy are:

1. Optical coherence tomography, developed by Fujimoto and co-workers (Tearney et al., 1997). This optical technique, useful for bioimaging, is discussed in detail in Chapter 7. This technique of imaging was adapted to allow high-speed visualization of tissue in a living animal with a catheter endoscope, 1 mm in diameter.

2. Diffuse-reflectance measurements, developed by Alfano and co-workers (Yang et al., 2001). This method was used to obtain the absorption spectra of malignant and normal human breast tissues. The absorption in the wavelength ranges of 275–285 nm and 255–265 nm, which are

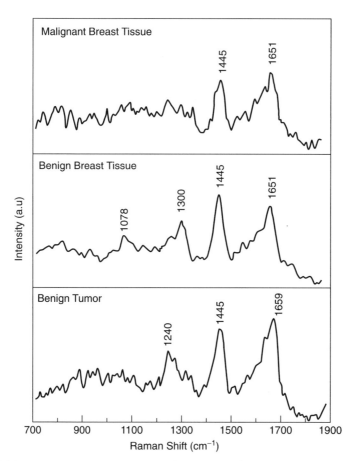

Figure 6.16. Raman spectra from normal (benign tissue), benign, and malignant breast tumors. (Reproduced with permission from Alfano et al., 1991.)

fingerprints of proteins and DNA components, revealed differences between the malignant, fibroadenoma, and normal breast tissues.

The applicability of autofluorescence and diffuse-reflectance spectroscopy for intraoperative detection of brain tumors was investigated in a clinical trail (Lin et al., 2001). The result suggested that brain tumors and infiltrating tumor regions can be effectively separated from normal brain tissues *in vivo* using a combination of these techniques.

6.8   SINGLE-MOLECULE DETECTION

Single-molecule detection is an exciting new frontier. The ability to detect a single molecule and study its structure and function provides the opportunity to probe properties that are not available in measurements on an ensemble. A new journal, *Single Molecules*, published by Wiley-VCH in 2000, is dedicated to single-molecule spectroscopy and detection.

Fluorescence has been the spectroscopic method for single-molecule detection. Single-molecule fluorescence detection has been successfully extended to biological systems (Ha et al., 1996, 1999; Dickson et al., 1997). Excellent reviews of the applications of single-molecule detection to bioscience are by Ishii and Yanagida (2000) and Ishijima et al. (2001). Single-molecule detection has been used to study molecular motor functions, DNA transcription, enzymatic reactions, protein dynamics, and cell signaling. The single molecule detection permits one to understand the structure–function relation for individual biomolecules, as opposed to an ensemble average property that is obtained by measurements involving a large number of molecules.

The single-molecule detection utilizes fluorescence labeling of a biomolecule or a biopolymer. An ideal choice will be a fluorescent marker with the quantum efficiency (defined in Chapter 4), close to 1 so that maximum emission can be observed. Recent advances in the photoelectric detection system (conversion of an emitted photon to an electrical signal) have pushed the limit to single-photon detection, thus greatly facilitating this field. It is to be pointed out that single molecule detection does not always imply the observation of an isolated single molecule. The observation generally utilizes two approaches:

1. Detection in the dilution limit where the sample volume under optical observation consists of only one molecule or biomolecule.
2. Detection where a single biomolecule is attached to a microscopic bead which can be observed and tracked.

Manipulation of these biomolecules can be conducted by using a glass microneedle (Ishijima et al., 1996). Many studies, however, have used laser trapping of microbeads by optical tweezers, where a single molecule is attached to a microbead. Laser tweezers are discussed in Chapter 14. Therefore, single-molecule detection studies using optical trapping will be discussed there.

Fluorescence probes used for single-molecule detection are fluorescence lifetime, two-photon excitation, polarization arrisotropy, and fluorescence resonant energy transfer (FRET). Single-molecule fluorescence lifetime measurements have been greatly aided by the use of a technique called *time-correlated single-photon detection* (Lee et al., 2001). Microscopic techniques such as confocal microscopy and near-field microscopy, discussed in Chapter 7, have provided the capability to perform single-cell imaging of stained cells or cell aggregates. A two-channel confocal microscope with separated polarization detection pathways has been used by Hochstrasser and co-workers at the Regional Laser and Biotechnology Laboratories (RLBL) at University of Pennsylvania to study time-resolved fluorescence arrisotropy of single molecules. Using this technique, Hochstrasser and co-workers simultaneously recorded single-molecule fluorescence in two-orthogonal polarizations, for free porphyrin cytochrome-c and Zn porphyrin cytochrome-c encapsulated in a glass (Mei et al., 2002). The fluorescence polarization anisotropy was shown to provide information on the orientational motions of these proteins and their interaction with the microenvironment surrounding them.

Zhuang et al. (2002) have studied the correlation between the structural dynamics and function of the hairpin ribozyme, a protein-independent catalytic RNA, by using fluorescence energy transfer. In this case they used the Cy3 dye as the energy donor and used the Cy5 dye as the energy acceptor, attached respectively to the 3′ and 5′ ends of the RzA strand of the ribozyme and studied fluorescence resonant energy transfer from Cy3 to Cy5. They found that this RNA enzyme exhibits a very complex structural dynamics with four docked (active) states of distinct stabilities and a strong memory effect. These observations would be difficult to obtain from an ensamble study because less stable conformational states are nonaccumulative. Thus, this result signifies the importance of single-molecule study in characterizing complex structural dynamics.

HIGHLIGHTS OF THE CHAPTER

- Photobiology deals with the interaction of light with complex, multistep biological processes, which can photo-induce physical, thermal, mechanical, and/or chemical effects.
- Light-absorbing components can be endogenous (cell and tissue constituents) or exogenous (photosensitizing dyes, staining components, etc).
- Transmission in biological specimen has an upper and lower cutoff of useable wavelength defined by scattering and absorptive losses.
- Most cells exhibit good transparency between 800 nm and 1300 nm.
- Light absorption in protein molecules is dictated by the characteristic absorption features of the constituent amino acids as well as the polypeptide bonds and disulfide linkages and by other chromophore that may be present.
- Absorption of light may cause radiative as well as nonradiative processes in endogenous molecules—that is, cellular constituents.
- Autofluorescence, a radiative process, is caused by the endogenous fluorophores—for example, NADH, flavins, tyrosine, porphyrins and lipopigments.
- Thermal nonradiative effects induced by light (e.g., protein denaturation and vaporization of water) are significant at the tissue level.
- Excited-state photochemical nonradiative processes include photoaddition, photofragmentation, photoxidation, photohydration, photoisomerization, and photorearrangement.

- Exogenous (added) photosensitizers can also induce photosensitization through photoaddition and photooxidation.
- Optical interactions in tissues occur in the form of refraction, reflection, absorption, and scattering.
- Scattering from tissues involves several different mechanisms: Rayleigh and Mie scattering (elastic) and Brillouin and Raman scattering (inelastic).
- Optical penetration depth δ, which is a measure of optical transparency, is the distance in a medium such that the light intensity becomes $1/e$ of the original intensity after traveling this distance.
- The experimental arrangement of double-integrating spheres can simultaneously determine the reflectance, absorption and scattering from a biological specimen.
- Interaction with light leads to radiative (autoflorescence) and nonradiative (photochemical, thermal, photoablation, plasma-induced ablation, and photodisruption) processes.
- Autofluorescence spectra provide a method of optical probing for detection of cancer, detection of metabolic changes, and so on.
- In photochemistry, recent focus is on nonlinear excitations such as two-photon absorption. It has the advantage of affecting the photochemistry at a much greater tissue depth than is possible in one-photon processes. This is advantageous in photodynamic therapy.
- Thermal effects include coagulation, vaporization, carbonization, and melting.
- Photoablation uses intense UV laser pulses (excimer laser) to photochemically decompose cellular and extracellular components and is very useful in tissue contouring.
- High-intensity irradiation leads to plasma formation, and shock waves are generated which disrupt the tissue structure (photodisruption). This has applications in minimally invasive surgery.
- The various methods of light delivery to specific regions of a biological specimen for *in vivo* spectroscopy are: free-space propagation with collimation, optical fiber delivery system e.g. medical endoscopes, articulated arm delivery system used for mid-infrared range, and hollow tube waveguides that offer some flexibility.
- Spectroscopic methods used for *in vivo* studies include electronic absorption through analysis of back-scattered light, Raman scattering, and fluorescence.
- *In vivo* spectroscopy and optical biopsy cover a broad spectrum of research such as study of cellular and tissue structures, biological functions, and early detection of cancer.

- The exciting new frontier of single-molecule detection and spectroscopy enables probing the structural and functional properties of individual biomolecules as opposed to ensemble averages. This has been used to study molecular motor functions, DNA transcription, protein dynamics, cell signaling, and so on.

REFERENCES

Adranov, A., and Frechet, J. M. J., Light-Harvesting Dendrimers, *Chem. Commun.* **18**, 1701–1710 (2000).

Agarwal, R., Krueger, B. P., Scholes, G. D., Yang, M., Yom, J., Mets, L., and Fleming, G. R., Ultrafast Energy Transfer in LHC-II Revealed by Three-Pulse Photon Echo Peak Shift Measurements, *J. Phys. Chem. B* **104**, 2908–2918 (2000).

Alfano, R. R., ed., Optical Biopsy IV, SPIE Proc. Vol. 4613, SPIE, Bellingham (2002).

Alfano, R. R., Liu, C. H., Sha, W. L., Zhu, H. R., Akins, D. L., Cleary, J., Prudente, R., and Celmer, E., Human Breast Tissue Studied by IR Fourier Transform Raman Spectroscopy, *Laser Life Sci.* **4**, 23–28 (1991).

Alfano, R. R., Pradham, A., Tang, G. C., and Wahl, S. J., Optical Spectroscopic Diagnosis of Cancer and Normal Breast Tissues, *J. Opt. Soc. Am. B.* **6**, 1015–1023 (1989).

Alfano, R. R., Tata, D. B., Cordero, J. J., Tomashefsky, P., Longo, F. W., and Alfano, M. A., Laser Induced Fluorescence Spectroscopy from Native Cancerous and Normal Tissues, *IEEE J. Quantum Electron.* **QE-20**, 1507–1511 (1984).

Brancoleon, L., Durkin, A. J., Tu, J. H., Menaker, G., Fallon, J. D, and Kellias, N., *In Vivo* Fluorescence Spectroscopy of Nonmelanoma Skin Cancer, *Photochem. Photobiol.* **73**, 178–183 (2001).

Carden, A., and Morris, M. D., Application of Vibrational Spectroscopy to the Study of Mineralized Tissues (Review), *J. Biomed. Opt.* **5**, 259–268 (2000).

Carden, A., Rajachar, R. M., Morris, M. D., Kohn, D. H., Ultrastructural Changes Accompanying the Mechanical Deformation of Bone Tissue: A Raman Imaging Study, *Calcif. Tissue Int.* **72**, in Press (2003).

Connelly, J. P., Müller, M. G., Hucke, M., Gatzen, G., Mullineaux, C. W., Ruban, A. V., Horton, P., and Holzwarth, A. R., Ultrafast Spectroscopy of Trimeric Light-Harvesting Complex II from Higher Plants, *J. Phys. Chem. B* **101**, 1902–1907 (1997).

Cossmann, P. H., Romano, V., Sporri, S., Alterman, H. J., Croitoru, N., Frenz, M., and Weber, H. P., Plastic Hollow Waveguides: Properties and Possibilities as a Flexible Radiation Delivery System for CO_2-Laser Radiation, *Lasers in Surgery and Medicine* **16**, 66–75 (1995).

Derbyshire, G. J., Bogden, D. K., and Unger, M., Thermally Induced Optical Property Changes in Myocardium at 1.06 Microns, *Lasers Surg. Med.* **10**, 28–34 (1990).

Dickson, R. M., Cubitt, A. B., Tsien, R. Y., and Meerner, W. E., On/Off Blinking and Switching Behavior of Single Molecules of Green Fluorescent Protein, *Nature* **388**, 355–358 (1997).

Frank, C. J., McCreery, R. L., Redd, D. C. B., Raman Spectroscopy of Normal and Diseased Human Breast Tissues, *Anal. Chem.* **67**, 777–783 (1995).

Grossweiner, L. I., and Smith, K. C., Photochemistry, in K. C. Smith, ed., *The Science of Photobiology*, 2nd edition, Plenum, New York, 1989, pp. 47–78.

Ha, T. J., Ting, A. Y., Liang, Y., Caldwell, W. B., Deniz, A. A., Chemla, D. S., Schultz, P. G., and Weiss, S., Single-Molecule Fluorescence Spectroscopy of Enzyme Conformational Dynamics and Cleavage Mechanism, *Proc. Natl. Acad. Sci. USA* **96**, 893–898 (1999).

Ha, T. J., Enderle, T., Ogletree, D. F., Chemla, D. S., Selvin, P. R., and Weiss, S., Probing the Interaction Between Two Single Molecules: Fluorescence Resonance Energy Transfer Between a Single Donor and a Single Acceptor, *Proc. Natl. Acad. Sci. USA* **93**, 6264–6268 (1996).

Hawes, C., Boevink, P., and Moore, I., Green Fluorescent Protein in Plants, in V. J. Allan, ed., *Protein Localization by Fluorescence Microscopy*, Oxford University Press, New York, 2000, pp. 163–177.

Ishii, Y., and Yanagida, T., Single Molecule Detection in Life Science, *Single Mol.* **1**, 5–16 (2000).

Ishijima, A., and Yanagida, T., Single Molecule Nanobioscience, *Trends in Biomedical Sciences* **26**, 438–444 (2001).

Ishijima, A., Kojima, H., Higuchi, H., Hasada, Y., Funatsu, T., and Tanagida, T., Multiple- and Single-Molecule Analysis of the Actomyosin Motor by Nanometer-Piconewton Manipulation with a Microneedle: Unitary Steps and Forces, *Biophys. J.* **70**, 383–400 (1996).

Jordan, P., Fromme, P., Witt, H. T., Klukas, O., Saenger, W., and Krau, N., Three-Dimensional Structure of Cyanobacterial Photosystem I at 2.5 Angstrom Resolution, *Nature* **411**, 909–917 (2000).

Katz, A., and Alfano, R. R., Optical Biopsy: Detecting Cancer with Light, in E. Sevick-Muraca, and D. Benaron, eds., *OSA TOPS on Biomedical Optical Spectroscopy and Diagnostics*, Vol. 3, Optical Society of America, Washington D.C., 1996, pp. 132–135.

Katz, A., and Alfano, R. R., Noninvasive Fluorescence-Based Instrumentation for Cancer and Precancer Detection and Screening, in G. E. Cohn, ed., *In Vitro Diagnostic Instrumentation*, *Proceedings of SPIE*, Bellingham, Vol. 3931, 2000, pp. 223–226.

Kennis, J. T. M., Gobets, B., van Stokkum, I. H. M., Dekkar, J. P., van Grondelle, R., and Fleming, G. R., Light Harvesting by Chlorophylls and Carotenoids in the Photosystem I Core Complex of *Synechococcus elongatus*: A Fluorescence Upconversion Study, *J. Phys. Chem. B* **105**, 4485–4494 (2001).

Kim, J. E., McCamant, D. W., Zhu, L., and Mathies, R. A., Resonance Raman Structural Evidence that the Cis-to-Trans Isomerization in Rhodopsin Occurs in Femtoseconds, *J. Phys. Chem. B* **105**, 1240–1249 (2001).

Kochevar, I. E., Photoimmunology, in J. Krutmann, and C. A. Elmets, eds., *Photoimmunology*, Blackwell Science, London, 1995, pp. 19–33.

Ksenzhek, O. S., and Volkov, A. G., *Plant Energetics*, Academic Press, San Diego, 1998.

Lee, M., Tang, J., and Hoshstrasser, R. M., Fluorescence Lifetime Distribution of Single Molecules Undergoing Förster Energy Transfer, *Chem. Phys. Lett.* **344**, 501–508 (2001).

Leung, F. W., Morishita, T., Livingston, E. H., Reedy, T., and Guth, P. H., Reflectance Spectrophotometry for the Assessment of Gastroduodenal Mucosal Perfusion, *Am. J. Physiol.* **252**, 6797–6804 (1987).

Lin, W.-C., Toms, S. A., Johnson, M., Jansen, E. D., and Mahadevan-Jansen, A., *In Vivo* Brain Tumor Demarcation Using Optical Spectroscopy, *Photochem. Photobiol.* **73**, 396–402 (2001).

Lodish, H., Berk, A., Zipursky, S. L., Matsudaira, P., Baltimore, D., and Darnell, J., *Molecular Cell Biology*, 4th edition, W. H. Freeman, New York, 2000.

Mei, E., Vanderkooi, J. M., and Hoshstrasser, R. M., Single Molecule Fluorescence of Cytochrome c, *Biophys. J.* **82** Part 1, 47a (2002).

Melkozernov, A. N., Excitation Energy Transfer in Photosystem I from Oxygenic Organisms, Photosynthesis Research, **70**, 129–153 (2001).

Niemz, M. H., *Laser–Tissue Interactions*, Springer-Verlag, Berlin, 1996.

Pan, D., Ganim, Z., Kim, J. E., Verhaeven, M. A., Lugtenburg, J., and Mathies, R. A., Time-Resolved Resonance Raman Analysis of Chromophore Structural Changes in the Formation and Decay of Rhodopsin's BSI Intermediate, *J. Am. Chem. Soc.* **124**, 4857–4864 (2002).

Pepperkok, R., and Shima, D., Fluorescence Microscopy of Living Vertebrate Cells in V. J. Allan, ed., *Protein Localization by Fluorescence Microscopy*, Oxford University Press, New York, 2000, pp. 109–132.

Reynolds, L., Johnson, C. C., and Ishijima, A., Diffuse Reflectance from a Finite Blood Medium: Applications to the Modeling of Fiber Optic Catheters (TE), *Appl. Opt.* **15**, 2059–2067 (1976).

Rol, P., Nieder, P., Durr, U., Henchoz, P. D., and Fankhauser, F., Experimental Investigations on the Light Scattering Properties of the Human Sclera, Lasers Light Ophthalmol. **3**, 201–212 (1990).

Shortreed, M., Swallen, S., Shi, Z.-Y., Tan, W., Xu, Z., Moore, J., and Kopelman, R., Directed Energy Transfer Funnels in Dendrimeric Antenna Supermolecules, *J. Phys. Chem. B* **101**, 6318–6322 (1997).

Stryer, L., *Biochemistry*, 4th edition, W. H. Freeman, New York, 1995.

Swallen, S. F., Kopelman, R., Moore, J., and Devadoss, C., Dendrimeric Photoantenna Supermolecules: Energetic Funnels, Exciton Hopping, and Correlated Excimer Formation, *J. Mol. Structure* (1999). special issue: R. Kuczkowski, and L. Laane, eds., *L. S. Bartell Festschrift, Molecules and Aggregates: Structure and Dynamics*, Vol. 485/486, pp. 585–597.

Tarnowski, C. P., Ignelzi, M. A., Wang, W., Taboas, J. M., Goldstein, S. A., and Morris, M. D., Earliest Mineral and Matrix Changes in Force-Induced Musculoskeletal Disease as Revealed by Raman Microspectroscopic Imaging, *J. Bone Mineral Res.*, submitted.

Tearney, G. J., Brezinski, M. E., Bouma, B. E., Boppart, S. A., Pitris, C., Southern, J. F., and Fujimoto, J. G., *In Vivo* Endoscopic Optical Biopsy with Optical Coherence Tomography, *Science* **276**, 2037–2039 (1997).

van Amerongen, H., and van Grondelle, R., Understanding the Energy Transfer Function of LHCII: The Major Light-Harvesting Complex of Green Plants, *J. Phys. Chem. B* **105**, 604–617 (2001).

Vo-Dinh, T., Allain, L. R., and Stokes, D. L., Cancer Gene Detection Using Surface-Enhanced Raman Scattering (SERS), *J. Raman Spectrosc.* **33**, 511–516 (2002).

Wagnieres, G. A., Star, W. M., and Wilson, B. C., *In Vivo* Fluorescence Spectroscopy and Imaging for Oncological Applications, *Photochemistry and Photobiology*. **68**, 603–632 (1998).

Wilson, B. C., and Jacques, S. L., Optical Reflectance and Transmittance of Tissues: Principles and Applications, *IEEE J. Quantum Electronics* **26**, 2186–2199 (1990).

Yang, Y., Celmer, E. J., Koutcher, J. A., and Alfano, R. R., UV Reflectance Spectroscopy Probes DNA and Protein Changes in Human Breast Tissues, *J. Clin. Laser Med. Surg.* **19**, 35–39 (2001).

Yang, X. Y., Gaspano, F. P., and DeLeo, V. A., 8-Methoxypsoralen-DNA Adducts in Patients Treated with 8-methoxypsoralen and Ultraviolet A Light, *J. Invest. Dermatol.* **92**, 59- 63 (1989).

Zellweeger, M., Grosjean, P., Goujon, D., Mounier, P., Van den Bergh, H., and Wagnieres, G., *In Vivo* Autofluorescence Spectroscopy of Human Bronchial Tissue to Optimize the Detection and Imaging of Early Cancers, *J. Biomed. Opt.* **6**, 41–51 (2001).

Zhuang, X., Kim, H., Pereira, M. J. B., Babcock, H. P., Walker, N. G., and Chu, S., Correlating Structural Dynamics and Function in Single Ribozyme Molecules, *Science* **296**, 1473–1476 (2002).

CHAPTER 7

Bioimaging: Principles and Techniques

Bioimaging using optical methods forms a major thrust of biophotonics. Optical bioimaging can be used to study a wide range of biological specimens, from cells to *ex vivo* tissue samples, to *in vivo* imaging of live objects. Optical bioimaging also covers a broad range of length scale, from submicron size viruses and bacteria, to macroscopic-sized live biological species. This chapter describes the basic principles and techniques used for optical bioimaging. Thus it is intended to provide the reader with the appropriate background for appreciating the various applications of optical bioimaging covered in the next chapter.

Optical bioimaging utilizes an optical contrast such as a difference in light transmission, reflection, and fluorescence between the region to be imaged and the surrounding region (background). The various optical principles involved and microscopic methods used to enhance these contrasts and utilize them for bioimaging are described.

Various types of fluorescence microscopic methods, currently in wide usage, are covered. An advantage offered by fluorescence microscopy is the use of laser beams to excite an illuminated point and scan the point of illumination to form the image. This is commonly called *laser scanning microscopy*. Confocal microscopy, which allows one to obtain images at different depths and thus reconstruct a three-dimensional image of a biological sample, is described.

Optical coherence tomography, which utilizes an interferometric method to enhance contrast in a reflection geometry and has emerged as a powerful technique for three-dimensional imaging of highly scattering biological media (such as a tooth), is discussed. Other types of microscopy described in this chapter are (i) near-field scanning microscopy (often abbreviated as NSOM), which allows one to obtain optical images at a resolution of ≤ 100 nm, much smaller than the wavelength of light itself, and (ii) total internal reflection fluorescence (TIRF) microscopy, which provides enhanced sensitivity to image and probe cellular environment close to a solid surface.

Introduction to Biophotonics, by Paras N. Prasad
ISBN: 0-471-28770-9 Copyright © 2003 John Wiley & Sons, Inc.

Other bioimaging techniques discussed are (i) spectral imaging, which provides information on spatial variation of spectra, (ii) fluorescence resonance energy transfer (FRET) imaging, which utilizies energy transfer from one fluorescent center to another to probe interactions, and (iii) fluorescence lifetime imaging microscopy (FLIM), which is used to obtain the spatial distribution of fluorescence lifetime and is a highly sensitive probe for the local environment of the fluorophore. These imaging methods offer multidimensional imaging to probe details of interactions and dynamical processes in biological systems.

Section 7.15 provides a discussion of nonlinear optical techniques used for bioimaging. These techniques have gained considerable popularity because of the ability to use short-pulse near-IR lasers that allow deeper penetration in biological materials with little collateral damage. Specifically covered in this section are second- and third-harmonic microscopies, two-photon microscopy and coherent anti-Stokes Raman scattering (CARS) microscopy.

The chapter concludes with a presentation of some future directions for further development in bioimaging. A list of commercial sources for various microscopes is also provided.

The contents of this chapter are developed using the following sources, which serve as excellent references on this subject:

Periasamy (2001): A comprehensive coverage of cellular imaging

Lacey (1999): Light microscopy in biology

Pawley (1995): Comprehensive coverage of confocal microscopy

Diaspro, ed. (2002): Coverage of foundations, applications, and advances in confocal and two-photon microscopy

Tearney and Bouma (2001): Coverage of optical coherence tomography

Pawslear and Moyer (1996): Coverage of theory, instrumentation, and applications of near-field microscopy

The following websites give extensive information on optical microscopy techniques including Java-based tutorials:

http://micro.magnet.fsu.edu
http://www.microscopyu.com
http://www.olympusmicro.com

The following periodic publications are convenient sources of updates in this field:

Journal of Bioimaging
Biophotonics International
Microscopy
BioTechniques
Journal of Biomedical Optics

7.1 BIOIMAGING: AN IMPORTANT BIOMEDICAL TOOL

Biomedical imaging has become one of the most relied-upon tools in healthcare for diagnosis and treatment of human diseases. The evolution of medical imaging from plain radiography (radioisotope imaging), to x-ray imaging, to computer-assisted tomography (CAT scans), to ultrasound imaging, and to magnetic resonance imaging (MRI) has led to revolutionary improvements in the quality of healthcare available today to our society. However, these techniques are largely focused on structural and anatomical imaging at the tissue or organ levels. In order to develop novel imaging techniques for early detection, screening, diagnosis, and image-guided treatment of life-threatening diseases and cancer, there is a clear need for extending imaging to the cellular and molecular biology levels. Only information at the molecular and cellular levels can lead to the detection of the early stages of the formation of a disease or cancer or early molecular changes during intervention or therapy.

The currently used medical techniques of x-ray imaging, radiography, CAT scans, ultrasound imaging, and MRI have a number of limitations. Some of these are:

- Harmful effects of ionizing radiations in the case of x-ray imaging and CAT scan
- Unsuitability of x-ray imaging for young patients and dense breasts, as well as its inability to distinguish between benign and malignant tumors
- Harmful radioactivity in radioisotope imaging
- Inability of MRI to provide specific chemical information and any dynamic information (changes occurring in real time response to a treatment or a stimulus)
- Inability of ultrasound to provide resolution smaller than millimeters as well as to distinguish between a benign and a malignant tumor

Optical imaging overcomes many of these deficiencies. Contrary to the perception based on the apparent opacity of skin, light, particularly in near-IR region, penetrates deep into the tissues as discussed in Chapter 6. Furthermore, by using a minimally invasive endoscope fiber delivery system, one can reach many organs and tissue sites for optical imaging. Thus, one can even think of an "optical body scanner" that a physician may use some day for early detection of a cancer or an infectious disease.

Optical imaging utilizes the spatial variation in the optical properties of a biospecie, whether a cell, a tissue, an organ, or an entire live object. The optical properties can be reflection, scattering, absorption, and fluorescence. Therefore, one can monitor spatial variation of transmission, reflection, or fluorescence to obtain an optical image. The use of lasers as an intense and convenient light source to generate an optical response, whether reflection, transmission, or emission, has considerably expanded the boundaries of optical imaging, making it a most powerful technique for basic studies as well as for clinical diagnostics. Some of the benefits offered by optical imaging are:

- Not being harmful
- Imaging from size scale of 100 nm (using near field, to be discussed in Section 7.11) to macroscopic objects
- Multidimensional imaging using transmission, reflection, and fluorescence together with spectroscopic information
- Imaging of *in vitro*, *in vivo*, and *ex vivo* specimens
- Information on cellular processes and tissue chemistry by spectrally resolved and dynamic imaging
- Fluorescence imaging providing many parameters to monitor for detailed chemical and dynamical information. These parameters are:
    Spectra
    Quantum efficiency
    Lifetime
    Polarization
- Ability to combine optical imaging with other imaging techniques such as ultrasound
- Sensitivity and selectivity to image molecular events

The area of optical imaging is very rich, both in terms of the number of modalities and with regard to the range of its applications. It is also an area of very intense research worldwide because new methods of optical imaging, new, improved, and miniaturized instrumentations, and new applications are constantly emerging.

7.2 AN OVERVIEW OF OPTICAL IMAGING

A number of methods based on the optical properties monitored are used for imaging. These methods are summarized in Table 7.1.

Transillumination microscopic imaging utilizes a spatial variation of absorption and scattering in the microscopic and macroscopic structures of tissues. A tissue is a highly scattering medium. As the light propagates through a tissue, the transmitted light is comprised of three components: unscattered (or coherently scattered), weakly scattered, and multiply scattered light. These different components can be visualized by taking an example of a short pulse of laser light propagating through a tissue, as illustrated in Figure 7.1 (Gayen and Alfano, 1996).

The coherently scattered light, called the *ballistic photons*, propagate in the direction of the incoming beam. They, therefore, travel the shortest path and emerge first from the tissue. Ballistic photons carry maximum information on

TABLE 7.1. Optical Methods of Imaging

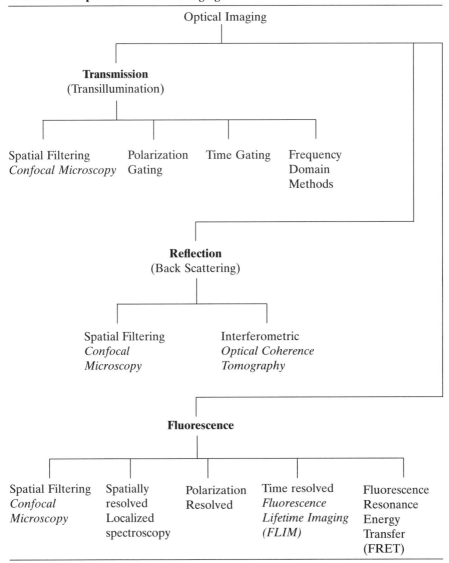

the internal structure of the tissue. The portions of light that scatter slightly more, but still in the forward direction, are called *snake photons* because of their wiggly trajectories in the forward direction. These photons are time-delayed with respect to the ballistic photons but still carry significant information on the scattering medium. However, most portions of the light beam undergo multiple scattering and travel long distances within the medium. They

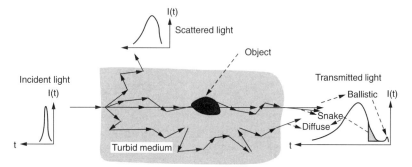

Figure 7.1. Propagation of a laser pulse through a turbid medium. (Reproduced with permission from Gayen and Alfano, 1996.)

emerge even later and are called *diffuse photons*. They carry little information on the microstructure of the tissue and have to be discriminated in order to image using ballistic and snake photons. Some of the commonly used methods to discriminate against the diffuse photons are listed in Table 7.1 and are briefly described below:

- *Spatial Filtering.* It is one of the simplest methods and relies on the fact that diffuse photons, undergoing multiple scattering, are more spread out and off-axis. Therefore, applying spatial filtering by using a transmitted light collection using an aperture (such as a small diameter fiber or a pinhole) provides rejection of a substantial amount of off-axis diffuse light. The most widely used microscope, a confocal microscope discussed later, uses a confocal aperture (pinhole) in the light collection path for spatial filtering. This confocal aperture in a confocal microscope is also used to enhance contrast and provide depth discrimination in reflection and fluorescence imaging.
- *Polarization Gating.* Here one utilizes a linearly polarized light. The transmitted ballistic and snake photons still retain much of the initial polarization, while the multiply scattered diffuse light are depolarized. Thus, by collecting the transmitted light through a polarizer allowing the transmission of light only with the polarizations parallel to the initial polarization, one can reject a significant portion of the diffuse light.
- *Time Gating.* This method utilizes a short laser pulse as the illumination source. The transmitted light is passed through an optical gate that opens and closes to allow transmission only of the ballistic and/or snake photons. Synchronization can be achieved by using a reference optical pulse that controls the opening and closing of the optical gate. A number of pulse gating techniques such as Kerr gate, nonlinear optical gate, and time-correlated single-photon counting are used. However, a detailed discussion of any of these methods is outside the scope of this book.

- *Frequency-Domain Methods.* In this method the time gating is transformed to intensity modulation in frequency domain (Lakowicz and Berndt, 1990). In this mode, the specimen is illuminated with an intensity-modulated beam from a CW laser, and the AC modulation amplitude and the phase shift of the transmitted signal are measured using methods such as heterodyning. One often uses the diffuse photon density wave description to analyze the transport of the modulated beam. In a sense, the situation here is analogous to the frequency domain measurement of fluorescence lifetime as described in Chapter 4, in which the temporal (time-resolved) information is obtained from the phase-shift information. The advantage of this method is that less expensive CW laser sources can be utilized. A limitation is that the readily available modulation frequency is only of a few hundred megahertz, which corresponds to time gating only with a few nanosecond resolution.

Reflection imaging collects the back-scattered light. Again, the coherently scattered light needs to be discriminated against the multiply scattered component. Two methods used are confocal and interferometric. The latter method has given rise to a very powerful microscopic technique called *optical coherence tomography* (OCT) for imaging of highly scattered tissues. This technique is discussed in detail in Section 7.9. In some cases, both confocal and OCT approaches are combined to enhance the discrimination against multiple-scattered light.

Fluorescence microscopy is the most widely used technique for optical bioimaging. It provides a most comprehensive and detailed probing of the structure and dynamics for *in vitro*, as well as *in vivo*, biological specimens of widely varying dimensions. The topic of fluorescence imaging is discussed in a separate section. Nonlinear optical methods have also recently emerged as extremely useful for bioimaging. Multiphoton-induced fluorescence microscopy is currently a very exciting new approach for bioimaging. Second-harmonic microscopy is also gaining popularity.

7.3 TRANSMISSION MICROSCOPY

7.3.1 Simple Microscope

A simple microscope is nothing but a single magnifying lens. The early design of a microscope had a single lens mounted on a metal plate with screws to move the specimen across the field of view and to focus its image. The concept of image formation by a lens is shown in Figure 7.2. A lens works by refraction and is shaped so that the light rays near the center are hardly refracted and those at the periphery are significantly refracted (Born and Wolf, 1999). A parallel beam of light passing through a convex lens is focused to a spot.

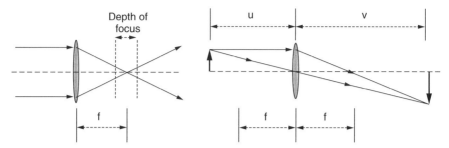

Figure 7.2. Ray tracing diagram showing the focusing action of a convex lens (*left*) and an image formation (*right*).

The distance from the center of the lens to the spot is known as the *focal length* of the lens (f). As shown in the figure, if an object is placed on one side of the lens at a distance u from it, a real image of that object is formed on the other side of the lens at a distance v. The image formed will be a magnified image, with magnification factor given by M = (image height) / (object height), which in turn is equal to the ratio of the image distance to the object distance (v/u). In a microscope, the object is placed in its focal plane and it forms a magnified image, which can either be observed by eyes or recorded by a camera. The object has to be placed close to the focal plane, within a short range known as the *depth of focus*, to obtain a sharp magnified image.

7.3.2   Compound Microscope

A compound microscope consists of a combination of lenses to significantly improve the magnification and functionality over a simple microscope. Figure 7.3 shows a compound microscope and its various components. Here, a magnified image of an object is produced by the objective lens that is again magnified by a second lens system (the ocular or eyepiece) for viewing. Thus, final magnification of the microscope is dependent on the magnifying power of the objective (lens) multiplied by the magnifying power of the eyepiece. Typical objective magnification powers range from 4× to 100×. Lower magnification objectives in a compound microscope are not commonly used because of spatial constraints of illumination (require special condensers for illumination), while higher magnification objectives are impractical due to their limited working distances.

Ocular magnification ranges are typically 8×–12×, though 10× oculars are most common. As a result, a standard microscope provides one with a final magnification range of ~40× up to ~1000×. Usually, a compound microscope contains many lenses to provide convenient illumination and to correct for different optical aberrations.

In a typical microscope, the objective lens projects an intermediate image of an object placed slightly off the front focal plane, onto a plane inside the

Modern Microscope Component Configuration

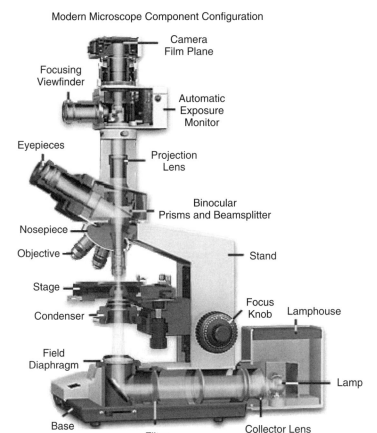

Figure 7.3. Schematic diagram of an upright transmission microscope. (Reproduced with permission from http://micro.magnet.fsu.edu/primer/anatomy/components.html.)

microscope. This intermediate image is then magnified and projected onto the retina by the eyepiece of the microscope. This type of microscope is called a *finite-tube-length microscope*, because it assumes a fixed path length between the objective and the eyepiece. But in most modern microscopes, a slightly different design is used to accommodate the introduction of different optical components, like a polarizer, inside the microscope, without affecting the image formation. In this design, the objective doesn't form an intermediate image, but an extra tube lens placed close to the eyepiece does the job of projecting the intermediate image. In this *infinity-corrected microscope*, there is a parallel beam of light in the space between the objective and the tube lens, whereby adding any other required optical component does not disturb the ray path. Figure 7.4 shows the optical ray path in these two variants of microscopes.

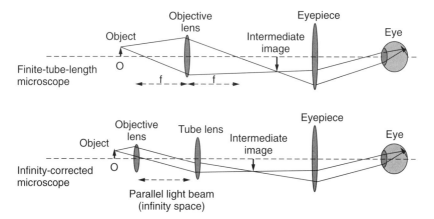

Figure 7.4. Schematic diagrams of optical ray paths in finite-tube-length microscope and in infinity-corrected microscope.

In a commonly used microscope, the sample plane is illuminated by a lamp through a set of collecting and condensing lenses and iris diaphragms. A proper illumination of the specimen observed under the microscope is critical for achieving a high-quality image through the eyepiece. The achievable resolution of a microscope depends not only on the objective lens used, but also on the way the sample is illuminated by its illumination system.

7.3.3 Kohler Illumination

One of the most commonly used illumination system in a transmission microscope is Kohler illumination, which provides an evenly illuminated field of view with a bright image, without glare and minimum heating of the specimen. Furthermore, as discussed below, it is important to have an illumination scheme where the sample is illuminated with a cone of light as wide as possible to achieve the best resolution possible. This feature is realized in the Kohler illumination scheme as shown in Figure 7.5. The light from the illuminating lamp, passing through a set of field diaphragms and lenses, simultaneously creates a uniformly illuminated field of view (parallel rays) while illuminating the specimen with a cone of light as wide as possible. The light pathways illustrated in Figure 7.5 are schematic representations of separate paths taken by the sample illuminating light rays and the image-forming light rays. Though in reality one cannot separate these two components, these diagrams help us to understand the process of uniform illumination and image formation.

As shown in Figure 7.5, the illuminating light ray path produces focused images of the lamp filament at the plane of the condenser aperture diaphragm, at the back focal plane of the objective, and at the eye point of the eyepiece.

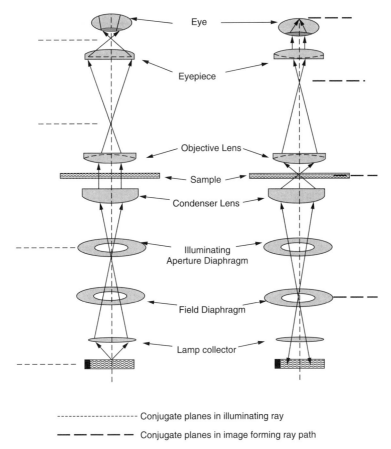

Eye

Eyepiece

Objective Lens

Sample

Condenser Lens

Illuminating
Aperture Diaphragm

Field Diaphragm

Lamp collector

------------------- Conjugate planes in illuminating ray

— — — — — Conjugate planes in image forming ray path

Figure 7.5. Schematic design for Kohler illumination.

These planes are called the *conjugate planes* of illuminating ray path. Conjugate planes in an optical system represent a set of planes such that an image focused on one plane is automatically focused on all other conjugate planes. The conjugate planes for image forming rays consist of the field diaphragm, the sample plane, the intermediate image plane and the retina of the eye. The field diaphragm and the condenser diaphragm are placed at the conjugate planes of image forming rays and illuminating rays, respectively. This allows independent controls over the angle, at which the sample is illuminated, and the intensity of illumination. A detailed description of Kohler illumination and Java-based tutorials on the effect of apertures in the microscopic illumination system can be found at the website http://micro.magnet.fsu.edu/primer /java/microscopy/transmitted/index.html.

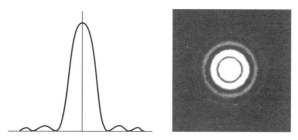

Figure 7.6. Intensity distribution and Airy disk formation.

7.3.4   Numerical Aperture and Resolution

The resolution of a microscope is its ability to distinguish between the smallest possible objects. This is directly related to the cone of light entering the objective from the sample. But, this optical resolution is limited by the diffraction of light occurring from the object, due to the wave nature of light. This principle can be understood by looking at a beam of light passing through a pinhole (Abbe, 1873; Born and Wolf, 1999). The image produced by the light passing through a pinhole and its intensity profile are shown below. This circular fringe pattern formation is known as the *Airy disk*. It looks like a negative target with a large bright central disk of light surrounded by a series of thin concentric circles of light of decreasing brightness as moving away from center (Figure 7.6). This effect is due to the diffraction of light emerging from the pinhole into multiple orders, which are represented by the concentric circles. The diffraction through pinholes has also been discussed in Chapter 2. While imaging small features in a sample using a microscope, a similar effect takes place. In order to get the entire information on these small features, the objective lens has to collect light of all these diffraction orders. Furthermore, the bigger the cone of light brought into the objective lens, the more of these diffraction orders can be collected by it, thus increasing the resolving power of the objective. As described below, this collection angle of the objective also determines the resolution of a microscope. This acceptance angle of light is quantified by a parameter called *numerical aperture* (NA) of the objective. The numerical aperture is defined as

$$NA = n\sin(\theta) \tag{7.1}$$

Here *n* is the refractive index of the medium from which the light rays enter the objective and θ is the maximum angle at which the light rays enter the objective as shown in Figure 7.7. The bigger the cone of light that can be brought into the lens, the higher its numerical aperture. From the expression shown above, it is clear that the maximum NA aperture an objective can achieve is the refractive index of the medium (since the maximum value of a

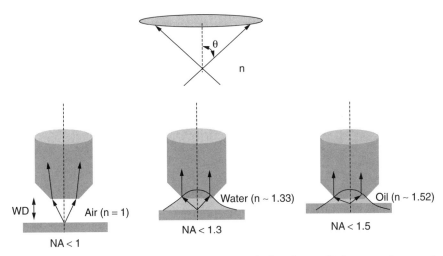

Figure 7.7. Numerical apertures for air, water, and oil as the media between the sample and the objective lenses.

sine function is 1). One way to improve the NA is to use an immersion medium with a higher refractive index than air. The most commonly used media for this purpose are water and oil. But, one thing to remember is that the use of higher NA objectives leads to a reduction of the working distance of the objective (distance between the objective and the sample). This behavior is also illustrated in Figure 7.7.

The resolution of an objective/microscope is defined as the distance, d, between two adjacent particles which still can be perceived as separate. Based on the limits of diffraction the resolution is given by Rayleigh's criteria (Born and Wolf, 1999) as

$$d = 1.22(\lambda/2\mathrm{NA})$$

Therefore, lenses with higher NA can give better resolution. In a transmission microscope, the numerical aperture of the objective, together with the NA of the condenser lens providing the illuminating light, determines the resolution. Thus, to achieve diffraction-limited resolution using a particular objective, the condenser must have an equal or higher NA. Furthermore, the magnification and the resolution of a microscope can be determined only by taking together into account the objective, the condenser, the eyepiece, and the illumination scheme used.

7.3.5 Optical Aberrations and Different Types of Objectives

The two major distortions or aberrations (Davidson and Abramowitz, 1999) in optical microscopy are (i) chromatic aberration which is due to the different refractive indices of the glass optical element (such as a lens) for different wavelengths and (ii) geometrical or spherical aberrations due to the shape of the lens, in which the rays from the edges of the lens don't get focused at the same point where the axial rays focus. Both these aberrations can be corrected by using lens doublets consisting of two lenses made of materials of different refractive indices. Good-quality objectives may contain multiple lenses to compensate for these errors. These corrected objectives are named *achromatic* and *aspheric* objectives. Another aberration in optical microscopy is due to the field curvature (curved image plane) of the objective lens that produces a curved image. New objectives made of special fluorite glass are available which correct for most of these aberrations. Thus one can choose from different types of objective lenses such as achromat, Plan-achromat, Plan-apochromat, Plan-Fluor, and so on, depending on the application and degree of aberration correction needed.

7.3.6   Phase Contrast Microscopy

Phase contrast microscopy is one of the most commonly used optical microscopic techniques in biology. Many of the unstained biological samples like cells don't introduce any amplitude changes (by absorption or scattering) in the transmitted light and hence are difficult to observe under normal bright-field microscopy. Phase contrast microscopy or dark-field microscopy provides enhancement of contrast.

In this technique (Zernike, 1942; Abramowitz, 1987a), the phase and the amplitude differences between undiffracted and diffracted light are altered to produce favorable conditions for interference and contrast enhancement. In phase specimens, the direct zeroth-order (undiffracted) light passes through or around the specimen undeviated. However, the light diffracted by the specimen is not reduced in amplitude, but is slowed by the specimen because of the specimen's refractive index or thickness (or both). This diffracted light, lagging behind by approximately 1/4 wavelength, arrives at the image plane shifted in phase from the undeviated light by 90°. Introduction of a phase plate that introduces an additional 1/4 phase difference between diffracted and undiffracted beams produces destructive interference between these two parts of the light. This interference can translate the phase difference into amplitude difference, which can be observed by eyes in a microscope. This is called *dark* or *positive phase* contrast, because the refractive object under observation appears dark in a bright background. For this, an annular ring is placed at the front focal plane of the condenser lens and a matching phase ring at the back focal plane of the objective, as shown in Figure 7.8.

7.3.7   Dark-Field Microscopy

Another technique commonly used for contrast enhancement is dark-field illumination (Abramowitz, 1987b; Davidson, 1999). In this case, the sample is

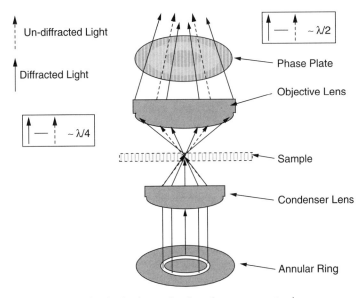

Figure 7.8. Optical schematics for phase contrast microscopy.

illuminated at an angle that cannot be accepted by the objective's aperture. In this case only the highly diffracted rays enter the objective. Hence only highly scattering or diffracting structures can be observed using this technique. Dark-field illumination requires blocking off the central light that ordinarily passes through and around the specimen, allowing only oblique rays from every angle to reach the specimen. This requires the use of special condensers that allow light rays emerging from the surface in all azimuths to form an inverted hollow cone of light with an apex centered in the sample plane. If no specimen is present and the numerical aperture of the condenser is greater than that of the objective, the oblique rays cross and all such rays miss entering the objective. In this case the field of view is dark. Because only the rays diffracted or refracted from the specimen reach the objective, this technique gives a high contrast image of the structures in the sample that diffract or refract the light.

7.3.8 Differential Interference Contrast Microscopy (DIC)

Differential interference contrast is a technique that converts specimen optical path gradients into amplitude differences that can be visualized as improved contrast in the image. This is accomplished by using a set of modified Wollaston prisms (Abramowitz, 1987b; Davidson, 1999). In this technique, living or stained specimens, which often yield poor images when viewed in bright-field illumination, are made clearly visible. Today there are several implementations of this design, which are collectively called *differential inter-*

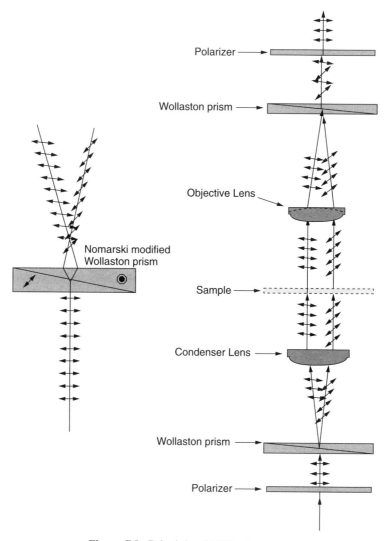

Figure 7.9. Principle of DIC microscope.

ference contrast (DIC), as shown in Figure 7.9. In transmitted light DIC, light from a lamp is passed through a polarizer located beneath the substage condenser, in a manner similar to polarized light microscopy. Next in the light path (but still beneath the condenser) is a modified Wollaston prism that is composed of two quartz wedges cemented together.

The plane-polarized light, oscillating only in one direction perpendicular to the propagation direction of the light beam, enters the Wollaston prism, which splits the light into two rays oscillating perpendicular to each other. The split beams enter and pass through the specimen where their wave paths are altered in accordance with the specimen's varying thicknesses, slopes, and refractive indices. When the parallel beams enter the objective, they are focused above the rear focal plane where they enter a second modified Wollaston prism that combines the two beams at a defined distance outside of the prism itself. As a result of having traversed the specimen, the paths of the two beams are not of the same length (optical path difference) for different areas of the specimen. After passing through another polarizer (analyzer) above the upper Wollaston beam-combining prism, these two beams interfere to translate the path difference introduced by the objects in the sample plane, into intensity difference. When a white light source from a lamp is used for imaging, each color will have a different optical path-length difference, thereby producing a color contrast. This results in observing the object details in pseudo—3-D and in color contrast.

7.4 FLUORESCENCE MICROSCOPY

Fluorescence microscopy has emerged as a major technique for bioimaging. Fluorescence emission is dependent on specific wavelengths of excitation light, and the energy of excitation under one photon absorption is greater than the energy of emission (the wavelength of excitation light is shorter than the wavelength of emission light). Fluorescence has the advantage of providing a very high signal-to-noise ratio, which enables us to distinguish spatial distributions of even low concentration species. To utilize fluorescence, one can use endogenous fluorescence (autofluorescence) or one may label the specimen (a cell, a tissue, or a gel) with a suitable molecule (a fluorophore, also called fluorochrome) whose distribution will become evident after illumination. The fluorescence microscope is ideally suited for the detection of particular fluorochromes in cells and tissues.

The fluorescence microscope that is in wide use today follows the basic "epi-fluorescence excitation" design utilizing filters and a dichroic beam splitter. The object is illuminated with fluorescence excitation light through the same objective lens that collects the fluorescence signal for imaging. A beam splitter, which transmits or reflects light depending on its wavelength, is used to separate the excitation light from the fluorescence light. In the arrangement, shown in Figure 7.10, the shorter-wavelength excitation light is reflected while the longer-wavelength emitted light is transmitted by the splitter.

With the advent of different fluorochromes/fluorophores, specifically targeting different parts of the cells or probing different ion channel processes (e.g., Ca^{2++} indicators), the fluorescence microscopy has had a major impact in biology (See Chapter 8). The development of confocal microscopy, discussed in Section 7.7, has significantly expanded the scope of fluorescence microscopy.

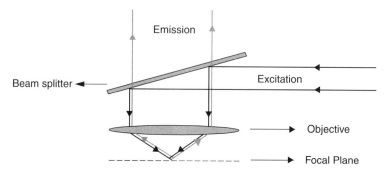

Figure 7.10. Basic principle of epi-fluorescence illumination. (See color figure.)

7.5 SCANNING MICROSCOPY

A primary problem with the fluorescence images that one observes or generates is that the out-of-focus regions of the sample appears as a "flare" in the object, reducing the signal-to-noise ratio substantially. Furthermore, during the imaging process, the entire sample is illuminated with high intensity excitation light, which can easily photooxidize (also known as photobleaching) the fluorochrome. As a means to eliminate both of these problems, it is possible to utilize a scanning optical microscope (Shepperd et al., 1978; Wilson et al., 1980) that permits observation of specimens at very high resolution, with comparatively low photooxidation of the fluorochrome.

A scanning optical microscope is designed to illuminate an object in a serial fashion, point by point, where a focused beam of light (from a laser) is scanned across the object rapidly in an X–Y raster pattern. The raster pattern is created by the repeated rotation of a beam deflecting galvanometric mirror assembly. Thus, a bright spot of light scans across an object and the image is generated point by point, in a raster format using a photomultiplier tube detection system. The intensity information is digitized and stored in a computer to generate the entire image of the scanned region. The resolution in scanning microscopy is limited by the spot size of the laser beam, which can approach the diffraction limit for the wavelength used.

Another approach to scanning microscopy is Nipkow disk microscopy (also known as tandem scanning microscopy). Instead of the point scanning techniques described above, Nipkow disk microscopy uses a spinning opaque disk perforated with multiple centrosymmetrical sets of holes (known as Nipkow disk) to illuminate the sample with multiple points (Petran et al., 1968, 1985). By rotating this disk, the entire sample area can be illuminated at high speeds, allowing real-time imaging or video rate imaging.

7.6   INVERTED AND UPRIGHT MICROSCOPES

The different types of transmission microscopy or fluorescence microscopy generally utilize a standard upright microscope in which light enters from the bottom and is viewed through the objective from the top. Alternatively, an inverted microscope can also be used in which the light source and the condenser are above the sample stage and the objective is below the stage.

For bioimaging, an inverted microscope offers certain advantages over an upright microscope. The main advantage of an inverted microscope is that gravity works in its favor. If the sample is something that will settle (or if the sample is at the bottom of a Petri dish), the settling will occur toward the objective in an inverted microscope, but away in an upright microscope. Thus settling objects are easier to image using an inverted microscope.

However, the design of an inverted microscope is more complex and the maximum magnification available is smaller than that for an upright microscope. All the options like phase contrast, DIC, or fluorescence imaging are available with an inverted microscope as well.

7.7   CONFOCAL MICROSCOPY

In a conventional wide-field microscope, thick specimens will produce an image that represents the sum of sharp image details from the in-focus region, combined with blurred images from all the out-of-focus regions. This effect does not significantly deteriorate images at low magnification ($10\times$ and below) where the depth of field is large. However, high-magnification objectives utilize high-numerical-aperture lenses that produce a limited depth of field, defined as the distance between the upper and the lower planes of the in-focus region. The area where sharp specimen focus is observed can be a micron or less at the highest numerical apertures. As a result, a specimen having a thickness greater than three to five microns will produce an image in which most of the light is contributed by the regions that are not in exact focus. The contribution from a blurred background reduces the contrast of the in-focus image.

Confocal microscopy overcomes this problem by introducing a confocal aperture (such as a pinhole) in the path of the image forming beam (fluorescence in case of fluorescence microscopy) to reject the out of focus contribution (Minsky, 1961).

In confocal microscopy, a point-like light source (laser) is focused by an objective onto a sample (Egger and Petran, 1967). The spatial extension of the focus spot on the sample is determined by the wavelength, the numerical aperture of the lens, and the quality of the image formation. The image spot is then focused through the same (or a second) lens onto an aperture (pinhole) and onto a detector. This pinhole is situated at a plane where the light from the

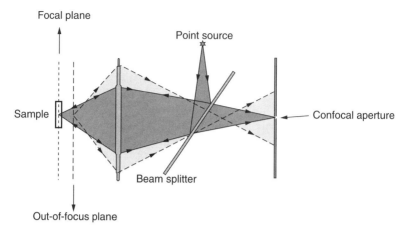

Figure 7.11. Ray path in confocal microscopy showing the out-of-focus rejection of the light from the sample by a confocal aperture.

in-focus part of the image converges to a point (i.e., at a conjugate focal plane). The principle utilized in a confocal microscope is shown in Figure 7.11.

Light from object planes above or below that of the focused image do not converge at the pinhole and hence is mostly blocked by it. Consequently, all out-of-focus optical background is removed from the image and the confocal image is basically an "optical section" of what could be a relatively thick object. The "thickness" of the optical section may approach the limit of resolution. However, in practice, the resolution in the z direction is somewhat greater: approximately 0.4–0.8 μm, depending on the excitation wavelength used. This value is dependent on multiple factors, such as the wavelength used and the size of the confocal aperture.

In order to build an image using the confocal principle, the focused spot of light is scanned across the specimen, either by scanning the laser beam (Brakenhoff, 1979; Brakenhoff et al., 1979) in a raster mode with a galvanometer mirror assembly (beam scanning) or by moving the sample stage (stage scanning) (Sheppard et al., 1978; Wilson et al., 1980). The stage scanning arrangement offers the advantage that the scanning beam is held stationary on the optical axis of the microscope, thus eliminating most aberrations introduced by lenses. For biological specimens, however, any movement of the specimen can cause wobble and distortion, resulting in a loss of resolution in the image. Therefore, bioimaging generally utilizes beam scanning with a galvanometer mirror assembly. To obtain a three-dimensional image, the focal plane is changed by translating the sample stage vertically, using a stepper motor or using a piezoelectric stage. The raster scanned data from each focal plane are obtained and stored in a computer, to reconstruct the 3D image of the sample. The principles of three-dimensional imaging in confocal microscopy has been discussed in detail by Gu (1996).

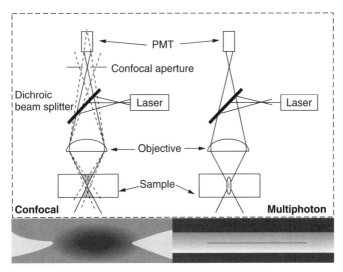

Figure 7.12. Confocal and multiphoton imaging. The bottom panel demonstrates the vertical cross section of the photo-bleached area in a sample. (Bottom panel reproduced with permission from Denk et al., 1995.)

Even though confocal microscopy provides a high-resolution optical sectioning capability, it has some inherent problems (Pawley, 1995). Since the confocal aperture reduces the fluorescence signal level, one needs a higher excitation power, which can increase the possibility of photobleaching. Furthermore, in single-photon confocal microscopy, linear excitation is used to obtain fluorescence. In this case the excitation of fluorescence occurs along the exciting cone of light, as shown in Figure 7.12, thus increasing the chance of photobleaching a large area. Another problem is that most fluorophores used for bioimaging are excited by one-photon absorption in the UV or blue light region. At these wavelengths, light is highly attenuated in a tissue, limiting the depth access. Some of these problems can be overcome (Konig, 2000) with the use of multiphoton excitation (also shown in Figure 7.12), which is described in the next section.

7.8 MULTIPHOTON MICROSCOPY

In multiphoton microscopy, a fluorophore (or fluorochrome) is excited by multiphoton absorption discussed in Chapter 5, and the resulting up-converted fluorescence (also discussed in Chapter 5) is used to obtain an image. Both two-photon and three-photon absorption-induced up-converted fluorescence have been used for multiphoton microscopy (Denk et al., 1990; Maiti et al., 1997). However, for practical reasons of three-photon absorption requiring extremely high peak power, only two-photon microscopy has emerged as a powerful technique for bioimaging. Two-photon laser scanning microscopy (TPLSM) can use a red- and near-infrared-wavelength short pulse (picoseconds and femtoseconds) laser as the excitation source and produce fluorescence in the visible range. Therefore, two-photon microscopy extends the range of dynamic processes by opening the entire visible spectral range for simultaneous multicolor imaging (Bhawalkar et al., 1997). Two-photon excitation, as discussed in Chapter 5, involves a simultaneous absorption of two laser photons from a pulsed laser source, to achieve fluorescence at the desired wavelength. The transition probability for simultaneous two-photon absorption is proportional to the square of the instantaneous light intensity, hence necessitating the use of intense laser pulses. It is preferable to use ultra-short laser pulses (picosecond or femtosecond pulses from mode-locked lasers), whereby the average power can be kept very low to minimize any thermal damage of the cell or biological specimen. A Ti: Sapphire laser (See Chapter 5), which produces very short (~100 fs) pulses of light around 800 nm (at a rate of ~80-MHz repetition rate), with a very large peak power (50 kW), has been a popular choice for two-photon microscopy.

Merits of TPLSM

In TPLSM, which utilizes a tightly focused excitation beam, the region outside the focus has much less chance to be excited. This eliminates the substantial "out-of-focus" fluorescence, often induced by one-photon excitation when used without a confocal aperture. TPLSM thus provides an inherent optical sectioning ability without using any confocal aperture. Two-photon excitation can also greatly reduce photobleaching, becasuse only the region at the focused point can be excited. This feature is derived from the fact that two-photon excitation is quadratically dependent on the intensity and hence highly localized at the focal point at which the intensity is greater.

Compared to short wavelength excitation in the UV-visible range, longer wavelength excitation with near-IR light can penetrate much deeper into a tissue because of less scattering or absorption at a longer wavelength (see Chapter 6). The elimination of UV also improves the viability of a cell (or tissue) and permits more scans to obtain a better 3-D image. From the instrumentation perspective, converting a conventional scanning laser microscope to TPLSM is straightforward. One only needs to change few optical elements. The dichroic mirror is to be replaced by a mirror to reflect near-IR or red excitation wavelength.

In addition to the sectioning ability, less photo damage and a better penetration ability, the resolution is another factor in evaluating the potential of TPLSM. As discussed earlier in this chapter, the image resolution is determined by the objective lens and the diffraction of light by the specimen (Rayleigh criteria). When comparing two-photon excitation with one-photon excitation of the same dye, two-photon excitation theoretically should yield worse resolution because it utilizes a significantly longer wavelength. However, other factors such as the out-of-focus fluorescence in one-photon excitation may come into play, affecting the achievable resolution. Generally, the strength of TPLSM does not lie in improvement of resolution, but in the other advantages described above.

7.9 OPTICAL COHERENCE TOMOGRAPHY

Optical coherence tomography (often abbreviated as OCT) is a new bioimaging technique that is rapidly growing in its applications (Tearney and Bouma, 2001). Already a number of clinical applications have been demonstrated in a widely diverse range of areas such as ophthalmology and dentistry (Huang et al., 1991; Brezinski et al., 1999; Schmitt et al., 1999). It is a reflection imaging technique similar to ultrasound imaging, except that light wave (usually in the near-IR to IR range) scattered from a specific tissue site is used to image. The sensitivity of the scattered light, as well as its selectivity from a specific back-scattering site, is achieved by using the interference between the back-scattered light and a reference beam. The OCT method of imaging is particularly suited for a highly scattering medium, such as a hard tissue. The interference between the propagating wavefronts of two light sources occurs when both wavefronts have well-defined coherence (phase relation) within the overlapping region. This well-defined coherence of a wavefront from a source is maintained within a distance called *coherence length*, as defined in Chapter 2. Therefore, if both the reference beam and the beam back-reflected from a scattering site are derived from the same light source, a well-defined interference pattern will be produced only if their path-length difference is within the coherence length. This behavior is shown in Figure 7.13.

A displacement of the reference beam produces the path-length difference between the light reflected from the reference mirror and the back-scattered ballistic photons (discussed in Section 7.2) from the scattering sample. In the case of a fully coherent source (such as a high-coherence laser source), the interference between the reference beam and the back-scattered beam can be maintained over a large path-length difference induced by reference mirror displacement. Thus no selectivity to back-scattering from a specific depth in the sample can be achieved in this case. The case on the right-hand side is for a low-coherence source (with a short coherence length). In this case the interference pattern between the reference beam and the back-scattered beam is produced only when their path difference is within the coherence length. In a three-dimensional scattering medium, at any given location of the reference beam mirror, only a given depth range (defined by the coherence length) of back-scattered light will interfere. Therefore, by scanning the reference mirror,

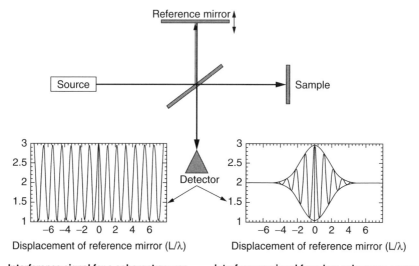

Figure 7.13. The interference signal as a function of the reference mirror displacement in case of a coherent source (e.g., laser) and a low-coherence source (e.g., SLD) are shown here.

one can achieve the depth discrimination. The interference patterns contain the information about the refractive index variation of the sample (tissue) which thus provides an optical image.

In OCT, the axial (depth) resolution is defined by the coherence length of the light source. The shorter the coherence length, the better the depth resolution. A broad-band light source will have a short coherence length (ΔL). Therefore, one often uses an incoherent but bright light source such as a super-luminescent diode (SLD) or a laser with a poor coherence (such as a femtosecond laser source with a broad-band width associated with it). An example of a bench-top OCT setup designed at our Institute for Lasers, Photonics, and Biophotonics (ILPB) is shown in Figure 7.14 (Xu et al., 1999). Here an SLD (8 mW at 850 nm with 20-nm bandwidth) is used as a low-coherence source. A polarizing beam splitter splits the beam into the reference and the sample arms. A combination of wave plates and polarizers allows the control of intensity of light in both arms. A phase modulator, introduced in one arm to modulate the signal derived from interference between the reference and sample beams, allows phase-sensitive detection of the signal using a lock-in amplifier. The sample is mounted on an XYZ stage, and the 3-D image of the sample can be obtained by using computer-controlled scanning and data acquisition. In this setup, the depth resolution was further enhanced by introducing a confocal aperture and a lens in the front of the detector, which further reduced the background interferences.

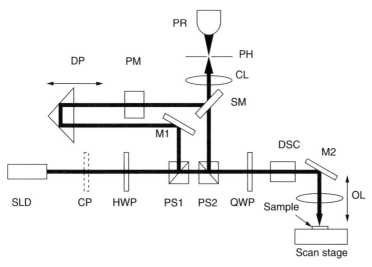

Figure 7.14. A table top OCT designed at the Institute for Lasers, Photonics, and Biophotonics (Buffalo) using an SLD light source. (Reproduced with permission from Xu et al., 1999).

The depth or axial resolution of this OCT is given by the FWHM of the round-trip coherence envelope, which is $\Delta L = 0.44 \lambda^2 / \Delta \lambda$ where $\Delta \lambda$ is the source bandwidth. The lateral resolution is given by the diffraction-limited spot size obtained by the focusing optics which is similar to that described in the previous section on confocal imaging. Fujimoto and co-workers (Swanson et al., 1993) developed a compact optical-fiber-based OCT setup as shown in Figure 7.15. In this arrangement, a dual-core fiber is used. One core of the fiber transmits the broad-band light source and splits it into the two arms: the sample probe and the reference. The other core of the fiber collects the back-scattered signal and reflected reference beam and combines to produce the interference.

Another variation of a fiber-based OCT has been used by Colston and others (Colston, 1998) for dental applications. The advantages of OCT and a comparison between OCT and confocal microscopy are presented below.

Advantages of OCT

High Resolution. Current OCT systems generally have resolutions of 4–20 µm compared to 110 µm for high-frequency ultrasound.

Real-Time Imaging. Imaging is at or near real time.

Catheter/Endoscopes. The fiber-based design allows relatively straight-forward integration with small catheter/endoscopes.

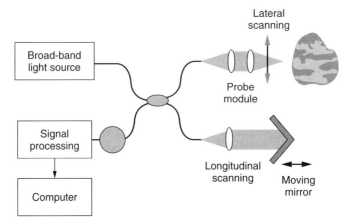

Figure 7.15. A fiber-based OCT. (Reproduced with permission from Swanson, et al., 1993).

Clinical Benefits

- Fiber-based OCT are compact, portable, and noncontact and can be combined with laser spectroscopy and Doppler velocimetry.

Comparison Between OCT and Confocal or Multiphoton Microscopy

Advantages

- Higher capability of imaging turbid medium (e.g., biological tissue).
- High depth resolution, even when the depth resolution of the objective is low.
- No need to use fluorescent stains or tags as in the case of confocal fluorescence or TPLSM.
- Compact compared to TPLSM setup (using SLD instead of Ti:sapphire laser as light source).

Disadvantage

- The depth resolution is lower than that for two-photon fluorescence or confocal microscopy with high magnification objective.

7.10 TOTAL INTERNAL REFLECTION FLUORESCENCE MICROSCOPY

Total internal reflection fluorescence microscopy, often abbreviated as TIRF microcopy, is best suited to image and probe a cellular environment within a

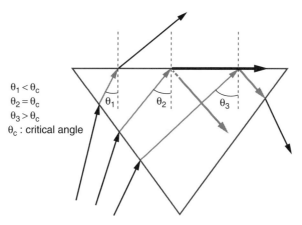

Figure 7.16. Principle of total internal reflection.

distance of 100 nm from a solid substrate. It relies on excitation of fluorescence in a thin zone of 100 nm from a solid substrate of refractive index higher than that of the cellular environment being imaged, by using the electromagnetic energy in the form of an evanescent wave. The concept of an evanescent wave can be understood by using the propagation of light through a prism of refractive index n_1 to the cellular environment of a lower refractive index n_2. At the interface, a refraction would occur at a small incidence angle. But when the angle of incidence exceeds a value θ_c, called the *critical angle*, the light beam is reflected from the interface as shown in Figure 7.16. This process is called *total internal reflection* (TIR). The critical angle θ_c is given by the equation

$$\theta_c = \sin^{-1}(n_2/n_1)$$

As shown in the figure, for incidence angle $>\theta_c$, the light is totally internally reflected back to the prism from the prism/cellular environment interface. The refractive index n_1 of a standard glass prism is about 1.52, while the refractive index n_2 of an intact cell interior can be as high as 1.38. The critical angle for these n_1 and n_2 parameters is 65°. For permeabilized, hemolyzed, or fixed cells, the n_2 value is that of an aqueous buffer which is 1.33, yielding a critical angle of 61°.

Even under the condition of TIR, a portion of the incident energy penetrates the prism surface and enters the cellular environment in contact with the prism surface. This penetrating light energy (or wave) is called an *evanescent wave* or an *evanescent field* (Figure 7.17). In contrast to a propagating mode (oscillating electromagnetic field with the propagation

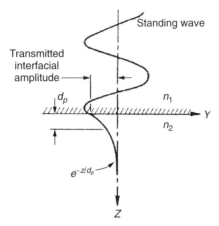

Figure 7.17. Evanescent wave extending beyond the guiding region and decaying exponentially. For waveguiding, $n_1 > n_2$, where n_2 is the refractive index of surrounding medium and n_1 is the refractive index of guiding region.

constant k, defined in Chapter 2, as a real quantity), an evanescent wave has a rapidly decaying electric field amplitude, with an imaginary propagation constant k. Therefore, its electric field amplitude E_z decays exponentially with distance z into the surrounding cellular medium of lower refractive index n_2 as

$$E_z = E_0 \exp(-z/d_p)$$

where E_0 is the electric field at the surface of the prism (solid substrate of higher refractive index). The parameter d_p, also called the *penetration depth*, is defined as the distance at which the electric field amplitude reduces to $1/e$ of E_0. The term d_p can be shown to be given as (Sutherland et al., 1984; Boisdé and Harmer, 1996).

$$d_p = \lambda \Big/ \left\{ 2\pi n_1 \left[\sin^2 \theta - (n_2/n_1)^2 \right]^{1/2} \right\}$$

Typically, the penetration depths d_p for the visible light are 50–100 nm. The evanescent wave energy can be absorbed by a fluorophore to generate fluorescence which can be used to image fluorescently labeled biological targets. However, because of the rapidly (exponentially) decaying nature of the evanescent field, only the fluorescently labeled biological specimen near the substrate (prism) surface generates fluorescence and can thus be imaged. The fluorophores that are further away in the bulk of the cellular medium are not excited. This feature allows one to obtain a high-quality image of the flu-

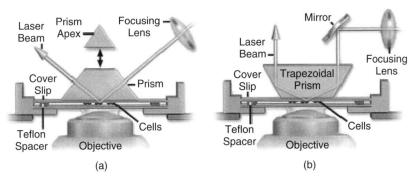

Figure 7.18. Inverted microscope TIP configuration. (Reproduced with permission from http://www.olympusmicro.com/primer/techniques/fluorescence/tirf/tirfconfiguration.html.)

orescently labeled biologic near the surface, with the following advantages (Axelrod, 2001):

- Very low background fluorescence
- No out-of-focus fluorescence
- Minimal exposure of cells to light in any other planes in the sample, except near the interface

The TIRF imaging offers a number of relative merits compared to the confocal microscopy. TIRF allows one to achieve a narrower depth of optical section (0.1 μm) compared to a typical value of 0.5 μm achieved in confocal microscopy. The illumination and hence the excitation are confined to a thin section (near the interface) in the case of TIRF, thus limiting any light-induced damage to cell viability. The TIRF microscopy is also much less expensive than the confocal microscopy, because one can use a standard microcope with TIRF attachment (or TIRF microscopy kits) available from a number of commercial sources.

Figure 7.18 shows two different prism based TIRF setups utilizing an inverted microscope. In Figure 7.18a, a prism is employed to achieve total internal reflection; the maximum incidence angle is obtained by introducing the laser beam from the horizontal direction. This arrangement is not compatible with conventional transmission imaging techniques. In Figure 7.18b, a trapezoidal prism is used and the incoming laser beam is vertical, so the total internal reflection area does not shift laterally when the prism is raised and lowered during specimen changes. In addition, transmission imaging techniques are compatible with this experimental design. Another approach is to utilize a hemispherical prism which permits continuous variation of the incidence angle over a wide range.

TIRF microscopy has been used for numerous applications that take advantage of the surface selectivity. Some of these are:

- Single-molecule fluorescence detection near a surface (Dickson et al., 1996; Vale et al., 1996; Ha et al., 1999; Sako et al., 2000)
- Study of binding of extracellular and intracellular proteins to cell surface receptors and artificial membranes (McKiernan et al., 1997; Sand et al., 1999; Lagerholm et al., 2000)

TIRF microscopy can be used with other optical imaging techniques such as fluorescence resonance energy transfer (FRET), fluorescence lifetime imaging (FLIM), fluorescence recovery after photobleaching (FRAP), and nonlinear optical imaging. FRET and FLIM are discussed in Sections 7.13 and 7.14. The TIRF microscopy can also utilize two-photon or multiphoton excitation of the fluorophores, similar to what was discussed above under two-photon laser scanning microscopy. Lakowicz and co-workers (Gryczynski et al., 1997) demonstrated two-photon excitation of a calcium probe Indo-1 using an evanescent wave.

7.11 NEAR-FIELD OPTICAL MICROSCOPY

Near-field optical microscopy is an optical technique that allows one to achieve a resolution of ≤100 nm, significantly better than permitted by the diffraction limit. As discussed in previous sections, the resolution of any optical imaging technique is limited by diffraction of light. The concept of using the near field for imaging was first discussed in 1928 by Synge, who suggested that by combining a subwavelength aperture to illuminate an object, together with a detector very close to the sample (<< one wavelength, or in the "near field"), high resolution could be obtained by a non-diffraction-limited process (Figure 7.19) (Synge, 1928). The implementation of this principle in practice (Ash, 1972; Pohl, 1984; Betzig and Trautman, 1992; Heinzelmann and Pohl, 1994) brought the field of near-field microscopy into existence. There are different variations

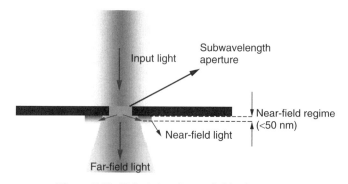

Figure 7.19. Principle of near-field microscopy.

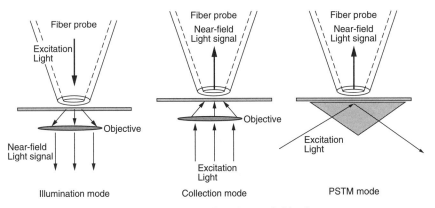

Figure 7.20. Different modes of near-field microscopy.

of this principle. One can illuminate the sample in the near field, but collect the signal in the far field or illuminate the sample in the far field while collecting the signal in the near field or do both in the near field. In almost all different methods, the most important component is the use of a subwavelength aperture that can be achieved by using a tapered optical fiber with a tip radius of <100 nm.

The most commonly used near-field probe consists of an optical fiber that is tapered and coated on the outside with a reflective aluminum coating. The tip of the fiber is typically about 50 nm. Light propagating through this fiber, either for excitation or for collection of emission, produces a resolution determined by the size of the fiber tip and the distance from the sample. The image is collected from point-to-point by scanning either the fiber tip or the sample stage. Hence the technique is called *near-field scanning microscopy* (NSOM) or *scanning near-field microscopy* (SNOM). Different modes of near-field microscopy are shown in Figure 7.20.

In illumination-mode NSOM, the excitation light is transmitted through the probe and illuminates the sample in the near field.

A typical setup used for near-field imaging is shown in Figure 7.21.

In collection-mode NSOM, the probe collects the optical response (transmitted or emitted light) in the near field. Another mode used in near-field imaging is photon scanning tunneling microscopy (PSTM) in which the sample is illuminated in a total internal reflection geometry using an evanescent wave (discussed in Section 7.10, but described here as due to photon tunneling); the emitted light is collected by a near-field optical probe.

The resolution in NSOM and PSTM is determined by two factors: the probe aperture (opening) size and the probe–sample distance. Because most samples exhibit some topography, it is important to keep the optical probe at a constant distance from the sample surface so that any change in the optical signal is attributed to a variation in the topographic feature, and not to variation in

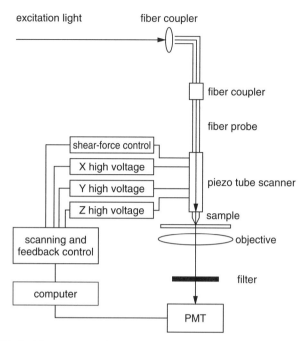

Figure 7.21. Typical instrumentation used for a near-field imaging setup. (Reproduced with permission from Shen et al., 2000.)

the probe–sample distance. A shear-force feedback technique can be used for distance regulation in cases of both conductive and nonconductive samples. In a shear-force feedback, the optical probe is attached to a tuning fork and oscillates laterally at its resonance frequency, with an amplitude of a few nanometers. As the probe approaches the sample surface, the probe–sample interaction dampens the amplitude and shifts the phase of the resonance. The change in the amplitude normally occurs over a range of 0–10 nm from the sample surface and is monotonic with the distance, which can be used in a feedback loop for distance regulation. The shear-force feedback can also be used to simultaneously obtain the topographic (AFM) image of the sample, to provide a monitoring reference for NSOM and PSTM. The applications of the near-field microscopy have ranged from single-molecule detection to biological imaging of viruses and bacteria (van Hulst, 1999; Hwang et al., 1998; Gheber et al., 1998; Subramanian et al., 2000).

7.12   SPECTRAL AND TIME-RESOLVED IMAGING

Discussions presented in the earlier sections of this chapter have focused on the types of microscopes used to provide spatial resolutions to probe structures down to subcellular levels. The optical probing methods utilized by them have been transmission, reflection or fluorescence to image structural details. Polarization characteristics have been used to enhance phase contrast for high-contrast imaging.

To probe biological functions, merely spatial imaging of structures is not sufficient. One needs to combine this information with spectrally resolved and time-resolved imaging in order to probe structure and dynamics that can provide useful information on biological functions. This combination of spatial, spectral, and temporal resolution, coupled with polarization discrimination, constitutes the emerging powerful field of multidimensional imaging. This section deals with spectral and time-resolved imaging. They are primarily used in conjunction with fluorescence detection. Thus, these methods of imaging can be used with epifluorescence, confocal, near-field, or TIRF microscopic techniques discussed above, which utilize fluorescence detection. Vibrational spectral imaging has also been gaining popularity in the form of Raman Imaging and CARS (coherent anti-Stokes Raman scattering) imaging. CARS imaging is described in Section 7.15.

7.12.1 Spectral Imaging

In fluorescence-based optical imaging described above, spatial distribution of fluorescence intensity is used to determine structure and organization at the cell or tissue level. In other words, the detection simply is of the intensity level of fluorescence. Spectral imaging provides spectral information on the spatial variation of fluorescence spectra. In other words, one can also obtain information on the fluorescence spectra at a given spatial location. This feature permits simultaneous use of more than one fluorescent marker and maps their distribution in various biological sites (tissue locations, cell organelles, etc.). Thus, spatial distribution and localization of a specific drug or biologic in a cell or a tissue can selectively be studied. Furthermore, by monitoring spectral shift in the emission maximum (or a change in the emission spectral profile) of a fluorescent marker, one can obtain information about the local environment (interaction and dynamics). This is a useful information in terms of understanding physiological changes occurring in response to a biological function or in elucidating the molecular mechanism of a drug–organelle interaction. Some of the spectral wavelength selection techniques are described below.

7.12.2 Bandpass Filters

In the simple case of imaging using more than one fluorescent probes, where their fluorescence spectra are widely separated, one can simply use a set of bandpass filters (green filters, red filters, etc.) in the path of fluorescence collection and detection. These bandpass filters allow transmission in a specific wavelength region, in which fluorescence can be detected, while rejecting the fluorescence from other regions. A number of choices exist such as interference transmission filters or dichroic mirrors. The simplicity and relatively low costs are the advantages offered by this method. The major disadvantage is that the system does not offer continuous tunability of wavelength selection and one has to use a mechanical method to introduce a specific filter in the fluorescence collection path (such as a rotating tray or filter wheel containing a set of filters).

7.12.3   Excitation Wavelength Selection

This method can be used when the fluorophores used have well-separated excitation spectra so that excitation at one wavelength generates fluorescence only from a particular fluorophore. A laser source providing a number of lasing lines (such as an argon–krypton laser) or a continuously tunable output over a broad range (such as a laser pumped OPO, described in Chapter 5), or a combination of lasers can be used for this purpose. The advantage, again, is the simplicity of the method, and there is no need to introduce any additional optical or mechanical element in the fluorescence collection (imaging) light path, thus providing minimal image degradation. However, this approach requires an achromatic objective lens to produce focusing of different excitation wavelengths at the same spot. Furthermore, this approach is limited because it involves the combination of only those fluorophores that have well-separated excitation spectra.

7.12.4   Acousto-Optic Tunable Filters

The use of acousto-optic tunable filters, often abbreviated as AOTF, for rapidly tunable multispectral imaging has been pioneered by Farkas and co-workers (Farkas, 2001; Wachman et al., 1997). AOTF are electronically controllable solid-state devices where a narrow spectral bandwidth is angularly deflected away (diffracted) from the incident beam by an acoustic frequency applied to AOTF. The central wavelength of the deflected light can be continuously tuned over a wide frequency range by the choice of the acoustic frequency, and the wavelength switching speed can be in tens of microseconds. The advantage of this approach is a fast switching speed, tunability of both the wavelength and the bandwidth, and control of the intensity of transmitted light.

   The mechanism of AOTF involves the modulation of the refractive index by elasto-optic (or acousto-optic) effect in a certain crystal (acousto-optically active) when an acoustic wave is generated inside the crystal. The acoustic field is electronically generated by applying an rf field to a piezoelectric transducer bonded to one of the faces of the crystal. The periodic modulation of the refractive index acts as a grating to diffract certain wavelengths at an angle from the incident beam direction.

   Image blur encountered using AOTF had limited the use of this technique in the past. Farkas and co-workers proposed a new transducer design, along

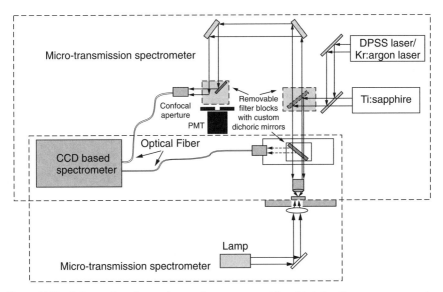

Figure 7.22. Schematics of experimental arrangement for obtaining fluorescence spectra from a specific biological site (e.g., organelle) using a CCD-coupled spectrograph. (Reproduced with permission from Pudavar et al., 2000.) (See color figure.)

with the use of two AOTFs in tandem, to provide out-of-band rejection that leads to an improvement of the image blur.

7.12.5 Localized Spectroscopy

Another approach is to use a spectrograph to analyze fluorescence from a specific point of the image plane (Masters et al., 1997; Wang et al., 2001). This approach, incorporated in the multiphoton confocal setup at our Institute for Lasers, Photonics, and Biophotonics is shown in Figure 7.22. The fluorescence is collected by the objective lens and, after the dichroic beam splitter, coupled into a fiber which also acts as a confocal aperture. The light is guided by the fiber to a spectrograph where the wavelengths are dispersed and detected by a CCD array. The CCD (charge coupled device) array is a multiarray detector where each array (pixel) detects a specific wavelength (or a narrow wavelength range). Therefore, the entire fluorescence spectrum can be simultaneously monitored.

7.13 FLUORESCENCE RESONANCE ENERGY TRANSFER (FRET) IMAGING

Fluorescence resonance energy transfer, abbreviated as FRET, is an example of spectral imaging that has emerged as a powerful technique for biomedical research. Its applications cover a broad range such as study of protein–protein interactions, calcium metabolism, protease activity, and high-throughput screening assays (Herman et al., 2001; Periasamy, 2001). The fundamental principle involves the use of Förster excitation energy transfer from an excited molecule of higher energy (donor) to another molecule of lower excitation energy (acceptor). This energy transfer, discussed in Chapter 4, occurs nonradiatively through dipole–dipole interaction, showing a distance dependence of R^{-6}. It is maximized when there is a significant overlap of the emission spectrum of the donor with the absorption spectrum of the acceptor.

Thus, the interactions of cellular components with each other (such as protein–protein interactions) can be studied and quantified by labeling the two components with two appropriately chosen fluorophores that act as an excitation donor and an excitation acceptor, respectively. In FRET spectral imaging, the donor is selectively excited and the quenching of its emission, concomitant with a gain in the fluorescence of the acceptor, indicates appreciable interaction between the donor and the acceptor labeled cellular components, leading to donor-to-acceptor excitation energy transfer. Another variation of FRET imaging utilizes lifetime where a considerable shortening of the fluorescence lifetime of the donor implies an efficient FRET process. The lifetime imaging is discussed below in a separate section.

A popular choice of fluorophores for studying subcellular interactions has been various mutants of green fluorescent proteins (GFP) which now offer a choice of fluorescence covering the entire visible range. Thus, both the donor and the acceptor fluorophores can be chosen from this family. GFP is discussed in Chapter 8.

FRET imaging involves measuring the intensities of the donor emission (I_D) and the acceptor emission (I_A) and obtaining a spatial distribution of the ratio I_A/I_D. This approach is also known as *steady-state FRET imaging*. An important consideration in getting a FRET image with a high signal-to-noise ratio is using optical means (as described in the above sections) to spectrally discriminate the donor and the acceptor absorption and emission spectra. A simple approach is to use narrow bandpass filters that allow selective excitation only of the donor so that any emission from the acceptor results from the FRET process. Also, the choice of the filters to separate the donor and the acceptor emissions is very important. This spectral discrimination using an appropriate combination of filters reduces any spectral bleedthrough background.

Another important factor to enhance the quality of FRET imaging is the appropriate choice of the donor and the acceptor optimal concentrations. These optimal concentrations can be determined in a systematic study of the FRET signal as a function of concentration of one component (e.g., donor) while keeping the concentration of the other component (acceptor) at a fixed value.

7.14   FLUORESCENCE LIFETIME IMAGING MICROSCOPY (FLIM)

Fluorescence lifetime imaging microscopy, often abbreviated as FLIM, provides a spatial lifetime map of a fluorophore within a cell or a tissue (Tadrous,

2000; Bastiaens and Squire, 1999). The use of fluorescence lifetime of a fluorophore as an imaging contrast mechanism offers a number of advantages over steady-state fluorescence microscopy. First, the fluorescence lifetime is a highly sensitive probe of the local environment of the fluorophore. The temporal resolution obtained in this modality of imaging provides an opportunity to study the dynamic organization of a living system. For FRET imaging described above, FLIM provides an advantage to measure the energy transfer only by measuring the donor fluorophore lifetime, which is significantly affected (reduced) by energy transfer to an acceptor. FLIM also has the advantage that the fluorescence lifetimes are independent of the fluorescence intensity, concentration, and, to a larger extent, photobleaching of the fluorophore. Furthermore, there may be cases where a fluorophore may exhibit similar spectra, but significantly different lifetimes in different environments, as the lifetime is a more sensitive probe of the environment. FLIM has been used for many different types of imaging experiments using both one- and two-photon excitations. These include imaging using multiple fluorophore labeling, quantitative imaging of ion concentrations, quantitative imaging of oxygen, and energy transfer efficiency in FRET (Periasamy et al., 1996; French et al., 1997; Bastiaens and Squire, 1999; Lakowicz et al., 1992).

The two methods of measurements of lifetime are (i) the time domain method using a pulse laser excitation and (ii) the frequency domain method using phase information. These methods have already been discussed in Chapter 4. The simplest case is a single exponential decay of fluorescence intensity $I(t)$ given as

$$I(t) = I_\text{o} \exp(-t/\tau)$$

where τ is the fluorescence lifetime. In such a case, the time domain FLIM imaging involves obtaining the fluorescence images by applying a pulse excitation at two different time delays t_1 and t_2, using gated detector for a duration of ΔT, as shown in Figure 7.23. The accumulated emitted photons during the ΔT periods are measured by integrating the signal using a CCD camera and given as D_1 and D_2 for the two time delays. The single exponential decay can be shown to yield the following relation for the fluorescence lifetime:

$$\tau = (t_2 - t_1)/\ln(D_1/D_2)$$

Using this equation, one can then obtain the FLIM image. An advantage offered by the time-resolved measurement is that the background noise due to scattering (such as Raman scattering) can be eliminated. Scattering occurs almost instantaneously. Hence using a time delay between the excitation pulse and opening of the electronic gate to collect fluorescence (such as t_1, in Figure 7.23), one can eliminate background scattering. In the case of a more complicated decay of multiexponential nature, one has to obtain the fluorescence

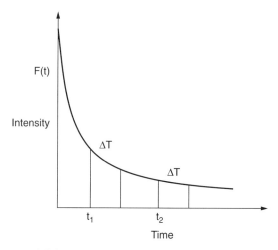

Figure 7.23. Exponential decay of fluorescence showing the sampling period T for two different delays t_1 and t_2.

signal at a number of delays to obtain the full decay curve and fit the curve to a double (or multiple) exponential.

In the frequency domain method, as described in Chapter 4, one utilizes a sinusoidally modulated light to excite fluorescence. Phase shift in fluorescence is measured to obtain lifetime information (Verneer et al., 2001). Measuring multiple lifetimes using a number of fluorophores simultaneously requires excitation with a sum of sinusoidally modulated signals with different frequencies. Thus, simultaneous detection of several fluorescently tagged biomolecules can be made and their individual interactions can be monitored. The fluorescence lifetimes of the fluorophores are in nanoseconds, enabling one to measure dynamical processes in cells and tissues. For example, activation state of proteins *in situ* can be investigated without any disruption of the cellular architecture.

FLIM is not only useful for cellular research, but is also an informative tool for the pharmacological and medical industries (Bastiaens and Squire, 1999). FLIM provides a detection platform for ultra-high throughput screening of drugs on their interactions with living cells. This technique is also well-suited to assess the early functional states of proteins implicated in the pathology of a diseased tissue.

7.15 NONLINEAR OPTICAL IMAGING

Nonlinear optical effects discussed in Chapter 5 can be used as a contrast mechanism for microscopic studies of biologics (Mertz, 2001). A clear advantage of using the nonlinear optical effects is a gain in the spatial resolution due to a higher-order dependence on the excitation intensity. A number of nonlinear optical effects have been used for bioimaging. A very popular method in bioimaging is that of two-photon excitation. The two-photon excitation has already been discussed in the Section 7.8, on multiphoton microscopy. This nonlinear optical method has also been used in TIRF and near field microscopy. There has also been reports of using three-photon processes for microscopy (Schrader et al., 1997; Wokesin et al., 1996). In this case, a simultaneous absorption of three IR photons produces an up-converted emission in the visible. Recently, we reported (He et al., 2002) highly efficient three-photon excitation of visible fluorescence ($\lambda^{max} = 550\,nm$) in a new fluorophore, abbreviated as APSS, by using a pump wavelength of $1.3\,\mu m$. The excitation is so efficient that one can create a population inversion to produce stimulated emission. At the wavelength of $1.3\,\mu m$, biological cells and tissues still exhibit very good transparency. Furthermore, emission at $550\,nm$ can also be more efficiently collected compared to emission in blue, generated by three-photon excitation at ~$1\,\mu m$. Therefore, the availability of highly efficient new three-photon fluorophores that can be excited in the IR around $1.3\,\mu m$ can lead to further expansion of applications of three-photon microscopy.

This section covers other nonlinear optical techniques. Specifically, second-harmonic generation, third harmonic generation and coherent anti-Stoke Raman scattering methods and microscopy based on them are described here.

7.15.1 Second-Harmonic Microscopy

In this approach, the nonlinear optical process of second-harmonic generation is used to generate image contrast. As discussed in Chapter 5, second-harmonic generation is a second-order nonlinear optical process, generated in a noncentrosymmetric medium (Prasad and Williams, 1991), whereby a second-harmonic output at the frequency of 2ν (or wavelength of $\lambda/2$) is generated from an input beam of frequency ν. Therefore, it is an up-conversion process, just like two-photon excited emission. Also, like a two-photon process, second-harmonic generation is quadratically dependent on the input intensity. However, the second-harmonic microscopy offers the following features, different from the two-photon microscopy.

- Since second-harmonic generation dos not involve any absorption of light (see Chapter 5), no thermal damage or photobleaching occurs if a pump beam of wavelength outside the absorption band is selectively chosen. In contrast, the two-photon microscopy involves a two-photon absorption of light.
- Second-harmonic generation shows a symmetry selection, occurring only in an asymmetric medium such as an interface or an electric-field-induced

noncentresymmetric environment. Therefore, the second-harmonic microscopy is more useful for probing structures and functions of membranes and the membrane potential induced alignment of dipolar molecules in a membrane. A two-photon process does not readily select and probe asymmetry.

- The second-harmonic signal is obtained at exactly half the wavelength of the pump laser beam and is thus easy to discriminate against the pump beam and any autofluorescence. The two-photon excited emission, in contrast, is a broad fluorescence band.
- The second-harmonic microscopy can be used with nonfluorescent samples and tissues.

The disadvantage of the second-harmonic microscopy is that it is not as versatile as the two-photon microscopy, since the signal is generated only in an asymmetric medium.

Second-harmonic generation was first reported by Fine and Hansen (1971) in nearly transparent tissues. Freund et al. (1986) described a cross-beam steering second-harmonic microscopy with a transmission geometry to obtain detailed variation of collagenous filaments in a rat tail tendon. Alfano and co-workers have used second-harmonic imaging to probe structures of animal tissues (Guo et al., 1996, 1997) and have used 100-fsec pulses at 625 nm to map subsurface structure of animal tissues by using second-harmonic generation tomography noninvasively. In the tomography approach (like in OCT described earlier), the second-harmonic signal is obtained in a back-reflection geometry to build a three-dimensional layered structure map near the surface of a tissue. They suggested that second-harmonic imaging can be implemented with fiber optics and adapted to endoscopy for morphological evaluation in cardiology, gynecology, and gastrointestinal applications.

Lewis and co-workers have shown that second-harmonic generation can be a powerful method to probe membrane structure and measure membrane potential with a single-molecule sensitivity (Bouevetch et al., 1999; Peleg et al., 1999; Lewis et al., 1999). They used this method to probe membrane proteins and obtain functional imaging around selective sties and at single molecule level in biological membranes. For this imaging they used a donor–acceptor-type dye structure, discussed in Chapter 5 for second-order nonlinear optical effect, that undergoes internal charge transfer and can bind and orient in a lipid bilayer. These molecules respond to membrane potential by an electrochromic mechanism in which the field due to the membrane potential is sufficiently large due to a change in the induced dipole moment of the dye. Thus they produce a change in the generated second-harmonic signal. The membrane potential variation can be induced by changes in extracellular potassium concentrations.

Moreaux et al. (2001) have used second-harmonic generation to measure membrane separation and have shown that this method can be used to measure separations over ranges not accessible by FRET.

7.15.2 Third-Harmonic Microscopy

Third-harmonic generation is a third-order nonlinear optical process in which a fundamental pump beam of frequency v (wavelength, λ) generates a coherent output at $3v$ (wavelength, $\lambda/3$). Thus, an input beam at the fundamental wavelength of 1064 nm in the IR will generate an output at ~355 nm in the UV. Again, this process, unlike three-photon absorption, does not involve light absorption in the medium. In contrast to second-harmonic generation, third-harmonic generation does not have any symmetry requirement and can occur both in bulk and at surfaces. The molecular structural requirement for an organic substance to efficiently produce third-harmonic is only that it has extended conjugation of π electrons (Prasad and Williams, 1991). Thus third-harmonic microscopy can be used for both interface and bulk imaging.

The third-harmonic generation has been shown to have monolayer sensitivity and was used to study conformational changes in monolayer films prepared by the Langmuir–Blodgett technique (Berkovic et al., 1987, 1988).

The third-harmonic microscopy has been used to image biological samples (Yelin and Sieberberg, 1999; Müller et al., 1998). Yelin and Sieberberg used a synchronously pumped OPO with a 130-fsec pulse output at 1.5 μm, at a repetition rate of 80 MHz as the fundamental pump source. They showed that even though a high peak power was needed to generate the third-harmonic signal, the use of ultra-short femtosecond pulses allows one to use low average power (50 nW in their experiment). Using third-harmonic generation for laser scanning microscopy, the image is collected point by point. Yelin and Sieberberg imaged live neurons in a cell culture and obtained detailed images of organelles. In view of the high peak power needed (hundreds of GW/cm^2) for third-harmonic generation, combined with the danger of damaging the specimen under illumination with such high intensity pulses, it is not yet apparent how wide an application this nonlinear technique will find.

7.15.3 Coherent Anti-Stokes Raman Scattering (CARS) Microscopy

Coherent anti-Stokes Raman scattering (abbreviated as CARS) is a third-order nonlinear optical process that can produce a vibrational transition. This nonlinear optical process has been discussed in Chapter 5. For CARS, two optical beams of frequencies v_p and v_s interact in the sample to generate an anti-Stokes optical output at $v_{AS} = 2v_p - v_s$ in the phase-matched direction (a specific direction). The signal has an electronic contribution (from the electronic third-order nonlinear optical response), but is resonantly enhanced if $v_p - v_s$ coincides with the frequency of a Raman active molecular vibration (see Chapter 4). The molecular vibration involved in a CARS signal enhancement can then be used as a contrast mechanism for bioimaging. Since the first attempt of Duncan et al. (1982), CARS microscopy has attracted a great deal of attention in recent years (Zumbusch et al., 1999; Muller et al., 2000; Potma et al., 2000; Hashimoto et al., 2000; Cheng et al., 2001; Volker et al., 2002). Xie and co-workers have made significant advances in the application of the laser scanning CARS microscopy to cell biology (Zumbusch et al., 1999; Cheng et al., 2001; Volkmer et al., 2001). The CARS microscopy provides a number of advantages, some of which are (Volker et al., 2001):

- Vibrational contrast in the CARS microscopy is inherent to the cellular species, thus requiring no endogenous or exogenous fluorophores that may be prone to photobleaching.
- CARS, being a coherent optical process (phase matching), offers much higher sensitivity than the spontaneous Raman process.
- CARS, being a nonlinear optical process, exhibits a nonlinear dependence on the pump intensity and, like in two-photon or other nonlinear microscopy described above, generates signals from the focal volume. This feature allows three-dimensional optical sectioning of thick samples to obtain a high-resolution three-dimensional image.
- CARS can provide chemical selectivity as different vibrational modes can be used for contrast.
- The CARS signal can be detected even in the presence of background autofluorescence, since a CARS signal is highly directional because of the phase-matching requirement.

CARS imaging is complicated by background signals derived from two sources: (i) the nonresonant electronic contributions that exist even when the Raman resonance condition is not met and (ii) electronic and Raman contributions from the solvent. The latter is particularly troublesome when using an aqueous medium as water generates a strong resonant CARS signal. Volker et al. (2001) have shown that by detecting CARS signal in the backward direction (which they call E-CARS) one can effectively suppress the solvent background and significantly increase the sensitivity of the CARS microscopy.

High-peak power lasers are needed to enhance the nonlinear optical process of CARS. At the same time it is necessary to maintain a narrow bandwidth of pulses in order to obtain good spectral resolution for selectively using a specific vibrational mode for resonance enhancement. Cheng et al. (2001) suggested the use of picosecond pulses instead of femtosecond pulses because the former allows one to achieve a better signal-to-background ratio (Cheng et al., 2001). Furthermore, to obtain and maintain overlap of two femtosecond laser pulses is much more difficult, in comparison with overlap to obtain and maintain of two picosecond pulses. The schematics of their experimental arrangement is shown in Figure 7.24. It utilizes two synchronized mode-locked Ti:sapphire lasers producing picosecond pulses at the 80-MHz repetition rate. The pump beam with frequency v_p is tunable from 690 to 840 nm, while the Stokes beam with frequency v_s is tunable from 770 to 900 nm. This arrangement allows one to cover the vibrational frequency range from 100 to $3400 \, cm^{-1}$. They used this arrangement to obtain the CARS image of unstained human epithelial cells. The Raman band at $1570 \, cm^{-1}$, arising from proteins and nucleic acid, was used.

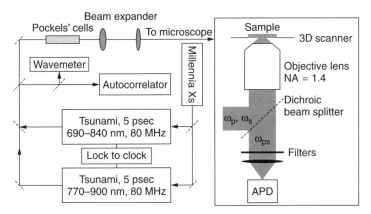

Figure 7.24. Schematics of a synchronized mode-locked picosecond Ti:sapphire laser system for backward detection CARS microscopy. Millenia is the diode-pumped Nd laser. Tsunami is the Ti:sapphire laser. (Reproduced with permission from Cheng, et al., 2001.)

7.16   FUTURE DIRECTIONS OF OPTICAL BIOIMAGING

7.16.1   Multifunctional Imaging

Each of the imaging techniques described in the previous sections has its own unique approach to imaging which can be suitable for obtaining certain biological information. But none of the techniques can be of universal use. For a comprehensive investigation of biological species and processes, one may require to use a combination of bioimaging methods, often at the same time. Integration of the various techniques is one of the evolving areas in bioimaging. Some modern confocal/multiphoton microscopes have the ability to provide simultaneous fluorescence, fluorescence lifetime imaging, and four-dimensional imaging (stack of three-dimensional images in different spectral region). Some of them can easily be adapted for polarization anisotropy imaging, Raman imaging, harmonic generation imaging, and so on, as well.

7.16.2   4Pi Imaging

Another technique, which is gaining popularity for high-resolution bioimaging is 4Pi imaging. In this technique, a standing wave is created in the sample plane, by the interference of two opposing wavefronts (Hell and Stelzer, 1992). In a 4Pi confocal microscope, two opposing high-NA objectives are used for illuminating and detecting the same point of a fluorescent sample. In such an arrangement with a coherent light source, an interference between the two light beams produce a standing wavefront, which in turn limits the volume of emission from the sample. In single-photon 4Pi imaging, even though the axial resolution is significantly improved, the lateral resolution is actually degraded (Martínez-Corral, 2002). But with the introduction of two-photon 4Pi imaging (Nagorni and Hell, 2001), good axial and lateral resolution can be achieved.

7.16.3   Combination Microscopes

Another important development in this area is the integration of near-field and far-field techniques (confocal or multiphoton imaging) to increase the dynamical range of imaging. Switching between the near-field and the far-field modes (e.g., from confocal) in the same instrument can give high-resolution small-area scans as well as large-area far-field images. Some commercial systems combining a near-field microscope with a confocal microscope are already available in the market.

7.16.4   Miniaturized Microscopes

Most of the instruments or setups used for the above-described techniques are desktop instruments. In order to use them as regular diagnostic tools in the field, there is a need for miniature instrumentation. Current research in bioimaging is concerned with exploiting the developments in the MEMS (micro-electro-mechanical devices) technology to miniaturize many of these imaging setups (Dickensheets and Kino, 1998). There are few reports of a compact confocal microscope (Dickensheets and Kino, 1998) or OCT (Bouma et al., 2000; Tearney et al., 1997). Another important aim of this miniaturization effort is to develop catheter-based imaging inside a human body.

7.17   SOME COMMERCIAL SOURCES OF IMAGING INSTRUMENTS

Confocal Microscopy:

Biorad:	http://www.biorad.com
Leica Microsystems, Inc.:	http://www.llt.de/
Nikon:	http://www.nikonusa.com/
Olympus:	http://www.olympusamerica.com/seg_section/seg_confocal.asp
Optiscan Inc.:	http://www.optiscan.com/
Carl Zeiss Inc.:	http://www.zeiss.de/us/micro/home.nsf

Optical Coherence Tomography:

LightLab Imaging:	http://www.lightlabimaging.com/
Advanced Ophthalmic Devices:	http://www.humphrey.com/Systems/prod&sol.html

Near Field Imaging:

Thermomicroscopes:	http://www.tmmicro.com/
Nanonics Imaging Ltd.:	http://www.nanonics.co.il
WITec Wissenschaftliche:	http://www.witec.de
Triple-O Microscopy GmbH:	http://www.triple-o.de

HIGHLIGHTS OF THE CHAPTER

- Optical imaging utilizes spatial variation in the optical properties such as transmission, reflection, scattering, and fluorescence of a cell, a tissue, an organ, or a living object to generate an optical contrast for obtaining an optical image of the specimen.
- The light transmitted through a tissue, which is a highly scattering medium, is comprised of three components: unscattered ballistic photons, weakly scattered snake photons, and multiply scattered diffuse photons.
- Information about the internal structure of a tissue is carried by ballistic and snakes photons. They are discriminated from the diffuse photons by spatial filtering, polarization-gating and time-gating, and frequency domain methods and are then used to obtain images.
- A transmission microscope utilizes the spatial variation of absorption and scattering in a tissue to obtain images. A common example is a compound microscope.
- The most common illumination method for transmission microscopy is Kohler illumination, which uses a specialized optical arrangement.
- The magnification, M, of a microscope is defined as the ratio of the image and the object dimensions.
- The numerical aperture, NA, is related to the cone of the angle θ and the refractive index n of the medium from which light enters the objective lens as $NA = n \sin \theta$.
- The resolution of a microscope, defined as the minimum resolvable distance between two adjacent spots, is determined by the numerical aperture according to $d = 1.22(\lambda/2NA)$.
- An objective lens with a higher NA also produces higher magnification, but provides shorter working distance between the specimen and the objective lens due to a tighter focus.
- The two types of distortions, called *optical aberrations*, encountered in optical imaging are (i) spherical aberrations, related to the shape of the lens in which rays refracted from the periphery of a lens do not focus at the same spot as those from near the center, and (ii) chromatic aberrations in which light of different wavelengths focus at different spots.
- Specially designed aspherical and achromatic lenses are commercially available which minimize the distortions.
- Phase-contrast microscopy utilizes changes in the phase of transmitted light, introduced by the biological sample, to obtain the image.
- Dark-field microscopy utilizes an angle of sample illumination at which only the highly diffracting structures can be imaged.
- Differential interference contrast microscopy (DIC) is based on optical interference techniques to convert the optical path difference, traveled by light passing through different parts of a specimen, to differences in intensities.

- Fluorescence microscopy utilizes either endogeneous fluorescence (autofluorescence) or fluorescence of an exogenous labeling (staining) fluorophore to obtain the image.
- Fluorescence microscopy provides the opportunity to utilize multiparameter control for imaging by using excitation wavelength, emission, lifetime and polarization selectivity. It is the most widely used method for bioimaging.
- Scanning microscopy which constructs an image using a serial, point-by-point illumination of the object provides the benefit of improved resolution with relatively low photodamage of the sample.
- Confocal microscopy is a popular imaging method, which utilizes a confocal aperture, such as a pinhole, to reduce the out-of-focus light from reaching the detector. It thus provides enhanced contrast and also the ability to obtain depth discrimination for three-dimensional imaging.
- Two-photon laser scanning microscopy (TPLSM) is gaining wide acceptance for fluoresence imaging. Here a two-photon excitation of the fluorophore using a near-IR pulsed laser source provides greater penetration in the tissue, more spatial localization, and less complication due to autofluorescence.
- Optical coherence tomography (OCT) is a refection imaging technique that utilizes back-scattered light from a tissue. Improved sensitivity in OCT is achieved by using interference between the back-scattered light and a reference beam.
- For a highly scattering dense medium, two-photon laser scanning microscopy and OCT are preferred techniques, with OCT having the advantage that no fluorescence labeling is required, but TPLSM generally provides better resolution.
- Total internal reflection fluorescence (TIRF) microscopy utilizes fluorescence excitation of the specimen, deposited on a solid surface, by an evanescent wave. This evanescent wave extends from the solid surface, when light is propagated on to the solid surface at a critical angle of total internal reflection. TIRF provides enhanced, sensitivity e to image and probes a cellular environment close to the solid surface.
- Near-field scanning optical microscopy (NSOM or SNOM) uses a tapered and metal-coated optical fiber with an optical opening of ~50 nm at the tip to excite a specimen and/or collect the transmitted, reflected, or fluorescence light signal, thus providing a resolution of <100 nm.
- Spectral imaging obtains information on spatial variation of spectra and provides information on the molecular mechanism of a biological function, a drug–organelle interaction, and so on.
- Fluorescence resonance energy transfer (FRET) imaging utilizes the ratio of fluorescence of an energy acceptor to that of the energy donor which is excited by light absorption and then transfers the excitation energy to the acceptor by the Forster energy transfer mechanism.

- FRET is useful in probing the interaction between cellular components such as protein–protein interactions or drug–binding-cell interactions.
- Fluorescence lifetime imaging (FLIM) maps the spatial distribution of the fluorescence lifetime. It serves as a sensitive probe of the local environment of a fluorophore and, thus, for the interactions and dynamics in a biological system.
- The nonlinear optical techniques of second-harmonic generation (SHG) and third-harmonic generator (THG) provide high spatial selectivity. Second-harmonic microscopy is very selective to interfaces and thus very suitable to probe interactions and dynamics at a membrane interface.
- The nonlinear optical techniques of coherent anti-Stokes Raman scattering (CARS) microscopy utilizes imaging by a coherent photon output generated at $2\nu_p - \nu_s$. The incident beams are at frequencies ν_p and ν_s, under the condition that $\nu_p - \nu_s$ corresponds to a Raman vibrational frequency of the molecule under illumination.
- CARS provides chemical information on the imaged region by mapping the spatial distribution of Raman vibrational spectra.

REFERENCES

Abbe, E., Beitrage zur Theorie des Mikroskops der microskopischen Wahrnehmung, *Schultzes Arch. Mikr. Anat.* **9**, 413–468 (1873).

Abramowitz, M., *Microscope Basics and Beyond*, Olympus Corporation Publishing, New York, 1987a.

Abramowitz, M., *Contrast Methods in Microscopy: Transmitted Light*, Olympus Corporation Publishing, New York, 1987b.

Abramowitz, M., *Fluorescence Microscopy: The Essentials*, Olympus America, New York, 1993.

Amos, W. B., White, J. G., Fordham, M., Use of Confocal Imaging in the Study of Biological Structures, *Appl. Opt.* **26**, 3239–3243 (1987).

Ash, E. A., Nicholls, G., Super-resolution aperture scanning microscope, *Nature* **237**, 510–513 (1972).

Axelrod, D., Total Internal Reflection Fluorescence Microscopy in A., Periasamy, ed. *Methods in Cellular Imaging*, Oxford University Press, Hong Kong, 2001, pp. 362–380.

Bastiaens, P. I. H., and Squire, A., Fluorescence Lifetime Imaging Microscopy: Spatial Resolution of Biochemical Processes in the Cell, *Trends Cell Biol.* **9**, 48–52 (1999).

Berkovic, G., Shen, Y. R., and Prasad, P. N., Third Harmonic Generation from Monolayer Films of a Conjugated Polymer, Poly-4-BCMV, *J. Chem. Phys.* **87**, 1897–1898 (1987).

Berkovic, G., Superfine, R., Guyot-Sinnoset, P., Shen, Y. R., and Prasad, P. N., A Study of Diacetylene Monomer and Polymer Monolayers Using Second- and Third-Harmonic Generation, *J. Opt. Soc. Am.* **135**, 668–673 (1988).

Betzig, E., and Trautman, J. K., Near-field Optics: Microscopy, Spectroscopy, and Surface Modification Beyond the Diffraction Limit, *Science* **257**, 189–195 (1992).

Bhawalkar, J. D., Swiatkiewicz, J., Prasad, P. N., Pan, S. J., Shih, A., Samarabandu, J. K., Cheng, P. C., and Reinhardt, B. A., Nondestructive Evaluation of Polymeric Paints and Coatings Using Two-Photon Laser Scanning Confocal Microscopy, *Polymer* **38**, 4551–4555 (1997).

Boisde, G., and Harmer, A., "Chemical And Biochemical Sensing With Optical Fibers And Waveguides", Artech House, Bosston-London, 1996.

Born, M., and Wolf, E., *Principles of Optics, 7th Edition*, Cambridge University Press, Cambridge, 1999.

Bouma, B. E., Tearney, G. J., Compton, C. C., and Nishioka, N. S., High-Resolution Imaging of the Human Esophagus and Stomach *In Vivo* Using Optical Coherence Tomography, *Gastrointest. Endosc.* **51**, 467–474 (2000).

Bouevitch, O., Lewis, A., Pinevsky, L., Wuskell, J. P., Loew, L. M., Probing Membrane-Potential with Nonlinear Optics. *Biophys. J.,* **65**, 672–679 (1993).

Bradbury, S., *An Introduction to the Optical Microscope*, RMS Handbook No.1, Oxford Scientific Publication, Oxford, 1989.

Brakenhoff, G. J., Imaging Modes of Confocal Scanning Microscopy, *J. Microsc.* **117**, 233–242 (1979).

Brakenhoff, G. J., Blom, P., and Barends, P., Confocal Light Microscopy with High-Aperture Immersion Lenses, *J. Microsc.* **117**, 219–232 (1979).

Brezinski, M. E., and Fujimoto, J. G., Optical Coherence Tomography: High-Resolution Imaging in Nontransparent Tissue, *IEEE J. Selected Topics in Quantum Electronics* **5**, 1185–1192 (1999).

Cheng, J.-X., Volkmer, A., Book, L. D., and Xie, X. S., An Epi-Detected Coherent Anti-Stokes Raman Scattering (E-CARS) Microscope with High Spectral Resolution and High Sensitivity, *J. Phys. Chem. B* **105**, 1277–1280 (2001).

Colston, Jr., B. W., Everett, J. M., Da Silva, L. B., Otis, L. L., Stroeve, P., and Nathel, H., Imaging of Hard- and Soft-tissue Structure in the Oral Cavity by Optical Coherence Tomography. *Appl. Opt.* **37**, 3582–3585, (1998).

Davidson, M. W., and Abramowitz, M., *Optical Microscopy*, Olympus America, New York, 1999.

Denk, W., Strickler, J. H., and Webb, W. W., "2-Photon Laser Scanning Fluorescence Microscopy, *Science* **248**, 73–76 (1990).

Denk, W., Piston, D., and Webb, W., Two-photon molecular excitation in Laser Scanning Microscopy, J. Pawley, ed., *Handbook of Biological Confocal Microscopy*, 2nd edition, Plenum Press, New York, 1995.

Diaspro, A., Corosu, M., Ramoino, P., and Robello, M., Adapting a Compact Confocal Microscope System to a Two-Photon Excitation Fluorescence Imaging Architecture, *Microsc. Res. Tech.* **47**, 196–205 (1999).

Diaspro, A., ed., *Confocal and Two-Photon Microscopy: Foundations Applications, and Advances*, John Wiley & Sons, New York, 2002.

Dickensheets, D. L., and Kino, G. S., Silicon-Micromachined Scanning Confocal Optical Microscope, *J. Microelectromech. Syst.* **7**, 38–47 (1998).

Dickson, R. M., Norris, D. J., Tzeng, Y.-L., and Moerner, W. E., Three-Dimensional Imaging of Single Molecules Solvated in Pores of Poly(acrylamide) Gels, *Science* **274**, 966–969 (1996).

Duncan, M. D., Reintjes, J., and Manuccia. T. J. Scanning Coherent Anti-Stokes Raman Microscope, *Opt. Lett.* **7**, 350–352 (1982).

Egger, M. D., and Petran, M., New Reflected Light Microscope for Viewing Unstained Brain and Ganglion Cells, *Science* **157**, 305–307 (1967).

Farkas, D. L., Spectral Microscopy for Quantitative Cell and Tissue Imaging in A. Periasamy, ed., *Methods in Cellular Imaging*, Oxford University Press, Hong Kong, 2001, pp. 345–361.

Fine, S., and Hansen, W .P., Optical Second Harmonic Generation in Biological Systems, *Appl. Opt.* **10**, 2350 (1971).

French, T., So, P. T. C., Weaver, D. J., Coehlo-Sampaio, T., Gratton, E., Voss, E. W., and Carrero, J., Two-Photon Fluorescence Lifetime Imaging Microscopy of Macrophage Mediated Antigen Processing, *J. Microsc.* **185**, 339–353 (1997).

Freund, I., Deutsch, M., and Sprecher, A., Connective Tissue Polarity-Optical 2nd Harmonic Microscopy, Crossed-Beam Summation, and Small Angle Scattering in Rat-Tail Tendon, *Biophys. J.* **50**, 69 (1986).

Gayen, S. K., and Alfano, R. R., Emerging Biomedical Imaging Techniques, Optics and Photonics News (1996).

Gheber, L. A., Hwang, J., and Edidin, M., Design and Optimization of a Near-Field Scanning Optical Microscope for Imaging Biological Samples in Liquid, *Appl. Opt.* **37**, 3574–3581 (1998).

Gryczynski, I., Gryczynski, Z., and Lakowicz, J. R., Two-Photon Excitation by the Evanescent Wave from Total Internal Reflection, *Anal. Biochem.* **247**, 69–76 (1997).

Gu, M., *Principles of Three-Dimensional Imaging in Confocal Microscopes*, World Scientific, Singapore, 1996.

Guo, Y., Ho, P. P., Tirkisliunas, A., Liu, F., and Alfano, R. R., Optical Harmonic Generation from Animal Tissues by the Use of Picosecond and Femtosecond Laser Pulses, *Appl. Opt.* **35**, 6810 (1996).

Guo, Y., Ho, P. P., Savage, H., Harris, D., Salks, P., Schantz, S., Liu, F., Zhadin, N., and Alfano, R. R., Second Harmonic Tomography of Tissues, *Opt. Lett.* **22**, 1323–1325 (1997).

Ha, T. J., Ting, A. Y., Liang, J., Caldwell, W. B., Deniz, A. A., Chemla, D. S., Schultz, P. G., and Weiss, S., Single-Molecule Fluorescence Spectroscopy of Enzyme Conformational Dynamics and Cleavage Mechanism, *Proc. Natl. Acad. Sci. USA* **96**, 893–898 (1999).

Hashimoto, M., Araki, T., and Kawata, S., Molecular Vibration Imaging in the Fingerprinting Region by Use of Coherent Anti-Stokes Raman Scattering Microscopy with a Collinear Configuration, *Opt. Lett.* **25**, 1768–1770 (2000).

He, G. S., Markowicz, P. P., Lin, T. C., and Prasad, P. N., Observation Of Stimulated Emission By Direct Three-Photon Excitation, *Nature* **415**, 767–770 (2002).

Heinzelmann, H., and Pohl, D. W., Scanning Near-Field Optical Microscopy, *Appl. Phys. A* **59**, 89–101 (1994).

Hell, S., and Stelzer, E. H. K., Properties of a 4PI Confocal Fluorescence Microscope, *J. Opt. Soc. Am.* 2159–2166 (1992).

Herman, B., Gordon, G., Mahajan, N., and Centonze, V., Measurement of Fluorescence Resonance Energy Transfer in the Optical Microscope, in A., Periasamy, ed., *Methods in Cellular Imaging*, Oxford University Press, Hong Kong, 2001, pp. 257–272.

Huang, D., Swanson, E. A., Lin, C. P., Shuman, J. S., Stinson, W. G., Chang, W., Hee, M. R., Flotte, T., Gregory, K., Puliato, C. A., and Fujimoto, J. G., Optical Coherence Tomography, *Science*, **254**, 1178–1181 (1991).

Hwang, J., Gheber, L. A., Margolis, L., and Edidin, M., Domains in Cell Plasma Membranes Investigated by Near-field Scanning Optical Microscopy, *Biophys. J.* **74**, 2184–2190 (1998).

König, K., Multiphoton Microscopy in Life Sciences, *Journal of Microscopy* **200**, 83–104 (2000).

Lacey, A. J., ed., *Light Microscopy in Biology—A Practical Approach*, 2nd edition, Oxford University Press, Oxford, 1999.

Lagerholm, B. C., Starr, T. E., Volovyk, Z. N., and Thompson, N. L., Rebinding of IgE Fabs at Haptenated Planar Membranes: Measurement by Total Internal Reflection with Fluorescence Photobleaching Recovery, *Biochemistry* **39**, 2042–2051 (2000).

Lakowicz, J. R., and Berndt, K., Frequency Domain Measurements of Photon Migration in Tissues, Chem. *Phys. Lett.* **166**, 246–252 (1990).

Lakowicz, J. R., Szymanski, H., and Nowaczyk, K., Fluorescence Lifetime Imaging of Calcium Using Quin-2, *Cell Calcium* **13**, 131–147 (1992).

Lewis, A., Khatchatouriants, A., Treinin, M., Chen, Z., Peleg, G., Friedman, N., Bouevitch, O., Rothman, Z., Loew, L., and Sheves, M., Second Harmonic Generation of Biological Interfaces: Probing the Membrane Protein Bacteriorhodopsin and Imaging Membrane Potential Around GFP Molecules at Specific Sites in Neuronal Cells of C. Elegans, *Chemical Physics* **245**, 133 (1999).

Maiti, S., Shear, J. B., Williams, R. M., Zipfel, W. R., and Webb, W. W., Measuring Serotonin Distribution in Live Cells with Three-Photon Excitation, *Science* **275**, 530–532 (1997).

Martínez-Corral, M., Effective Axial Resolution in Single-Photon 4Pi Microscopy, *Imaging and Microscopy* **4**, 29–31 (2002).

Masters, B. R., So, P. T. C., and Gratton, Multiphoton Excitation Fluorescence Microscopy and Spectroscopy of *In Vivo* Human Skin, *Biophys. J.* **72**, 2405–2412 (1997).

McKierman, A. M., MacDonald, R. C., MacDonald, R. I., and Axelrod, D., Cytoskeletal Protein Binding Kinetics at Planar Phospholipid Membranes, *Biophys. J.* **73**, 1987–1998 (1997).

Mertz, J., Nonlinear Microscopy, *C.R. Acad. Sci. Pants. T. 2 Series IV* 1153–1160 (2001).

Minsky, M., Microscopy Apparatus, United States patent 3013467, December 19, 1961 (filed November 7, 1957).

Moreaux, L., Sandre, O., Charpak, S., Blanchard-Depree, M., and Mertz, J., Coherent Scattering in Multi-Harmonic Light Microscopy, *J. Biophys. J.* **80**, 1568–1574 (2001).

Müller, M., Squier, J., De Lange, C. A., and Brakenhoff, G. J., CARS Microscopy with Folded BoxCARS Phasematching, *J. Microsc.* **197**, 150–158 (2000).

Müller, M., Sqier, J., Wilson, K. R., and Brakenhoff, G. J., 3-D-Microscopy of Transparent Objects Using Third-Harmonic Generation, *J. Microsc.* **191**, 266–274 (1998).

Nagorni, M., and Hel, S., Coherent Use of Opposing Lenses for Axial Resolution Increase in Fluoresence Microscopy, *J. Opt. Soc. Am.* **A18**, 36–48 (2001).

Pawley, J. B., ed., *Handbook of Confocal Microscopy*, 2nd edition, Plenum Press, New York, 1995.

Pawslear, M. A., and Moyer, P., *Near-Field Optics*: *Therapy, Instrumentation, and Applications*, Wiley, New York, 1996.

Peleg, G., Lewis, A., Linial, M., and Loew, L. M., Non-linear Optical Measurement of Membrane Potential Around Single Molecules at Selected Cellular Sites, *Proc. Acad. Sci.* **96**, 6700–6704 (1999).

Periasamy, A., Fluorescence Resonance Energy Transfer Microscopy. A Mini Review, *J. Biomed. Opt.* **6**, 287–291 (2001).

Periasamy, A., Wodnicki, P., Wang, X. F., Kwon, S., Gordon, G. W., and Herman, B., Time-Resolved Fluorescence Lifetime Imaging Microscopy Using a Picosecond Pulsed Tunable Dye-Laser System, *Rev. Sci. Instrum.* **67**, 3722–3731 (1996).

Petran, M., Hadravsky, M., Egger, M. D., and Galambos, R., Tandem Scanning Reflected-Light Microscope, *J. Opt. Soc. Am.* **58**, 661–664 (1968).

Petran, M., Hadravsky, M., and Boyde, A., The Tandem Scanning Reflected Light Microscope, *Scanning* **7**, 97–108 (1985).

Pohl, D. W., Denk, W., and Lanz, M., Optical Stethoscopy: Image Recording with Resolution λ/20, *Appl. Phys. Lett.* **44**, 651–653 (1984).

Potma, E. O., deBoij, W. P., and Wiersma, D. A., Nonlinear Coherent Four-Wave Mixing in Optical Microscopy, *J. Opt. Soc. Am B.* **17**, 1678–1684 (2000).

Prasad, P. N., and Williams, D. J., *Introduction to Nonlinear Optical Effects in Molecules and Polymers*, John Wiley & Sons, New York, 1991.

Pudavar, H. E., Kapoor, R., Wang, X., and Prasad, P. N., "Multi-Photon/confocal Localized Spectrometer: Applications in Biology and Material Science", Symposium on Bio-photonics and Nano Medicine, Buffalo, NY (2000).

Sako, Y., Miniguchi, S., and Yanagida, T., Single-Molecule Imaging of EGFR Signaling on the Surface of Living Cells, *Nature Cell Biol.* **2**, 168–172 (2000).

Schmitt, J. M., Optical Coherence Tomography (OCT): A Review, *IEEE J. Sel. Top. Quantum Electron.* **5**, 1205–1215 (1999).

Schrader, M., Bahlmann, K., and Hell, S. W., Three-Photon Excitation Microscopy: Theory, Experiment and Applications, *Optik* **104**, 116–124 (1997).

Shen, Y., Friend, C. S., Jiang, Y., Jakubczyk, D., Swiatkiewicz, J., and Prasad, P. N., Nanophotonics: Interactions, Materials and Applications, *J. Phys. Chem. B* **104**, 7577–7587 (2000).

Sheppard, C. J. R., Gannway, J. N., Walsh, D., and Wilson, T., Scanning Optical Microscopy for the Inspection of Electronic Devices, presented at Microcircuit Engineering Conference, Cambridge, 1978.

Smith, S., Cheong, H. M., Fluegel, B. D., Geisz, J. F., Olson, J. M., Kazmerski, L. L., and Mascarenhas, A., Spatially Resolved Photoluminescence in Partially Ordered GaInP$_2$, *Appl. Phys. Lett.* (1998).

Subramaniam, V., Kirsch, A. K., Jenei, A., and Jovin, T. M., Scanning Near-Field Optical Imaging and Spectroscopy in Cell Biology, in Gary Durack and J. Paul Robinson, eds., *Emerging Tools for Cell Analysis: Advances in Optical Measurement*, 2000, pp. 271–290.

Sutherland, R. M., Dahne, C., Place, J. F., and Ringrose, A. R., Immunoassays at a Quartz-liquid Interface: Theory, Instrumentation and Preliminary Application to the Fluorescent Immunoassay of Human Immunoglobulin G., *J. Immunol. Methods* **74**, 253–265 (1984).

Sund, S. E., Swanson, J. A., and Axelrod, D., Cell Membrane Orientation Visualized by Polarized Total Internal Reflection Fluorescence, *Biophys. J.* **77**, 2266–2283 (1999).

Swanson, E. A., Izatt, J. A., Hee, M. R., Huang, D., Lin, C. P., Schuman, J. S., Puliafito, C. A., and Fujimoto, J. G., In vivo Retinal Imaging by Optical Coherence Tomography. *Opt Lett.* **18**, 1864–1866 (1993).

Synge, E. H., A Suggested Method for Extending Microscopic Resolution into the Ultra-microscopic Region, *Philos. Mag.* **6**, 356–362 (1928).

Tadrous, P. J., Methods for Imaging the Structure and Function of Living Tissues and Cells: Fluorescence Lifetime Imaging, *J. Pathol.* **191**, 229–234 (2000).

Tearney, G. J., and Bouma, B. E., *Handbook of Optical Coherence Tomograpy*, Marcel Dekker, New York, 2001.

Tearney, G. J., Brezinski, M. E., and Bouma, B. E., *In Vivo* Endoscopic Optical Biopsy with Optical Coherence Tomography, *Science* **276**, 2037–2039 (1997).

Vale, R. D., Fanatsu, T., Pierce, D. W., Romberg, L., Harula, Y., and Yanagida, T., Direct Observation of Single Kinesin Molecules Moving Along Microtubules, *Nature* **380**, 451–453 (1996).

Van Hulst, N. F., Nearfield Optical Microscopy, in A. J. Lacey, ed., *Light Microscopy in Biology*, 2nd edition, Oxford University Press, New York, 1999.

Verneer, P., Squire, A., and Bastiaens, P. I. H., Frequency-Domain Fluorescence Lifetime Imaging Microscopy: A Window on the Biochemical Landscape of the Cell, in A. Periasamy, ed., *Methods in Cellular Imaging*, Oxford University Press, Hong Kong, 2001, pp. 273–294.

Volker, A., Cheng, J.-X., and Xie, X. S., Vibrational Imaging with High Sensitivity via Epidetected Coherent Anti-Stokes Raman Scattering Microscopy, *Phys. Rev. Lett.* **87**, 0239013–0239014 (2001).

Wachman, E. S., Niu, W.-H., and Farkas, D. L., AOTF Microscope for Imaging with Increased Speed and Spectral Versatility, *Biophys. J.* **73**, 1215–1222 (1997).

Wang, X., Pudavar, H. E., Kapoor, R., Krebs, L. J., Bergey, E. J., Liebow, C., Prasad, P. N., Nagy, A., and Schally, A. V., Studies on the Mechanism of Action of a Targeted Chemotherapeutic Drug in Living Cancer Cells by Two Photon Laser Scanning Microspectrofluorometry, *J. Biomed. Optics* **6**, 319–325 (2001).

Wilson, T., Gannaway, J. N., and Johnson, P., A Scanning Optical Microscope for the Inspection of Semiconductor Materials and Devices, *J. Microsco.* **118**, 390–394 (1980).

Wokesin, D. L., Centonze, V. E., Crittenden, S., and White, J., Three-Photon Excitation Fluorescence Imaging of Biological Specimens Using an All-Solid-State Laser, *Bioimaging* **4**, 208–214 (1996).

Xu, F. M., Pudavar, H. E., Prasad, P. N., and Dickensheets, D., Confocal Enhanced Optical Coherence Tomography for Nondestructive Evaluation of Paints and Coatings, *Opt. Lett.* **24**, 1808–1810 (1999).

Yelin, D., and Sieberberg, Y., Laser Scanning Third-Harmonic Generation Microscopy in Biology, *Opt. Express* **5**, 169–175 (1999).

Zernike, G. F., Phase Contract, A New Method for the Microscopic Observation of Transparent Objects, *Physica* **7**, 9 (1942).

Zumbusch, A., Holtom, G. R., and Xie, X. S., Three-Dimensional Vibrational Imaging by Coherent Anti-Stokes Raman Scattering, *Phys. Rev. Lett.* **82**, 4142–4145 (1999).

CHAPTER 8

Bioimaging: Applications

This chapter provides some examples of wide usage of optical bioimaging to investigate structures and functions of cells and tissues and to profile diseases at cellular, tissue, and *in vivo* specimen levels. For solely convenience purposes, wherever possible, examples presented here are from the work conducted at the Institute for Lasers, Photonics, and Biophotonics. This selection in no way minimizes extensive studies being conducted elsewhere and seminal contributions made by numerous research groups worldwide.

As described in Chapter 7, a broad range of optical methods and microscopic techniques are used for bioimaging. However, the contents of this chapter are more selective, focusing on fluorescence and optical coherence tomographic techniques. These techniques are, generally, the most widely used methods and will continue to receive the most attention for bioimaging. Fluorescence microscopy—in the form of confocal microscopy and, more recently, a variation of it, two-photon laser scanning microscopy—is the preferred method for cellular imaging. It has also shown promise for *in vivo* and *ex vivo* tissue imaging.

For fluorescence imaging, the use of exogenous fluorophores as labeling (staining) agents is widespread. Imaging, using the intrinsic autofluorescence, has also, to a limited extent, shown promise. This chapter starts with a discussion of the fluorophore characteristics needed for bioimaging. The criteria to select fluorophore for a particular application are presented. Developments in new fluorophores, particularly two-photon materials, are also presented.

Examples of bioimaging follow in subsequent sections. The organization of these sections is based on classification at the various levels of an organism. The first application discussed is cellular imaging where optical imaging to probe structures and functions of viruses, bacteria, and eukaryotic cells is presented. Then imaging at the tissue level is presented. Finally, *in vivo* imaging is discussed. This chapter concludes with a discussion of future directions for bioimaging.

For supplemental reading, books edited by Periasamy (2001) and Diaspro (2002) are useful. The reader is also encouraged to read the periodic publica-

Introduction to Biophotonics, by Paras N. Prasad.
ISBN: 0-471-28770-9   Copyright © 2003 John Wiley & Sons, Inc.

tions *Biophotonics International* and *Microscopy*, which feature highlights of new reports on bioimaging applications.

8.1 FLUOROPHORES AS BIOIMAGING PROBES

Both endogenous and exogenous fluorophores have been used for bioimaging. This section provides a description of some of these fluorophores.

8.1.1 Endogenous Fluorophores

A number of endogenous fluorophores (e.g., NADH, flavins) producing autofluorescence have proved extremely useful for bioimaging. While using exogenous fluorophores, the autofluorescence produced by the endogenous fluorophores gives rise to undesirable background. But in some cases, endogenous fluorophores prove to be useful for monitoring cellular processes. Some of them also exhibit sufficiently strong two-photon excited emission, making them useful for cellular bioimaging.

Some of the most common endogenous fluorophores are listed below. A detailed compilation of autofluorescent cellular components (endogenous fluorophores) is provided in a review by Billinton and Knight (2001).

Flavins: Flavins are derivatives of riboflavin (vitamin B_2), the most common of them being flavin mononucleotide (FMN) and flavin adenine dinucleotide (FAD). FMN, with its emission maxima at 530 nm, exhibits much brighter fluorescence efficiency than FAD. Intracellular riboflavin, flavin coenzymes, and flavoproteins show slightly shifted fluorescence (540–560 nm) compared to flavins (Billinton et al., 2001).

NAD(P)H: In the reduced protonated form, nicotinamide-adenine dinucleotide (NADH) and nicotinamide-adenine dinucleotide phosphate (NADPH) are fluorescent coenzymes, like FAD and FMN, and are crucial to many reactions in most types of cells. Together, they are often referred to as NAD(P)H. The excitation maximum of free NAD(P)H occurs at 360 nm with emission at 460 nm, while the protein-bound coenzyme emits around 440 nm. The fluorescence intensity due to NAD(P)H is reported to be between 50 and 100 times that of flavins (Aubin, 1979).

Lipofuscin: Lipofuscin is also called "age pigment" since its strong autofluorescence is seen more frequently with increasing age, in the cytoplasm of postmitotic cells. The excitation maximum of lipofuscin fluorescence is in the UV range of 330–390 nm. The emission maximum varies from blue (420 nm) to an orange–yellow color (540–560 nm), depending on its environment.

Elastin and Collagen: Structural proteins like collagen and elastin show green–yellow fluorescence with excitation around 480 nm.

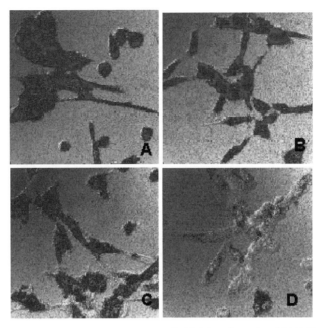

Figure 8.1. Confocal fluorescence images of RIF-1 cells using NADH emission at 426–454 nm when excited at 351 nm. The four images shown are (A) control without photosensitizer or light, (B) control with light irradiation but no photosensitizer, (C) control with photosensitizer but no light, and (D) cells with photosensitizer and irradiated with light. (Reproduced with permission from Pogue et al., 2001.) (See color figure.)

There are many other components like AGE (advanced glycation end-products), protoporphyrin, and lignin which are also classified as endogenous fluorophores. However, their fluorescence generally overlaps with those of different fluorescence stains used for bioimaging, and therefore are not as widely used.

An example of the use of an endogenous fluorophore used for imaging is provided by *in vivo* NADH fluorescence, monitored as an assay for the cellular damage in photodynamic therapy (Pouge et al., 2001). Photodynamic therapy is a cancer treatement, activated by light absorption in a photosensitizer, as will be discussed in Chapter 12. Figure 8.1 shows the confocal fluorescence images of NADH using emission at 426–454 nm from RIF-1 cells (fibrosarcoma tumor cells) when excited at 351 nm. A decrease in the magnitude of the fluorescence signal response to the increasing probe energy delivered, as seen by comparing image (D) with others, was used to monitor cell destruction by the action of photodynamic therapy.

8.1.2 Exogenous Fluorophores

There are many biological structures and processes which cannot be imaged or probed by using intrinsic fluorescence due to endogenous fluorophores.

For example, DNA does not exhibit any fluorescence. In these cases, labeling of these biological structures with exogenous fluorophores is needed for bioimaging. Some fluorophores can stain (disperse) throughout the cell, while others localize in a particular organelle. In some cases, a fluorophore can be chemically conjugated to derive the feature of targeting a specific biological site, a particular type of cancer cell, or an individual organelle in a cell. Thus, the specificity and sensitivity of a fluorescence probe can be used to derive valuable structural, biochemical, and biophysical information on cells and tissues (Harper, 2001).

In this section, selection criteria for exogenous fluorophores for bioimaging are discussed and examples of fluorophores used for targeting various organelles are presented. A more comprehensive discussion of currently available fluorophores for bioimaging can be found in the book *Principles of Fluorescence Spectroscopy* (Lakowicz, 1999), in the Handbook of Fluorescent Probes and Research Products (Haugland, 2002) and from the website of a company, Molecular Probes, which commercializes fluorophores (see Section 8.9). The fluorophores (fluorochromes), useful for flow cytometry, are also discussed in Chapter 11.

For bioimaging, the basic requirements for an ideal fluorophore are as follows:

- Dispersability (solubility) of the fluorophore in the biological medium to be probed
- Specific association with a target molecule, organelle, or cell
- High quantum efficiency of emission
- Environmental stability
- Absence of photobleaching

Photobleaching is a major problem with many organic fluorophores. Photobleaching generally refers to chemical degradation of a fluorophore, leading to the disappearance of fluorescence. This photodegradation may result from photochemistry in the excited state, photooxidation in the presence of oxygen, or thermal decomposition due to local heating by nonradiative processes following light absorption.

A practical consideration in the selection of a fluorophore is that it should be efficiently excited at the wavelengths of the common laser sources available for microscopy. Furthermore, the emission wavelength of the fluorophore should be compatible with the emission filters on the microscope.

Some situations may require multiple labeling with more than one fluorophore to simultaneously probe different target molecules/organelles. In such cases, an additional consideration is that these fluorescence spectra (peaks) be sufficiently separated so that appropriate cutoff (or narrow bandpass) filters can be used to discriminate their emission from each other. In some cases, such as in FRET imaging (see Section 8.5.5), it may be desirable that their excitation peaks are also well-separated, so that excitation wavelengths can be judiciously selected to excite only one fluorophore.

A wide array of fluorescence probes for most imaging applications as well as light sources most suitable to excite them are commercially available. Some of the commonly used fluorophores for bioimaging and their applications are shown in Table 8.1. The wavelengths of their one-photon excitation and resulting emission maxima are listed, along with convenient light sources used to excite them.

Using fluorescence labeling, bioimaging of a specific organelle or a site in a cell can be accomplished to study its structure and function. Fluorescent probes used for labeling specific sites and organelles in a cell can be divided into two categories:

1. Fluorophores targeting biological molecules, sites, or organelles without any prior coupling to a biomolecule. For example, some fluorophores in their commercially available form show selective staining of specific organelles. These features are also listed in Table 8.1, wherever applicable. Hence they can be conveniently used to probe the structure of an organelle and various biophysical and biochemical processes occurring in them.

2. Fluorophores that need to be conjugated to a biomolecule in order to acquire specificity for certain biological sites.

In the first case, the labeling characteristic is often derived from electrostatic and hydrophilic/hydrophobic interactions of a probe with the biomolecule or organelle of interest.

In the second case, fluorescent probes, which require conjugation with another biomolecule for selective staining, are often used for histological applications. They are chemically conjugated with oligonucleotides or proteins to allow targeting and imaging specific sites in cells. Examples are Alexa Fluor dyes, Cy dyes and Texas red, which are also listed in Table 8.1. In this case, high fluorescence quantum yield, low pH sensitivity, and high photostability are among the necessary demands on the fluorophore. Also, fluorescence of the probe should not be significantly reduced or quenched on conjugation to biological molecules. This approach has been used to obtain a multicolor image of bovine pulmonary artery epithelial cells that were stained with three different Alexa Fluor conjugates (see http://www.iwai-chem.co.jp/products/m-probes/alexa.pdf, a Molecular Probes website).

It is also worth noting that the same probe sometimes can be used for a number of purposes. For example, ethidium bromide, used as an intercalating dye for DNA studies, is also useful for monitoring cell viability, since dye permeation to the cells increases with a decrease in the cell viability.

TABLE 8.1. Some Commonly Used Fluorophores for Bioimaging

Fluorophore	Ex/Em	The Most Convenient Excitation Sources	Application
YOYO-1	491/509	Hg lamp, Ar ion laser (488 nm)	Nucleic acids labeling
Ethidium bromide	545/605	Hg lamp, green He–Ne (543 nm), Ti-sapphire laser (TPE)	DNA labeling, cell apoptosis
Acridine orange	490/530 (640)	Hg lamp, Ar ion laser (488 nm), Ti-sapphire laser (TPE)	Double (single) stranded nucleic acid labeling
Hoechst 33342	355/465	Ar—UV ion laser (351 nm), Hg lamp, Ti-sapphire laser (TPE)	A–T sequences of DNA labeling
DAPI	372/456	Ar—UV ion laser (364 nm), Hg lamp, Ti-sapphire laser (TPE)	A–T sequences of DNA labeling

Structure	Ex/Em	Light source	Application
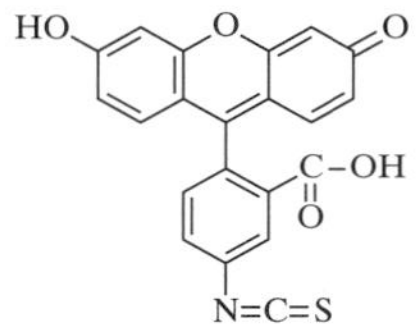 FITC	494/519	Ar ion laser (488 nm), Hg lamp	Conjugation with biomolecules (antibody labeling)
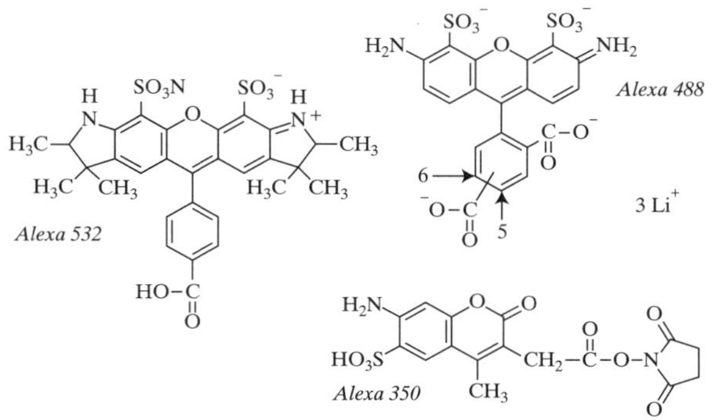 Alexa 488, Alexa 350, Alexa Fluor dyes	346/445	Ar–UV laser (351 nm), Ar laser (488 nm), Nd-YAG laser (532 nm), Hg lamp	Conjugation with biomolecules (antibody labeling)
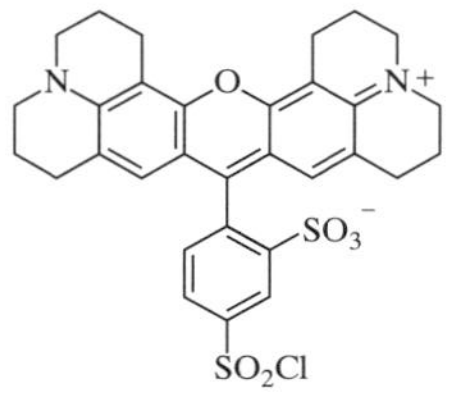 Texas red	596/620	Yellow He–Ne laser (594 nm), Hg lamp	Conjugation with biomolecules (antibody labeling)

Alexa 532

Alexa 488   3 Li⁺

Alexa 350

Alexa Fluor dyes

TABLE 8.1. (Continued)

Fluorophore	Ex/Em	The Most Convenient Excitation Sources	Application
	382/446	Ar–UV ion laser (351, 364 nm), Ar laser (457 nm, 488 nm, 514 nm), green He–Ne laser (543 nm), Hg lamp, Ti–sapphire laser (TPE)	Fluorescent proteins
	434/476		
	475/504		
	488/509		
	514/527		
	558/583		
BFP , CFP , GFP , EGFP , YFP , RFP chromophores			
Rhodamine 123	507/529	Ar laser (488 nm), Hg lamp, Ti–sapphire laser (TPE)	Mitochondria staining apoptosis

Structure	Ex/Em	Light source	Application
$CH_2NHCH_2CH_2N(CH_3)_2$ $CH_2NHCH_2CH_2N(CH_3)_2$ 4 HCl **Lysotracker blue**	373,394/400,422	Ar–UV ion laser (351, 364 nm) Ar laser (488 nm) Hg lamp,	Lysosomes staining
H₃C H₃C $CH_2CH_2-\overset{O}{\overset{\|}{C}}-NHCl$ **Lysotracker green**	503/509		
$CH_3(CH_2)_{12}CH=CHCHOH$ $(CH_2)_4-\overset{O}{\overset{\|}{C}}-\underset{\underset{CH_2OH}{\|}}{NHCH}$ H₃C H₃C **BODIPY FL C5-ceramide**	505/511	Hg lamp, argon laser (488 nm)	Golgi apparatus staining
Cl Cl $5\ NH_4^+$ CH₃ OCH_2CH_2O $N(CH_2CO^-)_2$ $(^-OCH_2)_2N$ **Fluo-3**	503 (506)/ No (526)	Hg lamp, argon laser (488 nm)	Calcium ions indicator
$(CH_3)_2N$ $\overset{}{C}-OH$ $HO-\overset{O}{\overset{\|}{C}}$ 5 6 **Carboxy SNARF-1**	548 (576)/ 587/635	Hg lamp, argon laser (488 nm, 514 nm), Nd–YAG laser (532 nm)	pH indicator

8.1.3   Organometallic Complex Fluorophores

Another group of highly fluorescent probes consists of organometallic complexes such as lanthanide chelates (Lakowicz, 1999). The organometallic fluorophores involve the inner f electrons of the rare-earth ions (lanthanides). The transition that involves an f orbital electron, being symmetry and, sometimes, spin forbidden, exhibits long lifetimes (0.5–3 msec). However, their emission spectra consist of very narrow lines, in contrast to broad emission bands for organics. The organometallic complexes often represent complexation between a central metal ion and the various electron donating organic molecules, called *ligands*. Generally, the excitation in these complexes is provided by the absorption into the organic ligand group, as the intrinsic absorption due to these rare-earth ions is very weak (extinction coefficients <10 L mol^{-1} cm^{-1}). The main advantage provided by rare-earth complexes is that the long lifetime of emission allows one to discriminate it from the generally short-lived (nanoseconds) autofluorescence from biological samples. This can be accomplished by electronic gating of the emission detection to an appropriate time delay following a laser pulse excitation. One can also enhance the sensitivity of detection by integrating the rare-earth emission signal over its long lifetime.

Other types of transition metal–ligand complexes (MLC) useful for bioimaging are those that involve a metal-to-ligand charge-transfer state (lifetime 10 nsec to 70 μsec), discussed in Chapter 2. These complexes are highly stable, with no significant dissociation of the ligands from the metal (Lakowicz, 1999). These MLCs can be conjugated with biomolecules and can also be intercalated in DNA. An example provided here (Figure 8.2) is a ruthenium complex with absorption and emission maxima near 450 nm and 520 nm.

Lanthanides have proved to be useful biological probes because they can substitute for calcium in many calcium-dependent proteins (Martin and

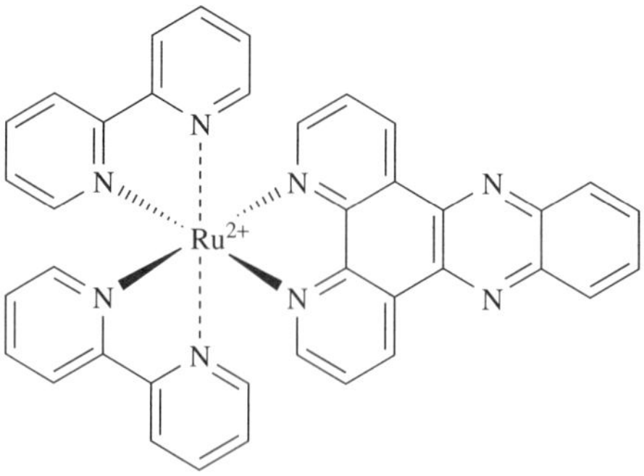

[Ru(bpy)$_2$(dppz)]$^{2+}$

Figure 8.2. Structure of a ruthenium complex; bpy stands for 2,2′-bipyridine, dppz stands for dipyrido(3,2-a:2′,3′-c)phenazine.

Richardson, 1979). The lifetime of their emission is strongly influenced by the surrounding water molecules. Hence the emission lifetime of a lanthanide bound to a protein can be used to calculate the number of bound water molecules around the calcium binding site.

8.1.4 Near-IR and IR Fluorophore

There has been a growing interest in long-wavelength, near-IR (NIR), and IR dyes (Lakowicz, 1999; Gayen et al., 1999). The principal reasons are threefold: (i) These NIR dyes require excitation near IR which produces practically no autofluorescence from any endogenous cellular components. Hence the sensitivity of detection, often limited by the autofluorescence background, is significantly improved; (ii) the longer excitation wavelength and the corresponding near-IR emission also produce reduced scattering in the tissue, and thus increase both the penetration depth and the efficiency of collection of emission; and (iii) commercially available, low-cost, and highly compact red, NIR, and IR diode lasers (e.g., 650 nm, 800 nm, 970 nm, etc.) can be used as convenient excitation sources for these dyes. The most common examples are the various cyanine dyes, often abbreviated as Cy3, Cy5, Cy7, and so on (in the order of increasing wavelengths). Some new dyes are shown in Figure 8.3.

A common problem with many of these dyes is their stability in biological fluids, which consist mainly of water. The ionic dyes tend to aggregate and form specific-type aggregates, such as the J aggregates discussed in Chapter 2. These aggregates have very different spectral characteristics, often resulting in self-quenching of emissions. One method used at our Institute to prevent aggregation and to improve dispersability is to isolate the dyes by encapsulation (in a nanobubble or a liposome) or to use chemical functionalization. Another problem with the NIR and IR dyes is their environmental and photochemical stability. These dyes are known to photobleach more readily than the dyes emitting in the visible.

8.1.5 Two-Photon Fluorophores

Availability of high-peak power pulse laser sources led to pioneering work by a number of groups on multiphoton processes in organic systems (Rentzepis et al., 1970; Fredrich and McClarin, 1980; Birge, 1986). Since the original work of Denk et al. (1990), the field of two-photon laser scanning microscopy has witnessed phenomenal growth. Much of the impetus has been derived from the availability of new fluorophores with considerably enhanced two-photon emission efficiency (Bhawalkar et al., 1996). In a number of cases, the efficiency of two-photon excited emission is so high that one can achieve population inversion and lasing at the up-converted wavelength (He et al., 1995). Figure 8.4 shows a chromophore APSS and its various derivatives, developed at the Insti-

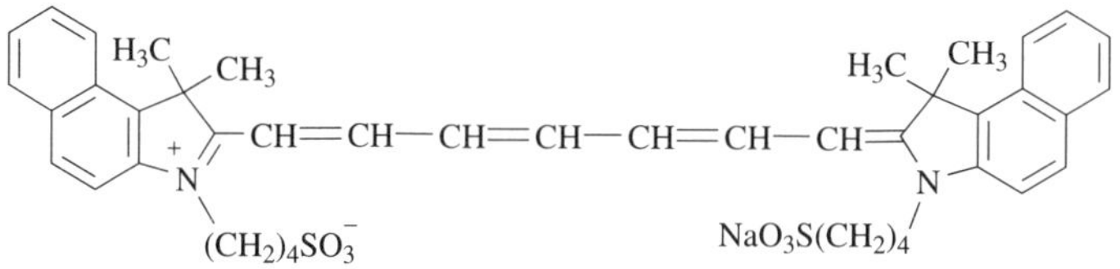

Commercially available indocyanine green: Absorption λ_{max}, 780 nm (water); fluorescence λ_{max}, 805 nm (water)

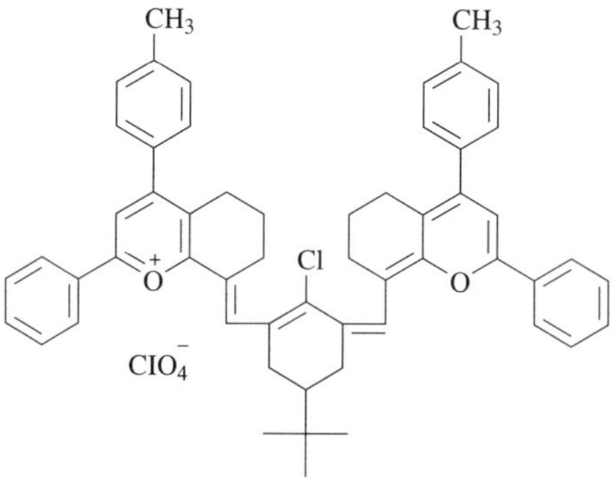

New IR dye[*]: Absorption λ_{max}, 1127 nm (dichloroethane); emission λ_{max}, 1195 nm (dichloroethane)

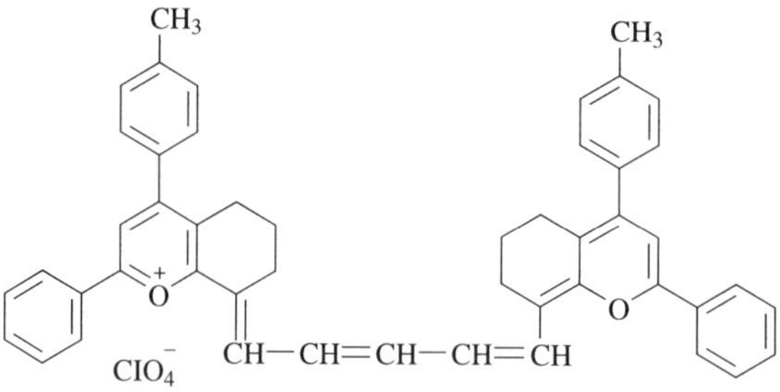

New IR dye[*]: Absorption λ_{max}, 1056 nm (dichloroethane); Emission λ_{max}, 1140 nm (dichloroethane)

Figure 8.3. Some new near-IR and IR dyes. * Developed at the Institute for Lasers, Photonics, and Biophotonics.

tute for Lasers, Photonics, and Biophotonics, which can efficiently be excited at 800 nm and emit in the green (~520 nm).

Another group of highly efficient two-photon dyes are the ionic dyes with stilbizolium structures, shown in Figure 8.5. Structure **1** contains a crown ether moiety. Our work revealed that by the appropriate choice of a cation incorporated in the crown ether cavity, along with that of the counterion X⁻, the two-photon emission peak can be varied over a 100-nm range. On the other

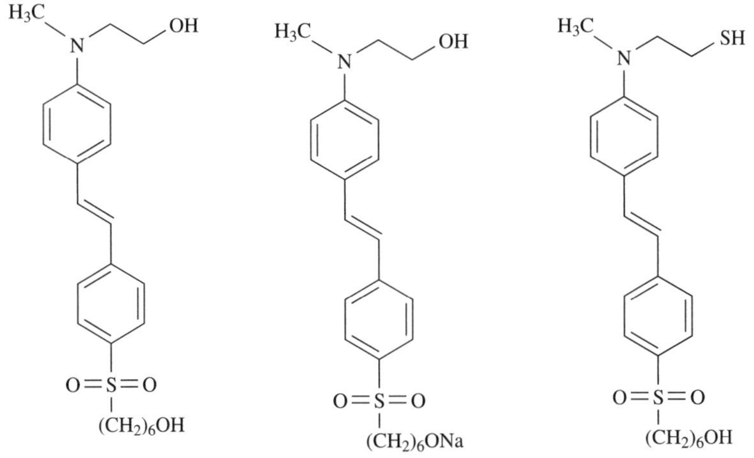

APSS | Water-soluble APSS | APSS-SH

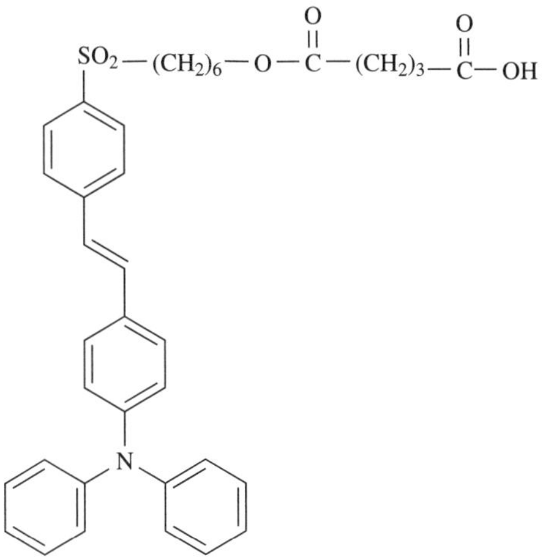

C625

Figure 8.4. Chromophore APSS and its various derivatives, developed at our Institute, which can very efficiently be excited at 800 nm and emit in the green (520 nm peak).

hand, the donor–acceptor-type dye, like structure **2**, shows a large variation in its two-photon property, depending on the nature of the donor (D), the substitution group (R′), and the counter anion X^-. Using these variations in a simple structure like structure **2**, we have achieved efficient two-photon emission, tunable across the entire visible spectral range.

A number of commercially available dyes also have two-photon absorption cross sections high enough to produce strong two-photon excited (TPE) emissions. For example, DAPI and Hoechst, which are widely used for fluorescence

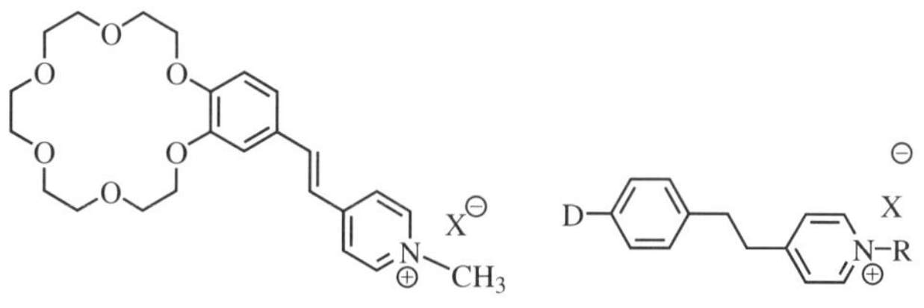

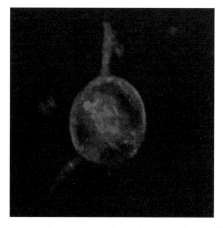

Figure 8.5. Examples of highly efficient two-photon active ionic dyes developed at the Institute for Lasers, Photonics, and Biophotonics.

Figure 8.6. Three-dimensional reconstruction of a retinal ganglion cell, imaged using two-photon confocal microscopy. The cell was stained with water-soluble APSS.

staining of nucleic acids, exhibit good TPE fluorescence and have been used for two-photon laser scanning microscopy. Specially designed fluorophores with high two-photon absorption cross sections, however, can improve imaging conditions because they permit the use of dye loading at relatively low concentration, whereby cell viability is maintained. Also, the use of a strong two-photon absorbing fluorophore requires excitation at lower intensities of laser beam.

As an example, two-photon imaging of a retinal ganglion cell using the water-soluble APSS is shown in Figure 8.6. This image was acquired at our Institute with a commercial Bio-rad confocal microscope, adapted for two-photon imaging. An 800-nm laser with a 90-fsec pulse width, 82-MHz rep rate, and 15-mW average power at the sample plane was used to excite the two-photon fluorescence. The fluorescence signal at around 490–550nm was collected using bandpass filters.

8.1.6 Inorganic Nanoparticles

Inorganic nanoparticles have also been used for bioimaging (Bruchez et al., 1998; Chan and Nie, 1998; Dahan et al., 2001). The two important types of nanoparticles are as follows:

- Semiconductor quantum dots composed of 2- to 4-nm-size nanoparticles, mainly of CdS and CdSe. The emission wavelength of these nanoparticles can be varied by changing their size. The larger the size, the longer the emission wavelength. They are excited by one-photon (linear) absorption. The advantage offered by the inorganic nanoparticles is that they do not photobleach. Furthermore, their surface can be functionalized with coupling groups to bond to biomolecules for selective staining. These nanoparticles are discussed in detail in Chapter 15, Section 15.3.
- Rare-earth-doped nanocrystals such as the Y_2O_3 nanocrystals containing Er^{3+} ions (Kapoor et al., 2000; Holm et al., 2002). These rare-earth-doped nanocrystals exhibit up-converted emission properties such as emission in the blue, green, or red region, when pumped by a small, readily available CW diode laser emitting at 974 nm. Hence they are also called *up-converting nanophores*. The up-conversion processes in these ions also involve multiphoton absorption. However, unlike the organic fluorophores discussed above, where the two-photon process involves a simultaneous absorption of two photons, the rare-earth ions in these nanocrystals exhibit stepwise multiphoton absorption. Hence the up-conversion produced by the rare-earth ions involves two or more sequential linear absorption, which only require a low-intensity CW laser. These up-converting nanophores are also discussed in detail in Chapter 15, Section 15.5.

8.2 GREEN FLUORESCENT PROTEIN

Green fluorescent protein (abbreviated as GFP) is a fluorescent protein, isolated from photogenic cells of the jellyfish, *Aequoria Victoria* (Chalfie and Kain, 1998). This fluorescent protein and its other fluorescent mutant forms deserve a unique place among the fluorophores used for bioimaging. Hence it is discussed in this separate section. In some sense, GFP falls between the categories of endogenous and exogenous fluorophores. Even though these fluorescent proteins are not naturally occurring in the cells or tissues being imaged, they are expressed in the cells and thus generated *in situ*.

GFP consists of 238 amino acids and has a cylindrical fold. The absorbing, and subsequently fluorescing, fluorophore unit is located within this very compact and folded structure that protects it from the bulk solvent to impart GFP a number of unique characteristics (Chalfie and Kain, 1998). Here are some of these characteristics:

- GFP is a very robust protein, resistant to denaturation.
- The intrinsic fluorescence of the protein is due to a chromophore, formed in the protein itself by the cyclization and oxidation of residues 65–67, Ser–Tyr–Gly (Periasamy, 2001).
- The encapsulation of the chromophore within the protein reduces non-radiative decays, induced by interaction with external environment, and leads to high quantum yield of fluorescence.
- GFP can be coupled (fused) to another protein (http://www.biochemtech.uni-halle.de).
- GFP is a noninvasive fluorescent marker for living cells, thus lending itself to a wide range of applications where GFP may function as a cell lineage tracer, a reporter of gene expression, or a measure of protein–protein interactions (http://pps99.cryst.bbk.ac.uk/projects/gmocz/gfp.htm).

GFP is generally introduced into a living cultured cell or into specific cells of a living organism as a GFP gene. This method involves recombinant DNA technology whereby the gene corresponding to GFP is inserted into a cell to synthesize GFP. In other words, the introduced GFP genes express GFP that emit green when irradiated and can be used to image the transfected cells and tissues containing them.

In yet another method, the gene for GFP can be fused to the gene for another protein of interest. This recombinant DNA encodes one long chimeric protein that contains both proteins in their entirety. The cells transfected with this recombinant DNA synthesize this chimeric protein. The green fluoresence from the chimeric protein can thus be used to image subcellular localization of the protein of interest.

Naturally occuring (wild-type) GFP is not optimal for some reporter gene applications, because it has quite a low extinction coefficient ($7000\,L\,mol^{-1}$ cm^{-1} at 475 nm) and a low expression level. A number of mutants of the GFP gene have been produced which exhibit enhanced emission and emit at different wavelengths. These variants of GFP include Blue FP(BFP), CyanFP(CFP), Yellow FP(YFP), and the fluorescence-enhanced variants of the above-mentioned variants (e.g., eGFP, eBFP, etc.). Their spectral characteristics are shown in Figure 8.7.

More recently, a red fluorescent protein (RFP or DsRed), isolated from a coral, has become available which extends the range of these fluorescent proteins for use in coexpression or in FRET imaging (Periasamy, 2001). The various variants of GFP, now covering a broad spectral range of emission, can be coexpressed in a living cell by again using recombinant DNA technologies. Using the methods discussed above, one fluorescent protein with appropriate absorption maximum can be fused to a protein or a protein segment to act as a donor. Another fluorescent protein with appropriate emission maximum to act as the energy acceptor can be fused to a different protein or segment. Then FRET imaging can be used to study protein–protein interactions or the

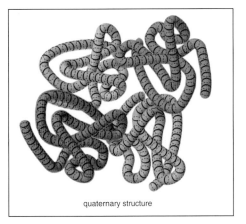
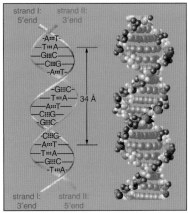

quaternary structure

strand I: 5'end   strand II: 3'end

-A∷T-
-T∷A-
-G∷C-
-C∷G-
-A∷T-

-G∷C-
-T∷A-
-A∷T-
-C∷G-
-G∷C-

-C∷G-
-A∷T-
-T∷A-
-G∷C-
-T∷A-

34 Å

strand I: 3'end   strand II: 5'end

Figure 3.15. *Left*: The quaternary structure of a protein consisting of four polypeptide chains (shown in four different colors). *Right*: The double-stranded structure of DNA in two representations: the base pairing through hydrogen bonding, shown on the left, and the space-filling model, shown on the right. (Reproduced with permission from Wade, 1999).

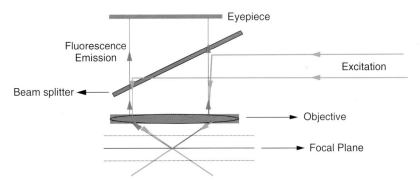

Eyepiece

Fluorescence Emission

Excitation

Beam splitter

Objective

Focal Plane

Figure 7.10. Basic principle of epi-fluorescence illumination.

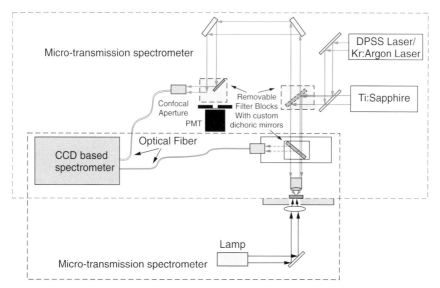

Figure 7.22. Schematics of experimental arrangement for obtaining fluorescence spectra from a specific biological site (e.g., organelle) using a CCD-coupled spectrograph. (Reproduced with permission from Pudavar et al., 2000.)

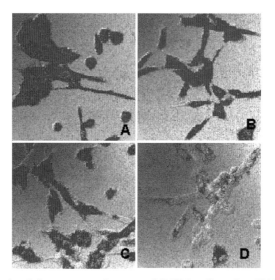

Figure 8.1. Confocal fluorescence images of RIF-1 cells using NADH emission at 426–454 nm when excited at 351 nm. The four images shown are (A) control without photosensitizer or light, (B) control with light irradiation but no photosensitizer, (C) control with photosensitizer but no light, and (D) cells with photosensitizer and irradiated with light. (Reproduced with permission from Pogue et al., 2001.)

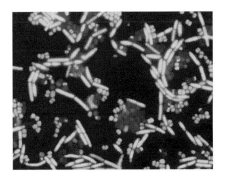

Figure 8.9. Confocal image of bacteria using multiple staining. Here live and dead *Micrococcus luteus* and *Bacillus cereus*, simultaneously stained with DAPI and SYTOX green nucleic acid probes, are imaged using a confocal microscope. (Reproduced by permission from http://www.probes.com/servlets/photo?fileid=g000651.)

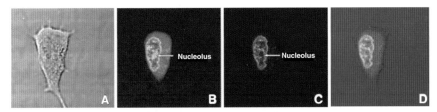

Figure 8.17. Two-photon laser scanning microscopic images of a KB cell stained with Hoechst 33342 and SYTO 43. (A) Transmission image; (B) SYTO 43 fluorescence image (excitation with 860 nm). (C) Hoechst 33342 fluorescence image (excitation with 750 nm). (D) Merged image of fluorescence and tramission images (Blue transmission, Green–Hoechst fluorescence, and Red–Syto 43 fluorescence) Hoechst stains exclusively the dsDNA sites; SYTO 43 labels both DNA and RNA. Arrow shows nucleolus, the major repository of RNA in the nucleus.

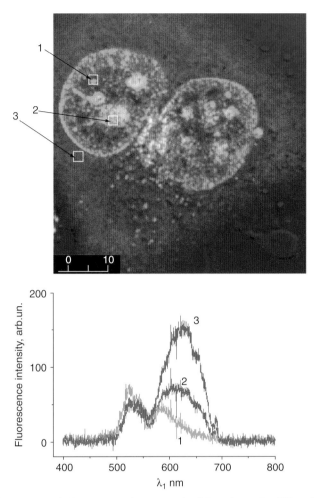

Figure 8.18. Top panel shows two-photon excited imaging of a KB cell with DNA (green pseudocolor) and RNA (red pseudocolor) staining with acridine orange. Bottom panel shows the fluorescence spectra obtained, using localized spectroscopy, from different locations in the cell.

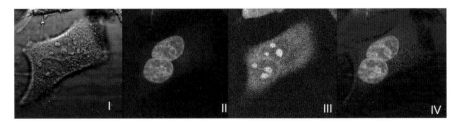

Figure 8.19. KB cell stained simultaneously with Hoechst 33342 and Cyan 40. See text for details.

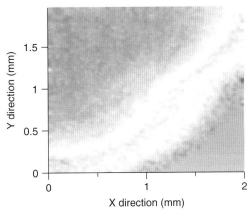

Figure 8.25. Optical coherence tomography image of cross section of human tooth. (Reproduced with permission from Rodman et al., 2002.)

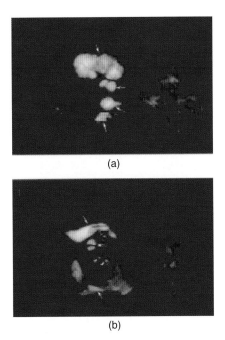

(a)

(b)

Figure 8.28. External images of B16F0-GFP colonizing the liver. A metastatic lesion of B16F0-GFP in the liver growing at a depth of 0.8 mm after portal vein injection was externally imaged through the abdominal wall of the intact nude mouse. (a) An external image of multilobe liver metastases of the B16F0-GFP cells (*large arrows*). (b) An external image of small liver metastatic lesions of approximately 1.5 mm in diameter (*small arrows*) and other larger metastatic lesions (*large arrows*). (Reproduced with permission from Yang et al., 2000.)

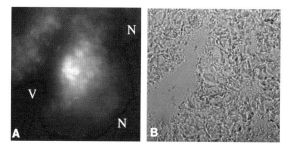

Figure 8.30. NIRF histology of tumor excised from an animal. Left image shows the NIRF acquisition, which is superimposed in red onto the correlative phase contrast microscopy image on the right. Vessels (V) and areas of necrosis (N) are labeled. Sections are unstained and unfixed to preserve fluorescence signal. (Reproduced with permission from Weissleder et al., 1999.)

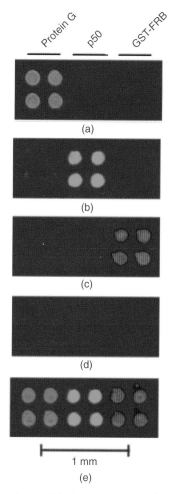

Figure 10.10. Detecting protein–protein interactions on glass slides. See text for details.

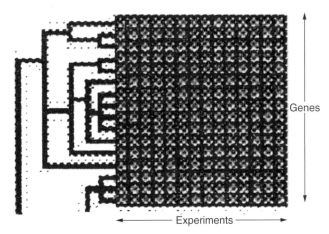

Figure 10.2. A representative of a cluster analysis of genes (on verticle axis) versus the experiments (horizontal axis). (Reproduced with permission from Dhiman et al., 2002.)

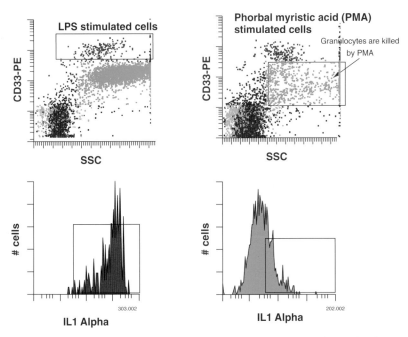

Figure 11.14. Response of cell populations to selected biological stimulus. *Left*: Stimulation with LPS (lipopolysaccharide). *Right*: Stimulation with PMA (phorbal myristic acid). (Courtesy of C. Stewart/RPCI.)

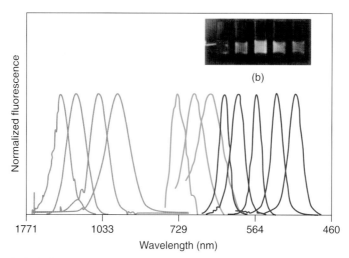

Figure 15.4. (A) Size- and material-dependent emission spectra of several surfactant-coated semiconductor nanocrystals in a variety of sizes. The first five from right represent different sizes of CdSe nanocrystals with diameters of 2.1, 2.4, 3.1, 3.6, and 4.6 nm (from right to left). The next three from right is of InP nanocrystals with diameters of 3.0, 3.5, and 4.6 nm. The IR emittersare InAs nanocrystals with diameters of 2.8, 3.6, 4.6, and 6.0 nm. (B) A true-color image of a series of silica-coated core (CdSe)-shell (ZnS or CdS) nanocrystal probes in aqueous buffer, all illuminated simultaneously with a handheld ultraviolet lamp. (Reproduced with permission from Bruchez et al., 1998.)

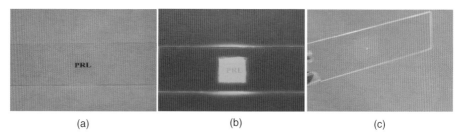

Figure 16.16. Thin film of dye-doped PHA: (a) Transmission, (b) UV-excited fluorescence, (c) two-photon-excited fluorescence.

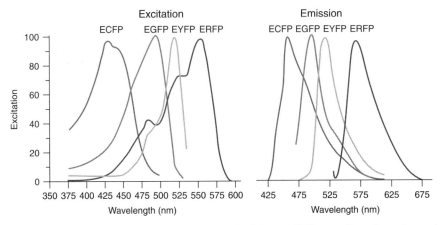

Figure 8.7. Excitation and emission spectra of wild-type GFP as well as the enhanced variants of GFP (eCFP, eGfp, eYFP, and eRFP). (Reproduced with permission from http://www.clontech.com.)

conformational dynamics of proteins of interest. Some specific applications of GFP and its variants are provided in later sections (e.g., see Figure 8.28).

8.3 IMAGING OF ORGANELLES

There is a diverse array of commercially available organelle probes, which are cell-permeant fluorescent stains that selectively associate with mitochondria, lysosomes, endoplasmic reticulum, and Golgi apparatus in live cells. Some examples of organelle specific dyes are given in Table 8.1. Many of these fluorescent probes can be used to investigate organelle structure and activity in live cells, with minimal disruption of cellular function.

Salvioli et al. (2000) have used an organelle-specific dye, MitoTracker™ Red CMXRos (MT-1), to selectively stain the mitochondria of live cells and study apoptosis. They have utilized this dye's sensitivity to the membrane potential to measure the changes in mitochondrial membrane ptotential, which is a very early indication of the beginning of apoptosis. In their study, they also have used a fluorescent probe (Fluorescin Isothiocyanate, FITC) conjugated to a protein, Annexin-V (ANX-V), to monitor apoptosis. Annexin-V has a high and selective affinity to phosphatidylserine (PS), an essential phospholipid present in all mammalian cellular membranes, which relocates from the inner leaflet of the cell membrane to the outer surface during apoptosis. An increase in the fluorescence of ANX-V (indicating the relocation of PS), as well as a decrease in the fluorescence of MT-1, indicates the initiation of apoptotic processes.

Figure 8.8 shows the confocal images of human HL60 promyelocytic cells stained with MT-1 and FITC-Annexin. In a healthy cell, the membrane-

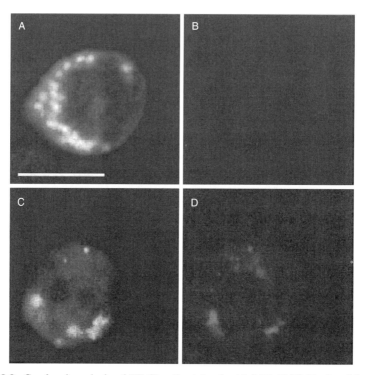

Figure 8.8. Confocal analysis of HL60 cells stained with MT ANX-V after 3 hours of treatment with STS (A–D). (A) Staining with MT; (B) no staining by ANX-V. Lower panels (C and D) are related to the same cell undergoing apoptosis. Note the functional heterogeneity of mitochondria, some of which still maintain a high membrane potential as seen in (C), even when the cell is positive to ANX-V binding (D), indicating the apoptosis. Scale bar = $10\,\mu$m. All the images have the same magnification. (Reproduced by permission from Salvioli et al., 2000.)

permeable MT-1 stains the mitochondria as seen in Figure 8.8A. At the same time, ANX-V doesn't show any fluorescence, because most of the PS is inside the cell membrane in the case of a healthy cell. But in apoptotic cells (apoptosis introduced by adding staurosporine, STS), a simultaneous decrease in the mitochondrial membrane potential and the relocation of PS to the outer surface of cell membrane results in a decrease in the MT-1 fluorescence (Figure 8.8C) and a simultaneous increase in the ANX-V fluorescence (Figure 8.8D).

Rhodamine R-123 is another dye that is commonly used for selective mitochondrial staining. The two-photon excited fluorescence image of a live KB cell stained with Rhodamine R-123 has been obtained by using two-photon laser scanning microscopy with 800-nm excitation from a mode-locked Ti:sapphire laser (Ohulchanskyy et al., unpublished results from the Institute for Lasers, Photonics, and Biophotonics).

Many nucleic acid specific dyes, like Hoechst 33342, are commonly used as counter stains for staining nucleus along with other organelle probes. Some applications of these nucleic acid probes are given in Section 8.5.4.

8.4 IMAGING OF MICROBES

Recent advances in optical imaging have dramatically expanded the capabilities of light microscopes and their usefulness in microbiology research to study the structure and functions of small microbes. Some of these advances include improved fluorescent probes, better cameras, new techniques such as confocal and near-field microscopy, and the use of computers in imaging and image analysis. These new technologies have now been applied to microbial investigations with great success (Fung and Theriot, 1998). This section provides selected examples of imaging of microbes that are comparable to or, in the case of viruses, significantly smaller than the wavelength of light used for imaging. The examples provided use both confocal and near-field microscopic techniques.

8.4.1 Confocal Microscopy

In the study of bacterial pathogenesis, confocal microscopy permits a sensitive detection of bacteria in much thicker sections than is possible by conventional immunohisto-chemistry. This allows for the use of much smaller, and more realistic, infectious doses to study pathogenesis *in vivo*. There are reports of using confocal microscopy to study *Salmonella* pathogenesis at much lower infectious doses than required previously (Richter-Dahifors et al., 1997). Confocal microscopy has also been used for studying the interaction of *S. typhimurium* with the host-cell actin cytoskeleton (Fu and Galan, 1998).

An example of bacterial imaging, using multiple staining and confocal microscopy, is provided here in Figure 8.9. In this case, a mixed population of live and dead *Micrococcus luteus* and *Bacillus cereus* are simultaneously stained with DAPI and SYTOX green nucleic acid stains. Bacteria with intact cell membranes are stained exclusively with the cell-permeant DAPI nuclear stain and exhibit blue fluorescence. On the other hand, cells with damaged membranes are stained with both fluorophores and exhibit green fluorescence.

This capability of imaging microbes can be further enhanced by using multiphoton microscopy (e.g., TPLSM). A second example provided here is that of the *Streptococcus gordonii* bacteria. These bacteria were stained metabolically with an efficient two-photon APSS-type fluorophore (water-soluble APSS), as shown in Figure 8.10, and were imaged at our Institute using TPLSM with 800-nm femtosecond pulses from a Ti:sapphire laser.

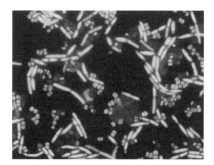

Figure 8.9. Confocal image of bacteria using multiple staining. Here live and dead *Micrococcus luteus* and *Bacillus cereus*, simultaneously stained with DAPI and SYTOX green nucleic acid probes, are imaged using a confocal microscope. (Reproduced by permission from http://www.probes.com/servlets/photo?fileid=g000651.) (See color figure.)

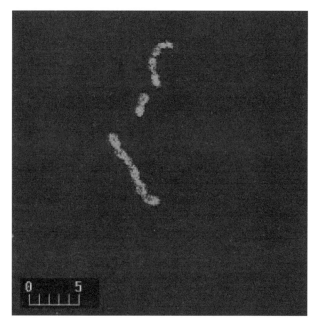

Figure 8.10. The *Streptococcus gordonii* bacteria stained with a two-photon dye (water-soluble APSS) and imaged using TPLSM with 800-nm femtosecond pulses from a Ti:sapphire laser as the excitation source.

8.4.2 Near-Field Imaging

Near-field scanning optical microscopy (NSOM) can provide images with a resolution less than the wavelength of light. Therefore, in principle, it is of great value in studies of biological structures such as viruses and bacteria, whose dimensions are often in submicrons. Conventional optical microscopy, with its inherent limitations on resolution, and electron microscopy (SEM and TEM), with rigid sample preparation requirements, cannot be used to image biological samples in *in vitro* conditions and with a nanometer-range resolution. Near-field scanning optical microscopy (NSOM) overcomes the diffraction limit of optical microscopy and allows optical images with a resolution of 10–100 nm (Dürig et al., 1986) without having to go through any special sample preparation, as is needed for electron microscopy. The principle and methods of near-field imaging are discussed in Section 7.11. Optical imaging of tobacco mosaic virus (TMV) particles is an example of the use of NSOM for the imaging of viruses. The NSOM image of TMV, shown in Figure 8.11 (Pylkki et al., 1993), is a transmission NSOM image, obtained on a Topometrix Aurora near-field microscope. The imaging was performed by attaching TMV to a silanated mica using glutaraldehyde. A 488-nm argon-ion laser was used as the light source. The image clearly shows the individual TMV particles with a separation of less than 30 nm.

As an example of imaging of bacteria, Figure 8.12 shows the NSOM image obtained at our Institute using fluororescently labeled bacteria. The *Porphyromonas gingivalis* bacteria were stained using the water-soluble APSS two-photon fluorophore (Figure 8.4) by incubating the bacterial suspension with

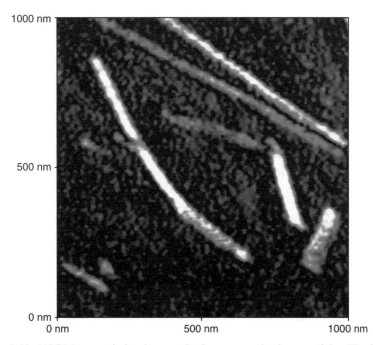

Figure 8.11. NSOM transmission image of tobacco mosaic virus particles. The imaged area is 1.4 μm × 1.4 μm. (Reproduced with permission from Pylkki. et al., 1993.)

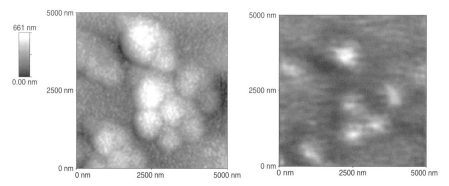

Figure 8.12. NSOM images (topography on left and emission NSOM on right) of oral bacteria (*Porphyromonas gingivalis*) stained with Water Soluble APSS.

dye solution. Near-field microscopy was used in the fluorescence collection mode. These images were obtained with 800-nm two-photon excitation. Due to the limitation of near-field propagation, the light collection is confined to a narrow layer of width in several nanometers. In this case, the fluorescence emitted by the dye is collected with the aid of an inverted microscope. It is processed together with the topographic signal obtained in the AFM mode. The optical resolution obtained is ~100 nm. A correlation between the topographic image and the near-field image shows the bacteria stained with the dye and the capability of NSOM to provide a detailed optical image of a bacteria.

8.5 CELLULAR IMAGING

8.5.1 Probing Cellular Ionic Environment

The activities of ions such as Ca^{2+}, Na^+, and H^+ play important roles in many cellular functions. Hence imaging and quantification of cellular distribution of these ions and their transport are of considerable significance. This important recognition is also reflected by the large number of ion-selective fluorophores available for bioimaging. These ion-selective or pH-sensing fluorophores exhibit a change in their fluorescence intensity upon binding with the respective ions. The same principle is used in the design of biosensors to detect this ion. The biodetection aspect of these ion-selective fluorophores is discussed in Chapter 9. The example provided in Figure 8.13 uses a calcium-selective fluorophore, fluo-3, to probe spontaneous intracellular Ca^{2+} fluctuations of neurons, developing *in vivo*. For this purpose, the spinal cord was dissected from a neurula-stage *Xenopus* embryo and stained with fluo-3. The regions of fluo-3 fluorescence, observed on the ventral side of the spinal cord, indicate areas of highest intracellular Ca^{2+} concentrations.

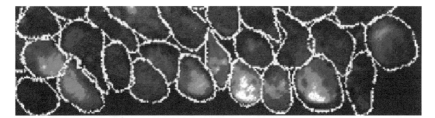

Figure 8.13. Fluorescence image of the spinal cord, dissected from a neurula-stage *Xenopus* embryo and stained with the Ca^{2+} binding fluo-3 fluorophore. The regions of bright fluo-3 fluorescence indicate areas of highest intracellular Ca^{2+} concentration. (Reproduced by permission from Gu et al., 1995.)

8.5.2   Intracellular pH Measurements

A knowledge of the intracellular pH in individual cells or cellular components is important for understanding many biological processes. There have been reports of measurements of intracellular pH from cell suspension (Bassnett et al., 1990; Buckler and Jones, 1990) or from individual cells. Using dyes like carboxy Snarf-AE-1, which is a long-wavelength fluorescent pH indicator commercially available from Molecular Probes, localized spectroscopy can be utilized to measure the pH of different compartments of a cell. Carboxy Snarf-1 can be excited at wavelengths in the range of 480–550 nm. The relative intensities of the two emission peaks, at 580 nm and 640 nm, are strongly dependent upon the pH of the dye environment (*Handbook of Fluorescent Probes and Research Products*, Molecular Probes, Inc.). A number of fluorescence measurement artifacts, such as those due to photobleaching, cell thickness, instrument instability, and leakage and nonuniform loading of the indicator, can be eliminated by using the ratio of these two peaks for a quantitative determination of the pH. It should be noted that photobleaching is also pH-dependent and can produce artifacts in the measurements.

In our study, Snarf-1 was loaded into U-937 cells (human promonocytocytic cell line) and the pH estimate was made using both multichannel imaging and spectroscopy. By calculating the ratio of the emission in the 640-nm range and in the 580-nm range, one can calculate the pH at a specific location, using a calibration curve. An *in situ* calibration was performed by using the ionophore, nigericin (N-1495), to equilibrate the intracellular pH with the extracellular medium of known pH. Thus by nigericin treatment and localized spectroscopy, a range of spectral ratio data are generated from cells with known intracellular pH varying from 6 to 8. These data are fitted with the modified Henderson–Hasselbalch equation for overlapping spectral components (Bassnett et al., 1990) to generate a calibration curve and to estimate the pK_a (defined as $-\log_{10}(K_a)$, where K_a is the dissociation constant of Snarf-1 inside the cells. This calibration curve and the resulting calculated pK_a value can be

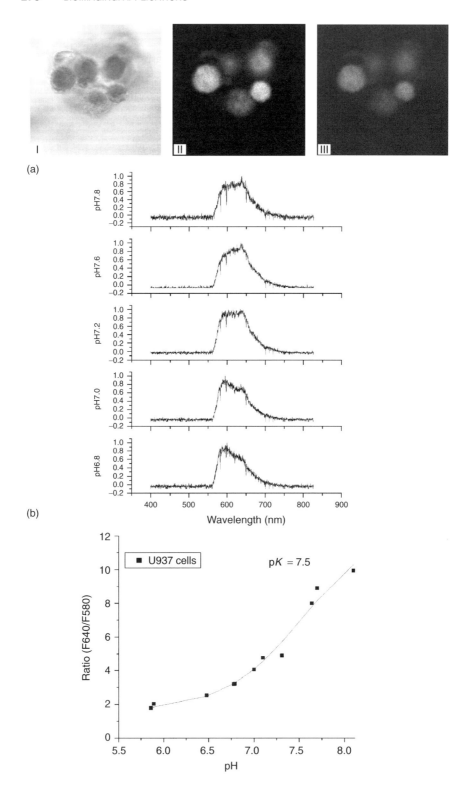

(a)

(b)

used to estimate the intracellular pH of U937 cells (each cell line needs its own calibration curve, because pK_a varies with different cell lines).

In the study presented here, the dye was excited at its pH-independent isosbestic excitation point, 532 nm, using a diode-pumped solid-state laser. For localized spectroscopy, the complete emission spectra from different compartments of a cell loaded with Carboxy Snarf were obtained. Figure 8.14 shows the multichannel image and a calibration plot obtained for Carboxy Snarf in the U-937 cell line.

8.5.3 Optical Tracking of Drug–Cell Interactions

Two-photon laser scanning microscopy can readily be used to monitor the cellular entry pattern of a drug, by coupling a two-photon fluorescent probe to the drug. In collaboration with Nobel Laureate Andrew Schally of Tulane Univeristy, our group used optical tracking of a fluorescently labeled chemotherapeutic agent to monitor its cellular uptake and to understand the mechanism of chemotherapy (Wang et al., 1999; Krebs et al., 2000).

Chemotherapy is commonly used in the treatment of cancers. However, the mechanism of action of many of these agents is not well understood. In the chemotherapeutic approach, pioneered by Dr. Schally and co-workers at Tulane University, drug targeting is used based upon selectivity of luteinizing hormone-releasing hormone analogues for specific binding sites in tumor tissues. An example is AN-152, developed in Schally's lab, which consists of an agonistic analogue of luteinizing hormone-releasing hormone, [D-Lys[6]]LH-RH, conjugated to a cytotoxic agent, doxorubicin, as shown in Figure 8.15 (Schally and Nagy, 1999). In order to utilize two-photon confocal microscopy to produce a three-dimensional image of the biological cellular process, a two-photon fluorophore probe (C625 of Figure 8.4) was attached to AN-152, with the help of which internalization of AN-152 was monitored. The results of this investigation visually showed the receptor-mediated entry of AN-152 into the cell cytoplasm and subsequently into the nucleus of a MCF-7 breast cancer cell (Figure 8.15). It was found that AN-152 entered the cell nucleus, supporting the assumption that the mechanism of antiproliferative activity of doxorubicin is due to its ability to intercalate into DNA and break the strands of double helix by inhibiting topoisomerase II.

Figure 8.14. Multichannel imaging and localized spectroscopy for pH measurement using carboxy Snarf-1 in U937 cells: (a) Channel 1 shows transmission image (I); Channel 2 shows yellow fluorescence (580 ± 25 nm) image (II), and Channel 3 shows red fluorescence (640 ± 25 nm) images (III). (b) The change in spectra with a change in the pH is shown in the top panel for the U-937 cells after pH equalization with the media using Nigericin treatment. The bottom panel shows the ratio of peaks at different pH, and it also shows the fit with the modified Henderson–Hasselbalch equation to estimate the pK_a inside these cell lines.

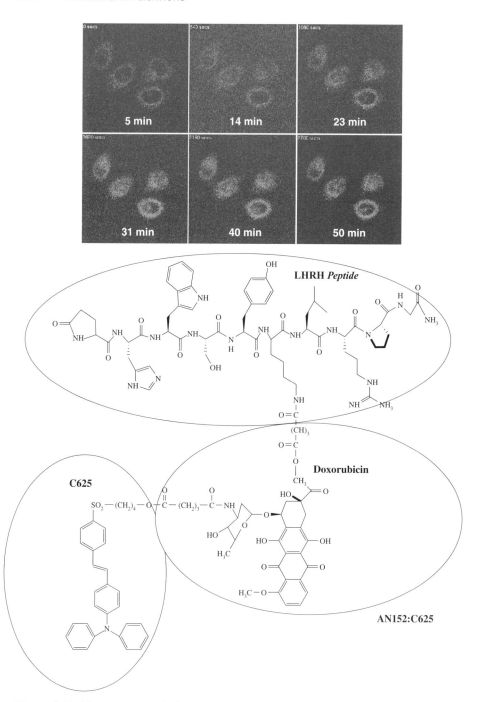

Figure 8.15. The structure of chemotherapeutic drug-carrier (LH-RH peptide)–dye conjugate AN152: C625. TPLSM images of MCF-7 cells show the intake of drug into cell over a time period of 50 minutes.

Localized spectroscopy was also used, in conjunction with imaging, to elucidate the mechanism of this targeting chemotherapeutic agent (AN-152) in living cancer cells (Wang et al., 2001). Two different two-photon fluorescent probes, with different fluorescence emission, were coupled to AN-152 and [D-Lys⁶]LH-RH separately. Multicolor fluorescence from the chemotherapeutic agent (AN-152) and the peptide (LH-RH) inside the different parts of the cell (nuclei and cytoplasm) were ratiometrically studied. AN-152 and the LH-RH peptide carrier showed different intracellular spectral profiles. These results are shown in Figure 8.16. Ratiometric studies showed that AN-152 could enter the nucleus more easily than LH-RH itself. This study confirmed the fact that the LH-RH peptide was helping the drug enter the cell, but the doxorubicin unit was responsible for entry into the nucleus and subsequent cell death. This study illustrates the utility of fluorescently labeled drugs for imaging and probing new cellular processes induced by drug interactions.

8.5.4 Imaging of Nucleic acids

Nucleic acid stains are usually cationic dyes that bind with the polyanionic nucleic acid, thus showing specificity for DNA and RNA. Furthermore, hydrophilic/hydrophobic interactions play an important role in (a) providing some dyes with the ability to intercalate between the nucleic acid bases and (b) providing other dyes with the ability to bind with double-stranded nucleic acids in grooves. In double-stranded DNA, the alignment of the strands is antiparallel and asymmetric along the axis. This asymmetry creates two different grooves on opposite sides of the base pairs, called major and minor grooves. Examples of intercalating dyes are phenanthridines (ethidium bromide, propidium iodide), acridines (acridine orange, ACMA), and numerous cyanine dyes (TO, YO, PO, JO, BO, LO, and their derivatives; SYTO and SYTOX dye families). The chemical conjugation of intercalating fluorophore and a linker carrying additional cationic charge provides both intercalating ability and the electrostatic mode of binding. The dimeric probes obtained in this way have extremely high binding efficiencies and sensitivities to nucleic acids. These features of the dimeric probes have led to their usage in a variety of DNA studies, including imaging at a single molecule level. Imaging of a single chain of 39-μm-long DNA molecules, stained with the dimeric dye YOYO-1 and attached to a 1-μm-diameter polystyrene bead, has been used to study the relaxation of a single DNA molecule stretched by a laser tweezer. Laser tweezer action is discussed in detail in Chapter 14. An example of this single DNA molecule imaging is shown in Figure 14.17.

Some of the nucleic acid stains, such as Hoechst dyes and DAPI, bind with double-stranded DNA in minor grooves. For these dyes, binding in the DNA minor grooves determines their selectivity not only to the double-stranded DNA (ds DNA)but also to the A–T sequences of the DNA molecule because of the narrow minor grooves of the A–T sequences in comparison with the wider minor grooves of the G–C sequences. Narrower minor grooves of the

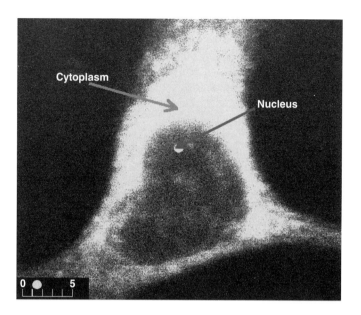

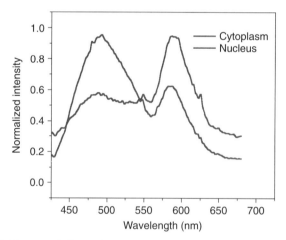

Figure 8.16. Top panel shows the two-photon fluorescence image of breast cancer cell treated with AN-152:C625 (drug labeled with dye C625) and [D-Lys⁶]LH-RH:TPR (the peptide carrier labeled with dye TPR). The bottom panel shows the localized spectra obtained from different parts of the cell as indicated by the arrows in the image.

A–T sequences provide a snag fit for the ribbon-like molecule of a groove-binding stain.

Nucleic acid stains can be used for staining nucleic acids inside cells, where the selection of the probe is determined by its ability to permeate through the membrane of the live cell. Cell-membrane-impermeant dyes (e.g., ethidium

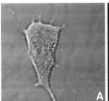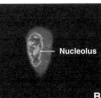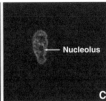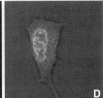

Figure 8.17. Two-photon laser scanning microscopic images of a KB cell stained with Hoechst 33342 and SYTO 43. (A) Transmission image; (B) SYTO 43 fluorescence image (excitation with 860 nm). (C) Hoechst 33342 fluorescence image (excitation with 750 nm). (D) Merged image of fluorescence and tramission images (Blue transmission, Green–Hoechst fluorescence, and Red–Syto 43 fluorescence) Hoechst stains exclusively the dsDNA sites; SYTO 43 labels both DNA and RNA. Arrow shows nucleolus, the major repository of RNA in the nucleus. (See color figure.)

bromide, SYTOX dyes) are involved in various studies of the cell apoptosis because of their increased permeability into apoptotic cells, caused by the compromised membranes in apoptotic cells. Consequently, the dyes are used as simple, single-step dead cell indicators.

In contrast, cell-membrane-permeant probes, such as SYTO dyes, are nucleic acid stains that passively diffuse through the membranes of live cells. These dyes are used to stain RNA and DNA in both live and dead eukaryotic cells, as well as in gram-positive and gram-negative bacteria. Hence, in terms of cell organelles, the SYTO dyes do not exclusively stain nuclei in live cells. In contrast, dyes like DAPI or Hoechst readily stain nuclei in live cells due to their selectivity to DNA concentrated in cell nucleus. These features of selective staining are demonstrated in Figure 8.17, which shows two-photon scanning microscopic images of KB cells (human oral epidermoid carcinoma cell line) stained with Hoechst 33342 and SYTO 43, obtained at our Institute. Hoechst 33342 is a cell-membrane-permeant, minor groove-binding DNA stain, showing a sequence-dependent DNA affinity as it binds to the A–T sequences.

The SYTO dyes do not exclusively stain nuclei in live cells. In contrast, dyes like DAPI or the Hoechst dyes at low concentrations readily stain nuclei in live cells and do not affect cell viability because of their high specificity to dsDNA.

Determination of the relative contents of DNA and RNA at definite sites inside a cell nucleus is still a challenging task for biologists. One approach to this problem is double staining with DNA- and RNA-selective stains. Though no true RNA-specific stains are commercially available, there are many specific fluorescent stains for double-stranded nucleic acids. A common approach, therefore, is to use a fluorescent stain, selective to the double-stranded DNA (e.g., Hoechst and DAPI), and another nucleic acid stain that does not show sufficient discrimination between double-stranded DNA and single-stranded RNA, existing in a cell nucleus.

Figure 8.17 shows a confocal image of a KB cell, stained with SYTO 43 and Hoechst, where this approach has been used. The observed difference between the SYTO 43 and the Hoechst fluorescence pattern reveals the distribution of RNA in the cell.

Another approach to differentiating DNA and RNA uses a single dye that shows a shift in its fluorescence spectra, when bound to the double-stranded DNA (dsDNA) and to single-stranded RNA. The cell-membrane-permeant intercalative dye acridine orange exhibits green fluorescence (530-nm peak), when staining dsDNA. However, when binding with the single-stranded RNA, it shows red fluorescence (640-nm peak), which is apparently associated with the formation of a dye aggregate on the RNA molecule. The image of KB cell stained with acridine orange is shown in Figure 8.18. In this case, two-photon excitation of acridine orange was used for imaging and localized spectroscopy (see Figure 7.22). Localized spectroscopy allowed ratiometric profiling of the "green" and the "red" components (indicating dsDNA and RNA, in this case) to estimate the DNA/RNA content in certain locations inside the nucleus (Figure 8.18).

It is important to note that the different conditions for the nucleic acids outside a cell and inside the cell (which includes an intracellular organization of RNA and DNA) can cause a difference in the binding efficiency of a probe to DNA and RNA in the two cases. For instance, a monomethine cyanine dye Cyan 40 (4-((1-methylbenzothiazolyliliden-2)methyl)-1,2,6-trimethylpyridinium perchlorate) does not show any significant preference in RNA staining versus DNA staining outside a cellular environment, but, as shown in Figure 8.19, it does show an apparent preferential binding with RNA inside a living cell (Ohulchanskyy et al., 2003).

An important technique enabling the detection and determination of spatial distribution of specific DNA or RNA sequences in the cytoplasm, nucleus, and chromosomes as well as in other organelles is that of fluorescence *in situ* hybridization (FISH) (Pinkel, 1999). Hence the imaging utilizing this technique is also referred to as FISH imaging (Kozubek, 2002). FISH imaging has proved to be of value in the analysis of the structures and functions of chromosomes and genomes. It has been used for the determination of the spatial and temporal expression of genes.

Hybridization, in the context of DNA and RNA, refers to nucleotide base pairing of two single-stranded nucleic acid chains (DNA or RNA). The FISH technique involves *in situ* hybridization of nucleic acids in the target cells or chromosomes, to be detected or imaged, with fluorescently labeled, single-stranded probe nucleic acids. This method is illustrated in Figure 8.20. First, the nucleic acid (DNA) in the target cell is made single-stranded, for example, by heating. Next, the cell is incubated with fluorescently labeled single-stranded probe nucleic acid molecules (DNA or RNA). *In situ* hybridization occurs between the target and the probe under the conditions of the matching of their base sequence for pairing (conditions of complementary base sequences). After hybridization, fluorescence imaging can be used to deter-

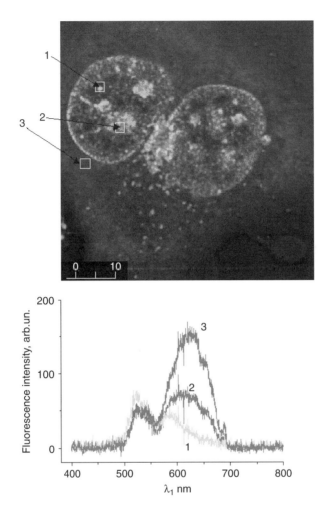

Figure 8.18. Top panel shows two-photon excited imaging of a KB cell with DNA (green pseudocolor) and RNA (red pseudocolor) staining with acridine orange. Bottom panel shows the fluorescence spectra obtained, using localized spectroscopy, from different locations in the cell. (See color figure.)

mine the number, intensity, and spatial distribution of each of the different-colored hybridization signals, probes, or segments fluorescing at different wavelengths.

Fluorescence labeling of the probe nucleic acid can involve two types of approaches, also shown in Figure 8.20. In one approach, called the *direct method*, the fluorescent label (fluorophore, represented by letter F in the figure) is directly attached to the end of the nucleic acid probe. In the other method, called the *indirect method*, the probe is modified chemically with molecules of biotin or digoxigenin (represented by the letter H in the figure).

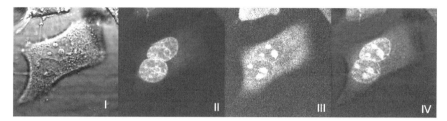

Figure 8.19. KB cell stained simultaneously with Hoechst 33342 and Cyan 40. (I) Light transmission image. (II) Fluorescence image with Hoechst excitation (λ_{ex} = 750 nm). (III) Fluorescence image with Cyan 40 excitation (λ_{ex} = 860 nm). (IV) Image generated by overlapping of I, II, and III images. Green pseudocolor marks Hoechst fluorescence, red pseudocolor represents Cyan 40, and blue pseudocolor means transmission. Scanning for both excitation wavelengths was performed in one focal plane. (Reproduced with permission from Ohulchansky et al., 2002.) (See color figure.)

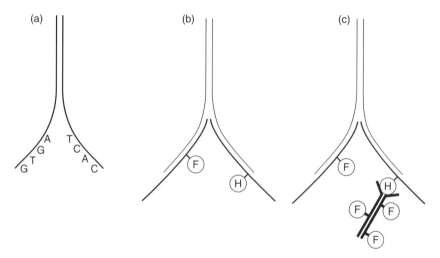

Figure 8.20. FISH Schematic to target DNA. (a) Double-stranded DNA (bold lines) is denatured to make them single-stranded. (b) The target is incubated with denatured probes (bold lines) that are labeled with fluorochromes. (c) Hapten-labeled probes need to be rendered visible using affinity reagents such as avidin or antibodies (bold lines) that carry fluorochromes. (Reproduced with permission from Pinkel, 1999.)

After hybridization with the target nucleic acid, fluorescently labeled affinity reagents are used to couple to the H group to render it fluorescent. Because the fluorophore number densities can significantly be increased using the indirect method, enhancement of sensitivity can be accomplished. The ideal length of the probe is 100–300 bp (base pairs), a shorter length resulting in lower stability of hybridization. On the other hand, larger-size probes pose difficulties in penetrating into the cell structure. The probes can be produced by cloning (recombinant DNA technique whereby DNA is inserted into a vector and amplified together inside an appropriate host cell). Synthetic oligonuclectides have also been used as probes. Multicolor FISH utilizes several types of probes simultaneously, which have been labeled with different fluorophores.

8.5.5. Cellular Interactions Probed by FRET/FLIM Imaging

FRET, as discussed in Chapter 7, is a phenomenon that occurs when two different chromophores with overlapping emission (donor) and absorption (acceptor) spectra are separated by a distance in the range 10–80 Å. Recently, the introduction of the green fluorescent protein (GFP) to FRET-based imaging has provided great impetus to the study of noncovalent molecular interactions inside cells (Jovin and Arndt-Jovin, 1989; Herman, 1989; Gadella et al., 1999; Bastiaens and Squire, 1999).

Mutagenic variants of GFP, such as CFP and YFP, have proven to be excellent donor–acceptor pairs for studying protein–receptor interactions in cells (Tsien, 1998; Heim et al., 1994, 1995; Tsien and Prasher, 1998). An example of FRET imaging is provided here using the CFP–YFP pair. The emission band of CFP overlaps well with the excitation/absorption band of YFP. CFP can be excited by one-photon abosorption (linear absorption) using either the 457-nm argon laser line or the 414-nm line from a krypton laser, while YFP is excited using the 514-nm line from an argon laser. The two-photon excitation of CFP can readily be accomplished at 800 nm with femtosecond pulses from a Ti:sapphire laser, while the two-photon excitation efficiency of YFP at 800 nm is very low (almost no signal from YFP transfected cells). In the results presented here, the COS cell (African green monkey kidney cells) lines were transfected with CFP- or YFP-labeled plasmid DNA vectors or with both of them. These plasmids, which are fragments of DNA carrying a specific gene, were designed to express a tumor necrosis factor receptor (TNFR), which is a membrane receptor. Individual receptor chains of TNFR, each tagged with CFP and YFP, can be detected as FRET from CFP-tagged receptor chain to YFP-tagged receptor chain (Chan et al., 2001).

The work reported here was done at the Institute for Lasers, Photonics, and Biophotonics, in collaboration with Sarah Gaffen from the Department of Oral Biology at SUNY at Buffalo. The cells expressing CFP and YFP were examined independently using 800-nm two-photon excitation. It was found that CFP could be two-photon excited with 800-nm excitation, while YFP expressing cells produced no fluorescence. Then the CFP and YFP coexpressing cells were examined with 800-nm excitation, in both the confocal imaging mode and the localized spectroscopy mode, to detect any FRET signal. As shown in Figure 8.21, the FRET signal from YFP was quite visible while the CFP signal was significantly quenched due to the energy transfer. The FRET signal clearly indicated significant protein–protein interactions in cells coexpressing both CFP and YFP. The spatial distribution of the FRET signal

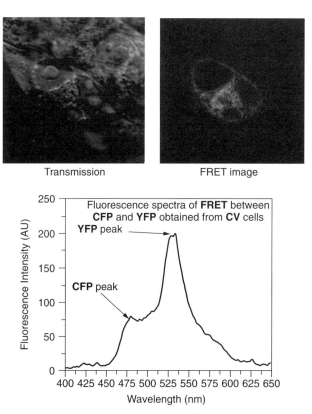

Figure 8.21. The image shown in the top left panel (red color) is the transmission image of a cell expressing both the CFP- and YFP-labeled receptors. The top right panel image is FRET image generated by exciting the cells with 800-nm femtosecond pulses and looking at the FRET emission from YFP at 525 nm. This image indicates the locations where CFP and YFP labeled receptors are in close proximity ($\sim$5–10 nm). In localized spectra (bottom), one can see both CFP and YFP emission peaks.

provided information on locations where the CFP- and the YFP-labeled receptors were in close proximity, within less than 10 nm.

As explained in Chapter 7, FLIM provides information on the spatial distribution of the fluorescence lifetime. The advantage of lifetime imaging, in comparison with steady-state fluorescence imaging, is that the absolute values of lifetimes are independent of the dye concentration, photobleaching, light scattering, or the intensity of excitation. At the same time, FLIM is sensitive to intracellular factors such as ion concentrations, polarity, binding to macromolecules, or FRET. The lifetime varies from fluorophore to fluorophore and also for the same fluorophore in different environments. For example, the lifetimes of many fluorophores are altered by the presence of ions such as Ca^{2+}, Mg^{2+}, Cl, pH, or K^+. This allows one to environmentally induce changes without any need to use wavelength-ratiometric probes. A combination of FRET and FLIM provides both the high spatial (nanometers) and the temporal (nanoseconds) resolution. Since only the donor fluorophore lifetime is measured, spectral bleed-through is not an issue in FRET–FLIM imaging. FLIM provides more quantitative information about the FRET phenomena than spectral analysis of the emitted light, because not all of the energy transferred from the donor probe results in the emission of photons from the acceptor probe.

Periasamy et al. (1999) used FLIM microscopy for imaging GFP in transfected cells. They showed that although the cell nucleus appeared brighter than the cytoplasm in fluorescence imaging, FLIM exhibited the same lifetime of the GFP fluorescence in both compartments. This result indicated that the GFP experienced similar molecular environments in both organelles.

Another application of lifetime imaging is in the identification of the nucleic acid (DNA or RNA) distribution in a cell. Nucleic acid stains can be expected to show significant changes in their lifetime, when bound to double-stranded or single-stranded nucleic acids. Hence, lifetime imaging of labeled DNA with or without FRET may provide new approaches to the analysis of the organization of intracellular DNA. Murata et al. (2000) reported the lifetime imaging of DNA in cell nuclei. They applied fluorescence lifetime imaging microscopy to study the spatial distribution of dsDNA-bound donors and acceptors in nuclei of fixed cells. In this study, cells were stained with a dsDNA label, Hoechst 33258, as the donor and 7-aminoactinomycin D (7-AAD) as the acceptor. The spatial variation of the lifetime of the Hoechst dye fluoresence in the presence of 7-AAD shows that one can identify the regions where FRET processes (indicated by a nonexponential decay) occurs. In some cells, addition of 7-AAD resulted in a spatially nonhomogeneous decrease in intensity. Lifetime distribution, represented in Figure 8.22, shows the Hoechst dye fluorescence that is associated with a specific time in cell cycle which relates to condensation of DNA.

8.6 TISSUE IMAGING

In terms of understanding biological processes, the main approaches are *in vitro, in vivo*, and *ex vivo*. The *in vitro* approach allows the study of individual cells, but cannot mimic the exact environment in which the real processes occur. The *in vivo* approach utilizes the study of cells and tissues in living organisms, such as a live animal or a human body. The *ex vivo* approach, in our context, refers to the study of a dissected or excised tissue specimen outside of the host organism. Optical methods of bioimaging of both *ex vivo* and *in vivo* tissues provide resolution and specificity to probe details of cellular processes in tissues (Cahalan et al., 2002).

The tissues can be divided into two broad classes: (i) soft tissues such as an organ and (ii) hard tissue such as bones. Optical imaging of both types of

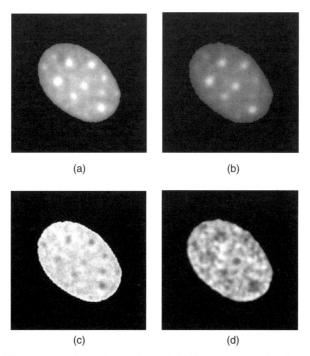

(a)   (b)

(c)   (d)

Figure 8.22. Fluorescence intensity, ratio, and lifetime images of cell nucleus stained with the Hoecht dye. (a) Fluorescence intensity image in the absence of 7-AAD. (b) Fluorescence intensity image in the presence of 7-AAD. (c) Ratio image in (b) over the image in (a). (d) Modulation lifetime in the presence of 7-AAD. This image is smoothed for better recognition of dark spots. The regions with high efficiency of energy transfer, which are shown as dark spots in image (c), correspond with short lifetimes in image (d). The short lifetime spots and long lifetime areas in (d) are about 1.2 ± 0.05 and 1.70 ± 0.06 nsec, respectively. (Reproduced with permission from Murata et al., 2000.)

tissues has been successfully performed. In the case of a soft tissue, optical imaging techniques work much better, because the light penetration is deeper and the scattering is less. Optical imaging of hard tissues are more of a challenge, because they are highly scattering media.

An example of soft tissue imaging is given here, where a study to evaluate the efficacy of drug delivery provided an ideal opportunity for corneal imaging. Targeted and more efficient delivery of agents for therapeutic applications is a priority for the pharmaceutical industry. Effective strategies are targeted to reduce the required dose, increase safety, and improve efficacy by focusing therapeutic agents into the desired site of action. Mucosal routes of drug delivery offer a number of logistical and biological advantages. Mucoadhesion is an important feature of topical sustained-release dosage forms, which may increase the duration or intensity of contact between drug molecules and the epithelium and thus provide control over the site and duration of drug release. In ocular drug delivery, one of the major problems is providing and maintaining an adequate concentration of a therapeutic agent in the pre-corneal area. Topical dropwise administration of ophthalmic drugs in aqueous solutions results in extensive drug loss, due to tear fluid and eyelid dynamics. It has been determined that only 1–3% of the administered dose penetrates the cornea and reaches intraocular tissues. A polyacrylic acid (PAA) nanopar-ticle formulation for ocular drug delivery has been developed at our Institute (De et al., 2003a). An *ex vivo* study utilizing dye-conjugated PAA nanoparti-cles was used to determine the adherence and penetration of the carrier par-ticles into the corneal tissue (De et al., 2003b). The two-photon laser scanning microscopy (TPLSM) images show optical sectioning done on an intact cornea incubated with the PAA nanoparticles that have been conjugated with a two-photon dye (AF-240, synthesized at Air Force Research Lab, Dayton, Ohio). The images shown in Figure 8.23 reveal penetration of the nanoparticles into the tissue and adherence to it, demonstrating the utility of the *ex vivo* imaging for targeted drug delivery. Additional studies indicate a controlled release of the therapeutic agent into the interior side of the cornea.

Imaging of tissue extracts, as in the case of biopsy, can be a valuable tool in cancer diagnosis or in studying the efficacy of a treatment. As shown in Figure 8.24, *ex vivo* imaging of a hamster cheek pouch tumor, after an injection of targeting chemotherapeutic agent AN-152 labeled with C625 (this dye–drug conjugate has been described in Section 8.5.3), shows a selective accumulation of the chemotherapeutic drug in the cancerous tissues. As seen from the images, the normal tissue does not show any significant fluorescence under the same imaging condition, whereas the tumor tissue shows strong fluorescence from the fluorescent tag. Details of this drug–dye conjugate's action at the cel-lular level and the tracking of its cellular pathways using TPLSM have already been described in Section 8.5.3.

In imaging of hard tissues, the scattering of light is a major problem. In mineralized tissue, such as bones and teeth, optical penetration is significantly reduced, resulting in only low-resolution images. As discussed in Chapter 7, optical coherence tomography (OCT) and two-photon laser scanning microscopy (TPLSM) can overcome this difficulty. Figure 8.25 shows the image of a human tooth obtained using a benchtop OCT, designed at our Institute (Figure 7.14). As can be seen, OCT can readily differentiate the structures of the tooth (e.g., dentin, cementum, and enamel). The significance of this imaging technique is its ability to detect incipient caries (demineralization of the enamel before involvement of the dentin), where remineralization, instead of the use of artificial restorations for treatment, can be implemented. Also, OCT could provide the ability to detect demineralization and microdefects underneath some restorative materials, thereby eliminating the need for x rays. Further-more, OCT could be designed to assess gingival pocket depth and visualize bone loss associated with periodontal disease in the oral cavity. This technology should lead to a significant advance in dental imaging, yielding the advantages of being safer, more versatile, and more cost effective (Rodman et al., 2002).

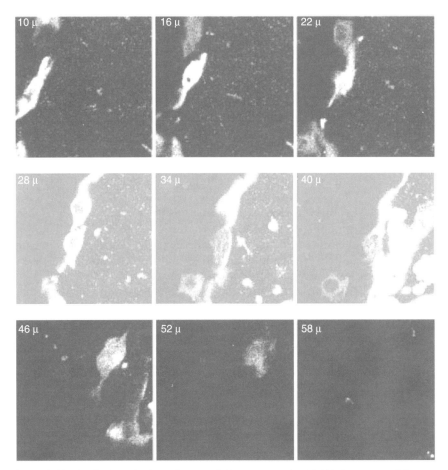

Figure 8.23. Optical sectioning of corneal tissue, treated with PAA nanoparticles conjugated with two-photon dye AF240 and imaged using TPLSM (6-μm step), showing the adhesion and penetration of the PAA particles used as the drug carrier. (Reproduced by permission from De et al., 2003b.)

Multiphoton imaging (e.g., TPLSM) provides another approach to improve penetration depth in hard tissues. An example of the use of this imaging method to image a tooth is shown in Figure 8.26 (Rodman et al., 2002). Here, an efficient two-photon fluorophore (APSS) was incorporated into a dental bonding agent (DBA) used in resin restorative procedures. After incubation with the doped DBA, the tooth was then optically sectioned (nondestructively) using TPLSM. In the particular case shown in Figure 8.26, computer-generated three-dimensional images were successfully used to image DBA that had penetrated ~50 μm into the dentinal tubules of the tooth. This obser-

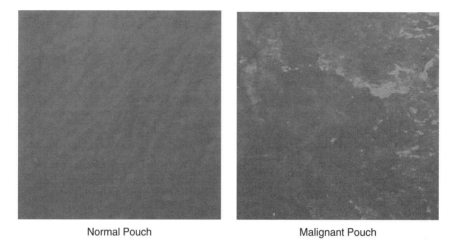

Normal Pouch   Malignant Pouch

Figure 8.24. *Ex vivo* imaging of extracted tumor tissue from a hamster cheek pouch showing the selective accumulation of drug–dye conjugate in tumor (two-photon imaging using 800-nm excitation). In this pseudocolored image, blue is the transmission signal while red shows two-photon fluorescence signal.

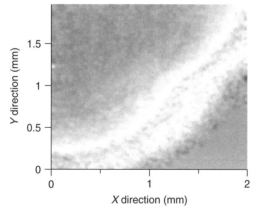

Figure 8.25. Optical coherence tomography image of cross section of human tooth. (Reproduced with permission from Rodman et al., 2002.) (See color figure.)

vation was verified by scanning electron microscopy, which involves a more labor-intensive process using hard tissue sectioning. Since the efficiency of any dental bonding agent is measured by its penetration efficiency in the dentinal tubules, TPLSM provides a nondestructive, rapid process for evaluation of DBA efficacy.

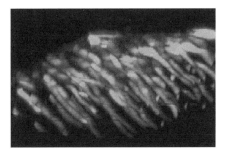

Figure 8.26. Three-dimensional reconstruction of dentinal tubule images, obtained using TPLSM. Extracted tooth was treated with dye-doped dental bonding agent and imaged using a confocal microscope with an 800-nm excitation from a Ti:sapphire laser. (Reproduced with permission from Rodman et al., 2002.)

8.7   *IN VIVO* IMAGING

Optical techniques like fluorescence imaging and optical coherence tomography, adapted for *in vivo* imaging, can complement other techniques such as MRI, CT, and ultrasound, for diagnosis and therapy. Imaging technologies that can provide the same kinds of cellular and *in vivo* molecular information, currently available only from *in vitro* techniques, would be very useful for molecular profiling of diseases and for monitoring their progression.

A major thrust of *in vivo* imaging is in the early detection of cancer. The *in vivo* imaging of tumors thus complements the *in vivo* spectroscopic techniques discussed in Section 6.7. It provides information on spatial distribution and localization of a tumor. Another area of major application of *in vivo* imaging is ophthalmology.

Examples of bioimaging for these applications are provided here.

In Vivo *Corneal Imaging.* Petran et al. (1968) successfully used a real-time direct view confocal microscope based on a spinning Nipkow disk (discussed in Chapter 7) to observe thin optical sections of an *ex vivo* animal cornea (Egger et al., 1969). Subsequently, a clinical confocal microscope, based on a Nipkow disk with an intensified video camera as a detector, was developed by Tandem Scanning Corporation, Inc. Webb utilized a scanning laser ophthalmoscope based on the confocal principle to image the retina (Webb, 1990, 1996; Webb et al., 1980, 1987).

The example provided in Figure 8.27 is that of *in vivo* imaging of a normal human cornea using a slit scanning confocal microscope (Böhnke and Masters, 1999), which utilizes line scanning instead of point scanning. This microscope uses a halogen lamp for illumination and uses a video camera as the detector. This study evaluated the effect of wearing contact lenses on the human cornea.

A normal human cornea ($\sim$500 μm thickness) has multiple layers consisting of epithelium (surface cells, intermediate cells, basal cells), Bowman's

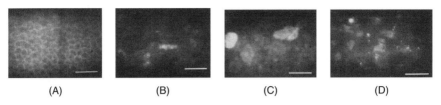

(A)   (B)   (C)   (D)

Figure 8.27. (A) Normal human cornea after 7 hours of wearing a 10-mm-diameter PMMA contact lens. (B) The same cornea with a section of the mid-stroma. (C) The same cornea, 1 hour after discontinuation of the contact lens. (D) A cornea after 15 years of wearing contact lenses. (Reproduced with permission from Böhnke and Masters, 1999.)

layer, anterior stroma, mid-stroma, posterior stroma, Descemet's membrane, and endothelium. Normally, the oxygen required for essential metabolism of the cornea is primarily derived from the atmosphere, via tears, and diffusion across the cornea's anterior surface. The oxygen requirement for each corneal layer is different and any disruption of this oxygen supply can create hypoxic conditions in the cornea. It is well known that wearing contact lenses changes the corneal oxygen consumption, creating hypoxic conditions. Due to hypoxia, the cornea may develop acidosis of the epithelium and stroma, which may cause tissue changes. Contact lens wear induces short-term and long-term changes that can be attributed to mechanical and hypoxic damage. Therefore, it is important to have a simple tool to monitor the changes that can occur over a long period of time. *In vivo* confocal microscopy can be an ideal tool for monitoring these tissue changes.

Figures 8.27A and 8.27B present images of a cornea after 7 hours of exposure to contact lens (Böhnke and Masters, 1999). This cornea was not previously exposed to a contact lens. The intercellular spaces of the basal epithelial cells are dilated with increased reflectivity, possibly at the anterior interface. As seen in Figure 8.27B, the keratocyte cell bodies become visible, in addition to the cell nuclei, due to a decrease in the refractive index caused by the stromal edema (accumulation of fluids in the stromal layer of cornea). This process is fully reversible. Figure 8.27C shows the same cornea as in Figures 8.27A and 8.27B, 1 hour after removal of the contact lens. The corneal epithelium shows an increased visibility of surface cells with signs of desquamation (scaling). Figure 8.27D shows a human cornea after 15 years of regular usage of soft contact lenses. Numerous small structures called microdots are present in all regions and layers of the corneal stroma. Some of these microdots may be deposited in the extracellular compartment, where they accumulate over time.

From their study, Böhnke and Masters (1999) concluded that long-term wearing of contact lenses causes irreversible deep stromal degeneration. *In vivo* confocal microscopy can also be useful in comparing the effect of different contact lenses and individual contact lens tolerance with high sensitivity. Thus, *in vivo* confocal microscopy can be used as a diagnostic tool for early detection as well as for long-term monitoring of corneal defects in ophthalmology.

The recent development of compact endoscopic confocal microscopes (e.g., Optiscan Inc.) has advanced the use of confocal microscopy for *in vivo* imaging. This kind of endoscopic tool, coupled with the monitoring of endogenous autofluorescence (described in a previous section) or of the fluorescence of drugs (e.g., Doxorubicin, the chemotherapeutic drug, shows fluorescence when excited with 480 nm), can be used for clinical diagnosis of cancers or to study the efficacy of drugs.

Imaging of Green Fluorescent Protein-Expressing Tumors and Metastases. The application of green fluorescent protein expression for the visualization of cancer invasion and micrometastasis in live tissues has been discussed by Chishima et al. (1997). Yang et al. (2000) used green fluorescent protein (GFP) expression in cancer cells as an effective tumor cell marker in conventional diagnostic dissections. The fluorescence imaging revealed tumor progression, allowing detection of metastases in exposed or isolated fresh visceral organs and tissues down to the single-cell level.

As shown in Figure 8.28, murine melanoma metastases in the mouse brain was imaged, in real time, using GFP expression (Yang et al., 2000). These experiments with the B16F0 melanoma utilized the tumor cells that were labeled by external GFP transduction and injected into the tail vein or portal vein of nude mice. The whole-body image as well as those of specific organs (such as the liver and skeleton) were obtained by using fluorescence (e.g., epifluorescence) microscopy. The size of the metastasis and micrometastasis determined the depth to which they could be imaged. Thus, a 60-μm tumor could be detected to a depth of 0.5 mm, while a 1.8-mm tumor could be imaged at even a larger depth of 2.2 mm.

Figure 8.28 shows the external images of B16F0-GFP colonizing the liver. The image clearly identifies the formation of metastatic lessions of B16F0-GFP in the nude mouse liver formed after portal vein injection. Images of multiple metastatic lessions in the liver can be seen at a depth of 0.8 mm through the abdominal wall of the intact mouse.

The work by Yang et al. (2000) illustrates that the tracking of cancer cells that stably express GFP *in vivo* can be a sensitive and rapid procedure. A major advantage provided by the GFP-expressing tumor cells is that imaging requires no preparative procedures. Therefore, this method is uniquely suited for visualizing live tissues using high GFP-expression tumor cells.

Retinal Imaging. An example of *in vivo* imaging is in angiography using an infrared dye, indocyanine green (ICG). Angiography is a diagnostic test used to image posterior ocular structures such as retinal or choroidal circulation. The test identifies leakage or damage to the blood vessels nourishing the retina (details of the eye are provided in Section 6.4.1). The ICG dye is injected into the patient's venous circulation. The dye travels through the blood stream to

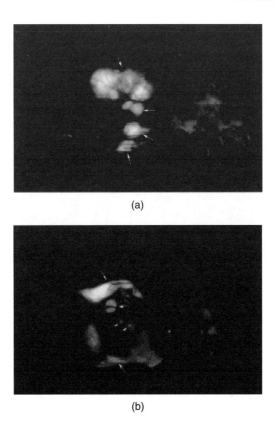

(a)

(b)

Figure 8.28. External images of B16F0-GFP colonizing the liver. A metastatic lesion of B16F0-GFP in the liver growing at a depth of 0.8 mm after portal vein injection was externally imaged through the abdominal wall of the intact nude mouse. (a) An external image of multilobe liver metastases of the B16F0-GFP cells (*large arrows*). (b) An external image of small liver metastatic lesions of approximately 1.5 mm in diameter (*small arrows*) and other larger metastatic lesions (*large arrows*). (Reproduced with permission from Yang et al., 2000.) (See color figure.)

the eye in about 15–20 seconds. Once the dye reaches the eye, it can be excited using an infrared excitation light (784 nm) to fluoresce at 820 nm. This fluorescence can be used to image "hot spots" such as subretinal neovascular membranes and leaking vessels of the choroid (the vascular layer of the eye between the retina and the sclera). The fine capillaries of the choroid grow upward beneath the basement membrane of the retina, and that growth is said to be a "subretinal neovascular membrane." The image can then guide the doctor to target the leak with the optimal laser modality, sealing the vessels using laser or photodynamic therapy (see Chapter 12). The ICG imaging procedure thus reduces the risk of damage to the surrounding retinal tissue.

Figure 8.29 is a representative example of the ICG angiogram which shows the deeper choroidal vessels that can be imaged readily by using the long-

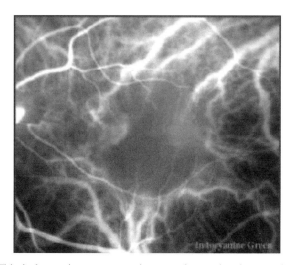

Figure 8.29. This indocyanine green angiogram shows the deeper choroidal vessels. The leaky vessels (choroidal neovascularization) are shown well here (Courtesy of The Macula Foundation, Inc.).

wavelength emission of ICG. The leaky vessels due to choroidal neovascularization (new blood vessel growth) are clearly visible in the image. This type of image using ICG is very useful as a diagnostic tool for age-related macular degeneration (described in Chapter 12), where the leakage is from the deeper choroidal vessels. Another dye, fluorescein, which is excited in the visible range and emits at ~520 nm, does not probe deeper choroidal blood vessel layers and is thus used for studying retinal circulation. Therefore, retinal imaging using fluorescein is useful to diagnose diabetic retinopathy and retinal occlusive diseases that primarily affect the retinal circulation.

Optical Mammography. Optical imaging may allow for an early assessment of cancer. An area of considerable current activity is optical mammography (also referred to as laser mammography) for the noninvasive detection of breast cancer. A prototype optical mammograph, relying on near-infrared fluorescence (NIR), has recently been reported (Weissleder et al., 1999; Bremer et al., 2002). In a mouse study, these researchers used *in vivo* imaging to noninvasively gauge the aggressiveness of breast tumors. This type of optical imaging takes advantage of the fact that tumors overexpress enzymes such as the protease cathepsin-B, which helps the cancer break down the surrounding tissues and invade the bloodstream. Weissleder et al. (1999) used a novel approach in which the autoquenched NIR fluorescence of an administered probe is generated (restored) by the lysosomal protease in tumor cells. The amount of light generated depends on the amount of protease in the tumor. For this purpose, aggressive and nonaggressive human breast tumors were

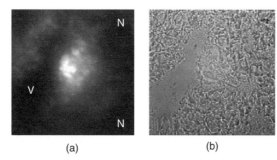

(a)   (b)

Figure 8.30. NIRF histology of tumor excised from an animal. Left image shows the NIRF acquisition, which is superimposed in red onto the correlative phase contrast microscopy image on the right. Vessels (V) and areas of necrosis (N) are labeled. Sections are unstained and unfixed to preserve fluorescence signal. (Reproduced with permission from Weissleder et al., 1999.) (See color figure.)

implanted into mice and the NIR probes were intravenously injected. These probes were bound to a long circulating graft copolymer consisting of poly-L-lysine and methoxypolyethylene glycol succinate. Following an accumulation of the NIR probe carrier in solid tumors, these macromolecules were cleaved by lysosomal proteases in tumor cells, generating fluorescence from the NIR probe. The *in vivo* imaging, shown in Figure 8.30, reveals a 12-fold increase in the NIR fluorescence signal, allowing the detection of tumors of submillimeter-sized diameters. In this study, the NIR fluorescence (NIRF) in tumors could be detected up to 96 hours (Weissleder et al., 1999). The simultaneous performance of autoradiography and NIRF microscopy using duly labeled probes confirmed that the NIRF signal originated from within the tumor cells.

Using the same technique, even whole-body images can be acquired noninvasively and tumorous and actively proliferating tissue sections can be identified, as shown in Figure 8.31. Time progression image studies showed that breast tumors were undetectable before and immediately after the injection of the NIR probe. But within 24 hours, all implanted tumors generated sufficient NIR fluorescence to image tumors of dimensions <300 μm. In their study, Weissleder et al. (1999) could achieve *in vivo* imaging up to a depth of 7–10 mm and projected that a depth resolution of 5–6 cm could be achieved by using their NIRF probe in more sensitive tomographic imaging systems such as OCT. This strategy could be used to detect the early stage of tumors *in vivo* and to probe specific enzyme activity.

Catheter-Based Endoscopic Optical Coherence Tomography. OCT, with its high resolution compared to MRI, is a powerful imaging technique for the diagnosis of a wide range of gastrointestinal pathologies. Here imaging can be performed *in situ* using an endoscope, allowing pathology of the gastrointestinal tract to be monitored on a screen and stored on a high-resolution

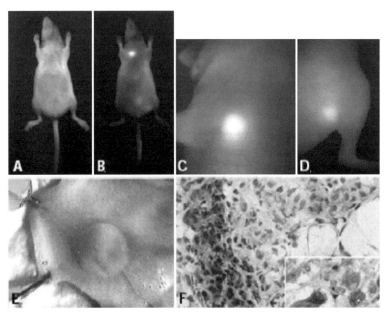

Figure 8.31. LX-1 tumor implanted into the mammary fat pad of a nude mouse. (A) Light image. (B) Raw NIRF image. Note the bright tumor in the chest. (C) High-resolution NIRF images of the chest wall tumor (2 mm). (D) High-resolution NIRF image of the additional thigh tumor (<0.3 mm). (E) Dissected tumor in the mammary pad. (F) Hematoxylin–eosin section of the NIRF positive tumor showing malignant and actively proliferating cells (magnification 200×; insert 400×). (Reproduced with permission from Weissleder et al., 1999.)

videotape (Fujimoto et al., 1999). Because OCT is fiber optic-based, it also allows imaging using a small catheter-based endoscopic accessory. *In vivo* endoscopic or catheter-based optical coherence tomographic imaging has been performed on a rabbit gastrointestinal tract, respiratory tract, and circu-latory system (Tearney et al., 1996, 1997a, 1997b). OCT is also emerging as a useful imaging technique in cardiology, capable of defining arterial structure on a micron scale.

The aorta is a large blood vessel that carries oxygen-rich blood from the heart to the rest of the body. Fujimoto et al. (1999) obtained *in vivo* aorta images in 12-week-old New Zealand white rabbits. Images were obtained after a manual 2–3 mL/sec saline flush of the aorta using extremely short (~10–15 fsec) and consequently broad bandwidth pulses (75 nm centered at 1280 nm from a Kerr lens mode-locked Cr^{4+}:Forsterite laser) were used as the light source. Figure 8.32 shows the results. A visual image of the entire aorta was seen at all times in the presence of the saline flush. However, in the absence of saline, no significant structural information was obtained. This result points to the attenuation of the optical signal in the presence of blood, thus requir-

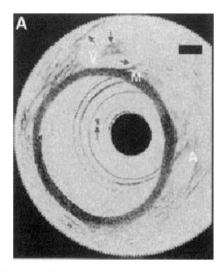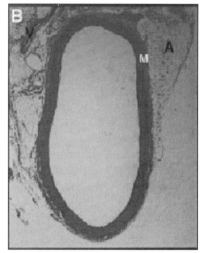

Figure 8.32. Aorta image after a saline flush and corresponding histology. (A) The media and surrounding supportive structures are clearly identified. (B) Histology has been included to confirm tissue identification. (Fujimoto et al., Heart (1999), 82, 128–133, reproduced with permission from BMJ publishers.)

ing the displacement of blood with saline. In this image, the media and the surrounding supportive structures are clearly identified. In the image A, the media "M" exhibits high back-scattering intensity; the adventitia "A," which consists primarily of loose connective tissue, exhibits low back-scattering intensity. A small clot adherent to the distal end of the catheter, which formed midway through imaging, is seen in many of the images (red arrow). A structure consistent with the inferior vena cava "V" is imaged through the wall of the aorta. The walls of the inferior vena cava are noted (green arrows), as well as a blood clot within (blue arrow). The bars represent 500 μm in all images. (The superior and inferior vena cava are veins that deliver oxygen-depleted blood from the upper part of the body, including the head, and the lower part of the body, respectively, to the right side of the heart.) This example demonstrates that high-resolution OCT imaging can be performed near the resolution of histology, which in the future could identify high-risk coronary plaques and provide guidance for interventional procedures.

8.8 FUTURE DIRECTIONS

The range of applications of optical bioimaging continues to expand at a rapid pace. Important developments in bioimaging techniques, which include miniaturization coupled with noninvasive/minimally invasive imaging, will open new frontiers. The development of new fluorophores providing enhanced selectivity, sensitivity, and ability to be activated *in situ* can lead to new capabilities to probe, in real time, *in vivo* physical and chemical processes.

In this section, some selected examples of projected areas of future developments are presented.

Near IR Imaging. Here opportunities lie in the development of new, stable and highly efficient fluorophores that can be excited by available near-IR lasers. New IR dyes, which can be excited at ~1200 nm with a Forsterite or new Cunyite laser, are of significant interest, because this wavelength provides significantly improved penetration, even in dense tissues.

New near-IR, fiber-based lasers will also be of significant value in the development of miniaturized imaging and light delivery systems.

Nanoparticle Platform. The use of nanoparticles, either for encapsulation of a fluorophore for enhanced luminescence efficiency or as a carrier to a specific biological site, is expected to be an area of considerable activity. The development of novel fluorescent nanoparticles, such as those of Si or Si/Ge, also holds promise for biomaging. Their luminescence wavelengths can be altered by varying the size of the nanoparticles when in the quantum confinement limit. At our Institute, Si nanoparticle emitters covering the entire visible range have already been accomplished (Swihart, Cartwright, and coworkers, unpublished). These nanoparticles can also be oxidized in a controlled way to produce a silica coating that can be surface functionalized and coupled to a carrier to target a specific site (organelle) to be imaged. Silicon nanoparticles are further discussed in Chapter 15.

In Situ *Fluorescence Activation of Probes.* As illustrated by the example of a near-IR probe that was activated from a nonfluorescent state to a highly fluorescent state by the action of an enzyme, the method of *in situ* generation or restoration of fluorescence is a very exciting prospect. It provides considerable enhancement of sensitivity and enzyme specificity either to probe various physiological and biochemical processes or to monitor drug action *in vivo*.

Real-Time* In Vivo *Imaging. Real-time *in vivo* imaging provides powerful capabilities to monitor a biological response to a stimulus. Thus, the progress and efficacy of a drug treatment or a drug action can be monitored in real time. Improvements are needed both in the instrumentation and in the development of fluorophores which do not interfere with the process to be monitored. The time resolution required may vary over many orders of magnitude, as discussed for various biological processes in Chapters 3 and 6.

Imaging of Microbes. As discussed in this chapter, near-field microscopy is well-suited for imaging of viruses and bacteria, since it provides the necessary resolution. Such studies have been limited, because NSOM in the traditional method of imaging requires immobilization (fixing) of a microbe. Imaging under physiological conditions requires imaging in a liquid environment. Although some reports of NSOM imaging in liquids have appeared, this area still needs further development (Keller et al., 1998). Another area of future development is the use of a chemically functionalized NSOM tip to selectively probe the structure and function of a specific biological constituent on the surface of a microbe.

8.9 COMMERCIALLY AVAILABLE OPTICAL IMAGING ACCESSORIES

Some suppliers of fluorescence imaging accessories are as follows:

Fluorescent Dyes: Molecular Probes http://www.probes.com/

Green Fluorescent Proteins: Clontech http://www.clontech.com/

Specialized Sample Chambers and Culture Dishes for Environment (e.g., Temperature) Control for Imaging: Bioptechs, Inc. http://www.bioptechs.com/

Beads, Dyes, Polymers, and Supplies for Microscopy: Polysciences http://www.polysciences.com/

Optical Filters for Fluorescence Microscopy: Chroma Technology Corp. http://www.chroma.com/ and Omega Filters http://www.omegafilters.com/

HIGHLIGHTS OF THE CHAPTER

- Optical bioimaging can be used to investigate structures and functions of cells and tissues and to profile diseases at cellular, tissue, and *in vivo* specimen levels.
- Fluorescence microscopy, a preferred method for cellular imaging, is also useful for *ex vivo* and *in vivo* tissue imaging. Both endogenous and exogenous fluorophores are useful for fluorescence bioimaging. Exogenous fluorophores can be used as such or chemically conjugated to target a specific organelle to be imaged.
- An ideal fluorophore for bioimaging has (i) dispersability in the biological medium to be probed, (ii) specific association with a target molecule, organelle, or cell, (iii) high quantum efficiency of emission, (iv) environmental stability, and (v) the absence of photobleaching.
- In addition to the usual fluorescent organic dyes, some other bioimaging fluorophores are: organometallic complexes, near-IR and IR fluorophores, highly efficient new two-photon absorbing fluorophores, and inorganic nanoparticle emitters.

- Another important class of fluorescent labels is fluorescent proteins (FP), which exhibit some unique features like resistance to denaturation and ease in coupling to other biomolecules. The fluorescent proteins are expressed in the cell and thus generated *in situ* by introducing an FP gene in a living organism.

- A wide variety of mutant variants of fluorescent proteins, producing fluorescence of varying wavelengths covering the entire visible spectral range and with enhanced emission, are available. They can be used for a wide range of applications such as a cell lineage tracer, reporter for gene expression or a measure of protein–protein interactions.

- Using selective fluorescent labels, bioimaging of a specific organelle can be accomplished to study its structure and function. An example is selective staining of mitochondria by a dye Mitotracker to monitor cell apotosis and measure mitochondrial membrane potential.

- Both confocal and near-field microscopy have been used for imaging of microbes such as viruses and bacteria.

- Near-field microscopy (NSOM) provides the resolution of <100 nm needed to study the structural details of viruses and bacteria. However, their applications to image viruses and bacteria have been limited, because traditional NSOM requires fixing them and they are not in their natural environment. Examples of NSOM images are that of a tobacco mosaic virus and a *Porphyromonas gingivalis* oral bacteria.

- Cellular imaging can be used to probe a cell's ionic environment, measure intracellular pH measurements, monitor drug–cell interactions, and determine nucleic acid distribution.

- Cellular imaging by two-photon laser scanning microscopy using a fluorescently labeled chemotherapeutic drug-carrier conjugate can optically track the cellular pathway and clarify the drug's mechanism of action. The bioimaging, together with localized (site-specific) fluorescence spectroscopy, indicates that the chemotherapeutic activity of the drug, doxorubicin, is due to its ability to intercalate into DNA and break the strands of the double helix by inhibiting topoisomerase II.

- Examples for confocal imaging of nucleic acid distribution and differentiation of RNA–DNA content are given. Some fluorescent probes are specific to double-stranded DNA, while others bind to both double- and single-stranded nucleic acids.

- Fluorescence *in situ* hybridization (FISH) enables the detection and determination of the spatial distribution of specific DNA or RNA sequences in the cytoplasm, nucleus, and chromosomes. FISH involves *in situ* hybridization of nucleic acids in the target cells or chromosomes to be detected or imaged, with fluorescently labeled single-stranded probe nucleic acids.

- Cellular interactions, such as protein–protein interactions, can be probed by fluorescent resonance energy transfer (FRET) and fluorescence life-time imaging microscopy (FLIM). A suitable energy donor and acceptor pair involves cyanofluorescent protein and yellow fluorescent protein.
- Tissue imaging using optical techniques can be achieved for soft and hard tissues under both *ex vivo* and *in vivo* conditions.
- Examples provided of imaging of soft tissues are (a) optical sectioning of a corneal tissue treated with a two-photon conjugated polyacrylic acid nanoparticles and (b) extracted tumor tissue of a hamster cheek pouch with selective accumulation of the drug: two-photon fluorophore conjugate. Both images are obtained by using two-photon laser scanning microscopy.
- Both two-photon laser scanning microscopy and optical coherence tomography have proved to be more suitable than other ways of optical bioimaging in highly scattering media such as hard tissues.
- Examples provided are (i) OCT images of a section of a human tooth displaying the ability to differentiate the various structures of a tooth and (ii) three-dimensional reconstruction of two-photon laser scanning images of dentinal tubules obtained by filling them with a dental bonding material containing a two-photon dye. A sharp image of dentinal tubules indicates a deep penetration by the dental bonding agent.
- *In vivo* imaging can be used at the level of tissue, organ, or entire live object (animal or human being).
- An example of optically sectioned corneal imaging with confocal microscopy provides evidence of irreversible deep stromal degeneration caused by long-term wearing of contact lenses.
- Another example of *in vivo* imaging utilizes GFP expressed in tumor cells to image tumor localization and growth in a live animal.
- Retinal imaging in angiography is an important example of *in vivo* imaging. It utilizes an infrared dye, indocyanine green, to image the structures in the back of the eye for finding leakage or damage to the blood vessels that nourish the retina.
- An exciting area of *in vivo* imaging is optical mammography, also known as laser mammography, for the noninvasive detection of breast cancer.
- A new approach in optical mammography is the use of a near-infrared dye administered in a self-quenched nonfluorescent state. The fluorescence of the dye is activated (restored) by an enzyme, overexpressed by tumors, thus enabling one to use fluorescence imaging for the study of tumor localization and growth.
- Catheter-based endoscopic optical coherence tomography (OCT) is emerging as a powerful approach for *in vivo* imaging of highly scattering tissues and organs. An application provided is gastrointestinal pathology.

- Future directions of research and development include (i) increased applications of near-IR imaging, thus opening opportunities for development of near-IR fluorophores and lasers, (ii) use of the nanoparticle approach for encapsulation and delivery to a specific biological size, (iii) *in situ* activation of a fluorescent probe in response to a stimulus or a drug, (iv) real-time *in vivo* imaging to monitor biological activities and (v) imaging of microbes much smaller than the wavelength of light in their natural environment.

REFERENCES

Aubin, J. E., Autofluorescence of Viable Cultured Mammalian Cells, *J. Histochem. Cytochem.* **27**, 36–43 (1979).

Bassnett, S., Reinisch, L., and Beebe, D. C., Intracellular pH Measurement Using Single Excitation–Dual Emission Fluorescence Ratios, *Am. J. Physiol.* **258**, C171–C178 (1990).

Bastiaens, P. I., and Squire, A., Fluorescence Lifetime Imaging Microscopy: Spatial Resolution of Biochemical Processes in the Cell, *Trends Cell Biol.* **9**, 48–52 (1999).

Bhawalkar, J. D., He, G. S., and Prasad, P. N., Nonlinear Multiphoton Processes in Organic and Polymeric Materials, *Rep. Prog. Phys.* **59**, 1041–1070 (1996).

Billinton, N., and Knight, A. W., Seeing the Wood Through the Trees: A Review of Techniques for Distinguishing Green Fluorescent Protein from Endogenous Autofluorescence," *Anal. Biochem.* **291**, 175–197 (2001).

Birge, R. R., Two-Photon Spectroscopy of Protein-Bound Fluorophores, *Acc. Chem. Res.* **19**, 138–146 (1986).

Böhnke, M., and Masters, B. R., Confocal Microscopy of the Cornea, *Prog. Retinal Eye Res.* **18**, 553–628 (1999).

Bremer, C., Tung, C. H., Bogdanov, A., Jr., and Weissleder, R., Imaging of Differential Protease Expression in Breast Cancers for Detection of Aggressive Tumor Phenotypes, *Radiology* **222**, 814–818 (2002).

Bruchez, M., Jr., Moronne, M., Gin, P., Weiss, S., and Alivisatos, A. P., Semiconductor Nanocrystals as Fluorescent Biological Label, *Science* **281**, 2013–2016 (1998).

Buckler, K. J., and Jones, V., Application of a New pH-Sensitive Fluoprobe CarboxyS-narf1 for Intracellular pH Measurement in Small, Isolated Cells, *Pflugers Arch.* **417**, 234–239 (1990).

Cahalan, M. D., Parker, I., Wei, S. H., and Miller, M. J., Two-Photon Tissue Imaging: Seeing the Immune System in a Fresh Light, *Nature Rev. Immunol.* **2**, 872–880 (2002).

Chalfie, M., and Kain, S., *GFP, Green Fluorescent Protein—Properties, Applications and Protocols*, Wiley-Liss, New York, 1998.

Chan, F. K., Siegel, R. M., Zacharias, D., Swofford, R., Holmes, K. L., Tsien, R. Y., and Lenardo, M. J., Fluorescence Resonance Energy Transfer Analysis of Cell Surface Receptor Interactions and Signaling Using Spectral Variants of the Green Fluorescent Protein, *Cytometry* **44**, 361–368 (2001).

Chan, W. C. W., and Nie, S., Quantum Dot Bioconjugates for Ultrasensitive Nonisotopic Detection, *Science* **281**, 2016–2018 (1998).

Chishima, T., Miyagi, Y., Wang, X., Yamaoka, H., Shimada, H., Moossa, A. R., and Hoffman, R. M., Cancer Invasion and Micrometastasis Visualized in Live Tissue by Green Fluorescent Protein Expression, *Cancer Res.* **57**, 2042–2047 (1997).

Dahan, M., Laurence, T., Pinaud, F., Chemla, D., Alivisatos, A. P., Sauer, M., and Weiss, S., Time-Gated Biological Imaging Using Colloïdal Quantum Dots, *Opt. Lett.*, **26**, 825 (2001).

De, T. K., Rodman, D. J., Bergey, E. J., Holm, B. A., and Prasad, P. N., Brimonidine Formulation in Polyacrylic Acid Nanoparticles for Ophthalmic Delivery, *J. Microencapsulation* (2003a), in press.

De, T. K., Rodman, D. J., Bergey E. J., Chung, S. J., Holm, B. A., and Prasad, P. N., Polyacrylic Acid Nanoparticles as Carrier of Brimonidine for Ophthalmic Delivery—An *Ex Vivo* Evaluation with Human Cornea, to be submitted, 2003b.

Denk, W., Strickler, J. H., and Webb, W. W., 2-Photon Laser Scanning Fluorescence Microscopy, *Science* **248**, 73–76 (1990).

Diaspro, A., ed., *Confocal and Two-Photon Microscopy—Foundations, Applications, and Advances*, Wiley-Liss, New York, 2002.

Dürig, U., Pohl, D. W., and Rohner, F., Near-Field Optical-Scanning Microscopy, *J. Appl. Phys.* **59**, 3318–3327 (1986).

Egger, M. D., Gezari, W., Davidovits, P., Hadravsky, M., and Petran, M., Observation of Nerve Fibers in Incident Light, *Experientia* **25**, 1225–1226 (1969).

Frederich, D. M., and McClarin, W. M., Two-Photon Molecular Electronic Spectroscopy, *Annu. Rev. Phys. Chem.* **31**, 559–577 (1980).

Fu, Y., and Galan, J. E., The *Salmonella typhimurium* Tyrosine Phosphatase SptP is Translocated into Host Cells and Disrupts the Actin Cytoskeleton, *Mol. Microbiol.* **27**, 359–368 (1998).

Fujimoto, J. G., Boppart, S. A., Tearney, G. J., Bouma, B. E., Pitris, C., and Brezinski, M. E., High Resolution *In Vivo* Intra-Arterial Imaging with Optical Coherence Tomography, *Heart* **82**, 128–133 (1999).

Fujimoto, J. G., Brezinski, M. E., Tearney, G. J., Boppart, S. A., Bouma, B. E., Hee, M. R., et al., Optical Biopsy and Imaging Using Optical Coherence Tomography, *Nat. Med.* **1**, 970–972 (1995).

Fung, D. C., and Theriot, J. A., Imaging Techniques in Microbiology, *Current Opin. Microbiol.* **1(3)**, 346–351 (1998).

Gadella, T.W., Jr., van der Krogt, G. N., and Bisseling, T., GFP-Based FRET Microscopy in Living Plant Cells, *Trends Plant Sci.* **4**, 287–291 (1999).

Gayen, S. K., Zevallos, M. E., Aerubaiee, M., and Alfano, R. R., Near-Infrared Laser Spectroscopic Imaging: A Step Toward Diagnostic Optical Imaging of Human Tissue, *Lasers Life Sci.* **98**, 187–198 (1999).

Graham, D. Y., Schwartz, I. T., Cain, G. D., and Gyorkey, F., Prospective Evaluation of Biopsy Number in the Diagnosis of Esophageal and Gastric Carcinoma, *Gastroenterology* **82**, 228–231 (1982).

Gu, X., and Spitzer, N.C., Distinct Aspects of Neuronal Differentiation Encoded by Frequency of Spontaneous Ca^{2+} Transients, *Nature* **375**, 784–787 (1995).

Harper, I. S. Fluorophores and Their Labeling Procedures for Monitoring Various Biological Signals, in A. Periasamy, ed., *Methods in Cellular Imaging*, Oxford University Press, Hong Kong, 2001, pp. 20–39.

Haugland, R. P., *Handbook of Fluorescent Probes and Research Products*, 9th edition, Molecular Probes, Inc., Eugene, Oregon, 2002.

He, G. S., Zhao, C. F., Bhawalkar, J., and Prasad, P. N., Two-Photon Cavity Lasing in Novel Dye Doped Bulk Matrix Rods, *Appl. Phys. Lett.* **67**, 3703–3705 (1995).

Heim, R., Cubitt, A. B., and Tsien, R. Y., Improved Green Fluorescence, *Nature* **373**, 663–664 (1995).

Heim, R., Prasher, D. C., and Tsien, R. Y., Wavelength Mutations and Posttranslational Autoxidation of Green Fluorescent Protein, *Proc. Natl. Acad. Sci. USA* **91**, 12501–12504 (1994).

Herman, B., Resonance Energy Transfer Microscopy, *Methods Cell Biol.* **30**, 219–243 (1989).

Holm, B. A., Bergey, E. J., De, T., Rodman, D. J., Kapoor, R., Levy, L., Friend, C. S., and Prasad, P. N., Nanotechnology in Biomedical Applications, *Mol. Cryst. Liq. Cryst.* **374**, 589–598 (2002).

Jovin, T. M., and Arndt-Jovin, D. J., Luminescence Digital Imaging Microscopy, *Annu. Rev. Biophys. Biophys. Chem.* **18**, 271–308 (1989).

Kapoor, R., Friend, C., Biswas, A., and Prasad, P. N., Highly Efficient Infrared-to-Visible Energy Up-Conversion in $Er^{3+}:Y_2O_3$, *Opt. Lett.* **25**, 338–340 (2000).

Keller, T. H., Rayment, T., and Klenerman, D., Optical Chemical Imaging of Tobacco Mosaic Virus in Solution at 60 nm Resolution, *Biophys. J.* **74**, 2076–2079 (1998).

Kozubek, M., FISH Imaging, in A. Diaspro, ed., *Confocal and Two-Photon Microscopy-Foundations, Applications and Advances*, Wiley-Liss, New York, 2002, pp. 389–429.

Krebs L. J., Wang, X., Pudavar, H. E., Bergey, E. J., Schally, A. V., Nagy, A., Prasad, P. N., and Liebow, C., Regulation of Targeted Chemotherapy with Cytotoxic Lutenizing Hormone Releasing Hormone Analogue by Epidermal Growth Factor, *Cancer Res.* **60**, 4194–4199 (2002).

Lakowicz, J. R., *Principles of Fluorescence Spectroscopy,* 2nd edition, Kluwer Academic/Plenum Publishers, New York, 1999.

Martin, R. B., Richardson, F. S., Lanthanides as Probes for Calcium in Biological Systems, *Q. Rev. Biophys.* **12**, 181–209 (1979).

Murata, S, Herman, P., Lin, H-J., and Lakowicz, J. R., Fluorescence Lifetime Imaging of Nuclear DNA: Effect of Fluorescence Resonance Energy Transfer, *Cytometry* **41**, 178–185 (2000).

Ohulchanskyy, T. Y., Pudavar, H. E., Yarmalok, S. M., Yashchuk, V. M., and Prasad, P. N., A Monomethine Cyanine Dye, Cyan 40 for Two-Photon Excited Fluorescence Detection of Nucleic Acids and Their Visualization in Live Cells, *Photochem. Photobiol.* (2003), in press.

Periasamy, A., Sharman, K. K., Ahuja, R. C., Eto, M., and Brautigan, D. L., Fluorescence Lifetime Imaging of Green Fluorescent Protein in a Single Living Cell, *SPIE Proc.* **3604**, 6–12 (1999).

Periasamy, A., ed., *Methods in Cellular Imaging*, Oxford University Press, Hong Kong, 2001.

Petran, M., Hadravsky, M., Egger, M. D., and Galambos, R., Tandem-Scanning Reflected-Light Microscopy, *J. Opt. Soc. Am.* **58**, 661–664 (1968).

Pinkel, D., Fluorescence *In Situ* Hybridization, in M. Andreef, and D. Pinkel, eds., *Introduction to Fluorescence In Situ Hybridization*, Wiley-Liss, New York, 1999, pp. 3–32.

Pogue, B. W., Pitts, J. D., Mycek, M. A., Sloboda, R. D., Wilmot, C. M., Brandsemal, J. F., and O'Hara, J. A., *In Vivo* NADH Fluorescence Monitoring as an Assay for Cellular Damage in Photodynamic Therapy, *Photochem. Photobiol.* **74**, 817–824 (2001).

Pylkki, R. J., Moyer, P. J., and West, P. E., Scanning Near-Field Optical Microscopy and Scanning Thermal Microscopy, *Jpn. J. Appl. Phys.* **33**, 3785–3790 (1994).

Rentzepis, P. M., Mitschele, C. J., and Saxman, A. C., Measurement of Ultrashort Laser Pulses by Three-Photon Fluorescence, *Appl. Phys. Lett.* **17**, 122–129 (1970).

Richter-Dahifors, A., Buchan, A. M. J., and Finlay, B. B., Murine Salmonellosis Studied by Confocal Microscopy: *Salmonella typhimurium* Resides Intracellularly Inside Macrophages and Exerts a Cytotoxic Effect on Phagocytes *In Vivo*, *J. Exp. Med.* **186**, 569–580 (1997).

Rodman, D. J., Bergey, E. J., Leibow, C., and Prasad, P. N., Biophotonics Perspectives for 21st Century, in H. Sasabe, ed., *Nanotechnology Toward the Organic Photonics*, Gootech Ltd., Chitose, Japan, 2002, pp. 29–40.

Salvioli, S., Dobrucki, J., Moretti, L., Troiano, L., Fernandez, M. G., Pinti, M., Pedrazzi, J., Franceschi, C., and Cossarizza, A., Mitochondrial Heterogeneity During Staurosporine-Induced Apoptosis in HL60 Cells: Analysis at the Single Cell and Single Organelle Level, *Cytometry* **40**, 189–197 (2000).

Schally, A. V., and Nagy, A., Cancer Chemotherapy Based on Targeting of Cytotoxic Peptide Conjugates to Their Receptors on Tumors, *Eur. J. Endocrinol.* **141**, 1–14 (1999).

Schmitt, J., Yadlowsky, M., and Bonner, R., Subsurface Imaging of Living Skin with Optical Coherence Microscopy, *Dermatology* **191**, 93–98 (1995).

Tearney, G. I., Boppart, S. A., Bouma, B. E., Brezinski, M. E., Weissman, N. I., Southern, I. F., et al., Scanning Single-Mode Fiber Optic Catheterendoscope for Optical Coherence Tomography, *Opt. Lett.* **21**, 543–545 (1996).

Tearney, G. I., Brezinski, M. E., Bouma, B. E., Boppart, S. A., Pitris, C., Southern, I. F., and Fujimoto, I. G., *In Vivo* Endoscopic Optical Biopsy with Optical Coherence Tomography, *Science* **276**, 2037–2039 (1997a).

Tearney, G. J., Brezinski, M. E., Southern, J. F., Bouma, B. E., Boppart, S. A., and Fujimoto, J. G., Optical Biopsy in Human Gastrointestinal Tissue Using Optical Coherence Tomography, *Am. I. Gastroenterol.* **92**, 1800–1804 (1997b).

Tsien, R. Y., The Green Fluorescent Protein, *Annu. Rev. Biochem.* **67**, 509–544 (1998).

Tsien, R. Y., and Prasher, D. C., Molecular Biology and Mutation of Green Fluorescent Protein, in S. K. Chalfie, ed., *GFP: Green Fluorescent Protein Strategies and Applications,* John Wiley & Sons, New York, 1998, pp. 97–118.

Wang, X., Pudavar, H. E., Kapoor, R., Krebs, L. J., Bergey, E. J., Liebow, C., Prasad, P. N., Nagy, A., and Schally, A. V., Studies of the Mechanism of Action of a Targeted Chemotherapeutic Drug in Living Cancer Cells by Two-Photon Laser Scanning Microspectrofluorometry, *J. Biomed. Opt.* **6**, 319–325 (2001).

Wang, X., Krebs, L. J., Al-Nuri, M., Pudavar, H. E., Ghosal, S., Liebow, C., Nagy, A. A, Schally, A. V., and Prasad, P. N., A Chemically Labeled Cytotoxic Agent: Two-Photon

Fluorophore for Optical Tracking of Cellular Pathway in Cshemotherapy, *Proc. Natl. Acad. Sci.* **96**, 11081–11084 (1999).

Webb, R. H., Scanning Laser Ophthalmoscope, in B. R. Masters, ed., *Noninvasive Diagnostic Techniques in Ophthalmology*, Springer-Verlag, New York, 1990, pp. 438–450.

Webb, R. H., Confocal Optical Microscopy, *Rep. Prog. Phys.* **59**, 427–471 (1996).

Webb, R. H., Hughes, G. W., and Delori, F. C., Confocal Scanning Laser Ophthalmoscope, *Appl. Opt.* **26**, 1492–1499 (1980).

Webb, R. H., Hughes, G. W., and Pomerantzeff, O., Flying Spot TV Ophthalmoscope, *Appl. Opt.* **19**, 2991–2997 (1980).

Weissleder, R., Tung, C. H., Mahmood, U., and Bogdanov, A., Jr., *In Vivo* Imaging of Tumors with Proteaseactivated Near-Infrared Fluorescent Probes, *Nat. Biotechnol.* **17**, 375–378 (1999).

Yang, M., Baranov, E., Jiang, P., Sun, F., Li, X., Li, L., Hasegawa, S., Bouvet, M., Tuwaijri, M. A., Chishima, T., Shimada, H., Moossa, A. R., Penman, S., and Hoffman, R. M., Whole-Body Optical Imaging of Green Fluorescent Protein-Expressing Tumors and Metastases, *Proc. Acad. Nat. Sci.* **97**, 1206–1211 (2000).

CHAPTER 9

Optical Biosensors

The field of biosensors has emerged as a topic of great interest because of the great need in medical diagnostics and, more recently, the worldwide concern of the threat of chemical and bioterrorism. The constant health danger posed by new strands of microbial organisms and spread of infectious diseases is another concern requiring biosensing for detecting and identifying them rapidly. Optical biosensors utilize optical techniques to detect and identify chemical or biological species. They offer a number of advantages such as the ability for principally remote sensing with high selectivity and specificity and the ability to use unique biorecognition schemes. The topic of optical biosensors is comprehensively covered in this chapter.

The objectives of this chapter are many. First, it describes the basic optical principles and the various techniques utilized in biosensing, which can be useful as a text for students or non-experts in this field. Second, the detailed coverage of the various optical biosensors, reported ongoing activities, and a list of commercially available optical biosensors can serve as a valuable reference source for researchers. Finally, some examples of opportunities for future developments, provided at the end of the chapter, are intended to stimulate the interest of a new researcher or one interested in expanding an ongoing research and development program in this field.

The two important components of biosensing, discussed in this chapter, are (i) a biorecognition element to detect chemical or biological species and (ii) a transduction mechanism which converts the physical or chemical response of biorecognition into an optical signal. The various types of biorecognition elements are discussed. This is followed by a coverage of the various principles of optical transduction and optical geometries utilized for biosensing. An important aspect of biosensing is to immobilize the biorecognition element to increase its local concentration in the sensor probe. The various physical and chemical methods utilized for this purpose are described.

The subsequent sections describe various types of optical biosensors that have been reported, some of which are already in practice. Specifically, these are fiber-optic biosensors, planar waveguide biosensors, evanescent wave

Introduction to Biophotonics, by Paras N. Prasad.
ISBN: 0-471-28770-9   Copyright © 2003 John Wiley & Sons, Inc.

biosensors, interferometric biosensors, and surface plasmon resonance (abbreviated as SPR) biosensors.

Some novel sensing methods reported recently are described in Section 9.9. Next is a discussion of future development opportunities in Section 9.10. The chapter concludes with Section 9.12, which provides a list of commercial available biosensors.

For further reading, suggested general references are:

Wolfbeis (1991): Covers fiber-optics-based chemical and biosensors

Boisdé and Harmer (1996): Covers optical fibers and waveguide-based sensors

Ramsay (1998): Covers commercial biosensors

Mehrvar et al. (2000): Covers trends and advances in fiber-optic biosensors

Ligler and Rowe-Taitt (2002): Provides comprehensive, up-to-date coverage of optical biosensors

9.1 BIOSENSORS: AN INTRODUCTION

Biosensors are analytical devices that can detect chemical or biological species or a microorganism. They can be used to monitor the changes in the *in vivo* concentrations of an endogenous specie as a function of a physiological change induced internally or by invasion of a microbe. Of even more recent interest is the use of biosensors to detect toxins, bacteria, and viruses because of the danger posed by chemical and biological terrorism. Biosensors thus find a wide range of applications:

• Clinical diagnostics
• Drug development
• Environmental monitoring (air, water, and soil)
• Food quality control

A biosensor in general utilizes a biological recognition element that senses the presence of an analyte (the specie to be detected) and creates a physical or chemical response that is converted by a transducer to a signal. The general function of a biosensor system is described in Figure 9.1. The sampling unit introduces an analyte into the detector and can be as simple as a circulator. The recognition element binds or reacts with a specific analyte, providing biodetection specificity. Enzymes, antibodies or even cells such as yeast or bacteria have been used as biorecognition elements. The principles of biorecognition are discussed in Section 9.2. Stimulation, in general, can be provided by optical, electric, or other kinds of force fields that extract a response as a result of biorecognition. The transduction process transforms the physical or chem-

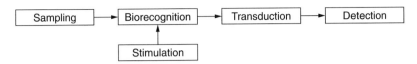

Figure 9.1. General scheme for biosensing.

ical response of biorecognition, in the presence of an external stimulation, into an optical or electrical signal that is then detected by the detection unit. The detection unit may include pattern recognition for identification of the analyte. In the most commonly used form of an optical biosensor, the stimulation is in the form of an optical input. The transduction process induces a change in the phase, amplitude, polarization, or frequency of the input light in response to the physical or chemical change produced by the biorecognition process. These processes are discussed in more detail in Section 9.2. Some of the other approaches use electrical stimulation to produce optical transduction (e.g., an electroluminescent sensor) or an optical stimulation to produce electrical transduction (e.g., a photovoltaic sensor).

The field of biosensors has been active over many decades. The earlier successes were sensors utilizing electrochemical response (Janata, 1989). This type of sensor still tends to dominate the current commercial market. However, progress in fiber optics and integrated optics (such as channel waveguides and surface plasmon waves) and the availability of microlasers (solid-state diode lasers) have made optical biosensors a very attractive alternative for many applications. An optical biosensor, in general, utilizes a change in the amplitude (intensity), phase, frequency or polarization of light created by a recognition element in response to a physiological change or the presence of a chemical or a biologic (e.g., microorganism). Enhancement of the sensitivity and selectivity of the optical response is achieved by immobilizing the biorecognition element (such as an antibody or an enzyme) on an optical element such as a fiber, a channel waveguide, or a surface plasmon propagation where light confinement produces a strong internal field or an evanescent (exponentially decaying; see Chapter 7, Section 7.7) external field. Thus, the main components of an optical biosensor are (i) a light source, (ii) an optical transmission medium (fiber, waveguide, etc.), (iii) immobilized biological recognition element (enzymes, antibodies or microbes), (iv) optical probes (such as a fluorescent marker) for transduction, and (v) an optical detection system.

Some of the advantages offered by an optical biosensor are:

- Selectivity and specificity
- Remote sensing
- Isolation from electromagnetic interference
- Fast, real-time measurements

- Multiple channels/multiparameters detection
- Compact design
- Minimally invasive for *in vivo* measurements
- Choice of optical components for biocompatibility
- Detailed chemical information on analytes

9.2   PRINCIPLES OF OPTICAL BIOSENSING

The two important principles involved in biosensing are biorecognition and optical transduction. They are discussed in this section together with the various geometries used for optical stimulation and collection of transduced optical response. A key step of immobilizing the biorecognition elements is discussed separately in Section 9.3.

9.2.1   Biorecognition

The biorecognition elements are biologics such as enzymes, antibodies, and even biological cells and microorganisms that selectively recognize an analyte. They are often immobilized to increase their local concentration near an optical sensing element and to allow them to be reused. Some of the molecular bioreceptors used for biorecognition in biosensitizing are described here.

Enzymes.   The use of an enzyme as a biorecognition element utilizes its selectivity to bind with a specific reactant (substrate) and catalyze its conversion to a product. This enzyme–substrate-catalyzed reaction, also discussed in Chapter 3, is often represented as

$$E + S \rightleftharpoons ES \rightarrow P$$

In addition to providing selectivity, the reaction of certain analytes/substrates with enzymes can also provide optical transduction by producing a product that absorbs at a different wavelength (change in absorption), or is fluorescent (fluorescence sensor). Alternatively, the product of the enzyme-catalyzed reaction can interact with a dye (an optical sensing element such as a fluorescence marker) to produce an optical response.

Antibodies.   Antibodies, as discussed in Chapter 3, are proteins that selectively bind with an antigen or hapten (analyte) because of their geometric (site) compatibility. Very often an antibody–antigen pair's selective association in terms of their conformational compatibility is represented as a lock (antibody) and key (antigen) combination, as shown in Figure 9.2. This specific physical association can also produce an optical response that can be intrinsic such as a change in the optical property of the antibody or the antigen

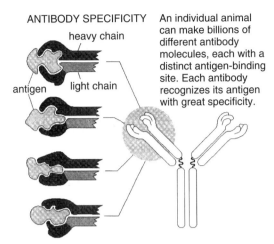

ANTIBODY SPECIFICITY

heavy chain

antigen   light chain

An individual animal can make billions of different antibody molecules, each with a distinct antigen-binding site. Each antibody recognizes its antigen with great specificity.

Figure 9.2. Schematic representation of antibody–antigen selective recognition. (Reproduced with permission from http://www.accessexcellence.org/AB/GG/antibodies.html)

as a result of association. Alternatively, an optical transducer (such as a fluorescent marker) can be used to tag the antibody or the antigen.

Lectins. Lectins are proteins that bind to oligosaccharides or single-sugar residues as well as to some glycoproteins such as immunoglobulins. Therefore, the lectins can act as biorecognition elements for these analytes. For example, concanavalin A in its A-form has been extensively used for its specific binding with α-D-mannose and α-D-glucose residues in a glucose sensor. In a glucose sensor utilizing concanavalin A, the lectin is immobilized on a sepharose film coated on the interior walls of a hollow fiber (Schultz et al., 1982; Boisdé and Harmer, 1996). Furthermore, it is liganded (conjugated) to dextran labeled with a fluorochrome, fluorescein-isothiocyanate (FITC). The glucose as an analyte diffusing through the hollow fiber displaces dextran from concanavalin A. The fluorescently labeled dextran then migrates to the area illuminated by light, being conducted through an inner solid optical fiber, to produce detectable fluorescence.

Neuroreceptors. These are neurologically active compounds such as insulin, other hormones and neurotransmitters that act as messengers via ligand interaction. They are also labeled with a fluorescent tag to produce an optical response through chemical transduction.

DNA/PNA. The specificity or complementary base pairing (that provides the basis for the DNA double-helical structure) can be exploited for recognition of base sequence in DNA and RNA (Kleinjung et al., 1998). An example is a DNA microarray (detailed coverage in Chapter 10) that consists of micropatterns of single-stranded DNA or finite-size oligonucleotides immobilized on a plate. They act as biorecognition elements by forming hydrogen bonds with a specific single-stranded DNA or RNA having a complementary base sequence. This process of base-pairing to form a double-stranded DNA is called *hybridization* (see 8.5.4). Another example of biosensing utilizing the hybridization in DNA is provided by a molecular beacon sensor, discussed below.

Recently, remarkable sequence specificity has been reported using peptide nucleic acids (PNAs) as biorecognition elements (Wang, 1998; Hyrup and Nielsen, 1996). The PNAs provide the advantage of a neutral backbone and correct interbase spacing to ensure that the PNAs bind to their complementary sequence with higher affinities and with specificity comparable to oligonucleotides.

9.2.2 Optical Transduction

Optical biosensing utilizes a rich variety of optical manifestations, in response to the presence of an analyte, created by the recognition element in the presence of an optical stimulation. Table 9.1 lists some of the principal optical manifestations (transduction) used for biosensing. Phase change produced by a change in the real part of the refractive index manifests itself as (i) a change of polarization of a linearly polarized light, (ii) a change in the propagation characteristics, particularly in relation to a light-confining geometry such as a fiber or a planar channel waveguide, or (iii) a change in the optical field distribution, particularly at an interface. All these manifestations have been used for optical biosensing, as described below.

Amplitude change derived from absorption, reflection, or other transmission loss mechanisms produces changes in the intensity of the sensing light. Frequency changes associated with biosensing utilize (a) fluorescence where

TABLE 9.1. Various Optical Manifestations Used for Biosensing

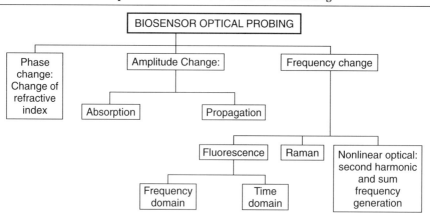

the optical signal generated is at a Stokes-shifted frequency from the exciting (absorbed) light, (b) Raman scattering, which is again Stokes-shifted, but now by vibrational excitation, (c) frequency shift by a nonlinear optical interaction mechanism such as second-harmonic generation. These effects were discussed in Chapters 4 and 5. For fluorescence sensors, one can use even an incoherent light source such as a light-emitting diode. However, for sensors utilizing Raman and second-harmonic generation, one needs a laser source.

9.2.3 Fluorescence Sensing

Direct Sensing. This type of sensing scheme utilizes a direct change in the fluorescence property as a result of the analyte binding with the biorecognition element (antibody or enzyme) or the production of a specie of a particular fluorescence property by a specific enzyme-catalyzed reaction. An example of this type of sensing is enzyme-catalyzed reactions that produce NADH. As discussed in Chapter 6, NAD^+ is nonfluorescent but NADH is fluorescent, with $\lambda_{ex}^{max} = 350\,nm$ and $\lambda_{em}^{max} = 450\,nm$. Therefore, dehydrogenase enzyme-catalyzed substrate reaction as

$$Substrate + NAD^+ \xrightarrow{\text{dehydrogenase}} Product + NADH$$

can be followed by monitoring the NADH fluorescence.

Indirect Sensing. Here an external dye, which may not be a part of the reaction but whose fluorescence property changes in response to biorecognition of an analyte, is used as a fluorescent tag for optical transduction.

An example of this type of biosensor is an ion-selective sensor that relies on the specific recognition of a specific ion. One example is that of a fluorescent dye conjugated to an enzyme, calmodulin, which recognizes and binds with Ca^{2+}. Figures 9.3 and 9.4 illustrate the principle and the response of Ca^{2+} binding to calmodulin on the fluorescence of the dye (producing a decrease in fluorescent intensity).

9.2.4 Fluorescence Energy Transfer Sensors

This scheme of fluorescence sensing involves an energy transfer that produces a change in the fluorescence of either the biorecognition element or the fluorescent marker (deSilva, 1997). The two main schemes used for biosensing utilizing this principle are shown in Figures 9.5 and 9.6.

A fluorescence resonance energy transfer (FRET) biosensor involves a donor and an acceptor group, with the electronic energy transfer between them being affected as a result of biorecognition. The biorecognition (such as antibody–antigen association) can lead to efficient electronic energy transfer from an excited donor group to an acceptor group that is highly fluorescent

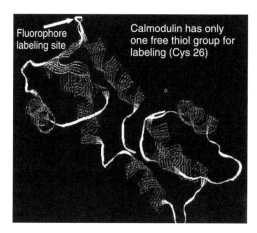

Figure 9.3. Calmodulin binding to Ca^{2+}, which produces conformational change in the enzyme structure. (Reproduced with permission from Watkins and Bright, 1998.)

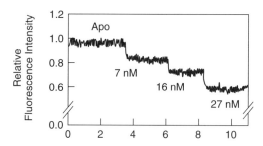

Figure 9.4. Decrease in fluorescence upon Ca^{2+} binding. (Reproduced with permission from Watkins and Bright, 1998.)

(or fluoresces at a different wavelength that is being detected). Alternatively, the facilitation or inhibition of energy transfer as a result of binding can affect the fluorescence of the donor group. In other words, the acceptor group acts as a fluorescence quencher. Another example of this type of sensing is the molecular beacon approach, discussed below. The energy level scheme dictates that the electronic excitation energy level of the donor is higher than that of the acceptor. Other requisites for an efficient fluorescence resonance energy transfer have been discussed previously in Chapter 7.

The scheme presented in Figure 9.6 utilizes the principle of photoinduced electron transfer. In this mechanism, the sensing unit consists of an electron donor group and an electron acceptor group. In the absence of the analyte, there is an efficient photoinduced electron transfer from the electron donor to the electron acceptor group when the acceptor is electronically excited. This results in quenching of the acceptor fluorescence. The appropriate energy level

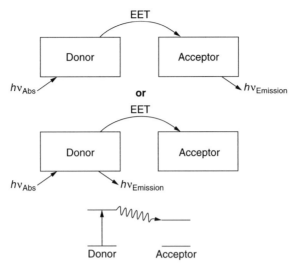

Figure 9.5. Fluorescence sensor utilizing fluorescence resonance energy transfer. EET represents electronic energy transfer. (Reproduced with permission from deSilva et al., 1997.)

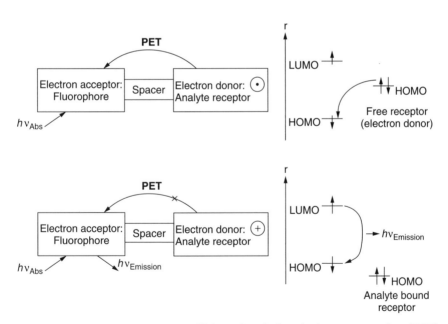

Figure 9.6. Fluorescence sensor utilizing photoinduced electron transfer (PET). (Reproduced with permission from deSilva et al., 1997.)

diagrams for this process are also shown in Figure 9.6. The HOMO (highest occupied molecular orbital, discussed in Chapter 2) energy level of the donor (when acting as a free receptor) is higher than that of the acceptor, permitting the photoinduced electron transfer when an electron is promoted from the HOMO to the LUMO (lowest unoccupied molecular orbital, discussed in Chapter 2) of the acceptor by optical absorption. When the electron donor group binds with an analyte or with a chemical product produced by the reaction of the analyte with the biorecognition element, it transfers the electron, thus transforming itself into a positively charged unit. Therefore, photoexcitation of the acceptor group now is unable to induce electron transfer from the donor to the acceptor group. Consequently, there is no quenching of the acceptor fluorescence. Therefore, the optical transduction here is the appearance of the acceptor fluorescence in the presence of an analyte. Figure 9.6 shows that the HOMO of the analyte-bound receptor (donor) is now lower than that of the electron acceptor.

9.2.5 Molecular Beacons

The molecular beacon approach is also based on the electronic energy transfer scheme between a fluorescent unit (Fl) and a fluorescent quencher (Q) (Tan et al., 2000). A molecular beacon consists of a loop and a stem. The loop structure involves a single-stranded oligonucleotide in a specific sequence. The stem usually consists of five to seven complementary base pairs. The two ends of the stem consist of a fluorophore (Fl) and a fluorescence quencher (Q). In the absence of the analyte, the stem is intact keeping the fluorophore and the quencher in close proximity and producing an efficient energy transfer, thus causing a quenching in the fluorescence of the fluorophore. In the presence of the analyte, the binding or biorecognition process forces the stem apart, thus increasing the distance between the fluorophore and the quencher sufficiently to inhibit the energy transfer. The result is restoration of the fluorophore fluorescence. This principle of operation is illustrated in Figure 9.7.

Molecular beacons have emerged recently as a new class of DNA, RNA, or PNA probes. Molecular beacons with a selected sequence of bases in the loop can be synthesized to detect the complementary DNA strand by hybridization (pairing up of complementary strands by hydrogen bonding), the hybridization forces the stem to open and restore the fluorescence of the fluorophore, as illustrated in Figure 9.8. In this study, a PNA–DNA hybrid probe was surface immobilized using biotin/streptavidin coupling. The hybridization with a single-stranded target DNA analyte opens the stem and produces fluorescence. A spectacular example of the molecular beacon approach is shown in Figure 9.9, where the molecular beacon loop consists of the oligonucleotide directed to the serine hydroxymethyltransferase pseudogene (SHMT-*ps1*). These studies confirmed those obtained using PCR that only primates possess this gene. As is clearly evident in Figure 9.9, fluorescent signal was obtained as the result of hybridization of the molecular beacon

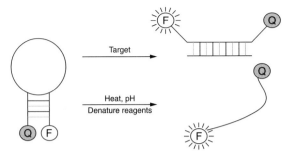

Figure 9.7. Molecular beacon approach for biosensing. Hybridization with the target DNA molecules of complementary sequence or unwinding with the increase of temperature, change of pH, or presence of denaturing agent produces an increase of fluorescence. (Reproduced with permission from Tan et al., 2000.)

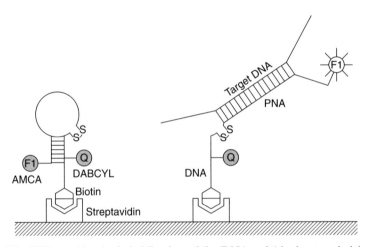

Figure 9.8. DNA probing by hybridization of the DNA unit/single-stranded (oligonucleotides) in the loop with a complementary sequence in the DNA analyte. (Reproduced with permission from Ortiz et al., 1998.)

probe with the target DNA obtained from primates (*H. sapiens, P. troglodytes, L. rosalia*). No signal was seen using nonprimate DNA (*V. varigegata, O. cunniculatus, O. aries*).

The molecular beacon approach has shown extremely high selectivity with single-base pair mismatch identification capability and suggests the prospect of studying biological processes in real time and *in vivo*.

9.2.6 Optical Geometries of Biosensing

A number of optical geometries have been used in the design of various optical biosensors. These geometries are listed in Table 9.2. The choice of any

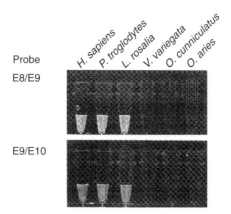

Figure 9.9. Molecular beacon fluorescence detection of pseudogene SHMT-*ps1* using two molecular beacon probes, E8/E10 and E9/E10. (Reproduced with permission from Devor, 2001.)

TABLE 9.2. Various Optical Geometries Used for Biosensors

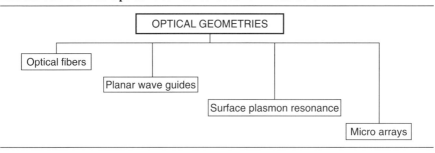

of these geometries is dependent on the nature of the analyte and the optical probing method used. A major consideration is enhancement of sensitivity and specificity. The guided wave geometries utilized in optical fibers and planar waveguide devices also provide an opportunity to use the evanescent waves that extend externally beyond the waveguiding region. As discussed in Chapter 7 (Section 7.7), the evanescent waves are nonpropagating optical fields whose strength decays exponentially as a function of distance away from the surface of the optical guiding region. The analyte/biorecognition element/optical probe at the interface between the biomedium and the guiding medium can interact with this evanescent field and produce an optical response. This evanescent wave can be utilized both for phase and amplitude modulation biosensors. The evanescent waves can be used to sense an analyte localized near the surface (by selectivity of the recognition element immobilized on the surface of an optical fiber, a planar waveguide or a surface plasmon resonance element). The surface plasmon geometry used for biosensing is discussed in Section 9.8. The microarray geometry is covered in detail in Chapter 10.

9.3 SUPPORT FOR AND IMMOBILIZATION OF BIORECOGNITION ELEMENTS

The biorecognition elements are normally immobilized on a solid support, although in some cases a membrane or a solid support is simply used for physical confinement of the biorecognition unit to increase its local concentration in the region of biodetection. The solid supports are usually a membrane, a polymer, a copolymer, or a glass such as sol–gel processed glass. A biore-cognition element is immobilized on this solid support, either by a physical method (such as adsorption) or by chemical attachment. In some approaches, a biorecognition element is entrapped in the volume of a matrix (solid support) with controlled porosity, in which case the solid support also provides selectivity toward an analyte of certain size compatible with its pore dimension.

In the case of evanescent wave sensing, discussed below, the surface of a fiber or a waveguide itself acts as a solid support for the biorecognition element. Also in some optical fiber sensors, the distal end of the fiber itself acts as the solid support.

9.3.1 Immobilization

The various physical or chemical methods used to immobilize a biorecognition element (an enzyme, antibody, etc.) are discussed extensively in the literature (Boisdé and Harmer, 1996; Kuswandi et al., 2001). A brief discussion of this topic is presented here.

Physical Methods. The simplest physical method is containment within semipermeable membranes. A number of optical fiber sensors have utilized this technique. Another method calls for adsorption on a solid support. Depending on the nature of the biomolecule, either ionic, hydrophobic, or even van der Waal's forces can be used for selective adsorption. The adsorption is facilitated by preactivation of the surface. A simple approach used for dye adsorption involves immersing a polar cross-linked polymer or copolymer placed at the end of the optical fiber into dye solution and then washing off the unadsorbed dye (Boisdé and Harmer, 1996). A number of pH sensors utilize this method. Another solid support utilizing adsorption for immobilization involves microspheres whose surfaces are preactivated to enhance adsorption of the biorecognition element or a dye on the surface. For example, the glass microspheres can be treated with a silane to make it hydrophobic, allowing protein adsorption. The advantage of using microspheres is in maximizing the available surface area for adsorption. However, if the microspheres are too small in size, they may require an additional porous membrane to contain them.

The advantage offered by the physical adsorption method is its simplicity. However, drawbacks are the nonspecific nature of physical adsorption, variation in the density of attachments and loss of biorecognition elements by leaching (desorption).

Ionic Binding. A biorecognition element can be immobilized on a solid support by electrostatic interactions between them. By adjusting the pH conditions, the biorecognition element (such as a protein) can be made polar or even made to carry a charge (become ionic). By appropriately selecting a polymer carrying an opposite charge, the biorecognition element can be immobilized by electrostatic attraction between opposite charges. Examples are polymers containing negatively charged sulfonic groups ($-SO_3^-$) or positively charged ammonium groups (NH_4^+).

Physical Entrapment. In this method, the biorecognition element or the sensor indicator (which changes its optical property on sensing) is entrapped within the body of a matrix such as polyacrylamide, polyvinyl alcohol, polyvinyl chloride, epoxy, sol–gel processed glass, or a Langmuir–Blodgett film. The two principal advantages offered by the entrapment method are (i) minimization of leaching of the biorecognition element from the matrix and (ii) minimization of any biofouling effect (the adherence of unwanted biologics) that is more manifested in surface immobilized sensing.

Furthermore, the use of a porous matrix such as a sol–gel processed glass, where the size of porosity can be tailored, offers the opportunity to introduce size-dependent specificity toward analytes.

Crucial requirements for this method to work are as follows:

- The immobilized biomolecule retains its affinity/activity in the entrapped form.
- The encapsulated structure is stable over time.
- The biorecognition molecule is accessible to the analyte.
- The entrapping matrix is optically transparent.

The sol–gel processing method to entrap various biorecognition elements has emerged as a powerful approach that meets all of these requirements. An excellent recent review on the application of this method for biosensing is by Rickus et al. (2002). It is a low-temperature, wet chemical method to produce inorganic glasses, with most of the focus being on silica. The procedure is well established and illustrated in Table 9.3.

This method offers tremendous flexibility with various approaches to entrap a biorecognition element. To date, a large variety of biomolecules and even entire cells have been successfully entrapped. Avnir, Braun, and their co-

TABLE 9.3. Sol–Gel Processing Scheme for Biosensing

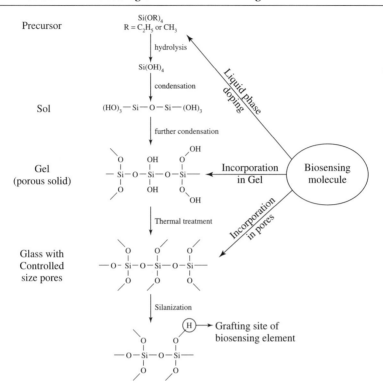

workers (1994) reported entrapping of an enzyme, alkalene phosphatase (AP), in a sol–gel monolith. Dunn, Valentine, and Zink (Ellerby et al., 1992; Yamanaka et al., 1994) reported the entrapment of copper–zinc superoxide desmutase, cytochrome C, and myoglobin in a sol–gel monolith. Shtelzer et al. (1992) entrapped trypsin and AP. Subsequently, groups of Avnir, Braun, Dunn, Valentine, and Zink reported encapulation of other enzymes like glucose oxidase (GO_x), aspertase, peroxidase, and urease. The review by Rickus et al. (2002) contains detailed references of these original contributions.

A disadvantage of this method is that the biomolecules are randomly oriented. Therefore, many of them may have their active sites buried in the matrix, thus unavailable.

Our group, in collaboration with Bright, reported the first successful intact entrapment of an antibody (anti-fluorescein) in active form within a sol–gel glass (Wang et al., 1993). More recent noteworthy contributions are reports of encapsulation of a whole cell by Pope et al. (1997) and Chia et al. (2000). The work cited above reports that by sol–gel entrapment, the enzymes retain their catalytic activities, antibodies retain their binding affinity and the cells retain their viability (Rickus et al., 2002).

Sol–gel processed glass containing the entrapped biostructures have been used in the forms of films, coatings, or monoliths. In recent biosensor work, the surface of an optical sensing element (fiber, waveguide, or a surface plasmon resonance device) has been coated by a sol–gel processed film.

Chemical Immobilization. This method involves formation of a covalent bond between the solid support or the optical sensor surface and a biore-cognition element such as a protein (enzyme or antibody). Often it involves modification of the support surface to introduce coupling groups such as —OH, —NH$_2$, —COOH, and —SH. An example is chemical immobilization of an enzyme on the surface of a fiber-optic sensor, as illustrated in Figure 9.10.

Another covalent attachment using a —COOH group utilizes reaction with the amine group to form an amide linkage, discussed in Chapter 3. This reaction is represented again here:

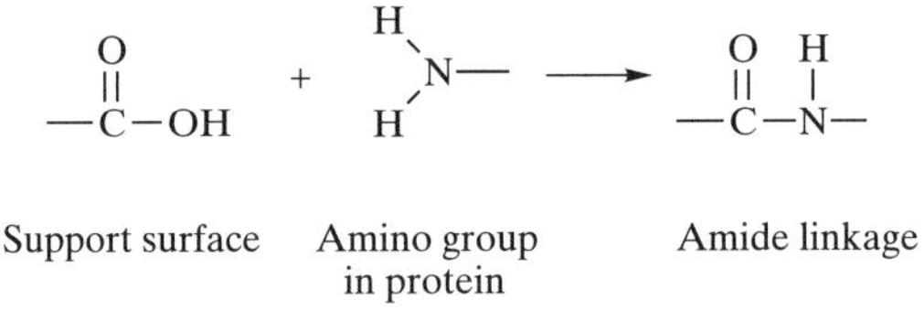

$$
\begin{array}{ccc}
\underset{\text{Support surface}}{\overset{\displaystyle O}{\underset{\displaystyle \parallel}{-\!C\!-\!OH}}} \;+\; \underset{\substack{\text{Amino group}\\\text{in protein}}}{\overset{\displaystyle H}{\underset{\displaystyle H}{\diagdown N-}}} \;\longrightarrow\; \underset{\text{Amide linkage}}{\overset{\displaystyle O\;\;H}{\underset{\displaystyle \parallel\;\;\;\mid}{-\!C\!-\!N-}}}
\end{array}
$$

An important method to immobilize an antibody and, at the same time, optimize its orientation so that it is accessible to an antigen utilizes avidin/biotin coupling (Rogers, 2000; Lowe et al., 1998). This avidin–biotin coupling scheme is represented in Figure 9.11.

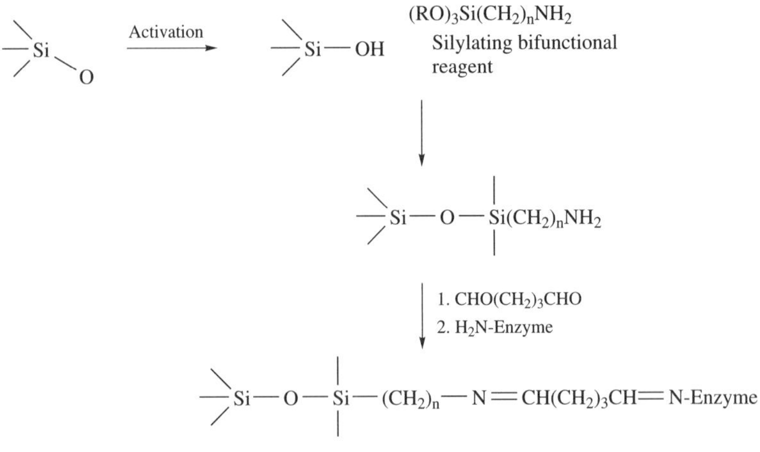

Figure 9.10. Chemical immobilization of an enzyme on an optical fiber surface using a bifunctional silylating group.

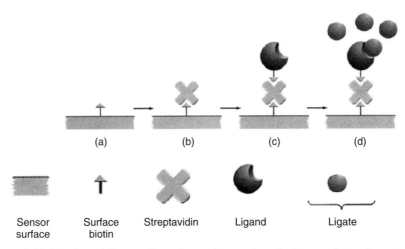

Figure 9.11. Biotin–avidin coupling scheme. (Reproduced with permission from Lowe et al., 1998.)

Another covalent attachment, frequently used in surface plasmon resonance (SPR) sensors, discussed in Section 9.8, requires immobilization of a biorecognition element on a gold surface. For this purpose, self-assembling of a monolayer formed from long-chain molecules with an —SH group at one end and an —NH$_2$ or a $-\overset{\text{O}}{\underset{\text{\large \|}}{\text{C}}}-$ group at the other end (Rogers, 2000) is used. The—SH group binds to the gold surface. The —NH$_2$ or $-\overset{\text{O}}{\underset{\text{\large \|}}{\text{C}}}-$ group at the other end can be used to couple to an enzyme or an antibody at multiple sites.

Cellulosic and polyacrylamide compounds, carboxylic-acid-modified polyvinyl chloride (PVC), and polystyrenes can be surface functionalized to bind with proteins (Boisdé and Harmer, 1996). For example, polystyrenes can be chloromethylated, sulfonated, and halogenated to bind with an indicator containing an —OH group.

9.4 FIBER-OPTIC BIOSENSORS

Fiber-optic biosensors are the most widely studied optical biosensors and have been a subject of extensive investigation over more than two decades. A number of excellent recent references describe fiber-optic based biosensors, their applications, and current status (Wolfbeis, 1991; Boisdé and Harmer, 1996; Mehrvar et al., 2000). Fiber-optic biosensors offer a number of advantages. Some of these are listed here:

- Optical fiber technology is now highly developed, providing optical fibers with many characteristics such as single-mode fibers, polarization preserving fibers, and multimode fibers. This topic was discussed earlier in Chapter 6 under light delivery systems. The availability of these well-defined characteristics has led to the application of optical fibers in biosensing based on all the principles listed in Table 9.1.

- Optical fibers also provide a number of convenient geometries such as a single core fiber, a dual core fiber, a Y-junction fiber, and fiber bundles, offering flexibility to make them compatible to a specific need. The single fiber configuration provides the advantage of compactness (small sample volume). It is also more efficient as the overlap between the incident light probe and the collected (returned) optical response is maximized. On the other hand, a fiber bundle yields a higher optical throughput, thus providing an opportunity to use inexpensive nonlaser sources and detection systems.

- Use of a longer-length fiber provides a gain in interaction length or surface area for multiple analyte detection. Using the evanescent wave coupling, one can utilize a longer interaction length of a fiber by simply increasing its length. One can also use different segments of the same optical fiber, by appropriate labeling, to probe different analytes, thus providing an opportunity for multianalyte detection.

- Optical fibers also offer compatibilities with catheters or endoscopes for *in vivo* biosensing. Thus, one can use minimally invasive optical biosensing methods to measure *in vivo* blood flow, glucose content, and so on.

A number of classification schemes have been used for fiber-optic biosensors. One scheme classifies fiber-optic sensors into extrinsic or intrinsic. In an extrinsic fiber-optic sensor, the optic fiber simply is used as a transmission channel to take light to and from the sensing elements. In an intrinsic sensor, the fiber itself acts as a sensing element (transduction) because one or more of the physical properties of the optical fiber changes in response to the presence of an analyte. Another scheme is based on whether a direct or indirect (indicator-based) sensing scheme is used. In the case of a direct fiber-optic sensor (sometimes abbreviated as FOS), the intrinsic optical properties of the analyte are measured, while in the case of an indirect sensor, optical properties (absorbance, fluorescence) of an immobilized indicator dye, label, or optically detectable bioprobe is monitored.

As discussed earlier, a fiber-optic biosensor can utilize an amplitude change, in which case it is called an *intensity-modulated sensor*. Alternatively, it can utilize a phase change, in which case it is called a *phase-modulated sensor*. A phase-modulated sensor utilizes interferometric techniques such as a Mach–Zehnder interferometer, which involves two fibers: a reference fiber and a sensing fiber. In the presence of an analyte, the basic optical parameter of the sensing fiber is changed, creating a phase difference between the light traveling through the two fibers, resulting in a change in the optical interference signal.

For intensity modulation even an incoherent light source such as a light-emitting diode (LED) can be used. In contrast, phase-modulated interferometric sensors require high-coherence single-mode lasers.

In its most basic form, an intensity-modulated fiber-optic sensor as well as a fluorescence fiber-optic sensor utilizes optical fibers of various types, the tip of which contains an immobilized biological recognition element such as an enzyme or an antibody. The different configurations of optical fiber geometries and immobilization scheme used are shown in Figures 9.12 through 9.14.

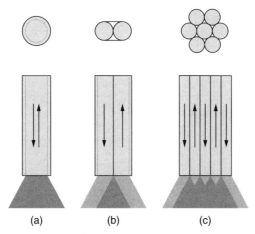

(a) (b) (c)

Figure 9.12. Different types of optical fiber configurations used for sensing. (Reproduced by permission of The Royal Society of Chemistry; Kuswandi et al., 2001.)

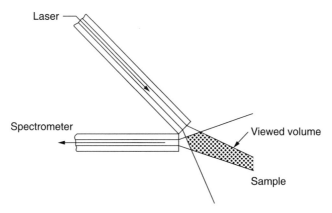

Figure 9.13. Double optical fiber terminal. (Reproduced with permission from CRC Press; Wolfbeis, 1991.)

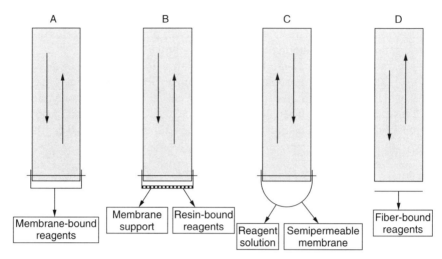

Figure 9.14. Various approaches of the sensing layers (immobilization of the recognition element) in fiber-optic sensing. (Reproduced by permission of The Royal Society of Chemistry; Kuswandi et al., 2001.)

The sensing layer containing the biorecognition element also can serve as a biochemical transduction system that produces a change in its optical property upon interaction with the analyte. As shown in Figure 9.14, the transduction reagent can be immobilized directly on a membrane held against the optical fiber (A), can be in a solid particulate form supported on the fiber end by a membrane (B), can be a liquid reagent confined near the fiber end by a semipermeable membrane (C), or can be bonded to the fiber itself (D).

A simple example is a urea sensor, depicted in Figure 9.15 (Abdel-Latif et al., 1990). Urea in the presence of urease, splits as ammonium and bicarbonate. Due to the production of ammonium, there is a change in the pH of the reagent. This results in a change in the spectral properties of the pH indicator. In the sensor shown above, urease and bromothyl blue are held at one of the fiber by a semipermeable membrane. The ammonium production changes the pH. The pH-sensitive dye changes its color when the pH of the surrounding environment changes. A similar pH-sensitive dye can be used for glucose sensing. In this case, a pH-sensitive dye along with a glucose biorecognition element (an enzyme: glucose oxidase) is held at one end of the optical fiber by a semipermeable membrane. The oxidation of glucose, which is catalyzed by glucose oxidase by the enzyme–substrate binding mechanism, consumes oxygen and produces protons (H^+), thereby changing the pH of the solution. The pH-sensitive dye changes its color (absorption) when the pH of the surrounding environment changes. Thus, by measuring the absorbance (absorption spectrum) using the light out (light returning in the optical fiber), one can get information on the glucose concentration.

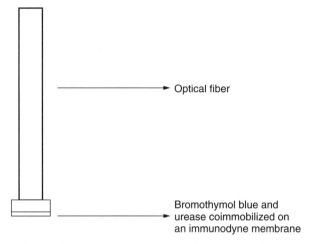

Optical fiber

Bromothymol blue and
urease coimmobilized on
an immunodyne membrane

Figure 9.15. A fiber-optic urea sensor utilizing a pH-sensitive dye. (Reproduced with permission from Abdel-Latif et al., 1990.)

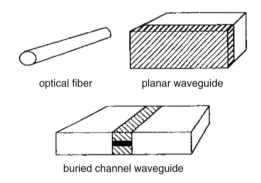

optical fiber   planar waveguide

buried channel waveguide

Figure 9.16. Typical examples of optical waveguides.

9.5 PLANAR WAVEGUIDE BIOSENSORS

Like optical fibers, planar waveguides are media in which the propagation of an optical waveguide is confined in a dimension comparable to the wavelength of light. Planar waveguides were discussed in Chapter 7 in the section on total internal reflection fluorescence (TIRF) imaging. However, for the sake of clarity, they are represented here in Figure 9.16.

As pointed out earlier (Chapter 6), a fiber is a waveguide in which the optical propagation is confined in two dimensions. In a planar waveguide the confinement is in one dimension (the thickness of the film). The film guiding the wave, again, is typically of the dimensions of ~1 μm.

A channel waveguide actually produces two-dimensional confinement (height and width) and is quite analogous to a fiber. The three important techniques for coupling light into a planar waveguide are (i) prism coupling, (ii) grating coupling, and (iii) end-fire coupling.

An excellent recent review on planar waveguide sensors is provided by Sapsford et al. (2002) and by the book by Boisdé and Harmer (1996). Just like an optical fiber sensor, a planar waveguide utilizes immobilization of the biorecognition element on its surface. Most planar waveguide sensors are evanescent wave sensors, described in Section 9.6. The advantage of a planar waveguide sensor is that it allows the immobilization of multiple biorecognition elements, thus providing the prospect for multianalyte detection using a single substrate. This approach utilizes patterns of immobilized biomolecules. A number of techniques have been used to create such a patterned structure (Blawas and Reichert, 1998; Sapsford et al., 2002). Photolithography has been used to produce patterns of protein (Bhatia et al., 1992, 1993). Another approach utilizes photopatterning of a polymer surface by photoablation (Schwarz et al., 1998). Ink jet printing technology has also been used to pattern antibodies or the protein, avidin, in 200-μm-diameter spots on the surface of polystyrene films (Silzel et al., 1998).

Most planar waveguide sensors have been used in the fluorescence detection mode using evanescent wave excitation as described in the Section 9.6. When the dimensions of the waveguide are comparable to optical wavelength (≤1 μm), the wave propagation is described in terms of concepts of integrated optics in which a continuous field distribution along the propagation path exists. In such a case, the waveguide is often referred to as an *integrated optical waveguide* (IOW). If the dimensions (width) of the waveguide are considerably thicker (100 μm), classical ray-optics describing total internal reflection of rays from the boundaries of the waveguide is used. Therefore, the waveguide is often referred to as an *internal reflection element* (IRE). In this case, the fluorescence sensing method utilizing an immobilized biorecognition system on the surface of the waveguide is referred to as total internal reflection fluorescence (TIRF) sensing. TIRF has also been discussed in Section 7.10 of Chapter 7 in the context of bioimaging. In the TIRF sensing, the waveguide surface produces a series of sensing "hot spots" along the planar surface from where the light beam is reflected. These discrete regions of high intensity can be used as sensing regions. However, it may be preferable to have a uniform field distribution achieved by reducing the waveguiding dimensions.

Many different kinds of materials have been used for waveguides. They include silica glass, polystyrene, and Ta_2O_5 (Sapsford et al., 2002). Depending on the material used, different surface chemistry approaches have been used to immobilize a molecule on the surface of a waveguide. In the case of a silica glass, silanization has been used. The avidin–biotin binding approach has been extensively used in general for various waveguides. These methods have been described in Section 9.3. More details are provided in the review by Sapsford et al. (2002).

As discussed earlier, a principal advantage of using a planar waveguide geometry is patterning for simultaneous multichannel multianalyte detection. In these approaches, a patterned array of a series of biorecognition elements is immobilized on the surface of a planar waveguide. Various analytes, fluorescently labeled with different fluorophores, are flowed over the surface of the waveguide. Then the pattern of fluorescent biorecognition:analyte complexes is detected. Image analysis software then can be used to correlate the position of a particular fluorescence signal with the identity of a specific analyte. An approach using patterning of captured biomolecules using flow cells is schematically represented in Figure 9.17, which is taken from the work of Feldstein et al. (1999). In this approach, a multichannel flow cell was pressed onto a planar waveguide surface and each channel was filled with a solution of the biomolecule. Then the sample and a fluorescent-tagged antibody were passed over the waveguide surface perpendicular to the immobilized biomolecule channel using another flow cell. Further advances have been made recently using automated fluidic systems and automated image analysis programs to develop a fully automated array sensor (Feldstein et al., 2000; Rowe-Taitt et al., 2000a).

Wadkins et al. (1998) used a scheme, shown in Figure 9.18, that used patterned antibody channels. They demonstrated the detection of *Y. pestis* F1 in clinical fluids such as whole blood, plasma, urine, saliva, and nasal secretion.

In another approach, Zeller et al. (2000) developed a TIRF system in which the planar waveguide consisted of multiplanar single pad sensing units. Each of the single pads had its own laser light input, coupling of fluorescence emission to the detector, and background suppression. In one example, they demonstrated a two-pad sensing device in which one pad was modified with mouse IgG while the other was modified with rabbit IgG. Other work in this direction is by Silzel et al. (1998), Plowman et al. (1999), and Rowe-Taitt et al. (2000b).

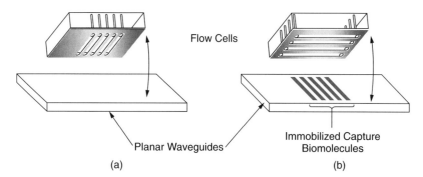

Figure 9.17. Patterning of capture biomolecules using flow cells. (Reproduced with permission from Feldstein et al., 1999.)

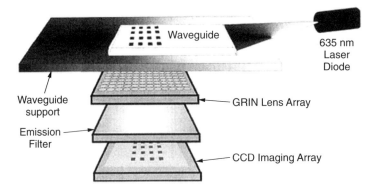

Figure 9.18. Array biosensors developed by Ligler, Golden, and co-workers at the Naval Research Laboratory. (Reproduced with permission from Wadkins et al., 1998.)

Hug et al. (2001) have proposed another method, which they call optical waveguide lightmode spectroscopy (OWLS), that is based on measurements of the effective refractive index of a thin layer above a waveguide. This thin layer can be due to whole cells. The effective refractive index of this adsorbate layer is dependent on the nature of adhesion and the cell and determines the coupling angle of light of a given polarization (TE or TM) into the planar waveguide. They used this approach to monitor the adhesion behavior of anchorage-dependent cells such as fibroblasts.

9.6 EVANESCENT WAVE BIOSENSORS

Evanescent wave sensors utilize the interaction with the electromagnetic field (evanescent wave) that extends away from the surface of the light guiding medium, whether a planar waveguide, a channel waveguide, or a fiber (Rowe-Taitt and Ligler, 2002; Boisdé and Harmer, 1996). In other words, these sensors rely on the light that is not confined within the waveguide itself, but penetrates into the surrounding medium of lower refractive index (cladding or air or into a surface immobilized biorecognition element) and thus senses the chemical environment on the surface of the waveguide (or fiber). In contrast to a propagating mode (oscillating electromagnetic field with the propagation constant k, defined in Chapter 2, as a real quantity), an evanescent wave has a rapidly decaying electric field amplitude, with an imaginary propagation constant k. The topic of evanescence has already been covered in detail in Chapter 7, Section 7.7.

Evanescent wave biosensing can utilize a number of optical transduction mechanisms, as illustrated in Table 9.4. In frequency conversion techniques, the fluorescence excitation has been used extensively both in a planar waveguide and in a fiber geometry. In this sensing scheme, the fluorescence is gen-

TABLE 9.4. Evanescent Wave Sensing

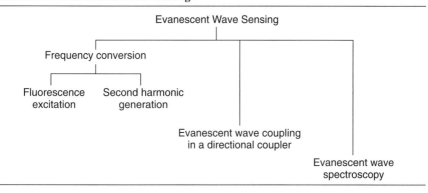

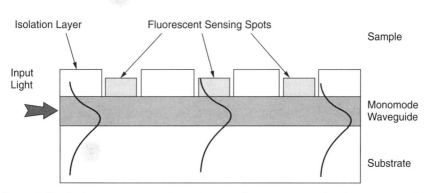

Figure 9.19. A fluorescence sensing scheme using a monomode planar waveguide. (Reproduced with permission from http://barolo.ipc.uni-tuebingen.de/projects/riana/ summary/james.html).

erated from the analyte (antigen) specifically binding with a biorecognition element (antibody) which is immobilized on the surface of the waveguide or a fiber. Alternatively, the biorecognition element can be entrapped in a sol–gel film coated on the surface of the waveguide or the optical fiber. Even though the fluorescence is radiated isotropically in all directions, it is the fluorescence from the molecules close to the surface which couples into the waveguide (or fiber) and is detected for sensing.

A fluorescence sensing scheme using a monomode planar waveguide is shown in Figure 9.19. The idea is to excite with the evanescent field as well as to detect the fluorescence. Isolation layers (windows) are drawn on the surface of the waveguide, leaving a certain area of the waveguide surface exposed. This is achieved by rf spluttering with silica. The areas left exposed on the surface form the sensing spots of the sensor. These sensing spots form the regions of interaction with the analyte. If appropriate fluorophores are positioned in these sensing spots, then the evanescent wave protuding from the waveguide will excite the fluorophores to induce fluorescence.

In the case of a cladded fiber-optic probe, the amount of evanescent power is related to the fraction (f_e) of light power, P_{clad}, in the cladding region compared to the total power, P_t. It is defined as

$$f_e = (P_{clad}/P_t) = 1 - (P_{core}/P_t)$$

where P_{core} is the power in the core of the fiber. A crucial factor determining f_e is the V number of the optical fiber, which is defined as

$$V = (2\pi r/\lambda)(n_1^2 - n_2^2)^{1/2}$$

where r is the radius of the optical fiber, n_1 is the refractive index of the fiber, and n_2 is the refractive index of the surrounding medium or cladding layer. The fraction f_e decreases with the increasing V number (i.e., for a greater difference between n_1 and n_2 or larger r). In contrast, the efficiency of coupling of fluorescence emission from the surface back into the fiber increases with an increase in the V number (Thompson, 1991).

In order to enhance the interaction of the evanescent wave with the sensing layer containing a fluorescent marker, optical fibers with unclad, partially clad, and D-shaped forms have been used (Rowe-Taitt and Ligler, 2002). The D fibers are fibers in which the cladding from one-half of the fiber is removed, exposing that half to the sensing layer. In a partially cladded region, a major problem is V-number mismatch between the cladded region and the uncladded region. This can happen if the refractive index of the cladding layer is different from the medium surrounding the declad sensing region. This V mismatch creates light loss; particularly the fluorescence emission from the decladded sensing region is not guided into the core but enters the cladding layer and is therefore not transmitted to the detector. Approaches used to reduce the V-number mismatch are based on decreasing the radius of the fiber. The various geometries used for this purpose are shown in Figure 9.20.

Evanescent wave coupling sensors involve coupling between two channel waveguides or fibers that are close enough so that their evanescent fields overlap and couple them. The coupling is analogous to two coupled oscilla-

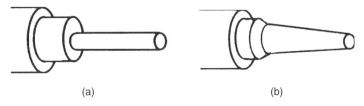

(a)   (b)

Figure 9.20. (A) Step-tapered core fiber. (B) Continuously tapered fiber.

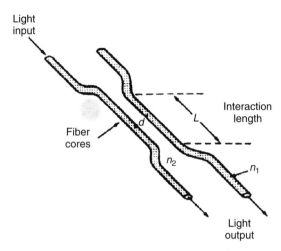

Figure 9.21. Evanescent wave coupled fiber-optic sensor. (Reproduced with permission from CRC Press; Wolfbeis, 1991.)

tors. It is, in fact, a phase sensing device that is highly sensitive to a change in the refractive index of the region between the two guides. An example of an evanescent wave coupled sensor is shown in Figure 9.21, where a pair of optical fibers are simply brought together.

When light is launched in one fiber, the overlap of the evanescent wave field with that of the adjacent fiber leads to a power transfer into the second fiber. The power transfers back and forth between the two fibers with a periodicity determined by the coupling constant between them that is strongly dependent on the refractive index. Therefore, when the refractive index between them changes as a result of biosensing the power transfer conditions change, resulting in a change of the intensity of light exiting one of the fibers. One specific case is when the length of the coupling region is half of a characteristic length called the *beat length*; then the light launched in one fiber is completely transferred to the other fiber. The biorecognition element in this case is immobilized in the region of evanescent wave overlap. The analyte binding changes the refractive index, thus changing the coupling condition whereby the same length now does not meet the condition of complete transfer. Therefore, the power transfer to the second fiber decreases.

In evanescent wave spectroscopic sensors, the interaction of the evanescent wave with the sensing layer is used to get spectroscopic information on the analyte binding (Boisdé and Harmer, 1996). The spectroscopic information can be on the IR (vibrational) or UV-visible (electronic) absorption band or the Raman spectroscopic transitions. Recent studies have used near-IR and FT–IR spectroscopic approaches. Silver halide fibers have been used for obtaining spectral information in the region 2–20 μm. Also, uncladded chalcogenide fibers and sapphire fibers have been utilized.

9.7 INTERFEROMETRIC BIOSENSORS

An interferometric biosensor utilizes interference between the light from a waveguiding channel with a sensing layer on its surface, and that from a reference channel. In this type of sensing, one utilizes the information on phase change introduced by binding of an analyte with a biorecognition element that is immobilized on the surface of the sensing waveguide channel. This phase change is detected by an interferometric technique by creating an interference between the two beams. The most commonly used technique for biosensing has been that of a Mach–Zehnder interferometer. Figure 9.22 shows the schematics of such a biosensor used at our Institute. The sample arm exhibits specific binding with the biological analyte, based on the specificity of the immunoglobulin (IgG) that is immobilized on its surface. Control of nonspecific binding is provided on the reference arm by using immobilized IgG that is not selective to the analyte of interest. An optical beam I_{in} is split into two parts, which travel through two arms (channel waveguides or fibers)—that is, the reference arm and the sample arm of the interferometer. If the two arms are not exactly identical, the phase-shift introduced in the two arms are different (because of a difference in the propagation time). The sample arm forms the sensing area of the interferometer. Therefore, when the two beams from these two arms are recombined near the output port, the output intensity I_{out} is modulated due to interference between them. If the relative phase shift between the reference and the sample arms is $\Delta\Phi$, the output intensity is given as

$$I_{out} = I_{in}(1 + M\cos\Delta\Phi)$$

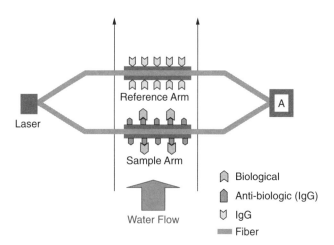

Figure 9.22. Schematics of a Mach–Zehnder interferometer biosensor.

where M is the modulation factor and $\Delta\Phi$ is the relative phase shift between the two arms. If the initial conditions are adjusted so that the relative phase shift $\Delta\Phi$ is zero, the binding of an analyte to the sensing layer on the sample arm channel waveguide introduces an additional phase shift $\Delta\Phi_{sens}$, given as (Heideman and Lambeck, 1999)

$$\Delta\Phi_{sens} = (2\pi/\lambda)L_{int}\Delta N$$

where L_{int} is the "interaction length" of the guided wave with the analyte, λ is the wavelength, and ΔN is the change in the effective index in the evanescent field.

Integrated-optic Mach–Zehnder interferometric sensors utilizing a planar waveguide geometry have been used for many applications. Some examples are glucose sensor (Liu et al., 1992), immunosensor (Brecht et al., 1992), and sensors for pesticide determination (Schipper et al., 1995).

9.8 SURFACE PLASMON RESONANCE BIOSENSORS

Surface plasmon resonance (often abbreviated as SPR) sensors are, perhaps, the most extensively utilized optical biosensors that are also commercially sold by a number of companies as discussed below in Section 9.11. There are a number of excellent reviews on this subject (Liedberg et al., 1995; Schuck, 1997; Homola et al., 1999; Myszka and Rich, 2000). The SPR technique has been utilized for a variety of biosensing methods, from biochemical detection such as of glucose and urea, to immunosensing for immunoassays (for protein hormones, drugs, steroids, immunoglobulins, viruses, whole bacteria, and bacterial antigens), to DNA binding assays, to real-time kinetics of drugs binding to therapeutic targets. Spangler et al. (2001) have compared the performance of a commercial SPR sensor with that of a quartz crystal microbalance for detection of *E. coli* heat-labile enterotoxin.

In principle, the SPR technique is an extension of evanescent wave sensing, described in Section 9.6, except that a planar waveguide is replaced by a metal–dielectric interface. Surface plasmons are electromagnetic waves that propagate along the interface between a metal and a dielectric material such as organic films (Wallis and Stegeman, 1986). Since the surface plasmons propagate in the frequency and wave-vector ranges for which no light propagation is allowed in either of the two media, no direct excitation of surface plasmons is possible. The most commonly used method to generate a surface plasmon wave is attenuated total reflection (ATR).

The Kretschmann configuration of ATR is widely used to excite surface plasmons (Wallis and Stegeman, 1986). This configuration is shown in Figure 9.23. A microscopic slide is coated with a thin film of metal (usually a 400- to 500-Å-thick gold or silver film by vacuum deposition). Then a biosensing layer containing an immobilized biorecognition element can be coated on the metal

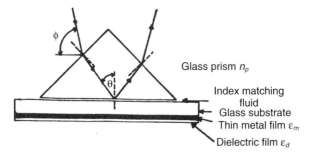

Figure 9.23. Kretschmann (ATR) geometry used to excite surface plasmons (Prasad, 1988).

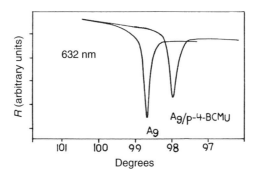

Figure 9.24. Surface plasmon resonance curves. The left-hand-side curve is for just the silver film (labeled Ag); the right-hand-side curve shows the curve (labeled Ag/p-4-BCMU) shifted on the deposition of a monolayer Langmuir–Blodgett film of poly-4-BCMU on the silver film (Prasad, 1988).

surface. The microscopic slide is now coupled to a prism through an index-matching fluid or a polymer layer. A p-polarized laser beam (or light from a light-emitting diode) is incident at the prism. The reflection of the laser beam is monitored. At a certain θ_{sp}, the electromagnetic wave couples to the interface as a surface plasmon. At the same time, an evanescent field propagates away from the interface, which extends to about 100 nm above and below the metal surface. At this angle the ATR signal drops. This dip in reflectivity is shown by the left-hand-side curve in Figure 9.24. The angle is determined by the relationship

$$k_{sp} = kn_p \sin\theta_{sp}$$

where k_{sp} is the wave vector of the surface plasmon, k is the wave vector of the bulk electromagnetic wave, and n_p is the refractive index of the prism. The surface plasmon wave vector k_{sp} is given by

$$k_{sp} = (\omega/c)[(\varepsilon_m \varepsilon_d)/(\varepsilon_m + \varepsilon_d)]^{1/2}$$

where ω is the optical frequency, c is the speed of light, and ε_m and ε_d are the relative dielectric constants of the metal and the dielectric, respectively, which are of opposite signs. In the case of a bare metal film, ε_d (or square of the refractive index for a dielectric) is the dielectric constant of air and the dip in reflectivity occurs at one angle. In the case of metal coated with the sensing layer, this angle shifts. Upon binding with an analyte, a further shift of the SPR coupling angle occurs. Figure 9.24 also shows as an illustration the shift in the coupling angle on deposition of a monolayer Langmuir–Blodgett film of a diacetylene, poly-4-BCMU. The shifted SPR curve curve is shown on the right-hand side in Figure 9.24 (Prasad, 1988).

In this experiment one can measure the angle for the reflectivity minimum, the minimum value of reflectivity, and the width of the resonance curves. These observables are used for a computer fit of the resonance curve using a least-squares fitting procedure with the Fresnel reflection formulas, which yields three parameters: the real and the imaginary parts of the refractive index and the thickness of the sensing layer. The experiment involves the study of angular shift (change in θ_{sp}) as a function of analyte binding.

From the above equations, one can see that the change $\delta\theta$ in the surface plasmon resonance angle (the angle corresponding to minimum reflectivity; for simplicity the subscript sp is dropped) caused by changes $\delta\varepsilon_m$ and $\delta\varepsilon_d$ in the dielectric constants of the metal and the covering film, respectively, is given by (Nunzi and Ricard, 1984)

$$\cot\theta\,\delta\theta = (2\varepsilon_m\varepsilon_d(\varepsilon_m + \varepsilon_d))^{-1}(\varepsilon_m^2 + \delta\varepsilon_d + \varepsilon_d^2\delta\varepsilon_m)$$

Since $|\varepsilon_m| \gg |\varepsilon_d|$, the change in θ is much more sensitive to a change in ε_d (i.e., of the sensing layer) than to a change in ε_m. Therefore, this method appears to be ideally suited to obtain $\delta\varepsilon_d$ (or a change in the refractive index) as a, function of analyte binding to the sensing layer. Another way to visualize the high sensitivity of SPR to variations in the optical properties of the dielectric above the metal is to consider the strength of the evanescent field in the dielectric, which is an order of magnitude higher than that in a typical evanescent wave sensor utilizing an optical waveguide as described above. The magnitude of the change in θ can be quantitatively related to the amount of analyte binding or to the extent of a chemical change in the sensing layer.

In an SPR sensor, the change $\delta\varepsilon_d$ (and hence $\delta\theta$) (such as in antibody–antigen reactions) that can be induced is independent of wavelength. However, in some cases such as for various immobilized chromophores, the change in ε_d is at specific wavelengths. In SPR biosensors the immobilized probe is usually attached to a sensor chip with a thin layer of metal. In an SPR sensor, the sensing response is a change in the refractive index of the sensing layer containing ligands (e.g., antibodies) upon analyte binding which is measured as a change, $\delta\theta$, in the coupling angle. In commercial SPR sensors, this change in the coupling angle is measured by a CCD or photodiode array using a convergent light beam, rather than scanning the angle as described above. This arrangement, as shown in Figure 9.25, permits real-time monitoring of the ligand–analyte binding to obtain kinetics of association and dissociation. To get this information, the sample solution containing the analyte flows over the sensor chip containing the ligand. During the association phase, the analyte binds with the ligand immobilized on the sensor chip, generating an increase in response (amount of shift of the coupling angle). The magnitude of the response ($\delta\theta$) levels off over the time as an equilibrium condition between the free and the bound analyte is reached. To monitor dissociation, the flow switches to that of a running buffer which washes out the analyte (leading to dissociation of it from the ligand). During the dissociation, the magnitude of the response decreases. The generated response curve for the association and dissociation cycle is often called a *sensorgram*. The association process is also shown in Figure 9.25 as an inset on the right-hand side.

A wide variety of surface chemistries have been used to provide functionality to minimize nonspecific binding of ligands to the gold surface. Some of these are (Homola et al., 1999)

- Streptavidin monolayer immobilized onto a gold film with biotin which can further be functionalized with biotinylated biomolecules.

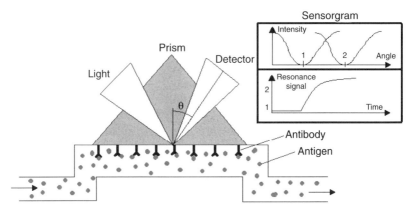

Figure 9.25. Surface plasmon resonance sensor schematic utilizing a CCD or photodiode array (*left*). The inset on the right-hand side shows the sensorgram. The top in the inset shows a shift in the SPR curve from 1 to 2 upon binding with the analyte. The bottom curve is obtained by monitoring the SPR signal at the shifted coupling angle as a function of time when the analyte is introduced. (Reproduced from http://chem.ch.huji.ac/il/~eugeniik/spr.htm.)

- A self-assembled monolayer (SAM) of thiol molecules such as 16-thiohexadecanol. The thiol group attaches to the gold surface. The tail can be bound to the ligands forming a monolayer of ligand molecules.
- An SAM layer covalently bonded to a dextran layer using epichlorohydrin. After treating dextran with iodoacetic acid, the resulting carboxylic group can be used to immobilize the ligands.
- Gold surface coated by a plasma polymerized thin film onto which the ligands can be immobilized via an amino group.

SPR sensors offer several distinct advantages, such as:

- No labeling (such as by a fluorescent marker) required, thus allowing for the analysis of a wide range of biomolecular systems.
- Real-time monitoring permitted, thereby providing rapid and quantitative information on kinetics of binding.
- Small amounts of materials required for typical analysis.

9.9   SOME RECENT NOVEL SENSING METHODS

Photonic Crystals Sensors. Photonic crystals are ordered dielectric arrays that diffract light at wavelengths determined by the lattice spacing between the arrays and the average refractive index of a structure (Carlson and Asher, 1984; Asher, 1986; John, 1987; Yablonovitch, 1987). One example of a photonic crystal is a closely packed colloidal array as shown in Figure 9.26. This crystal was prepared at our Institute, using 200-nm polystyrene spheres. These spheres were floated, as a suspension, over a patterned template where they settled to form a highly ordered array (Markowicz and Prasad, unpublished).

Figure 9.26. Close packing of colloidal nanospheres to form a photonic crystal of close-packed colloidal array. (*Left*) Atomic force microscope (AFM) image of the surface layer. (*Right*) Scanning electron microscope (SEM) image of a cross section (Markowicz and Prasad, unpublished).

Asher and co-workers in their pioneering work have used this colloidal array to propose a novel chemical and biosensing scheme (Holtz and Asher, 1997; Holtz et al., 1998; Lee and Asher, 2000; Reese et al., 2001). They used a three-dimensional periodic structure of colloidal crystal arrays (CCA) of highly charged polystyrene spheres of diameter 100 nm. Electrostatic interactions between these charged spheres lead the spheres to self-assemble into a body-centered or a face-centered cubic structure. This periodic CCA produces diffraction of light according to Bragg's law:

$$m\lambda = 2\,nd\,\sin\theta$$

Here m is the diffraction order, λ is the wavelength of light in vacuum, n is the refractive index of the system, d is the diffracting plane spacing (separation between the centers of the spheres), and θ is the Bragg glancing angle. Furthermore, microcavity resonance producing strong scattering within the sphere from its boundary can also be produced under the condition that $\lambda = 2nd$. The effect of both this microcavity resonance and the Bragg diffraction is that the propagation of a certain wavelength matching the above resonance condition (diffraction and scattering) is highly attenuated, thus the crystal acting as a narrow band filter for this wavelength. This wavelength region of high attenuation is often called the stop-gap region. For polystyrenes of the dimensions ~200 nm, the stop-gap region is in the visible around ~500 nm, thus imparting an intense color to the CCA, based on the strong attenuation (filtering) of this visible wavelength of light.

For chemical and biochemical sensing, Asher's group fabricated a CCA of polystyrene spheres (of diameters ~100 nm) polymerized within a hydrogel that swells and shrinks reversibly in the presence of certain analytes such as metal ions or glucose. For this purpose, the hydrogel contains a molecular-recognition group that either binds or reacts selectively with an analyte. The result of the recognition process is a swelling of the gel owing to an increase in the osmotic pressure which in turn leads to a change in the value d in the above equation. As a result, the diffraction and scattering conditions changes to a longer wavelength. Results by Asher's group suggest that a mere change of 0.5% in the hydrogel volume shifts the diffracted wavelength by ~1 nm.

For the detection of metal ions such as Pb^{2+}, Ba^{2+}, and K^+, Holtz and Asher (1997) copolymerized 4-acryloaminobenzo-18-crown-6 (AAB18C6) into polymerized crystalline colloidal array (PCCA). This crown ether was chosen because of its selective binding ability with Pb^{2+}, Ba^{2+}, and K^+. In these cases, the gel swelling mainly results from an increase in the osmotic pressure within the gel.

For glucose sensing, Asher and co-workers (Holtz and Asher, 1997; Holtz et al., 1998) attached the enzyme, glucose oxidase (GO_x), to a polymerized crystalline colloidal array (PCCA) of polystyrene. For this purpose, the PCCA was hydrolyzed and biotinylated. This PCCA was polymerized from a solution containing ~7 wt% polystyrene colloidal spheres, 4.6 wt% acrylamide (AMD) and 0.4 wt% *N,N'*-methylenebisacrylamide (bisAMD), with water constituting the remaining fraction. This hydrogel was hydrolyzed in a solution of NaOH and was then biotinylated with biotinamidopentylamine, which was attached using a water-soluble carbodiimide coupling agent. Avidinated glucose oxidase was then directly added to the PCCA. A glucose solution prepared in the air causes PCCA to swell and produce a red shift in the diffraction wavelength as shown in Figure 9.27.

Optical Sensor Array and Integrated Light Source. A new multianalyte detection scheme that utilizes an optical sensor array and integrated light source (OSAILS) has recently been introduced by Bright and co-workers (Cho and Bright, 2001). They utilized microwells of dimensions approximately 250 μm machined directly into the light-emitting diode (LED) face. The individual microwells are filled with a sol–gel precursor solution containing an oxygen-sensitive emitter [tris(4,7-diphenyl-1,10-phenanthroline)ruthenium(II) dication, $[Ru(dpp)_3]^{2+}$]. The xerogel (porous gel) forms within individual

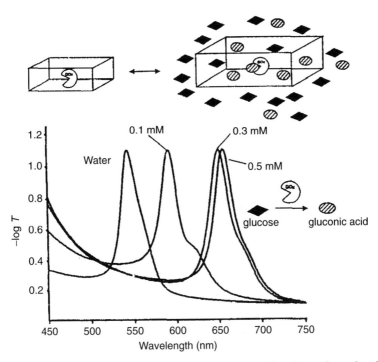

Figure 9.27. Visible extinction spectra showing how diffraction depends on the glucose concentration for a 125-μm-thick PCCA glucose sensor. The ordinate is given as −log *T*, where *T* is the transmittance. The PCCA expands for concentrations between 0.1 and 0.5 mM glucose. (Reproduced with permission from Holtz and Asher, 1997.)

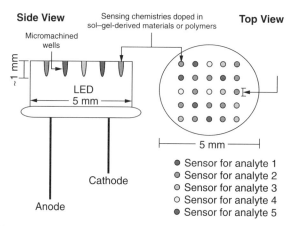

Figure 9.28. The schematics of an optical sensor array and integrated light source (OSALLS). (Reproduced with permission from Cho and Bright, 2001.)

microwells. A schematic of an OSAILS is shown in Figure 9.28. The OSAILS is then placed within a flow cell holder and can be powered by a low-voltage dc power supply or a battery. The LED light output is used directly to excite the emitter immobilized within the microwell-entrapped xerogel. The fluorescence output from the array of microwells is collected by a charge-coupled device (CCD).

Hybrid Transduction Biosensors. All of the above sensing schemes utilize optical transduction in which biosensing produces an optical response. Another area of well-developed sensors utilizes electrical transduction (Janata, 1989; Ramsay, 1998). A number of schemes have recently utilized a hybrid optical–electrical or electrical–optical transduction mechanism that may offer some unique capabilities by combining the electrical and optical effects. One earlier example is a light addressable potentiometric sensor (LAPS) in which specific locations on a silicon-nitride-coated sensor are made sensitive to pH by illumination from an LED array (Hafemen et al., 1988). This approach was illustrated for monitoring of enzymes that produce protons. Other suggested applications of this type of sensing include measurement of DNA, pathogenic bacteria, anticholinesterase drugs, and pesticides.

Electroluminescence in which an electrical stimulation (charge carrier injection) produces fluorescence has also been proposed for sensing in which the sensing function produces a change in the electroluminescence efficiency (Leca and Blum, 2000).

Time Domain Sensing. Time domain methods have also received considerable attention recently as pulse and modulated laser sources become more readily available at an affordable cost. One example is a method utilizing optical time domain reflectometry (OTDR) in which a light pulse is transmitted down a fiber and the back-scattered light produces an echo signal (Dakin, 1991). The signal propagation time depends on the speed of light (refractive index of the medium) as well as on the length of transmission in the fiber. Any variation in light attenuation/absorption, scattering, and its spatial distribution produces a change in the echo response. The whole fiber length can act as a distributed sensor. An example is a microbend fiber-optic sensor with a water-sensitive polymer (hydrogel) deposited onto a central support (Michie et al., 1995). In the presence of water, the swelling of the hydrogel exerts a microbending force on the fiber, producing attenuation, which is sensed by OTDR. This type of sensing scheme can also be adapted for distributed pH measurements.

Surface-Enhanced Raman Sensors. Vo-Dinh and co-workers at Oak Ridge National Laboratory have developed biosensors that are based on surface-enhanced Raman scattering (abbreviated as SERS) (Isolo et al., 1998; Vo-Dinh et al., 1999). The SERS effect, described in Chapter 4, utilizes an enhancement of Raman spectra of species deposited on a metal surface. The Oak Ridge group has utilized the SERS technique as gene probes for DNA detection and DNA trapping, utilizing nanostructured metallic substrates as biosensor platforms.

9.10 FUTURE DIRECTIONS

The area of biosensors is rapidly growing worldwide. In recent years, it has received a great deal of attention because of the danger posed by chemical and bioterrorism. The needs cover a wide range, from point detection, to environmental monitoring, to *in vivo* monitoring. Opportunities for future development are also manyfold and multidisciplinary. Some of these future directions are briefly described here.

Multianalyte Detection. Multianalyte detection will continue to be a major focus of future development. Different methods of patterning efficient and mutually compatible biorecognition elements, as well as coupling them separately to an array of light sources and an array of detectors are opportunities for chemists, biomedical researchers and engineers. Various imprint technologies will be of value in patterning. An important consideration will also be the capabilities of these sensors for real time continuous monitoring. Here the long-term stability of the pattern of immobilized biorecognition elements and fluidic consideration will play an important role.

New Biorecognition Molecules. Development of new biorecognition elements enhancing selectivity and sensitivity of sensing is another important direction. Here an important avenue of approach may be combinatorial biochemistry. Some recent examples of new biorecognition elements showing promises are single-chain Fv antibodies, ligand binding oligonucleotides, also known as *aptamers* (Kleinjung et al., 1998), and phage-displayed peptides (Goldman et al., 2000). The use of a whole biological cell or virus as a biorecognition element is another exciting new direction.

Fluidics. To increase the capabilities of real-time monitoring and to obtain information on kinetics of binding and dissociation, key areas of development are improved fluidics and signal sampling rates. Microfluidic devices may play an important role. Screening of large areas will place additional demands.

In Vivo *Sensors.* There is a growing need for *in vivo* monitoring to detect infections and diseases as well as to determine in real time the efficacy of a drug and the effectiveness of a specific treatment or response to a particular medical procedure. The two important factors are:

1. Miniaturization of all sensor components to produce a small, implantable sensor that is minimally invasive. Efforts are also underway to produce smart patches or band-aids that can generate sensor response in a noninvasive manner.
2. Biocompatibility of sensors in the case of an implant. An important consideration in the design of the biosensing materials is whether the sensor induces any change in the host tissue and/or the host tissue induces any change in the sensor.

From a miniaturization perspective, recent development in microlasers, vertical cavity lasers (VCSELS), micro-optics, microelectromechanical systems (MEMS), and photodetector arrays will play important roles.

Chemical Identification Biosensors. From the perspective of threat of new toxins and new microbes, sensors have to be developed which detect these new species and provide detailed chemical information on their structures and identification of chemical functional groups that are known to be toxic and a health hazard. This type of chemical information can be obtained from a detailed mass-spectrometric analysis or spectral analysis (such as vibrational bands from IR or Raman spectra). This is, again, an exciting opportunity to produce new designs of all optical or hybrid optical sensing which can respond to chemical properties.

Data Processing, Pattern Recognition, and Automation. There is also an exciting opportunity for smart software and algorithm development for background variation compensation, data mining, correlation and recognition. Automation is a major need in producing commercializable biosensors for multianalyte detection, computer control of fluidics, screening of a large number of samples, continuous monitoring, and increased reproducibility of assays.

New Applications. Broadening the range of applications of biosensors is another area of opportunity. Some areas of growth are food processing, water quality control, and drug screening.

9.11 COMMERICALLY AVAILABLE BIOSENSORS

A number of manufacturers commercialize optical biosensors. Table 9.5 lists some of these biosensors. More details on them can be found from the websites of the manufacturers, which are also listed in the table.

HIGHLIGHTS OF THE CHAPTER

- Biosensors are analytical devices that can detect bioactive chemicals, biological species, or microorganisms.
- Optical biosensors offer the advantage of real-time monitoring and remote sensing with no electrical interference.
- An optical biosensor consists of (i) a sampling unit that introduces an analyte, (ii) a biorecognition element that recognizes the presence of the analyte, (iii) a transducer that converts the resulting response of the biorecognition element into an optical signal, and (iv) a detector that responds to the optical signal.
- Biorecognition elements are biological species such as enzymes, antibodies, and microorganisms, optimally having an extremely high specificity for the analyte of interest. The recognition elements are often immobilized spatially to increase their concentration near the optical sensing element.
- Optical transduction utilizes changes in optical properties such as phase, amplitude, and frequency, manifested because of the selective binding of an analyte with the biorecognition element.
- A commonly used method is based on fluorescence detection, which utilizes (i) direct sensing where the fluorescence properties of the analyte are altered or (ii) indirect sensing in which the fluorescence properties of an external dye are changed, when the analyte interacts with the biorecognition element.
- Fluorescence energy transfer sensors utilize a change in the electronic energy transfer between a donor and an acceptor group, caused when the analyte interacts with the biorecognition element.
- Optical sensors based on photoinduced electron transfer between an electron donor and an electron acceptor utilize a change in fluorescence,

TABLE 9.5. Optical Biosensor Manufacturers, Instruments, and Website Information

Manufacturer	Instrument	Sensing Method	Website
Biacore AB	BIACORE 1000, 2000, 3000, X, J, Quant, 8 channel prototype, S51	SPR	www.biacore.com
Affinity Sensors	IAsys, IAsys Plus, IAsys Auto+	Evanescent wave	www.affinity-sensors.com
IBIS Technologies	IBIS I, IBIS II	SPR	www.ibis-spr.nl
Nippon Laser Electronics	SPR670, SPR Cellia	SPR	www.nle-lab.co.jp/English/ZO-HOME.htm
Texas Instruments	Spreeta	SPR	www.ti.com/sc/docs/products/msp/control/spreeta
Analytical μ-Systems	BIO-SUPLAR 2	SPR	www.micro-systems.de
AVIV Instruments	PWR Model 400	Plasmon waveguide resonance	www.avivinst.com
Farfield Sensors Ltd.	*Ana*Light Bio250	Planar waveguide interferometric	www.farfield-sensors.co.uk
Luna Innovations	Fiber optical prototype	Fiber optic using long period gratings	www.lunainnovations.com
ThreeFold Sensors	Label-free prototype	Evanescent wave fiber-optics fluorescence	http://ic.net/~tfs
Graffinity	Plasmon Imager	Wavelength-dependent SPR	www.graffinity.com
Leica	Prototype	SPR	www.leica-ead.com
Prolinx	OCTAVE	SPR	www.prolinxinc.com
HTS Biosystems	SPR array	SPR	www.htsbiosystems.com
Quantech Ltd	FasTraQ SPR array	SPR	www.quantechltd.com
SRU Biosystems	BIND	Resonant diffraction grating surface	www.srubiosystems.com

caused when an analyte interacts with the biorecognition element affecting electron transfer.

- A molecular beacon consists of a loop of specifically sequenced single-stranded oligonucleotide and a stem made up of five to seven complementary base pairs that at the two ends consist of a fluorophore and a fluorescence quencher. In the presence of an analyte the two ends move farther apart, thus restoring the fluorescence of the fluorophore.

- Immobilization of the biorecognition element to enhance its local concentration can be brought about by physical methods of containment within semipermeable membranes, a selective adsorption of the element on a solid support, electrostatic interactions, and physical entrapment within a matrix.

- Chemical methods of immobilization involve the actual chemical bonding of the solid surface to the biorecognition element via reactions that usually involve coupling groups.

- Fiber-optic biosensors are widely used because of their convenient geometry—for example, longer interaction length and compatibility with instruments used for *in vivo* biosensing.

- Planar waveguides offer the advantage of simultaneously using several different biorecognition elements to provide multianalyte detection capability.

- Evanescent wave biosensors are based on interaction of the analyte species with the electromagnetic wave, which extends from the surface of the light-guiding materials.

- Evanescent wave coupling between two channel waveguides or fibers in close enough proximity, introduced by the overlap of their evanescent fields, can detect binding of the analyte by changes in the refractive index between the waveguides. The result is a change in the transfer of optical power from one channel to another.

- Interferometric biosensors are based on the changes that occurs in the interference between light from a sensing waveguide and that from a reference, when an analyte is present on the sensing waveguide.

- Surface plasmon resonance (SPR) biosensors utilize surface plasmons that propagate when the light is directed at a certain angle called the *critical angle* along the interface between a metal and a dielectric medium.

- The change in the critical angle, on analyte binding to biorecognition elements immobilized on the metal film surface, provides biosensing.

- Some examples of novel sensing elements are photonic crystals, optical sensor arrays integrated with light sources, hybrid transadation biosensors, time domain sensors, and surface-enhanced Raman sensors.

REFERENCES

Abdel-Latif, M. S., Suleiman, A. A. and Guilbault, G. G., Fiber-Optic Sensors: Recent Development, *Anal. Lett.* **23**, 375–399 (1990).

Asher, S. A., Crystalline Colloidal Narrow Band Radiation Filter, U.S. patent 4,627,689 (1986).

Avnir, D., Braun S., Lev O., and Ottolenghi M., Enzymes and Other Proteins Entrapped in Sol-Gel Materials, *Chem. Mater.* **6**, 1605–1614 (1994).

Bhatia, S. K., Hickman J. L., and Ligler, F. S., New Approach to Producing Patterned Biomolecular Assemblies, *J. Am. Chem. Soc.* **14**, 4432–4433 (1992).

Bhatia, S. K., Teixeira, J. L., Anderson, M., Shriner-Lake, L. C., Calvert, J. M., Georger, J. H., Hickman, J. J., Ducley, C. S., Schoen, P. E., and Ligler, F. S., Fabrication of Surfaces Resistant to Protein Adsorption and Application to Two-Dimensional Protein Patterning, *Anal. Biochem* **208**, 197–205 (1993).

Blawas, A. S., and Reichert, W. M., Protein Patterning, *Biomaterials* **19**, 595–609 (1998).

Boisdé, G., and Harmer, A., *Chemical and Biochemical Sensing with Optical Fibers and Waveguides*, Artech House, Norwood, MA, 1996.

Brecht, A., Ingenhoff, J., and Gauglitz, G., Direct Monitoring of Antigen–Antibody Interactions by Spectral Interferometry, *Sensors and Actuators* **B6**, 96–100 (1992).

Carlson, R. J., and Asher, S. A., Characterization of Optical Diffraction and Crystal Structure in Monodisperse Polystyrene Colloids, *Appl. Spectrosc.* **38**, 297–304 (1984).

Chia, S., Urano, J., Tamanoi, F., Dunn, B., and Zink, J. I., Patterned Hexagonal Arrays of Living Cells in Sol–Gel Silica Film, *J. Am. Chem. Soc.* **122**, 6488–6489 (2000).

Cho, E. J., and Bright, F. V., Optical Sensor Array and Integrated Light Source, *Anal. Chem.* **73**, 3289–3293 (2001).

Dakin, J. P., Distributed Optical Fiber Sensor Systems, Chapter 15 in B. Culshaw, D. P. Dakin, and A. M. Morwood, eds., *Optical Fiber Sensors*, Vol. 2, Artech House, Norwood, MA, 1991, pp. 575–598.

deSilva, A. P., Gunaratne, H. Q. N., Gunnlaugsson, T., Huxley, A. J. M., McCoy, C. P., Rademacher, J. T., and Rice, T. E., Signaling Recognition Events with Fluorescent Sensors and Switches, *Chem. Rev.* **97**, 1515–1566 (1997).

Devor, E. J., Use of Molecular Beacons to Verify that the Serine Hydroxymethyl-transferase Pseudogene *SHMT-ps1* Is Unique to the Order Primates, *Genome Biol.* **2**, 1–5 (2001).

Ellerby, L. M., Nishida, C. R., Nishida, F., Yamanaka, S. A., Dunn, B., Valentine, J. S., and Zink, J. I., Encapsulation of Proteins in Transparent Porous Silicate Glasses Prepared by Sol–Gel Method, *Science* **255**, 1113–1115 (1992).

Feldstein, M. J., Golden, J. P., Rowe, C. A., MacCraith, X., and Ligler, F. S., Array Biosensor: Optical and Fluidic Systems, *J. Biomed. Microdevices* **1:2**, 139–153 (1999).

Feldstein, M. J., MacCraith, B. D., and Ligler, F. S., U.S. patent 6,137,117 (2000).

Goldman, E. R., Pazirandeh, M. P., Mauro, J. M., King, K. D., Frey, J. C., and Anderson, G. P., Phage-Displayed Peptides as Biosensor Reagents, *J. Mol. Recognition* **13**, 382–387 (2000).

Hafeman, D. G., Parce, J. W., and McConnell, H. M., Light Addressable Potentiometric Sensor for Biochemical Systems, *Science* **240**, 1182–1185 (1988).

Heideman, R. G., and Lambeck, P. V., Remote Opto-Chemical Sensing with Extreme Sensitivity: Design, Fabrication and Performance of a Pigtailed Integrated Optical Phase-Modulated Mach–Zehnder Interferometer System, *Sensors and Actuators* **B**, 100–127 (1999).

Holtz, J. H., and Asher, S. A., Polymerized Colloidal Crystal Hydrogel Films as Intelligent Chemical Sensing Materials, *Nature* **389**, 829–832 (1997).

Holtz, J. H., Holtz, J. S. W., Munro, C. H., and Asher, S. A., Intelligent Polymerized Crystalline Colloidal Arrays: Novel Chemical Sensor Materials, *Anal. Chem.* **70**, 780–791 (1998).

Homola, J., Yee, S., and Gauglitz, G., Surface Plasmon Resonance Sensors: Review, *Sensors and Actuators* **B54**, 3–15 (1999).

Hug, T. S., Prenosil, J. E., and Morbidelli, M., Optical Waveguide Lightmode Spectroscopy as a New Method to Study Adhesion of Anchorage-Dependant Cells as an Indicator of Metabolic State, *Biosens. Bioelectron.* **16**, 865–874 (2001).

Hyrup, B., and Nielsen, P. E., Peptide Nucleic Acids (PNA): Synthesis, Properties and Potential Applications, *Bioorg. Med. Chem.* **4**, 7–14 (1996).

Isolo, N. R., Stokes, D. L., and Vo-Dinh, T., Surface-Enhanced Raman Gene Probe for HIV, *Anal. Chem.* **70**, 1352–1356 (1998).

Janata, J., *Principles of Chemical Sensors*, Plenum Press, New York, 1989.

John, S., Strong Localization of Photons in Certain Disordered Dielectric Superlattices, *Phys. Rev. Lett.* **58**, 2486–2489 (1987).

Kleinjung, F., Klussmann, S., Erdmann, V. A., Scheller, F. W., Furste, J. P., and Bier, F. F., Binders in Biosensors: High-Affinity RNA for Small Analytes, *Anal. Chem.* **70**, 328–331 (1998).

Kuswandi, B., Andres, R., and Narayanaswamy, R., Optical Fiber Biosensors Based on Immobilized Enzymes, *Analyst* **126**, 1469–1491 (2001).

Leca, B., and Blum, L. J., Luminol Electrochemiluminescence with Screen-Printed Electrodes for Low-Cost Disposable Oxidase-Based Optical Sensors, *Analyst* **125**, 789–791 (2000).

Lee, K., and Asher, S. A., Photonic Crystal Chemical Sensors: pH and Ionic Strength, *J. Am. Chem. Soc.* **122**, 9534–9537 (2000).

Liedberg, B., Nylander, C., and Lundstrom, I., Biosensing with Surface Plasmon Resonance, How It All Started, *Biosens. Bioelectron.* **10**, i–ix (1995).

Ligler, F. S., and Rawe-Taitt, C. A., eds, *Optical Biosensors: Present and Future*, Elsevier, Amsterdam, 2002.

Liu, Y., Hering, P., and Scully, M. O., An Integrated Optical Sensor for Measuring Glucose Concentration, *Appl. Phys. B, Photophys. Laser Chem.* **B54**, 18–23 (1992).

Lowe, C. R., Chemoselective Biosensors, current opinion in *Chemical Biology* **3**, 106–111 (1999).

Lowe, P. A., Clark, J. H. A., Davies, R. J., Edwards, P. R., Kinning, T., and Yeung, D., New Approaches for Analysis of Molecular Recognition Using IAsys Evanescent Wave Biosensors, *J. Mol. Recognition* **11**, 194–199 (1998).

Lukosz, W., and Tiefenthaler, K., Sensitivity of Integrated Optical Grating and Prism Couplers as (Bio) Chemical Sensors, *Sensors and Actuators* **15**, 273–284 (1988).

Mehrvar, M., Bis, C., Scharer, J. M., Moo-Young, M., and Lerong, J. H., Fiber-Optic Biosensors—Trends and Advances, *Anal. Sci.* **16**, 677–692 (2000).

Michie, W. C., Culshaw, B., MacKenzie, I., Konstantakis, M., Graham, N. B., Moran, C., Santos, F., Bergqvist, E., and Carlstrom, B., Distributed Sensor for Water and pH Measurements Using Fiber Optics and Swellable Polymeric Systems, *Opt. Lett.* **20**, 103–105 (1995).

Myszka, D. G., and Rich, R. L., Implementing Surface Plasmon Resonance Biosensors in Drug Discovery, *Pharm. Sci. Technol. Today* **3**, 310–317 (2000).

Nunzi, J. M., and Ricard, D., Optical Phase Conjugation and Related Experiments with Surface Plasmon Waves, *Appl. Phys.* **B35**, 209–216 (1984).

Ortiz, E., Estrada, G., and Lizardi, P. M., PNA Molecular Beacons for Rapid Detection of PCR Amplicons, *Mol. Cell. Probes* **12**, 219–226 (1998).

Plowman, T. E., Durstchi, J. D., Wang, H. K., Christensen, D. A., Herron, J. N., and Reichert, W. M., Multiple-Analyte Fluoroimmunoassay Using an Integrated Optical Waveguide Sensor, *Anal. Chem.* **71**, 4344–4352 (1999).

Pope, E. J. A., Braun, K., and Pehrson, C. M., Bioartifical Organs I: Silica Gel Encapsulated Pancreatic Islets for the Treatment of Diabetes Mellitus, *J. Sol–Gel Sci. Technol.* **8**, 635–639 (1997).

Prasad, P. N., Design, Ultrastructure, and Dynamics of Nonlinear Optical Effects in Polymeric Thin Films, in P. N. Prasad and D. R. Ulrich, eds., Nonlinear Optical and Electroactive Polymers, Plenum Press, New York, 1988, pp. 41–67.

Ramsay, G., ed., *Commercial Biosensors, Applications to Clinical, Bioprocess and Environmental Samples*, John Wiley & Sons, New York, 1998.

Reese, C. E., Baltusavich, M. E., Keim, J. P., and Asher, S. F., Development of an Intelligent Polymerized Crystalline Colloidal Array Colorimetric Reagent, *Anal. Chem.* **73**, 5038–5042 (2001).

Rich, R. L., and Myszka, D. G., Survey of the 1999 Surface Plasmon Resonance Biosensor Literature, *J. Mol. Recognition* **13**, 388–407 (2000).

Rickus, J. L., Dunn, B., and Zink, J. I., Optically Based Sol–Gel Biosensor Materials, in F. S. Ligler and C. A. Rowe-Taitt, eds., *Optical Biosensors: Present and Future*, Elsevier, Amsterdam, 2002, pp. 427–456.

Rogers, K. R., Principles of Affinity-Based Biosensors, *Mol. Biotechnol.* **14**, 109–129 (2000).

Rowe-Taitt, C. A., Golden, J. P., Feldstein, M. J., Cas, J. J., Hoffman, K. E., and Ligler, F. S., Array Biosensor for Detection of Biohazards, *Biosens. Bioelectron.* **14**, 785–794 (2000a).

Rowe-Taitt, C. A., Hazzard, J. W., Hoffman, K. E., Cras, J. J., Golden, J. P., and Ligler, F. S., Simultaneous Detection of Six Biohazardous Agents Using a Planar Waveguide Array Biosensor, *Biosens. Bioelectron.* **15**, 579–589 (2000b).

Rowe-Taitt, C. A., and Ligler, F. S., Evanescent Wave Fiber Optic Biosensors, in F. S. Ligler and C. A. Rowe-Taitt, eds., *Optical Biosensors: Present and Future*, Elsevier, Amsterdam, 2002, pp. 57–94.

Sapsford, K., Rowe-Taitt, C. A., and Ligler, F. S., Planar Waveguides for Fluorescence Biosensors, in F. S. Ligler and C. A. Rowe-Taitt, eds., *Optical Biosensors: Present and Future*, Elsevier, Amsterdam, 2002, pp. 95–122.

Schipper, E. F., Kooyman, P. H., Heideman, R. G., and Greve, J., Feasibility of Optical Waveguide Immunosensors for Pesticide Detection: Physical Aspects, *Sensors and Actuators* **B24/25**, 90–93 (1995).

Schuck, P., Use of Surface Plasmon Resonance to Probe the Equilibrium and Dynamic Aspects of Interactions Between Biological Macromolecules, *Annu. Rev. Biophys. Biomol. Structure* **26**, 541–566 (1997).

Schultz, J. S., Mansouri, S., and Goldstein, I. J., Affinity Sensor: A New Technique for Developing Implantable Sensors for Glucose and Other Metabolites, *Diabetes Care* **3**, 245–253 (1982).

Schwarz, A., Rossier, J. S., Roulet, E., Mermod, N., Roberts, M. A., and Girault, H. H., Micropatterning of Biomolecules on Polymer Substrates, *Langmuir* **14**, 5526–5531 (1998).

Shtelzer, S., Rappoport, S., Avnir, D., Ottolenghi, M., and Braun, S., Properties of Trypsin and of Acedphosphatase Immobilized in Sol–Gel Glass Matrices, *Biotechnol. Appl. Biochem.* **15**, 227–235 (1992).

Silzel, J. W., Cercek, B., Dodson, C., Tsay, T., and Obrenski, R. J., Mass-Sensing, Multianalyte Microarray Immunoassay with Imaging Detection, *Clin. Chem.* **44**, 2036–2043 (1998).

Spangler, B. D., Wilkinson, E. A., Murphy, J. T., and Tyler, B. J., Comparison of the Spreeta® Surface Plasmon Resonance Sensor and a Quartz Crystal Microbalance for Detection of *Escherichia Coli* Heat-Labile Enterotoxin, *Anal. Chim. Acta* **444**, 149–161 (2001).

Tan, W., Fang, X., Li, J., and Liu, X., Molecular Beacons: A Novel DNA Probe for Nucleic Acid and Protein Studies, *Chem. Eur. J.* **6**, 1107–1111 (2000).

Thompson, R. B., Fluorescence-Based Fiber-Optic Sensors, in J. R. Lakowicz, ed., *Topics in Fluorescence Spectroscopy,* Vol. 2, 1991, pp. 345–365.

Vo-Dinh, T., Stokes, D. L., Griffin, G. D., Volkan, M., Kim, V. J., and Simon, M. I., Surface-Enhanced Raman Scattering (SERS) Method and Instrumentation for Genomics and Biomedical Analyses, *J. Raman Spectrosc.* **30**, 785–793 (1999).

Wadkins, R. M., Golden, J. P., Pritsiolas, L. M., and Linger, F. S., Detection of Multiple Toxic Agents Using a Planar Array Immunosensor, *Biosens. Bioelectron.* **13**, 407–415 (1998).

Wallis, R. F., Stegeman, G. I., eds., *Electromagnetic Surface Excitations*, Springer-Verlag, Berlin, 1986.

Wang, J., DNA Biosensors Based on Peptide Nucleic Acids (PNA) Recognition Layers. A Review, *Biosens. Bioelectron* **13**, 757–762 (1998).

Wang, R., Nasang, V., Prasad, P. N., and Bright, F. V., Affinity of Antifluorescein Antibodies Encapsulated Within a Transparent Sol–Gel Glass, *Anal. Chem.* **65**, 2671–2675 (1993).

Watkins, A. N., and Bright, F. V., Effect of Fluorescent Reporter Group Structure on the Dynamics Surrounding Cysteine-26 in Spinach Calmoldulin: A Model Biorecognition Element, *Appl. Spectrosc.* **52**, 1447–1456 (1998).

Wolfbeis, O. S., Fiber-Optic Chemical Sensors and Biosensors, *Anal. Chem.* **72**, 81R–89R (2000).

Wolfbeis, O. S., ed., *Fiber Optic Chemical Sensors and Biosensors,* Vol. 1, CRC Press, Boca Raton, FL, 1991.

Yamanaka, S. A., Nguyen, N. P., Ellerby, L. M., Dunn, B., Valentine, J. S., Zink, J. I., Encapsulation and Reactivity of the Enzyme Oxalate Oxidase in a Sol-Gel Derived Glass, *J. Sol-Gel Sci. and Tech.* **2**, 827–829 (1994).

Yablonovitch, E., Inhibited Spontaneous Emission in Solid-State Physics and Electronics, *Phys. Rev. Lett.* **58**, 2059–2062 (1987).

Zeller, P. N., Voirin, G., and Kunz, R. E., Single-Pad Scheme for Integrated Optical Fluorescence Sensing, *Biosens. Bioelectron.* **15**, 591–595 (2000).

CHAPTER 10

Microarray Technology for Genomics and Proteomics

Microarray technology provides a powerful tool for high-throughput rapid analysis of a large number of samples. This capability has been of significant value in advancing the fields of Genomics, Proteomics, and Bioinformatics, which are at the forefront of modern structural biology, molecular profiling of diseases, and drug discovery. Biophotonics has played an important role in the development of microarray technology, since optical methods are used for detection and readout of microarrays.

This chapter on microarray technology follows Chapter 9 on biosensors because, in a true sense, it is a natural extension of biosensing. It utilizes a micropatterned array of biosensing capture agents for rapid and simultaneous probing of a large number of DNA, proteins, cells, or tissue fragments. Since many disciplines have contributed to the development of microarray technology, this chapter is written to cover the various multifaceted aspects such as the fabrication of microarrays, the immobilization of capturing/biorecognition elements, and scanning and readout of a vast amount of data.

Four types of microarrays are covered here: DNA microarrays, protein microarrays, cell microarrays and tissue microarrays. Among them, the most developed are DNA microarrays, also known as *biochips* or by a trade name (e.g., GeneChip®). They are widely used in clinical laboratories around the world. Protein microarrays form an emerging technology, because the emphasis is shifting to proteomics with the recognition that it is the proteins that need to be catalogued and analyzed to understand biological complexity and functions. Cell microarrays and tissue microarrays are relatively new developments.

The chapter introduces the principles utilized in these various microarray technologies. The various methods of fabrications of each of these four microarrays are discussed. Some selected examples of applications of microarrays are provided.

A discussion of future directions is provided to identify some areas of opportunities. Finally, the chapter concludes with a list of some companies

Introduction to Biophotonics, by Paras N. Prasad.
ISBN: 0-471-28770-9   Copyright © 2003 John Wiley & Sons, Inc.

commercializing the various microarrays. This list demonstrates that the microarray technology is already being perceived to be a growing field of applications and thus an expanding business opportunity. This list is also intended to be of help to new users of microarray technologies as well as to researchers entering this field.

For further reading on DNA microarray technology, the books suggested are:

Rampal (2001):   *DNA Arrays*: *Methods and Protocols*
Knudsen (2002):   *Analysis of DNA Microarray Data*
Schena (2000):   *Microarray Biochip Technology*
Palzkill (2002):   *Proteomics*

In addition, an entire supplemental issue of *Nature Genetics* is dedicated to microarrays [1999, Vol. 21 (Suppl.)]

10.1 MICROARRAYS, TOOLS FOR RAPID MULTIPLEX ANALYSIS

Microarray technology is a powerful, universally applicable analytical tool for rapid and simultaneous detection and analysis of a large number of biological assemblies with high sensitivity and precision. Thousands of DNA and RNA species can be simultaneously probed using a DNA microarray to provide detailed insight into cellular phenotyping and genotyping. Phenotyping is the process of identifying cells with specific markers. When antibodies are used, the process is called *immunophenotyping*. When genetic constructs are used, it is called *genotyping*.

Microarray technology provides valuable input that aids in understanding the molecular basis of health and disease and thus accelerates drug discovery. One can readily envision development of individual custom tailored treatment plans to replace a one-size-fits-all approach to health care (Friend and Stoughton, 2002). Microarray technology is an extension of biosensing techniques and utilizes a micropatterned array of biosensors. These arrays allow rapid and simultaneous probing of a large number of DNA, proteins, cells, or tissue fragments. The important steps of a microarray technology are shown in Scheme 10.1.

By comparing the flow chart presented in Scheme 10.1 and that in Chapter 9 on biosensors, one can readily see two new additions to the hardware: (i) microarray fabrication and (ii) scanning/readout of microarrays. Detecting and analyzing a large number of bioassemblies simultaneously and rapidly has been of considerable value to high-throughput projects such as the Human Genome Projects, where thousands of genes and their products need to be characterized. For example, using a DNA microarray, tens of thousands of genes can be tracked or detected. Microarray technology has practically rev-

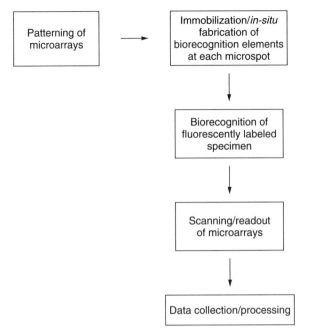

Scheme 10.1

olutionized the field of genome research, making possible studies ranging from gene expression monitoring and transcription profiling for drug target identification, to large-scale identification of single nucleotide polymorphisms (SNP) which results from a base variation at a single nucleotide position.

Microarray technology is rapidly advancing the frontier of molecular understanding of disease and speeding the development of new molecular diagnostics and drug discovery. It is also guiding researchers in the design of better crops. Microarray technology has also generated significant attention from the private sector, thereby stimulating investment and producing a rapid growth in new industries.

The two complementary areas of considerable current interest for molecular analysis of genome structure and function leading to molecular diagnostics are Genomics and Proteomics described in Table 10.1.

Genomics, derived from *genome*, is a term used to refer to an organism's complete set of genetic information. Genomics was originally used to describe the mapping, sequencing, and analyzing of genomes. This aspect of genomics now is frequently termed *structural genomics*, which encompasses the construction of high-resolution genetic, physical, and transcript maps of an organism, including its complete DNA sequence. Another aspect of genomics, called *functional genomics*, deals with the study of gene functions seeking a complete understanding of the gene function of a biological system.

TABLE 10.1. Genomics and Proteomics for Molecular Analysis of Genomes, Along with Their Expression, Transcription, and Function

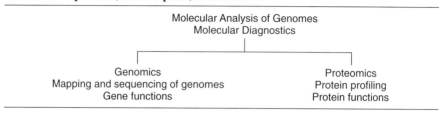

Molecular Analysis of Genomes
Molecular Diagnostics

Genomics	Proteomics
Mapping and sequencing of genomes	Protein profiling
Gene functions	Protein functions

To date, genomes of many organisms have been completely sequenced, and more are in progress. The current status of genome mapping can be found at the websites www.tigr.org or genomes@ncbi.nlm.nih.gov. However, molecular profiling of diseases or drug action requires more than a catalogue of all the genes or DNA sequence. Rather, it is the understanding of how the genes work together in determining functions of cells and organisms. This is the subject of functional genomics, which is now receiving increasingly more attention.

Genomics monitors expressed genes whose translation ultimately yields the respective proteins (Chapter 3). Recently, the focus of biology has shifted considerably with the recognition that it is the proteins that need to be catalogued and analyzed in order to understand biological complexity and functions, thus giving rise to the field of proteomics. Analogous to genomics, structural proteomics deals with the identification, profiling, and quantification of proteins while functional proteomics involves a study of protein structure, localization, modification, interactions, activities, and functions.

Two important general areas of proteomics are high throughput: (i) protein profiling dealing with determination of the abundance, modification, and subcellular localization of proteins in a given cell or tissue and (ii) determination of protein function.

An interesting way of looking at genomics and proteomics is that while genomics provides the biological equivalent of a chemist's Periodic Table, proteomics is biological analogy of the encyclopedia of reactions known to chemistry. Genomics and proteomics complement each other in providing a molecular description of a biological specie and the effect of environment, diseases, and drugs on it.

The various types of biological microarrays are listed in Table 10.2. Of these the DNA microarray technology is most developed and in wide usage all around the world. DNA microarrays consist of single-stranded DNA fragments, oligonucleotides, or RNA as DNA capture agents, with immobilization occurring at 5- to 150-μm size spots. As described in Section 10.8, a number of companies sell DNA microarrays. The term *gene chip* or *biochip* is also often used to describe a DNA microarray because the approaches used to fabricate DNA microarrays often involve processing analogous to that used to produce

TABLE 10.2. Various Biological Microarrays

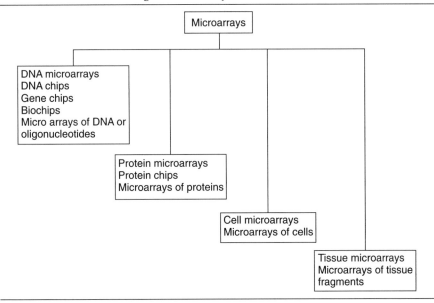

semiconductor microchips. Since its commercial availability in 1996, DNA microarrays have now become a major tool for genomics and drug discovery. The DNA microarrays have also led scientists to explore the operations of normal cells in the body and understand molecular aberrations underlying medical disorders. The subject of DNA microarrays is extensively covered by a number of books (Schena, 1999, 2000; Rampal, 2001) and a large number of reviews (Lockhart and Winzeler, 2000; Wang, 2000; Epstein et al., 2002; Dhiman et al., 2002; Van Hal et al., 2000; Epstein and Butoa, 2000; Hedge et al., 2000).

Protein microarrays or protein chips utilize microarrays of immobilized fusion proteins (proteins that fuse with other proteins) or antibodies (Habb et al., 2001). However, the protein chips currently are not sufficiently robust for high throughput studies (MacBeath and Schreiber, 2000; Arenkov, 2000; Fang et al., 2002). Although protein chip technology is not as well developed as the gene chip technology, their improvement is inevitable because it is the proteins which mediate nearly all cellular functions. Proteins also constitute the vast majority of pharmaceutical targets. Thus, protein chips are destined to be valuable tools for determining the molecular basis of disease and the mechanistic basis for drug action and toxicity (pharmacoproteomics). Proteins also can provide identification and validation of new biomarkers for disease and diagnosis, as well as for monitoring of drug efficacy and safety (Schweitzer and Kingsmore, 2002).

Cell microarrays are a relatively new development that utilize live cells expressing a c-DNA of interest (Ziauddin and Sabatini, 2001; Blagoev and Pandey, 2001). Cell microarray consists of mammalian cells cultured on a glass slide which is printed in defined locations with DNAs. Cells growing on the top of the DNA printed area are transfected, within a lawn of nontransfected cells. Thus by printing sets of complementary DNAs cloned in expression vectors, microarrays can be made whose features are clusters of live cells that express a defined c-DNA at each location. This approach provides an expression cloning system for the discovery of gene products that alter cellular physiology. The cells at a specific spot, by absorbing the DNA, form distinct cell clusters, each manufacturing the particular protein encoded in the absorbed DNA. By screening transfected cell microarrays expressing 92 different c-DNAs, Ziauddin and Sabatini (2001) were able to identify proteins involved in tyrosine kinase signaling, apoptosis, and cell adhesion, as well as identify proteins with distinct subcellular distributions. The cell microarray technology offers the benefit of rapid functional characterization of molecular and phenotyping effects produced by specific genes on live cells.

Tissue microarrays developed in the laboratory of Kallioniemi (Kononen et al., 1998) provides a new high-throughput tool for the study of gene dosage and protein expression patterns in a large number of individual tissues. This tool can provide a rapid and comprehensive molecular profiling of cancer and other diseases without exhausting limited tissue resources. The tissue microarrays are fabricated by taking core needle biopsies of preexisting paraffin-embedded tissues and reimbedding them in an arrayed "master" block. Thus, tissues from hundreds of specimens can be represented on a single paraffin block that can be analyzed by a number of techniques such as *in situ* fluorescence hybridization FISH which has been discussed in Chapter 8 (Bubendorf et al., 1999; Moch et al., 1999; Mucci et al., 2000; Perrone et al., 2000; Camp et al., 2000). The major advantages offered by the Tissue Microarray technology are as follows: (i) It enables the study of cohort of cases by simply analyzing a few master slides, (ii) all specimens are processed at one time using identical conditions, and (iii) it reduces the amount of archival tissue required for a specific study, thus preserving remaining tissue specimen for other studies. A major concern limiting wide acceptance of the tissue microarray technology is that it reduces the amount of tissue analyzed from a whole tissue section to a 0.6-mm-diameter disk that may not be representative of the protein expression patterns of the entire tumor. Regardless of these concerns, a number of companies are commercializing tissue microarrays. A list of current companies producing various microarrays is provided in Section 10.8. Tissue arrays are currently used for large-scale epidemiology studies aimed at identifying new diagnostic and prognostic markers for tumors.

To conclude this section, the interest in the field of microarray technology is increasing at a rapid pace as is evidenced by the rapid growth of publication in this field. Major developments can be expected both in the area of microarray technologies and in new applications of this technology.

TABLE 10.3. Websites Listing Links and Publications on Microarrays

http://brownlab.stanford.edu

http://www.microarrays.org/

http://resresources.nci.nih.gov/tarp/

http://bioinformatics.phrma.org/microarrays.html

http://industry.ebi.ac.uk/~alan/MicroArray/

www.rii.com/publications/default.htm

www.biologie.ens.fr/en/genetiqu/puces/links.html#news

Some of the websites that may provide updates on new developments in this field are listed above in Table 10.3.

10.2 DNA MICROARRAY TECHNOLOGY

The DNA microarrays have 5- to 150-µm size spots, depending on the method of fabrication used, on a solid support such as a glass strip. At these sites, fragments of single-stranded DNA ranging from 20 to 1000 or even more bases are attached. These arrays identify DNA sequences of a gene in a sample by using fluorescently labeled m-RNA or c-DNA. c-DNA (complementary DNA) is a single-stranded DNA that is produced from m-RNA by a process called *reverse transcription*, while the regular transcription process, described in Chapter 3, produces m-RNA from the DNA of the nucleus. The fluorescently labeled m-RNAs or c-DNAs are then hybridized with the immobilized DNA fragments on the chip. The chips are subsequently scanned with high-speed fluorescence detectors.

The two main pieces of hardware for microarray technology are (i) the microarray slide spotter and (ii) the microarray scanner. The website http://ihome.cuhk.ed.hk/~b400559/array provides an excellent account of the various options available. The two main approaches used to fabricate DNA microarrays are described in Table 10.4.

10.2.1 Spotted Arrays

The two major stages in spotted arrays approach are (i) printing of microarrays and (ii) the sample preparation. The important steps in each are described below along with their schematic representations in Figure 10.1. The main steps involved in the printing of microarrays (schematics in Figure 10.1) are

TABLE 10.4. Two Main Approaches for DNA Microarray Technology

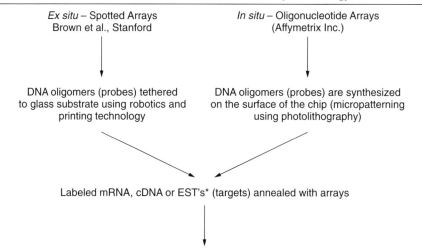

Ex situ – Spotted Arrays
Brown et al., Stanford

In situ – Oligonucleotide Arrays
(Affymetrix Inc.)

DNA oligomers (probes) tethered
to glass substrate using robotics and
printing technology

DNA oligomers (probes) are synthesized
on the surface of the chip (micropatterning
using photolithography)

Labeled mRNA, cDNA or EST's* (targets) annealed with arrays

Laser scanner applied for fluorescence detection of positive pairing of probe and target
(scanner usually employs confocal detection)

*EST's – Expressed sequence tags

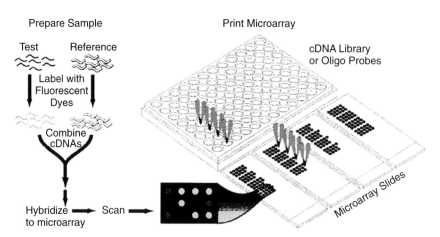

Figure 10.1. The spotted array technology. (Reproduced with permission from Knudsen, 2002.)

- Coating of a glass slide with polylysine
- Use of robot to spot probes (c-DNA or oligonucleotides) from a microliter plate on a glass slide (or capillary printing)
- Blocking of remaining exposed amines of polylysine with succinic anhydride
- If DNA is double-stranded, denaturing it to produce single strand

The DNA spots are in the range of 50–150 µm in diameter; thus, a 3.6-cm^2 chip can bear 10,000 spots. Other methods, besides nonspecific binding to polylysine, have also been used to fix probes to the surface. A variety of printing techniques have been utilized to produce spotted arrays of DNA (Rose, 2000). They are used to deliver a small volume of the target DNA on the solid surface to create the spotted micropatterns. The technologies for printing the spotted arrays fall in two distinct categories: (i) noncontact, where drops are ejected from a dispenser onto the surface; the common noncontact method involves an ink-jet printing technology; and (ii) contact printing, which involves a direct contact between the printing head and the solid support. These contact printing mechanisms involve solid pins, capillary tubes, tweezers, split pins, and ink stamps. These are described in detail by Rose (2000). The main steps involved in the sample preparation are:

- Isolate m-RNA and amplify by the well-known technique of polymerase chain reaction (PCR), which produces multiple copies of specific fragments of DNA.
- Convert m-RNA to c-DNA by reverse transcription, a process that is the reverse of the regular transcription process producing m-RNA from DNA (see Chapter 3, Section 3.7).
- Label the sample c-DNA with red (Cy5) and control (reference) with green (Cy3) fluorescent dyes.
- Hybridize with the microarray probes (hybridization is discussed in Chapter 8).
- Wash away unhybridized materials.
- Scan the microarray.
- Analyze the data.

As represented in the schematics (Figure 10.1), the test and the reference are labeled by two different fluorescent markers (Cy3 and Cy5). Then the two labeled c-DNA are combined and hybridized to the microarray. The microarray is scanned with a laser scanner or by using a confocal microscope. Using an image analysis software, signal intensities for each dye at each microarray spot are determined and log(Cy5/Cy3) ratios are obtained. A positive log(Cy5/Cy3) ratio indicates a relative excess of the transcript in the Cy5-labeled sample, while a negative log(Cy5/Cy3) ratio is indicative of a relative excess of the transcript in the Cy3-labeled c-DNA levels of gene expression relative to the reference. The data are then analyzed by cluster analysis and displaced in a format where red boxes represent the positive log(Cy5/Cy3) values, green boxes represent the negative log(Cy5/Cy3) values, and the black boxes indicate near zero values of log(Cy5/Cy3). A typical display is shown in Figure 10.2.

Current capabilities permit printing of more than 31,000 elements on a microscope slide.

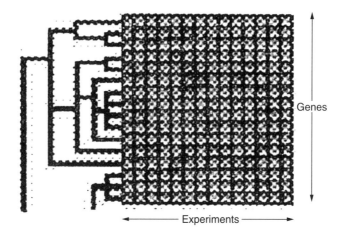

Figure 10.2. A representative of a cluster analysis of genes (on verticle axis) versus the experiments (horizontal axis). (Reproduced with permission from Dhiman et al., 2002.) (See color figure.)

10.2.2 Oligonucleotide Arrays

This technology, also called by a trade name Genechip® and introduced by Affymetrix (*source*: www.affymetrix.com), involves *in situ* synthesis of oligonucleotides of known sequence in a site-specific arrangement on a substrate. This approach can produce hundreds of thousands of different oligonucleotide probes packed at an extremely high density. These oligonucleotides are up to 25 nucleotides (25-mer) long. A schematic of the photolithographic process used is shown in Figure 10.3.

The process of fabrication of a Genechip® combines combinatorial DNA synthesis chemistry with photolithographic techniques adapted from the semiconductor industry (McGall and Fidanza, 2000). The photolithographic process, like for semiconductor chip manufacturing, utilizes ultraviolet light through holes in masks to deprotect photolabile groups and subsequently direct parallel and stepwise synthesis of oligonucleotides with a specific sequence. When using a fused silica or a planar glass substrate, its surface is first covalently modified using a silane reagent to provide hydroxyalkyl groups which serve as the initial synthesis sites. These sites are then extended with linker groups, protected with a photolabile-protecting group, such as 5′-(α-methyl-6-nitropiperonyloxycarbonyl), abbreviated as MeNPOC, which can be activated at specific spatial locations by UV light exposure for addition of nucleoside phosphoramidite monomers, also containing the photolabile group at the 5′ (or 3′) position. The photodeprotection is induced by the ~350-nm wavelength of the UV light from a commerical photolithographic exposure system.

Repetition of the cycle of changing the mask, deprotecting by photolithography, and adding a nucleotide to ~70 times can allow the synthesis of a com-

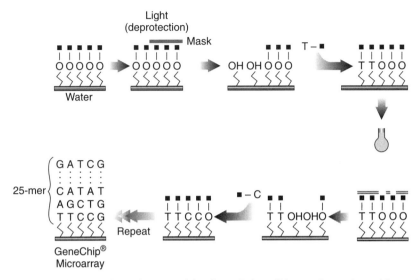

Figure 10.3. The use of a unique combination of photolithography and combinatorial chemistry to manufacture Genechip® arrays. (Reproduced with permission from www.affymetrix.com.)

plete array of thousands of 25-mer oligonucleotides in parallel. Generally, 20 pairs of oligonucleotides are arrayed to represent each gene. Each nucleotide matching the gene (perfect match, PM) is paired with a second mismatch oligonucleotide differing only by a central nucleotide (mismatch, MM). MM oligonucleotides serve to detect nonspecific and background hybridization, important for quantifying weakly expressed m-RNAs. The maximum achievable microarray density is determined by the spatial resolution provided by the photolithographic process. Typical dimensions for each array are $24\,\mu m \times 24\,\mu m$ on a 1.6-cm^2 chip. This method has been used to display 65,000–400,000 DNA oligonucleotides on a 1.6-cm^2 glass surface (Lockhart et al., 1996).

The advantage offered by the Affymetrix Genechip approach is that the microarrays are very uniform. However, as opposed to the spotted array technology discussed earlier where both the sample and the control are hybridized to the same chip using different fluorescent markers, the Genechip approach can handle only one fluorescent marker at a time, thus requiring two chips to compare a sample and a control.

10.2.3   Other Microarray Technologies

A number of new microarray technologies are emerging. It is not possible to describe all of them in this monograph. The best way to keep up with the development in this area is to visit the websites listed in Section 10.1. Here, two specific examples are briefly described.

MAGIChip™ Technology. MAGIChip™, an acronym for microarrays of gel-immobilized compounds on a chip, utilize glass substrates on which an array of polyacrylamide gel pads are produced (Zlatanova and Mirzabekov, 2001). The size of these pads can range from $10\mu m \times 10\mu m \times 5\mu m$ to $100\mu m \times 100\mu m \times 20\mu m$. The overall scheme of the MAGIChip™ technology is presented in Figure 10.4. The array of the polyacrylamide gel pads is created by photopolymerization of the acrylamide monomer, spread over the glass substrate and exposed to UV light through a patterned mask. The unpolymerized acrylamide monomer (in the dark region) is then washed away. Each individual gel pad then acts as a separate test tube for localizing DNA probes, because the surrounding hydrophobic glass surface prevents exchange of sample solution among the pads. The acrylamide has also been copolymerized with oligonucleotides to combine the polymerization step with the step of probe immobilization (Vasiliskov et al., 1999).

The Flow-Thru Chip™. This technology provides a three-dimensional biochip platform, with the benefit of enhancing the surface area and thus increasing the capture rate (Steel et al., 2000). This concept of a 3D Flow-Thru Chip™ is illustrated in Figure 10.5. This technology utilizes a uniformly porous substrate. Three types of porous substrates have been utilized: glass capillary arrays, electrochemically etched porous silicon, and metal oxide filters. The probe molecules are immobilized on the walls of the pores (microchannels) of the substrate. A spot in the array may contain several discrete microchannels in which a single probe may be immobilized. This technology is still in an early stage of development.

10.3 PROTEIN MICROARRAY TECHNOLOGY

The general schematic of a protein microarray is shown in Figure 10.6 (Mitchell, 2002). In a protein microarray, a glass slide serving as a substrate for a protein chip is printed with thousands of protein probes. A biological sample is subsequently spread over the chip, and any binding is detected. The detection methods used with protein chips can be of many different types as shown in Figure 10.6. The optical methods involving fluorescence and surface plasmon resonances have already been presented in Chapter 9.

Although the projected market for a protein chip is expected to surpass that for a gene chip and already a number of companies are producing protein chips, the production of a protein chip and its successful utilization requires more stringent conditions.

The fundamental differences between DNA and proteins contribute to difficulties in a simple extension of the DNA microarray technology to be used for fabrication of protein arrays. Some of the important differences between DNA and proteins from this perspective are listed below:

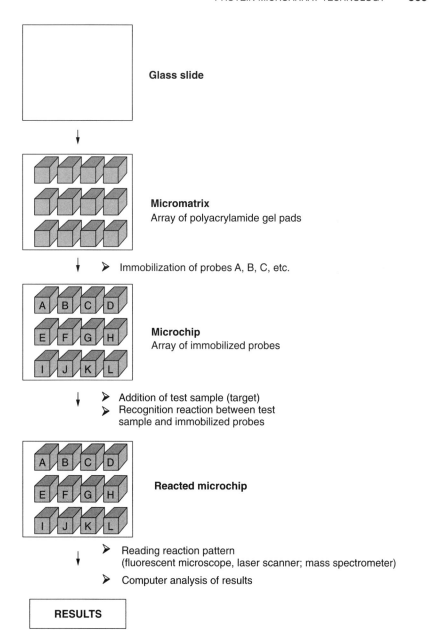

Figure 10.4. Overall scheme of the MAGIChip™ technology. (Reproduced with permission from Zlantanova and Mirzabekov, 2001.)

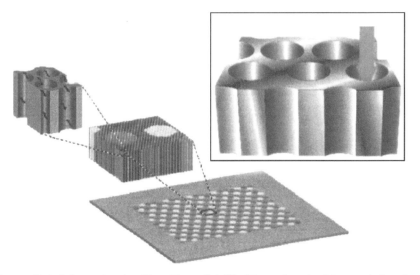

Figure 10.5. Schematic of a Flow-Thru Chip™. (Reproduced with permission from Steel et al., 2000.)

- While DNA involves a double-stranded rigid structure, a protein's function is determined by a complex set of three-dimensional structures (primary, secondary, tertiary, and quarternary, as discussed in Chapter 3) which determine its function and which are extremely sensitive to the environment. Thus DNA molecules are robust, which can be dried and rehydrated to restore their functions. Proteins, on the other hand, are unstable and easily denatured at solid–liquid and liquid–air interfaces.

- The principle for biorecognition used in the DNA microarray involves hybridization of the single-stranded c-DNA or m-RNA with a complementary strand, which is highly specific. In the case of protein microarrays, one often utilizes a wide array of immobilized antibodies. Unlike DNA capture, which is high-affinity, the current antibody production capability produces low-affinity capture antibodies that severely compromise the validity of the interpretation of the result.

- While DNA microarray technology can utilize PCR methods to amplify detection, no equivalent of PCR exists for proteins.

- Antibodies, being generally glycosylated, have large surface areas for interactions and thus exhibit significant cross-reactivity between target proteins. The large surface area occupied on a small protein spot also promotes denaturation.

Protein chips are difficult to handle. Consequently, protein chips have yet to find wide acceptance and have not matched the projected market demands.

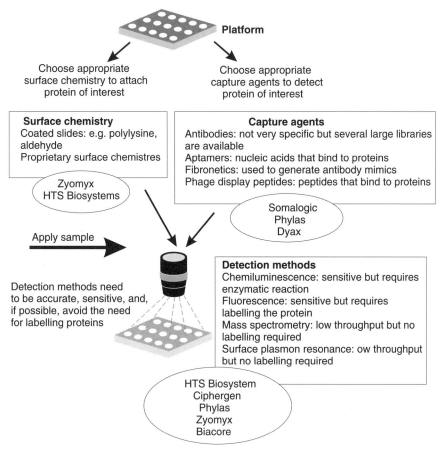

Figure 10.6. A general schematic presenting the different stages, and the components involved in the construction of protein chips. (Reproduced with permission from Nature, Mitchell, 2002.)

Some of the major challenges for a protein chip technology are listed below (Mithcell, 2002).

- Appropriate surface chemistry to immobilize proteins of a widely diverse range and to retain their biologically active secondary and tertiary structure.
- Identification and isolation of a suitable capturing agent (e.g., antibody) for the protein of interest.
- A suitable detection method with desired sensitivity and range of operation to measure the degree of protein binding.
- The ability to extract the detected protein from the chip and analyze it further, if needed.

- A large dynamic range of detection to cover the wide range exhibited by concentrations of various proteins.

Printing of Protein Microarrays. Just as in the case of a DNA microarray, a protein microarray is produced on a glass or a silicon substrate that has been treated with an aldehyde or other agent to immobilize the protein capturing agent such as an antibody (MacBeath and Schreiber, 2000). The reaction of the aldehyde group with the amino group of the antibody (or protein) to form a Schiff base linkage is used for immobilization. In this scheme, following the attachment of the proteins to the substrate, the unreacted aldehyde groups are quenched to minimize nonspecific binding by immersing the substrate in a buffer solution containing bovine serum albumin (BSA). The schematic of this process is shown in Figure 10.7.

In the case of printing of peptides or very small proteins, BSA obscures the molecules of interest. In such cases, the scheme presented in Figure 10.8 is utilized. Here, first a molecular layer of BSA is attached to a glass substrate. Then BSA is activated with N,N'-succinimidyl carbonate to produce active residues

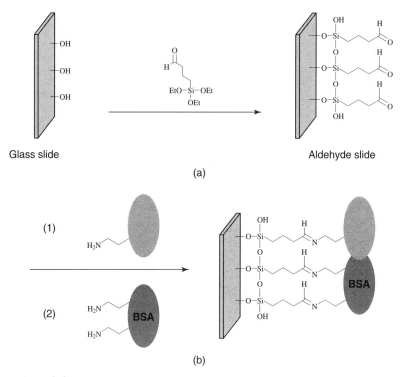

Figure 10.7. (A) Aldehyde group attachment to the substrate; (B) immobilization of a protein at an array site and subsequent quenching of the unreacted site with BSA. (Reproduced with permission from http://cgr.harvard.edu/macbeath/research/protein_microarrays.)

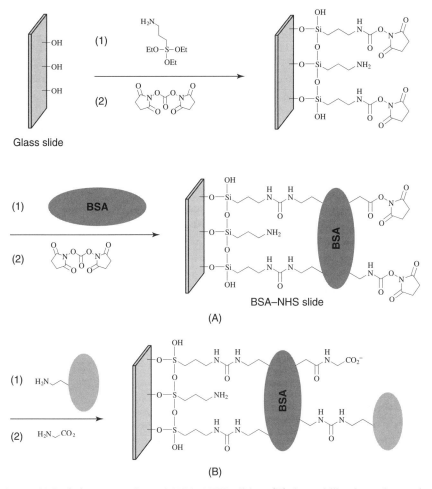

Figure 10.8. (A) Preparation of BSA–NHS slides; (B) immobilization of proteins and subsequent quenching with glycine. (Reproduced with permission from http://cgr.harvard.edu/macbeath/ research/protein_microarrays.)

that can react with surface amines on the proteins (step (2) in (A)). The unreacted sites on the substrate are then quenched with glycene (step (2) in (B)).

A number of other approaches have been utilized to produce protein microarrays. They include the use of layers of aluminum or gold, hydrophilic polymers, and polyacrylamide gels for immobilization of capture agents. The polyacrylamide gel pad method has already been discussed in Section 10.2. Each method requires appropriate chemistry to orient each molecule in the same direction and to create the necessary hydrophilic environment for proteins.

An approach being pursued by Zyomy of Hayward, California utilizes photolithography to etch miniature wells on the surface of a silicon chip. The proteins or antibodies are immobilized in the flow chambers on the chip to maintain an aqueous solution environment.

Another approach used by Large Scale Biology of Vacaville, California utilizes hundreds of thin plastic rods, each doped with a particular antibody and bundled together in a sheaf. Chips in the form of micrometer-thin slices are then produced from this sheaf by cutting them transversely.

Recently, Lahiri and co-workers (Fang et al., 2002) have reported the fabrication of a membrane–protein microarray. They utilized surface modification with γ-aminopropylsilane (GAPS) and produced microspots on GAPS by printing vesicular solutions of G-protein-coupled receptors (GPCRs) using a quill pin printer.

Capture Agents. Immobilized antibodies have been frequently used as capture agents. Protein microarrays have been an array of different antibodies that are monoclonal, polyclonal, antibody fragments, or synthetic polypeptide ligands. Other capture agents used are aptamers (single-stranded nucleic acids that complex with proteins) and oligonucleotides that bind specifically to proteins. Light-sensitive "photoaptamers" are being used by SomaLogic of Boulder, Colorado. These photoaptamers capture proteins and covalently cross-link with them when exposed to UV light. The oligonucleotides and aptamers offer the advantage that the same technology used to print the microarrays for m-RNA expression can be used.

Amplification. The technique of rolling circle amplification (RCA) (Lizardi et al., 1998) has been used to increase the sensitivity for multiplexed arrays on antibody microarray (Schweitzer et al., 2000). An adaptation of RCA is used where the 5' end of an oligonucleotide primer is attached to an antibody. Consequently, the presence of a circular DNA, DNA polymerase, and nucleotides allows rolling circle replication to generate a concatamer of circle DNA sequence copies attached to the antibody. The concatamer, which are tandem arrays of monomeric DNA molecules with complementary ends, can then be detected by hybridization with fluorescently labeled complementary oligonucleotide probes. This method of amplification has been reported to yield a 100- to 1000-fold improvement over detection using simply fluorescently labeled antibodies or streptavidin (Schweitzer, 2000).

Detection. Fluorescence detection is the widely used method. However, this necessitates labeling proteins with a fluorochrome. The risk is that the binding with the fluorochrome may alter the ability of the protein to bind (interact) with the immobilized capture agent. Another method is based on the use of the surface plasmon resonance technique (SPR) discussed in Chapter 9. The microarrays for this detection are fabricated by immobilizing the test proteins or antibodies on a metal (gold)-coated glass chip. Biacore is marketing SPR-based protein chips. Ciphergen of Fremont, California utilizes laser evapora-

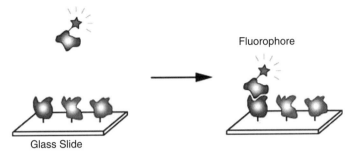

Fluorophore

Glass Slide

Figure 10.9. Schematic of the specific binding between an immobilized antibody and the fluorescently labeled protein. (Reproduced with permission from http://cgr.harvard.edu/macbeath/research/ protein_microarrays.)

tion of the captured protein spot into a benchtop time-of-flight mass spectrometer to analyze the protein.

Protein Microarray in Action. As an illustration of the use of protein microarray, Figure 10.9 shows the schematic of three different types of antibodies immobilized on a glass slide. One type of antibody selectively binds with the fluorescently labeled protein that can then be detected. This approach was used to probe protein–protein interactions between three pairs: (i) protein G and IgG, (ii) p50 and IKBα, and (iii) FRB domain of FKBPI2, the last pair requiring a small molecule, rapamycin, to enhance interaction. The results are shown in Figure 10.10.

10.4 CELL MICROARRAY TECHNOLOGY

One of the latest developments in the area of microarray technology is that containing live cells (Ziauddin and Sabatini, 2001; Blagoev and Pandey, 2001). A schematic for producing such microarrays is shown in Figure 10.11. Further details of the cell array technology can be found from the website of Sabatini's group (http://staff.wi.mit.edu/sabatini_public/reverse_transfection/ frame.htm). The step of coating of a cationic lipid on top of the DNA–gelatin microarray produces a DNA–lipid complex. When a suspension of transfectable live cells (such as 293 or COS cells) are added to a culture dish containing the DNA–gelatin patterned microarray plate, the cells take up DNA and produce c-DNA at each location in the microarray by "reverse transfection" because it is the cells that are added to the DNA–lipid complex and not the other way around used for conventional transfection. In this process, the cells grow by dividing two or three times in the process of creating a microarray with features consisting of clusters of transfected cells (Ziauddin and Sabatini, 2001). After proteins of interest are expressed within the

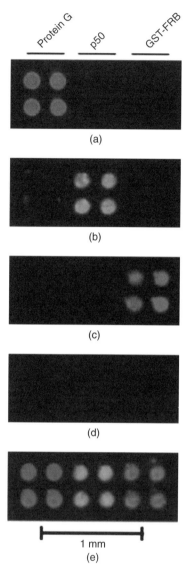

Figure 10.10. Detecting protein–protein interactions on glass slides. (A) Slide probe with BODIPY-FL-IgG; (B) slide probed with Cy3-I-kappa-B-alpha; (C) slide probed with Cy5-FKBP12 and 100 nM rapamycin; (D) slide probed with Cy5-KBP12 and no rapamycin; (E) slide probed with a mixture of BODIPY-FL-IgG, Cy3-I-kappa-B-alpha, Cy5-FKBP12, and 100 nM rapamycin. In all panels, BODIPY-FL, Cy3, and Cy5 fluorescence were false-colored blue, green, and red, respectively. (Reproduced with permission from American Association for the Advancement of Science, MacBeath and Schreiber, 2000.) (See color figure.)

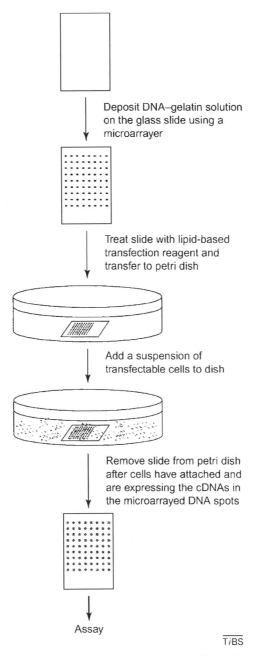

Deposit DNA–gelatin solution on the glass slide using a microarrayer

Treat slide with lipid-based transfection reagent and transfer to petri dish

Add a suspension of transfectable cells to dish

Remove slide from petri dish after cells have attached and are expressing the cDNAs in the microarrayed DNA spots

Assay

T*i*BS

Figure 10.11. Schematic of various steps in producing a live cell microarray. (Reproduced with permission from Blagoev and Pandey, 2001.)

cells, the microarray plate is then removed from the culture dish; the cells are fixed and the expressed protein is analyzed by incubation with fluorescently labeled antibodies and subsequent visualization using a fluorescence microscope.

As illustrated in Figure 10.12, the live cell microarray technology offers a number of possibilities. First, many types of c-DNA can be transfected. Second, use of spotted peptide fusion or small interfering RNAs (si-RNAs) allows live cells to be transfected by them. Furthermore, by varying the cell line being transfected, one can use knockout lines or cell lines stably transfected with a molecule or reporter of interest. Finally, cytokines and chemical inhibitors can be added to the medium during the stage the cells are attaching to the microarray to study transcriptional regulation.

Although cell-based microarrays overcome many of the deficiencies of a protein microarray and are also more flexible, the limited availability of c-DNAs for large-scale experiments is the bottleneck in wide acceptance of this technology. Since this technology relies on the ability of the cell to transfect, only cells with high transfection efficiency are optimal for this method. On the positive side, the microarrays with spotted DNAs are stable for a month or more and thus, like DNA microarrays, can be mass-produced.

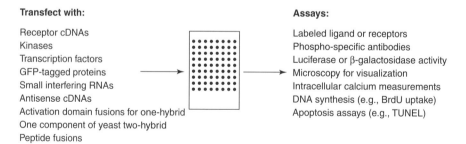

Agents that can be added to the medium:

Cytokines and growth factors
Chemical activators or inhibitors
BrdU
^{3}H thymidine
Fura-2

Transfect with:

Receptor cDNAs
Kinases
Transcription factors
GFP-tagged proteins
Small interfering RNAs
Antisense cDNAs
Activation domain fusions for one-hybrid
One component of yeast two-hybrid
Peptide fusions

Assays:

Labeled ligand or receptors
Phospho-specific antibodies
Luciferase or β-galactosidase activity
Microscopy for visualization
Intracellular calcium measurements
DNA synthesis (e.g., BrdU uptake)
Apoptosis assays (e.g., TUNEL)

Overlay slide with:

Cells stably expressing a promoter reporter construct
Knockout cells
Cells expressing a receptor subunit
Cells expressing the other component of yeast two-hybrid

TiBS

Figure 10.12. The flexibility and possibilities offered by live cell microarrays. (Reproduced with permission from Blagoev and Pandey, 2001.)

10.5   TISSUE MICROARRAY TECHNOLOGY

This technology currently offers a broad range of tissue microarrays ranging from ~50 times (low-density) to ~1000 times (high-density). The microarray is prepared by obtaining tissue biopsy samples from morphologically representative regions of a regular formalin-fixed paraffin embedded tissue block (Kononen et al., 1998). Core tissue biopsies of 0.6-mm diameter and 3- to 4-mm height are obtained and arrayed into a new recipient paraffin block of dimensions 45 mm × 20 mm with the help of a commercially available tissue microarray instrument.

A 5- to 8-μm section of the resulting tissue microarray block is cut with the help of a microtome. An adhesive-coated tape system is often used to assist in the sectioning of the tissue array and transferring it to a glass slide. The major advantages offered by the tissue microarray technology are (i) it enables the study of cohort of cases by simply analyzing a few master slides, (ii) all specimens are processed at one time using identical conditions, and (iii) it reduces the amount of archival tissue required for a specific study, thus preserving remaining tissue specimen for other studies. A major concern, limiting wide acceptance of the tissue microarray technology, is that it reduces the amount of tissue analyzed from a whole tissue section to a 0.6-mm-diameter disk that may not be representative of the protein expression patterns of the entire tumor. Regardless of these concerns, a number of companies are commercializing tissue microarrays. A list of current companies producing various microarrays is provided in Section 10.8. Tissue arrays are currently used for large-scale epidemiology studies aimed at identifying new diagnostic and prognostic markers for tumors.

10.6   SOME EXAMPLES OF APPLICATION OF MICROARRAYS

Since the DNA microarrays are the most advanced state of development and applications, this section provides applications of microarray technology using them.

Molecular Profiling of Tumor. The microarray technology is emerging as a powerful diagnostic tool in identifying molecular signatures of cancer and other diseases. It provides a rapid screening method to systematically explore changes in expression accompanying progression of cancer or a disease. It is particularly useful in the cases of tumors where various subclasses are morphologically indistinguishable. This technique has been used for many types of cancer—for example, breast cancer, diffuse large B-cell lymphoma, leukemia, colon adenocarcinoma, and ovarian cancer.

DNA array technology provides a method for rapid genotyping, facilitating the diagnosis of diseases for which a gene mutation has been identified. It also assists in the diagnosis of diseases for which known gene expression biomarkers of a pathologic state, or signature genes, exist. Signature genes are genes that are constitutively expressed in a normal or diseased cell or tissue that can serve as an identifier of that cell or tissue. As disease-related gene abnormalities are identified, the design of custom arrays will increase, tailoring sequence number and features to answer the type of question that is to be addressed. Custom-designed DNA variation detection arrays will be used to scan the genome and detect single nucleotide polymorphisms (SNPs). The SNPs that are identified can be used in designing further genotyping chips for performing association and linkage analysis. The DNA microanalysis of a patient's primary tumor can be useful in improving the patient's clinical treatment (Shoemaker and Linsley, 2002).

Golub et al. (1999) used microarray data to distinguish between two similar diseases, namely, acute myloid leukemia (AML) and acute lymphoblastic leukemia (ALL). They chose these cancers to validate the microarray method, as the two forms are already distinguishable by a number of methods. They applied a 6817-gene Affymetrix chip to 38 bone marrow samples and were able to identify 50 genes whose expression most distinguished AML from ALL. Golub and his colleagues are now using DNA chips to investigate if gene expression patterns can predict which patients will respond to standard chemotherapy (Steinberg, 2000). DNA microarrays were used by Alizadeh et al. (2000) to distinguish between two previously undifferentiated forms of diffuse large B-cell lymphoma. Welsh et al. (2001) used DNA microarray analysis to elucidate molecular markers for ovarian cancer. More recent reports on the use of DNA microarrays for assessing ovarian cancer gene expressions are by Haviv and Campbell (2002). Dhanasekaran et al. (2001) identified prognostic markers for prostate cancer. A great deal of activities utilizing microarray technology have also focused on breast cancer (Monni et al., 2001; Hedenfalk et al., 2001).

Elucidation of Biochemical Pathways. DNA microarrays are also playing a significant role in the elucidation of many complex biochemical pathways (Shoemaker and Linsley, 2002). Expression profiling has been used to monitor signaling pathways. This was accomplished by monitoring, in parallel, changes in expressions of thousands of genes affected by proximal signaling events. Ideker et al. (2001) used a combination of expression profiling and proteomic analysis to systematically analyze the metabolic galactose utilization pathway from yeast, following systematic perturbations by genetic deletion.

DNA microarrays have been successfully used to elucidate the interaction between bacteria and their hosts. The effect of a commensal bacterium on gene expression in the intestine was reported by Hooper et al. (2001). Microarrays were used by Huang et al. (2001) and Boldrick et al. (2002) to investigate the effects of pathogenic microorganisms on gene expression by dendritic cells and unfractionated peripheral blood mononuclear cells, respectively. They found massive and unique gene-expression changes triggered by different organisms.

Live cell arrays may provide valuable information on the signal transduction pathways. DNA microarrays may not be suitable for most signaling pathways because they involve post-translational modifications such as phosphorylation. In addition, because protein complexes are formed in stimulated cells, they cannot easily be studied in *in vitro* cell-free systems. Using live cell microarrays, expressions of hundreds of c-DNAs under various stimuli can be studied.

Applications to Neurobiology. A major current goal of neurobiology is an understanding of the role of anatomical organization in determining behavior, memory, cognition and various other neurological processes. An important step toward this goal is an understanding of differences in the molecular composition of different brain regions associated with distinct functional properties. DNA microarrays offer the ability to simultaneously monitor levels of nearly every transcript in the genome.

A number of studies have recently been reported on the application of DNA microarrays in the identification of transcripts specific for different brain regions (Dent et al., 2001; Zhao et al., 2001; Sandberg et al., 2000). DNA microarray has also been applied to, study the molecular basis of mental retardation in Fragile X Syndrome (Brown et al., 2001).

Drug Development. The presence of alternate gene forms or atypical expression of a gene involved in drug action or metabolism can manifest as resistance to therapy or as an atypical response to therapy. Pharmacogenomic and toxicogenomic studies correlate the genetic profile of individual patients and the individual response to a drug or toxin, respectively. The information obtained from these studies can be used to design arrays that will assist in the selection of custom and rational drug therapy. The identification of signature genes or biomarkers indicative of a disease process can identify candidate targets for therapeutic intervention.

Arrays assist in the identification of sentinel genes that demonstrate altered expression in a given cell or tissue type in response to drug or toxin exposure. Creating profiles of sentinel genes associated with drugs sharing a common mechanism allows potential new therapies to be rapidly screened for similar activities. This facilitates the selection of compounds for further investigation and may reduce the need for animal testing. An *in vitro* screen for potential toxicity has the potential to reduce drug-screening costs, prevent human suffering, and reduce product liability.

Analysis of sentinel genes can assist in determining the mechanism of action of a drug or toxin. Given that there is a multitude of events triggered by the initial action of a drug, screening thousands of genes at one time can identify multiple potential drug effectors. This allows robust hypotheses of drug mechanism to be formed and tested in subsequent investigations.

The valuable assistance provided by rapid screening using microarray technology can lead to accelerated FDA approval. FDA has established an accelerated review that allows approval based on evidence of the effect of the candidate drug on a surrogate endpoint, a disease-specific biomarker, or a signature gene. Once a drug has been shown to affect the expression of a surrogate endpoint in the desired manner, the FDA can grant marketing approval. Prior to this juncture, it has been difficult to identify disease biomarkers due to limited technology and individual patient variability. DNA array analysis allows multiple sequences to be searched for the presence of a suitable biomarker or a group of biomarkers. Testing the effects of a drug on a group of biomarkers may compensate for differences that reflect genetic variability.

Genetically Modified Food. Quality control of food is a major concern of the public, especially in the plant biotechnology world where genetically modified foods are involved. Microarray technology offers numerous applications. It can be used for screening of unintended changes in expression of a large number of genes in an unbiased manner. For example, a c-DNA library for tomatoes is used for microarray analysis on genetically modified tomatoes to observe differences in expression. The microarray technology can also be used for functionality assessment of food components.

10.7   FUTURE DIRECTIONS

The DNA microarray technology has reached a sufficiently mature level. Major opportunities in the development of microarray technologies are in the areas of protein microarrays and live cell microarrays. As described above, in their respective sections, there are numerous challenges. For protein microarrays, major breakthroughs in the area of improving their stability and providing appropriate amplification scheme to improve sensitivity in order to cover a wide range of concentrations of various proteins have to occur in order for them to find wide acceptance.

For live cell microarrays, a major limitation is the availability of c-DNAs for large-scale experiments. Recently a new type of microarray consisting of microbeads have been described by Brenner et al. (2000). It provides gene expression analysis by massive parallel signature sequencing (abbreviated as MPSS) on microbead arrays. This approach combines non-gel-based signature sequencing within vitrocloning of millions of templates on separate 5μm-diameter microbeads which were then assembled as a planar array in a flow cell at a density greater than 3×10^6 microbeads/cm^3. A fluorescence-base signature sequencing method was used to simultaneously analyze sequences of the free ends of the cloned templates on each microbead.

Another area of future development of technology is that of a multitask chip that will allow capture, separation, and quantitative analyses of proteins on one chip.

A growing trend is also the shift of focus from production of microarrays to enhancing the scope of applications and improving experimental design and analysis. In order for this widening of scope to occur, development at many fronts has to occur. Some of these are (van Berkum and Holstege, 2001)

- Improved clone collections
- Commercial availability of cheaper arrays
- Ability to use small amount of RNA
- Improved sensitivity and reproducibility
- Improved and validated analysis techniques
- Accepted universal standards

10.8 COMPANIES PRODUCING MICROARRAYS

The followings are just some of the companies that are selling or developing microarrays:

DNA Microarrays

Affymetrix, Santa Clara, California (www.affymetrix.com)

Agilent Technologies, Palo Alto, California (www.agilent.com)

Amersham Biosciences, Piscataway, New Jersey (www.apbiotech.com)

Axon Instruments, Union City, California (www.axon.com)

BioDiscovery, Marina del Rey, California (www.biodiscovery.com (software))

Clontech, Palo Alto, Califorina (www.clontech.com)

Genomic Solutions, Ann Arbor, Michigan (www.genomicsolutions.com)

Mergen, San Leandro, Califorina (www.mergen-ltd.com)

Motorola Life Sciences, Northbrook, Illinois (www.motorola.com/lifesciences/)

Nanogen, San Diego, California (www.nanogen.com)

PerkinElmer, Boston, Massachusetts (www.perkinelmer.com)

Virtek Vision International, Ontario, Canada (www.virtekbiotech.com)

Protein Microarrays

Biacore International Uppsala, Sweden (www.biacore.com)

Biosite Diagnostics, San Diego, California (www.biosite.com)

Ciphergen Biosystems, Inc., Fremont, California (www.ciphergen.com)

Large Scale Biology, Germantown, Maryland (www.lsbc.com)

PerkinElmer, Boston, Massachusetts (www.perkinelmer.com)

Cell Microarrays

Akceli, Cambridge, Massachusetts (www.akceli.com)

Tissue Microarrays

Beecher Instruments (www.beecherinstruments.com)
Inndrenex (subsidiary of BioGenex) (www.biogenex.com)

HIGHLIGHTS OF THE CHAPTER

- Microarray technology is used for rapid and simultaneous detection of a large number of biologics, with high throughput, sensitivity, and precision.
- Microarrays are micropatterned arrays of biosensing capture agent which allow rapid and simultaneous probing of a large number of DNA, proteins, cells, or tissue fragments.
- The microarray technology has significantly accelerated the development of two exciting frontiers: (i) genomics dealing with mapping and sequencing of genomes and determining gene functions and (ii) proteomics covering the structure, localization, modifications, interactions, and functions of proteins.
- While genomics is the biological equivalent of a chemists periodic table, proteomics is the biological analogy of the encyclopedia of reactions known to chemists.
- The four different types of biological microarrays are (i) DNA microarrays to identify DNA sequences of a gene (ii) protein microarrays to capture and identify proteins of interest, (iii) cell microarrays in which cells are transfected by growing them on the top of printed microarrays containing a specific DNA at a defined printed location, (iv) tissue microarrays consisting of micro tissue fragments taken by core biopsy needle and reimbedded in an array marker block.
- The various steps of microarray technology are (i) patterning of microarrays, (ii) immobilization/*in situ* fabrication of biorecognition elements at the microarray locations, (iii) biorecognition of fluorescently labeled specimens, (iv) scanning/readout of microarrays, and (v) data collection and processing.
- The two main approaches to fabricate DNA microarrays are (i) spotted arrays in which single stranded c-DNA or finite-size nucleotide oligomers (oligonucleotides) are covalently bonded by printing (spotting) on a polylysine film on glass and (ii) oligonucleotide arrays, also called Genechip®, which utilizes photolithography to produce *in situ* synthesis of oligonucleotides of known sequence in a site-specific arrangement.

- Other approaches for DNA microarrays are (i) MAGIChips™, which utilizes an array of polyacrylamide gel pads to localize DNA probes, and (ii) the Flow-Thru Chip™, which utilizes a porous substrate matrix to provide a three dimensional biochip platform.
- Analytes detected by DNA microarrays are single-stranded DNA fragments, m-RNA, produced from the sample cell by the transcription process, or c-DNA (complementary DNA) produced from m-RNA by reverse transcription.
- To enhance the sensitivity of detection, a well-known chemical process called the *polymerase chain reaction* (PCR) is used to create multiple copies of m-RNA from a specific fragment of the parent DNA in the specimen.
- The DNA microarrays use fluorescence detection of the fluorescently labeled c-DNA or m-RNA sample.
- Protein microarrays are usually fabricated by using a substrate, covalently bonded to an aldehyde which then is coupled to a protein capturing agent (e.g., an antibody) to immobilize it.
- Other approaches to produce protein microarrays are use of layers of aluminum or gold, hydrophillic polymers, and polyacrylamide gels to immobilize the capture agent.
- The capture agents used in protein microarrays are various antibodies, synthetic polypeptide ligands, aptamers, and oligonucleotides.
- A technique called *rolling circle amplification* (RCA) can be used to increase the sensitivity of multiplexed arrays on antibody microarrays.
- For protein microarrays, the detection techniques used are primarily fluorescence. Surface plasmon resonance has also been used.
- Protein microarrays have not found as wide an acceptance as have the DNA microarrays, because of the difficulties in handling them, some of which are as follows: (i) Proteins are not as robust as DNA and thus readily denature, (ii) protein microarrays usually have to be kept in water to reduce denaturation, (iii) protein detection is not as specific as DNA, and (iv) no equivalent to PCR exists for proteins to enhance sensitivity.
- Fabrication of cell microarrays starts with a coating of a cationic lipid on top of the DNA–gelatin microarray producing a DNA–lipid complex. Then a suspension of transfectable live cells is added to a culture dish containing the DNA–gelatin patterned microarray plate.
- Cells take up DNA and produce c-DNA by reverse transcription at each location in the cell microarray. After proteins of interest are expressed within the cells, they are analyzed by incubation with fluorescently labeled antibodies.
- Tissue microarrays are prepared by obtaining tissue biopsy samples from morphologically representative regions of a regular formalin-fixed, paraffin-embedded tissue block and reimbedding them in an arrayed

"master block." Thus, tissues from hundreds of specimens can be represented on a single paraffin block for subsequent analysis.

- Applications of microarrays are highly diverse, covering fundamental areas such as (a) molecular profiling of diseases and tumors and (b) elucidation of biochemical pathways. Some examples of applications are in drug development and in functionality assessment of genetically modified food.
- Major opportunities in the development of microarray technology are in the areas of protein microarrays and live cell microarrays.
- Another future technology area is the development of a multitask chip that will allow capture, separation, and quantitative analysis of proteins on one chip.

REFERENCES

Alizadeh, A. A., Eisen, M. B., Dairs, R. E., Ma, C., Lossos, I. S., Rosenwald, A., Boldrick, J. C., Sabet, H., Tran, T., Yu, X., Powell, J. I., Yang, L., Marti, G. E., Moore, T., Hudson, J., Lu, L., Lewis, D. B., Tibshirani, R., Sherlock, G., Chan, W. C., Greiner, T. C., Weisenberger, D. D., Armitage, J. O., Warnke, R., and Staudt, L. M., Distinct Types of Diffuse Large B-Cell Lymphoma Identified by Gene Expression Profiling, *Nature* **403**, 503–511 (2000).

Arenkov, Protein Microchips Use for Immunoassay and Enzymatic Reactions, *Anal. Biochem.* **278**, 123–131 (2000).

Blagoev, B., and Pandey, A., Microarrays Go Live-New Prospects for Proteomics, *Biochem. Sci.* **26**, 639–641 (2001).

Boldrick, J. C., Alizadeh, A. A., Diehm, M., Dudoit, S., Liu, C. L., Belcher, C. E., Botstein, D., Staudt, L. M., Brown, P. O., and Relman, D. A., Stereotyped and Specific Gene Expression Programs in Human Innate Immune Responses to Bacteria, *Proc. Natl. Acad. Sci. USA* **99,** 972–977 (2002).

Brenner, S., Johnson, M., Bridgham, J., Golda, G., Lloyd, D. H., Johnson, D., Luo, S., McCurdy, S., Foy, M., Ewan, M., Roth, R., George, D., Eletr, S., Albrecht, G., Vermaas, E., Williams, S. T., Moon, K., Burcham, T., Palla, M., DuBridge, R. B., Kirchner, J., Fearon, K., Mao, J., and Corcoran, K., Gene Expression Analysis by Massively Parallel Signature Sequencing (MPSS) on Microbead Arrays, *Nature Biotech.* **18,** 630–634 (2000).

Brown, V., Jin, P., Ceman, S., Darnell, J. C., O'Donnell, W. T., Tenenbaum, S. A., Jin, X., Feng, Y., Wilkinson, K. D., and Keene, J. D., Microarray Identification of FMRP-Associated Brain m-RNAs and Altered m-RNA Translational Profiles in Fragile X Syndrome, *Cell* **107**, 477–487 (2001).

Bubendorf, J., Kononen, J., Koivisto, P., et al., survey of Gene Amplification During Prostate Cancer Progression by High Throughput Fluorescence in situ Hybridization on Tissue Microarrays, *Cancer Res.* **59**, 803–806 (1999).

Camp, R. L., Charette, L. A., and Rimm, D. L., Validation of Tissue Microarray Technology in Breast Carcinoma, *Laboratory Investigation* **80**, 1943–1948 (2000).

Dent, G. W., O'Dell, D. M., and Eberwine, J. H., Gene Expression Profiling in the Amygdala: An Approach to Examine the Molecular Substrates of Mammalian Behavior, *Physiol. Behav.* **73**, 841–847 (2001).

Dhanasekaran, S. M., Barrette, T. R., Ghosh, D., Shah, R., Varambally, S., Kurachi, K., Pienta, K. J., Rubin, M. A., and Chinnaiyan, A. M., Delineation of Prognostic Biomarkers in Prostate Cancer, *Nature* **412**, 822–826 (2001).

Dhiman, N., Bomilla, R., O'Kane, D., and Poland, G. A., Gene Expression Microarrays: A 21st Century Tool for Directed Vaccine Design, *Vaccine* **20**, 22–30 (2002).

Epstein, C. B., and Butoa, R. A., Microarray Technology—Enhanced Versatility, Persistent Challenge, *Curr. Opin. Biotechnol.* **11**, 36–41 (2000).

Epstein, J. R., Biran, I., and Walt, D. R., Fluorescence-Based Nucleic Acid Detection and Microarrays, *Anal. Chim. Acta* **21827**, 1–34 (2002).

Fang, Y., Frutos A. G., and Lahiri J., Membrane Protein Microarrays, *J. Am. Chem. Soc.* **124**, 2394–2395 (2002).

Friend, S. H., and Stoughton, R. B., The Magic of Microarrays, *Sci. Am.* February, 44–49 (2002).

Golub, T. R., Slonim, D. K., Tamayo, P., Huord, C., Gasenbeck, M., Mesirov, J. P., Coller, H., Loh, M. L., Downing, J. R., Caligiuri, M. A., Bloomfield, C. D., and Lander, E. S., Molecular Classification of Cancer: Class Discovery and Class Prediction by Gene Expression Monitoring, *Science* **286**, 531–537 (1999).

Haab, B. B., Dunham, M. J., and Brown, P. O., Protein Microarrays for Highly Parallel Detection and Quantitation of Specific Proteins and Antibodies in complex solutions, *Genome Biol.*, **2**, research 0004.1–0004.13 (2001).

Haviv, I., and Campbell, I. G., DNA Microarrays for Assessing Ovarian Cancer Gene Expression, *Mol. Cell. Endocrinol.* **191**, 121–126 (2002).

Hedenfalk, I., Duggan, D., Chen, Y., Radmacher, M., Bittner, M., Simon, R., Meltzer, P., Gusterson, B., Esteller, M., and Kallioniemi, O. P., Gene Expression Profiles in Hereditary Breast Cancer, *N. Engl. J. Med.* **344**, 539–548 (2001).

Hedge, P., Qi, R., Abernathy, K., Gay, C., Dhasap, S., Gaspard, R., Hughes, J. E., Snesrud, E., Lee, N., and Quackenbush, J., A Concise Guide to c-DNA Microarray Analysis, *BioTechniques* **29**, 548–562 (2000).

Heng, Z., Global Analysis of Protein Activities Using Proteome Chips, *Science* **293**, 2101–2105 (2001).

Hooper, L. V., Wong, M. H., Thelin, A., Hansson, L., Falk, P. G., Gordon, J. L., Molecular Analysis of Commensal Host–Microbial Relationships in the Intestine, *Science* **291**, 881–884 (2001).

Huang, Q., Liu, D., Majewski, P., Schulte, L. C., Korn, J. M., Young, R. A., Lander, E. S., and Hacohen, N., The Plasticity of Dendritic Cell Responses to Pathogens and Their Components, *Science* **294**, 870–875 (2001).

Ideker, T., Thorsson, V., Ranish, J. A., Christmas, R., Buhler, J., Eng, J. K., Bumgarner, R., Goodlett, D. R., Aebersold, R., and Hood, L., Integrated Genomic and Proteomic Analyses of a Systematically Perturbed Metabolic Network, *Science* **292**, 929–934 (2001).

Karlstrom, A., and Nygren, P., Dual Labeling of a Binding Protein Allows for Specific Fluorescence Detection of Native Protein, *Anal. Biochem.* **295**, 22–30 (2001).

Knudsen, S., *A Biologist's Guide To Analysis of DNA Microarray Data*, John Wiley & Sons, New York (2002).

Kononen, J., Bubendorf, L., Kallioniemi, A., Bartund, M., Schrami, P., Leighton, S., Tarhorst, J., Mihatsch, M. J., Sauter, G., and Kallioniemi, O. P., Tissue Microarrays for High Throughput Molecular Profiling of Tumor Specimens, *Nat. Med.* **4**, 844–847 (1998).

Lizardi, P., Huang, X., Zhu, Z., Bray-Ward, P., Thomas, D., and Ward, D., Mutation Detection and Single-Molecule Counting Using Isothermal Rolling Circle Amplification, *Nat. Genet.* **19**, 225–232 (1998).

Lockhart, D. J., Dong, H., Byrne, M. C., Follettie, M. T., Gallo, M. V., and Chee, M. S., Expression Monitoring by Hybridization to High-Density Oligonucleotide Arrays, *Nat. Biotechnol.* **14**, 1675–1680 (1996).

Lockhart, D. J., and Winzeler, E. A., Genomics, Gene Expression and DNA Arrays, *Nature* **405**, 827–836 (2000).

MacBeath, G., and Schreiber, S. L., Printing Proteins as Microarrays for High-Throughput Function Determination, *Science* **289**, 1760–1763 (2000).

McGall, G. H., and Fidanza, J. A., Photolithographic Synthesis of High-Density Oligonucleotide Arrays, in J. B. Rampal, ed., *DNA Arrays,* Humana Press, Totowa, NJ, 2001, pp. 71–101.

Mitchell, P., A Perspective on Protein Microarrays, *Nat. Biotechnol.* **20**, 225–229 (2002).

Moch, H., Schraml, P., Bubendorf, L., Mirlacher, M., Kononen, J., Gasser, T., Mihatsch, M. J., Kallioniemi, O. P., and Sauter, G., High-Throughput Tissue Microarray Analysis to Evaluate Genes Uncovered by c-DNA Microarray Screening in Renal Cell Carcinoma, *Am. J. Pathol.* **154**, 981–986 (1999).

Monni, O., Barlund, M., Mousses, S., Konenen, J., Sauter, G., Heiskanen, M., Paavola, P., Avela, K., Chen, Y., Bittner, M. L., Comprehensive Copy Number and Gene Expression Profiling of the 17q23 Amplicon in Human Breast Cancer, *Proc. Natl. Acad. Sci. USA* **98**, 5711–5716 (2001).

Mucci, N. R., Akdas, G., Manely, S., Rubin, M. A., Neuropendocrine Expression in Metastatic Prostate Cancer: Evaluation of High Throughput Tissue Microarrays to Detect Heterogeneous Protein Expression, *Hum. Pathol.* **31**, 406–414 (2000).

Palzkill, T., *Proteomics*, Kluwer Academic Publishers, Hingham, MA, 2002.

Perrone, E. E., Theohanis, C., Mucci, N. R., Hayasak, S., Taylor, J. M., Cooney, K. A., and Rubin, M. A., Tissue Microarray Assessment of Prostate Cancer Tumor Proliferation in African-American and White Men, *J. Natl. Cancer Inst.* **92**, 937–939 (2000).

Rampal, J. B., ed., *DNA Arrays: Methods and Protocols*, Humana Press, Totowa, NJ, 2001.

Rose, D., Microfluidic Technologies and Instrumentation for Printing DNA Microarrays, in M. Schena, ed., *Microarray Biochip Technology,* Eaton Publishing, Natick, MA, 2000, pp. 19–38.

Sabatini, D., and Ziauddin, J., Microarrays of Cells Expressing Defined cDNA, *Nature* **411**, 107–110 (2001).

Sandberg, R., Yasuda, R., Pankratz, D. G., Carter, T. A., Del Rio, J. A., Wodicka, L., Mayford, M., Lockhart, D. J., and Barlow, C., Regional and Strain-Specific Gene

Expression Mapping in the Adult Mouse Brain, *Proc. Natl. Acad. Sci. USA* **97**, 11038–11043 (2000).

Schena, M., *DNA Microarrays—A Practical Approach,* Oxford, University Press, Oxford, 1999.

Schena, M., ed., *Microarray Biochip Technology*, Eaton Publishing, Natick, MA, 2000.

Schweitzer, B., and Kingsmore, S. F., Measuring Proteins on Microarrays, *Curr. Opin. Biotech*, **13**, 14–19 (2002).

Schweitzer, B., Wiltshire, S., Lambert, J., O'Malley, S., Kukanskis, K., Zhu, Z., Kingsmore, S. F., Lizardi, P. M., and Ward, D. C., Immunoassays with Rolling Circle DNA Amplification: A Versatile Platform for Ultrasensitive Antigen Detection, *Proc. Natl. Acad. Sci., USA* **97**, 10113–10119 (2000).

Shoemaker, D. D., and Linsley, P. S., Recent Developments in DNA Microarrays, *Curr. Opin. Microbiology*, **5**, 334–337 (2002).

Steel, A., Torres, M., Hartwell, J., Yu, Y-Y., Ting, N., Hoke, G., and Yang, H., The Flow-Thru Chip™: A Three-Dimensional Biochip Platform, in M. Schena, ed., *Microarray Biochip Technology*, Eaton Publishing, Natick, MA, 2000 pp. 87–117.

Steinberg, D., DNA Chips Enlist in War on Cancer, *The Scientist* **14**, 1–7 (2000).

Van Berkum, N. L., and Holstege, F. C. P., DNA Microarrays: Raising the Profiles, *Curr. Opin. Biotech.* **12**, 48–52 (2001).

Van Hal, N. L., Vorst, O., van Houwelingen, A. M., Kok, E. J., Peijnenburg, A., Aharoni, A., and van Tuness, A. J., The Applications of DNA Microarrays in Gene Expression Analysis, *J. Biotechnol.* **78**, 271–280 (2000).

Vasiliskov, A V., Timofeev, E. N., Surzhikov, S. A., Drobyshev, A. L., Shick, V. V., and Mirzabekov, A. D., Fabrication of Microarray of Gel-Immobilized Compounds on a Chip by Copolymerization, *Biotechniques* **27**, 592–606 (1999).

Wang, J., Survey and Summary: From DNA Biosensors to Gene Chips, *Nucleic Acids Res.* **28**, 3011–3016 (2000).

Welsh, J. B., Zarrinkar, P. P., Sapinoso, L. M., Kern, S. G., Behling, C. A., Monk, B. J., Lockhart, D. J., Burger, R. A., and Hampton, G. M., Analysis of Gene Expression Profiles in Normal and Neoplastic Ovarian Tissue Samples Identifies Candidate Molecular Markers of Epithelial Ovarian Cancer, *Proc. Natl. Acad. Sci. USA*, **98**, 1176–1181 (2001).

Zhao, X., Lein, E. S., He, A., Smith, S. C., Aston, C., and Gage, F. H., Transcriptional Profiling Reveals Strict Boundaries Between Hippocampal Subregions, *J. Comp. Neurol.* **441**, 187–196 (2001).

Zhu, H., Bilgin, M., Bangham, R., Hall, D., Casamayor, A., Bertone, P., Lan, N., Jansen, R., Bidlingmaier, S., Houfek, T., Mitchell, T., Miller, P., Dean, R., Gerstein, M., and Snyder, M., Global Analysis of Protein Activities Using Proteome Chips, *Science. Computers and Science* **293**, 2101–2105 (2001).

Ziauddin, J., and Sabatini, D. M., Microarrays of Cells Expressing Defined c-DNAs, *Nature* **411**, 107–110 (2001).

Zlatanova, J., and Mirzabekov, A., Gel-Immobilized Microarrays of Nucleic Acids and Proteins, in J. B. Rampal, ed., DNA Arrays: Methods and Protocols, Humana Press, Totowa, NJ, 2001, pp. 17–38.

CHAPTER 11

Flow Cytometry

A flow cytometer is an optical diagnostic device which is used in research and clinical laboratories for disease profiling by measuring the physical and/or chemical characteristics of cells. Flow cytometry is also suitable for rapid and sensitive screening of potential sources of deliberate contamination, an increasing source of concern of bioterrorism. It is also emerging as a powerful technique for agriculture research and livestock development. This chapter introduces the principle of flow cytometry describing the various steps involved in its operation. The various components of a flow cytometer are described.

Fluorescence is the generally used optical response, which is discussed in relation to the flow cytometric applications. Criteria selection of fluorophores (often called *fluorochromes* by the cytometry community), used to stain cells for analysis by cytometry, is discussed. An important part of flow cytometry is the manipulation of a large amount of data and their presentation for analysis. This subject is also covered.

This chapter also presents selected examples of applications of flow cytometry, which are in current usage. They cover both clinical and research. For the convenience of a reader who may be interested in acquiring a flow cytometer, this chapter lists some of the commercial sources providing this instrument. The chapter concludes with a discussion of future directions in flow cytometry, which a researcher interested in entering this field may find useful. For further reading on flow cytometry, the following books are recommended:

Givan (2001): An excellent book introducing the basics of flow cytometry.

Shapiro (1995): A very authoritative book with a comprehensive coverage of flow cytometry.

Stewart and Nicholson (2000): An edited book covering various aspects of clinical applications of immunophenotyping.

Nunez (2001): A book covering research applications of flow cytometry.

Introduction to Biophotonics, by Paras N. Prasad.
ISBN: 0-471-28770-9   Copyright © 2003 John Wiley & Sons, Inc.

Ormerod (2000): An edited book dealing with practical aspects of flow cytometry.

11.1   A CLINICAL, BIODETECTION, AND RESEARCH TOOL

The term *cytometry* refers to the measurement of physical and/or chemical characteristics of cells or, in general, of any biological assemblies (Shapiro, 1995; Givan, 2001; Stewart et al., 2002). In flow cytometry, such measurements are made while the cells, the biological assemblies, or microbeads (as calibration standards) flow in suspension, preferably in a single file, one by one, past a sensing point. The sensing is conducted by using an optical technique where the beam from a light source interacting with each individual cell, a bioassembly, or a microbead produces scattering or fluorescence. The optical response is used to determine cellular features and organelles, providing counts and ability to distinguish different types of cells in a heterogeneous population. Though the fluorescence detected can be autofluorescence, generally the cells or the intracellular products are tagged with special fluorescently labeled antibodies or dyes that bind to cellular components and are capable of producing fluorescence. Flow cytometry yields measurement of various optical responses that provide a set of properties, also called *parameters*, that provide unique characteristics of a specific type of cell. The identification and quantification of a particular type of cells can be then used to correlate with a specific pathological condition that can be used to identify a specific disease or microbial invasion.

Another name used for flow cytometry and thus applied interchangeably is fluorescence-activated cell sorting (FACS), which emphasizes the utilization of fluorescence detection and the ability of the instrument to sort cells that meet specific measured criteria. In a sense, both flow cytometry and optical microscopy perform similar functions—that is, to look at microscopic objects. Like a microscope, a flow cytometer incorporates a light source, an illumination optics, and a light collection optics. However, while in a regular microscope with point detection (such as in a confocal microscope), the light (laser) beam moves (scanned) to detect and image cells, the cells are moving (flowing in a single file) in a flow cytometer.

Furthermore, incorporation of the cell sorting feature in a flow cytometer by using electrical or mechanical methods allows one to collect cells with one or more specific characteristics. This feature thus allows one to isolate pure population of viable cells with more homogeneous characteristics from within a mixed heterogeneous population of cells. The process of diverting a particular type of cell, after measuring its identifying features, will be described later. In addition to isolating a pure population, the cell sorting capability also allows for further biochemical analysis of the selected cells, or other desired processing such as cell culture.

Although a flow cytometer measures optical response from one individual cell at a time, progress in fluidics and detection, together with rapid data acquisition and processing software, can now readily provide the ability to detect and analyze up to 75,000 cells per second. Thus, the two significant advantages offered by a flow cytometer over a traditional microscope are its rapid throughput rate at which each cell is interrogated and the ability to sort selected populations while maintaining viability. Thus real-time monitoring of many biological events can be achieved.

The current market for flow cytometry is near one billion dollars worldwide, showing the wide usage of this instrument, even through primary application has been on eukaryotic cells. The usage of flow cytometry for the study of prokaryotic cells is only beginning to emerge. An instrument capable of accurately measuring the properties of microbes, whether bacterial, fungal, or viral, can be expected to significantly expand the current market. The impact will be felt in many diverse areas, such as the pharmaceutical industry, microbiology applications, and agriculture. As pharmaceutical research moves more toward target-directed intervention and biological intervention and away from drug screening with intact animals, flow cytometry offers a cost-effective method for the development and testing of agents by the pharmaceutical industry.

Biological organisms are now being recognized as offensive weapons against population centers. Flow cytometry will likely be a method of choice for the rapid and sensitive screening of potential sources of deliberate contamination. New generation flow cytometers can significantly impact on the ability of physicians and clinicians to quickly determine microorganism infection in humans, animals, food supplies, and water supplies with a compact high-performance cytometer.

Flow cytometry is also beginning to impact agricultural research and livestock development. Selection of sex type in feed animals, or the ability to selectively sort male and female sperm, will have an enormous economic impact on food supply. The ability to select desired animal gender, with flow cytometric techniques, and to use artificial insemination ensures desired selection of offspring gender and physical traits. This capability dramatically reduces breeding and maintenance costs. This application is expected to grow substantially in coming decades.

In addition to clinical and biodetection applications of flow cytometry, its ability as a valuable research tool is also rapidly expanding. It has proved to be a powerful technique for cell cycle analysis where measurements of DNA content can be used to provide a great deal of information about the cell cycle. Cell division, apoptosis, and necrosis (Darzynkiewicz et al., 1997; also discussed in Chapter 3) can be studied. Furthermore, metabolic characteristics such as calcium flux, mitochondrial activity, cellular pH, and free radical production in live cell populations can be probed and quantified in real time. Flow cytometry using fluorescent tagged protein or a reporter gene can readily be used to measure gene expression in cells transfected with recombinant DNA. Thus using flow cytometry one can readily measure the following (*source*: http://facs.stanford.edu/5minuteguide.htm): (i) expression of proteins and (ii) transfection efficiency. In addition, transfection assays can be combined with staining and sorting for other markers. Also, flow cytometry can be used to purify transfected cells for further analysis or use. Other cellular functions studied by flow cytometry are (*source*: http://www.cyto.purdue.edu/flowcst) (i) phagocycosis, (ii) intracellular cytokines, (iii) oxidative burst, and (iv) membrane potential.

Molecular cytometry is a relatively new area of application of cytometry which is attracting a great deal of attention from the research community (Nunez, 2001). It utilizes flow cytometry to obtain information on cell-to-cell variations in molecular parameters being investigated. Fluorescence resonance energy transfer (FRET), discussed in Chapter 7, can also be used in flow cytometry to determine if two protein markers are closely associated on the cell surface or inside the cell. Another technique being used in combination with flow cytometry to obtain molecular (or submolecular) information is fluorescence *in situ* hybridization (also abbreviated as FISH and discussed in Chapter 8).

Another active area of research is detecting or assessing the activities of microorganisms in a wide variety of samples such as milk, bean, river water, biosolids, and biofilms.

The various applications of flow cytometry are further summarized in Tables 11.1, 11.2, and 11.3. The field of flow cytometry bridges many disciplines involving biologists, physicians, organic chemists, laser physicists, and optical

TABLE 11.1. Clinical Applications

- HIV monitoring
- Leukemia or lymphoma immunophenotyping
- Organ transplant monitoring
- DNA analysis for tumor ploidy and SPF
- Primary and secondary immunodeficiency
- Hematopoietic reconstitution
- Paroxysmal nocturnal hemoglobinuria

TABLE 11.2. Research Applications

- Multiplexing immunoassays
- Multiparameter immunophenotyping
- Measurement of intracellular cytokines
- Signal transduction pathways
- Cell cycle analysis
- Measuring cellular function

TABLE 11.3. Molecular Flow Cytometry

- Multiplexing oligonucleotide assays
- Measuring gene expression
- In situ hybridization
- Drug discovery

and fluidic engineers, as well as software and hardware engineers for data acquisition systems. It holds tremendous opportunities for future development, ranging from basic research at the molecular and cell biology level, to functional genomics and proteomics, to new applications, to new designs of flow cytometers with considerably expanded capabilities. Areas of future development are detailed in Section 11.6. This wide diversity of flow cytometry is clearly reflected at various flow cytometry centers around the world as well as at numerous conferences which attract a truly multidisciplinary participation. The rapidly growing interest of the scientific and clinical community is also evident from the popularity of a number of courses offered on cytometry which cover principles, applications, and hands-on training. Detailed information about these courses can readily by obtained from various websites, some of which are:

National Flow Cytometry Resource:
http://www.lanl.gov/orgs/ibdnew/DTIN/open/UsrFac/userfac36.html

The International Society for Analytical Cytometry: http://www.isac-net.org

Clinical Cytometry Society: http://www.cytometry.org

Purdue University Cytometry Laboratories: http://www.cyto.purdue.edu

Roswell Park Cancer Institute Laboratory of Flow Cytometry:
http://rpciflowcytometry.com

Cancer Research UK FACS Laboratory:http://science.cancerresearchuk.org/

Major journals specializing in research reports using cytometry are:

Cytometry: *The Journal of the International Society for Analytical Cytology*
*Clinical Cytometry: A Publication of the Clinical Cytometry Society and The
    International Society for Analytical Cytology*
The Journal of Immunological Method

11.2 BASICS OF FLOW CYTOMETRY

11.2.1 Basic Steps

The basic key steps in the operation of a flow cytometer are illustrated by the following block diagram:

Block diagram: basic steps in the operation of a flow cytometer:

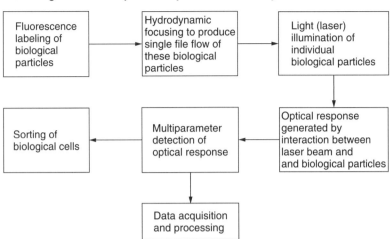

Block Diagram: Basic Steps in the Operation of a Flow Cytometer

11.2.2   The Components of a Flow Cytometer

A flow cytometer consists of a light source, which in all modern flow cytometers is a laser, and illumination optics to focus the laser beam on to a flowing biological cell or a polystyrene microsphere, which is fluorescently labeled. The scattered laser light and the fluorescence response are separated and focused onto photodetectors. A special electronics processes the optical response and controls a sorter if it is provided. These components of a flow cytometer are described in detail here.

Light Source. A flow cytometer may use a single excitation wavelength or a number of excitation wavelengths from different laser sources. The laser beams can be coaxial or separated so that one or more interrogation point occurs. The critical requirements for a laser in flow cytometry are power stability, high-quality beam characteristics, and low-level high-frequency noise.

In older versions of flow cytometers, the blue-green wavelength region was covered by an argon ion laser (Chapter 5), which provides excitation wavelengths of UV (350–365 nm) as well as 488 nm and 514.5 nm. The red region is covered by a helium neon laser (632 nm). The current trend, however, is to replace these bulky and inefficient gas lasers by compact solid-state lasers. As discussed in Chapter 5, diode pumped solid-state lasers are available at 488 nm and 532 nm. The red (635–670 nm) and blue (~405 nm) regions are covered by a number of diode lasers. Another new prospect is the use of near IR lasers (700–800 nm) to excite IR dyes, whereby the problem due to autofluorescence interference is considerably reduced.

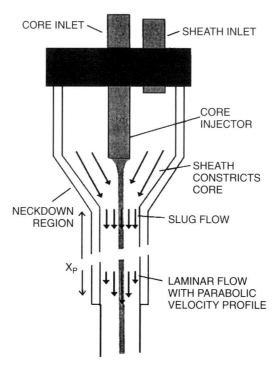

Figure 11.1. A typical flow cell diagram. (Reproduced with permission from Shapiro, 1995.)

Flow Cell. The flow cell is designed to hydrodynamically focus the sample stream. A typical flow cell design is shown in Figure 11.1. A cell suspension is introduced through a core inlet which has an inner diameter of 20 μm and is surrounded by a larger (~200 μm) stream of flowing saline (sheath liquid). Therefore, in this arrangement, a core stream of the cell suspension is injected into the center of the sheath stream. During the flow, the sheath fluid produces hydrodynamic focusing of coaxial flow of the core fluid, whereby the two streams maintain their relative positions and do not mix significantly and move at the velocity of the sheath fluid. The hydrodynamic focusing produces the flow of cells in a single file (one cell at a time). The core and sheath streams are driven by syringe pumps or by sources of pressure that deliver a known volume of sample per unit time with minimum pulsation. From the sample flow rate, one can easily derive the cell count per unit volume.

The rate of cell flow is significantly influenced by the sheath flow rate, the differential pressure (the pressure difference between the sheath and core), and their concentration in the sample tube. Generally, the sheath pressure, which determines the velocity of the fluid passing the laser beam, is held constant. As the differential pressure is reduced, a stronger hydrodynamic focusing of the core produces increased resolution because the core diameter becomes smaller. On the other hand, if the differential pressure is increased, the core diameter becomes larger and more cells are "pushed" into the system. If the sheath pressure is decreased, the velocity is decreased so that the time spent by a cell in the illuminating beam is increased. This may be useful if the amount of light collected from cells is a limiting factor.

To maintain good stability, the core stream requires a velocity of at least one meter per second. At this speed, a cell would traverse its own diameter (<20μm) in a few microseconds. An important consideration of the core diameter is also the width of the illuminating beam. The core diameter needs to be considerably less than the beam diameter to keep the cells at the center of the light beam in order to ensure uniform illumination. To accomplish this, beam-shaping optics are used so the laser beam is not circular.

Illumination Optics. Optical elements between the laser and the sample are referred to as illumination optics and are used to shape and focus the laser beam. Except for certain diode lasers, most laser sources produce circular beams that can even be a true Gaussian intensity distribution (see Chapter 5). However, the use of such a circular Gaussian beam producing a circularly focused spot is not very desirable for flow cytometry. The beam shaping optics most frequently used in current flow cytometers utilize a pair of anomorphic prisms or two crossed cylindrical lenses that provide an elliptical spot of 10–20μm in dimension parallel to the direction of cell flow and 60μm in dimension perpendicular to the flow dimensions. The advantage of the use of such an elliptical beam over circular beams of either 20-μm dimension or 60-μm dimension are illustrated by Figure 11.2. The elliptical beam provides a wider illumination field across the width of the flow so that the optical response (fluorescence or scattering) does not fluctuate if cells stray from the center of the beam. In other words, it provides a considerable side-to-side tolerance. At the same time, the smaller (20μm) dimension of the elliptical beam parallel to the flow direction allows cells to pass in and out of the light illumination quickly and avoids simultaneous illumination of more than one cell at one time. This optical focusing also improves the power density distribution across the cell stream. In a multilaser system, each beam can be focused at the same spot using an elliptical lens system, which is achromatic and aspherical (see Chapter 7). Alternatively, each beam can be elliptically focused at different points along the stream, in which case a cell moves through each beam sequentially. The latter configuration provides for the resolution of dyes that have the same emission spectra but different excitation spectra.

The alignment of the laser beam with respect to the stream of cells is crucial; a proper alignment ensures that the core of the stream is uniformly illuminated. The region of intersection between the stream and the laser beam is often called the *analysis point, observation point,* or *interrogation point.* A poor alignment between the laser beam and the stream may be due to misalignment of the illumination optics (focusing lenses) or from shifts in the fluid stream due to bubbles or partial obstruction.

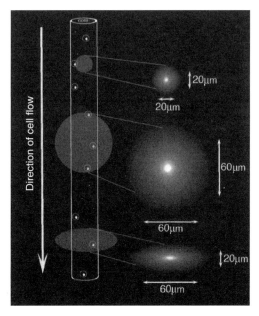

Figure 11.2. Illumination of cells flowing in a flow cytometer using focused laser beams of different profiles: (a) The focused circular beam of diameter 20 μm; (b) focused circular beam of diameter 60 μm; (c) focused elliptical beam of dimensions 20 μm × 60 μm. (Reproduced with permission from Givan, 2001.)

Collection Optics. The collection optics consists of a train of optical ports to separate and collect various optical responses produced by illumination of each flowing cell. These optical responses, discussed in more detail in the next subsection, are (i) forward scattering count (FSC), (ii) side scattering count (SSC), and (iii) various fluorescence signals at different wavelengths collected at 90° to the laser beam. The number of optical responses detected (hence the number of photodetectors used) define what is known as the number of measured parameters in a flow cytometer. Figure 11.3 shows a typical schematic of collection optics in a five-parameter cytometer: two scattering channels and three fluorescence channels (FL1 PMT, FL2 PMT, and FL3 PMT—in Figure 11.3) to monitor forward scattering, side scattering, and fluorescence signals in three different spectral regions.

For collection of forward-scattered count (FSC), a beam stop in the form of an obstruction element/slit is used first in front of the beam passing through the flowing fluid to block the direct laser beam. The low-angle scattered light, which constitutes FSC, is generated usually at a bend angle of 0.5° and is not blocked by the slit; it is focused by a collecting lens at a photodetector such as a photodiode. The side-scattered signal (SSC) and fluorescence are collected at an angle of 90° to the illuminating laser beam and the flowing stream. For this purpose the 90° light signal is processed by an arrangement of various

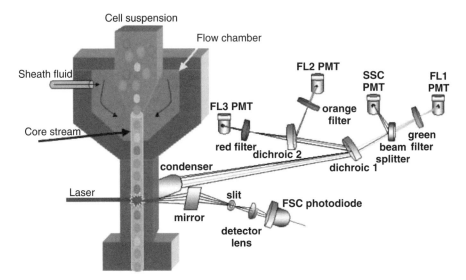

Figure 11.3. Schematic of a five-parameter flow cytometer showing details. (Courtsey of C. Stewart/RPCI.)

dichroic mirrors. A specific dichroic mirror, when oriented at 45° to the light signal, reflects fluorescence of a given wavelength while transmitting light of other wavelengths. As illustrated in Figure 11.3, the fluorescence signals of three different wavelengths (red, orange, and green) are separated by placing the dichroic mirrors in the path of the light collected at 90° to the illuminating beam. This arrangement also separates the side-scattered (SSC) signal. The beams after being reflected from a specific dichroic mirror pass through wide bandpass filters to further discriminate the desired fluorescence from other light. Then they are focused, through a pinhole for further spatial filtering to reduce stray light, onto the photodetector. In the case of light collection for the side scatter, the beam passes through a narrow bandpass filter (±5 nm), centered at the laser wavelength, before going to the photodetector. Since the magnitude of the side-scattered signal is considerably greater than that of the fluorescence, often one may have to use a beam attenuator or even an optical glass plate (to reflect only 1% of the light) of light into the photodetector. The wide bandpass filters are helpful to discriminate two fluorescence signals that may overlap.

The type of collection optics shown in Figure 11.3 are designed for particular fluorescence wavelengths being detected. In other words, the types of fluorochromes which can be used for fluorescence tagging of the cells are fixed. Ideally, one may wish to use a different combination of fluorochromes, depending on the biological samples being analyzed. In such a case, a fixed set of dichroic mirrors is not appropriate and one has to use a dispersion element such as a prism or a grating.

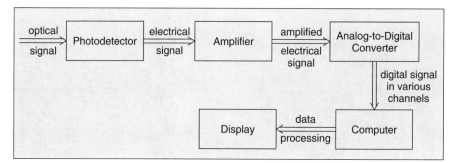

Figure 11.4. Block diagram showing necessary detection and electronics for flow cytometry.

Detection and Electronics. The detection system and electronics used to acquire and process optical response are shown in Figure 11.4 as a block diagram. In most flow cytometers, the photodetectors used for flourescence detection are photomultiplier tubes. For the forward-scattered signal (FSC), which is significantly stronger than fluorescence, a much cheaper photodiode that has considerably reduced sensitivity can be used. An electrical preamplifier following the photodetector is required because a dc offset voltage that establishes zero baseline is used to account for steady-state stream fluorescence. This stream fluorescence is the result of fluorochromes remaining in solution. Subsequent amplifiers can be either logarithmic or linear. The logarithmic amplifier allows one to process signals over a wide range of intensities, while a linear amplifier restricts sensitive measurements to signals in a small linear range. The logarithmic amplifiers also have an "offset" control to select an intensity range to be analyzed without changing the amplification. For further details the readers are referred to a number of books (Shapiro, 1995; Givan, 2001).

Logarithmic amplifiers are normally used for the analysis of fluorescence signals from cells stained with surface markers, because these cells often exhibit a great range of variation in fluorescence intensities. Linear amplifiers may be useful for analysis of forward- and side-scatter signals as well as for low-intensity fluorescence and for narrow-band fluorescence.

A recent trend has been to replace the conventional photomultipliers with more compact solid-state photodetectors such as avalanche photodiodes (APD) (http://usa.hamamatsu.com/cmp-detectors/apds/default.htm) or new miniaturized photomultipliers (Shapiro, 1995). The APDs do not quite match the sensitivity of the PMTs in the visible range, but are the detector of choice in the infrared. They also provide the benefit of being small and all solid state.

As mentioned above, another recent trend in the design of collection optics and detection is to use dispersion optics for polychromatic detection. In this schematic the orthogonal optical response, consisting of side scatter and fluorescence from various fluorochromes, used to stain cells, is dispersed using a prism or a grating. This dispersed signal can then be detected using a linear array of photodetectors such as that provided by a charge-coupled device often abbreviated as CCD. In this case, each element of the detector array detects a narrow spectral wavelength of light (Vesey et al., 1994). Another approach for polychromatic detection has utilized a Fourier transform or interferometric spectrometer (Buican, 1987, 1990). However, this type of detection is considerably more expensive. Optical fiber technology has also been used to deliver light to the detector.

After being amplified by the amplifier, the electrical signal from the photodetector is then fed into an analog-to-digital converter (ADC). The ADC plays the role of converting a continuous distribution of electrical response into a group of discrete (digital) signals that can be conveniently displayed in various forms as discussed below in a separate section (Givan, 2001). The ADCs in a flow cytometer are divided into a number of discrete channels (usually 1024), each channel representing a certain specific light intensity range. Therefore, the signal from a specific cell can be recorded in a particular channel depending on the intensity (digital count level) of that signal. The intensity or count level of each channel can be set by adjusting the photodetector (PMT) voltage and the amplifier gain. The full range of channels should encompass the full range of intensities relevant to a particular experiment.

While the forward-scattered light (FSC) is often used to set the threshold level to trigger the ADC to accept data, any parameter can be used. The threshold defines the minimum brightness of any signal (digital count registered by ADC) used to trigger all ADCs for the parameters that are measured. The scatter threshold is helpful in avoiding problems derived from debris or electronic noise in the system.

The signal from the ADC is then fed into a computer (PCs) and processed using various software packages to display data in appropriate forms as discussed in Section 11.4.

Cell Sorter. Most commercial flow cytometers are not equipped with a cell sorter capability. However, for the sake of completeness the designs of cell sorting units used in flow cytometry are briefly discussed here. The two types of cell sorting devices used in flow cytometry are schematically represented in Figure 11.5.

In the electrostatic sorting device, the cells of a specific type, after passing the interrogation point, are charged and electrostatically deflected to a collection point. It involves the following steps:

1. Hydrodynamic focusing in a nozzle after passing through the interrogation point (illumination zone).
2. Fluorescence from a selective fluorescently tagged cell providing an appropriate trigger to vibrate the nozzle by a transducer and break

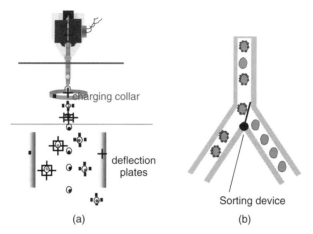

Figure 11.5. Schematics of two types of cell sorting devices for flow cytometry. (a) electrostatic sorting, (b) mechanical sorting (Courtsey of C. Stewart and M. Casstevens).

the cell stream into droplets ejected into air. This process ensures that the droplets will contain the specific cells.

3. Fluorescence response from a selective fluorescently labeled cell providing an electronic trigger that is time delayed to charge this cell as it reaches the charging collar.

4. Charged droplets containing selectively the specific cell population deflected by an electrostatic field from plates held at high voltage (3000 V).

5. Various collection devices such as tubes, plates, and so on, placed at appropriate location to collect a specific type of charge (positive, negative or neutral) and thus selected population of cells.

In the electrostatic sorting method, as the cell intercepts the laser beam and a particular fluorescence characteristic of a specifically labeled cell is detected, the sort logic board of the cytometer electronics makes a decision based on the user-defined criteria whether this type of cell is to be sorted or not. The cytometer then waits until the cell has traveled the distance between the illumination point and the nozzle break-off point (called the *drop delay*) to charge the droplet with a specific charge (positive or negative). These charged droplets are sorted out on the basis of their charge by attraction toward the plate of opposite polarity.

The electrostatic sorting method can be operated at rates up to 50,000 cells per second. The necessary time delay to trigger the vibrating nozzle and the charging collar can be determined from the flow rate of the cells.

The second method utilizes a mechanical gate that swings back and forth to direct a particular type of cell into a desired pathway. While this method is

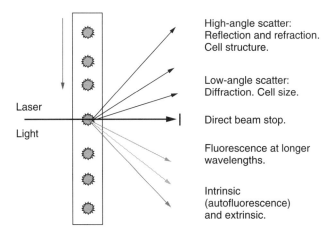

High-angle scatter:
Reflection and refraction.
Cell structure.

Low-angle scatter:
Diffraction. Cell size.

Laser

Direct beam stop.

Light

Fluorescence at longer
wavelengths.

Intrinsic
(autofluorescence)
and extrinsic.

Figure 11.6. The various optical response generated by interaction of laser light with the flowing cell. (Reproduced with permission from http://www.uwcm.ac.uk/study/medicine/haematology/cytonetuk/introduction_to_fcm/optics.htm

considered to be more gentle on cells, it has only a maximum rate of 300 sorted cells per second.

11.2.3 Optical Response

As described in Section 11.2.2, the optical response generated by interaction of a cell with the laser beam at the illumination point consists of forward-scattered light and side-scattered light that are of the same wavelength as the exiting beam. In addition, absorption either by the cellular components or by fluorochromes staining a cell produces fluorescence signals shifted in wavelength from that of the exciting beam. These detected parameters provide information about the type of the cell as well as its structure and function. They are illustrated in Figure 11.6.

The forward-scattered signal (FSC) that is generated within a few degrees from the incident laser beam is often related to the cell size and is used to determine the cross-sectional area and volume of the cell. However, a cell with a very different refractive index from its surrounding can also produce a FSC signal. For example, a dead cell appears smaller than the corresponding living cell because its refractive index is more like the surrounding stream due to the leaky outer membrane. The dead cells therefore bend less light into the FSC detector than the corresponding living cells.

The high-angle scattered signal are often collected at 90°C as a side-scattered count (SSC). This has also been discussed in the previous subsection. This signal is produced by reflection and refraction from the variation in cell structure. Therefore, SSC is related to the cell's surface texture and internal structure. SSC is sometimes also referred to as a granularity signal because it provides information on the granularity of the cell. For example, granulocytic blood cells that have granular and irregular nuclei produce a significantly more intense SSC signal compared to that from more regular lymphocytes or erythrocytes.

Fluorescence labeling allows one to selectively label (stain) a specific subpopulation of cells in order to investigate their cell structures and functions. For example, immunofluorescence, used for immunophenotyping discussed later in Section 11.5.1, involves staining of cells with antibodies which are conjugated to fluorochromes. This staining can be used to label antigens on the cell surface. Alternatively, antibodies can also be directed at targets in cytoplasm. The two approaches used for immunofluorescence are: (i) direct immunofluorescence in which an antibody is directly conjugated to a fluorochrome, thus the cells are stained in a single step; and (ii) indirect immunofluorescence in which the primary antibody to cell surface antigen is not labeled. Instead a second antibody specific to the primary antibody is conjugated to a fluorochrome.

Modern flow cytometry utilizes a large array of monoclonal antibodies which are specific for various antigen proteins on the cell surface (Givan, 2001). The antigens defined by these antibodies are characterized by a CD number designation where CD stands for "cluster of differentiation," which defines a particular protein differentiating one type of cells from another. The CD numbers now range from CD1 to over 200, as the number of characterized antigens have steadily grown. These CD number designations are used to specify an antibody that is specific to the antigen.

In the direct staining of cells, they are incubated with a monoclonal antibody that has been conjugated to a fluorochrome. This procedure takes only 15 to 30 minutes of cell incubation with antibody at 4°C, followed by several washes to remove antibodies that are weakly bound or bound nonspecifically. In the indirect staining method, the cells are incubated with a nonfluorescent monoclonal primary antibody. Then after washing to remove any weakly bound antibody, a second incubation with an antibody conjugated to a fluorochrome is used to react with the primary antibody of the first layer. The advantage of this indirect staining process is the cheaper cost of unconjugated primary antibodies. Another advantage offered by this sequential layer approach is that it can be extended to more than two layers with each layer producing amplification of fluorescence. The disadvantage of this indirect staining process is that it is considerably time-consuming and that complications due to nonspecific binding is significantly increased with each added layer.

Another optical parameter monitored can be the polarization of the scattered light, which provides information on the birefringence produced by the cell structure such as that of eosinophil granules. Furthermore, pulse shape analysis of the scattered signal can be used to get information on the cell shape.

11.3   FLUOROCHROMES FOR FLOW CYTOMETRY

A wide variety of fluorochromes have been used in flow cytometry. The number of fluorochromes still keep on growing, as new applications of flow cytometry dictate the need for novel fluorochromes to be developed. The choice of fluorochromes is primarily dictated by specific applications as well as by the laser excitation source available on the flow cytometer. For example, for phenotyping, fluorochromes have been preferred because they can easily be conjugated to antibodies and are excitable by the popular argon ion laser.

Table 11.4 lists many of the fluorochromes and their common abbreviations, if any, used in flow cytometry. It also lists the excitation and emission wavelengths, along with their typical applications. For most applications, it is desirable to use more than one fluorchrome so that one can conduct multi-parameter analysis of the specimen. For this application, it is desirable to have a flow cytometer with more than one laser excitation source to offer a wide choice in selecting the fluorochromes. However, in the case where the flow cytometer has only one laser such as an argon ion laser providing 488-nm excitation, one can select a dye pair such that each dye has a different amount of Stokes shift (separation between the excitation wavelength and the emission peaks as discussed in Chapter 4). With such a pair, even though the excitation is provided at the same wavelength, the emission spectra from the two dyes are well separated in two different regions. Figure 11.7 represents the excitation and emission spectra of some fluorochromes.

The two most commonly used fluorochromes for dual color flow cytometry are (i) fluorescein, often abbreviated as FITC because it is the fluorescein isothiocyanate form that is used for conjugation with specific antibodies for phenotyping application, and (ii) phycoerythrin, abbreviated as PE, which is derived from red sea algae. As can be seen from Figure 11.7, both these dyes can be excited at 488 nm. However, their fluorescence peaks are well separated; while FITC emits in the green (~520 nm), the emission from PE is of orange color (~575 nm).

Another approach used to separate the emissions of two fluorochromes is that of a tandem fluorochrome. Here one utilizes a combination of two dyes, one absorbing efficiently at the excitation wavelength and then exciting another chemically attached dye by Förster energy transfer, which then emits at a wavelength considerably red shifted. An example of a natural tandem dye is PerCP (peridinin chlorophyll protein), which is a carotenoid:chlorophyll complex. Here the carotenoid unit absorbs at 488 nm and transfers energy to chlorophyll, which emits at ~670 nm. Hence the tandem dye exhibits a large apparent Stokes shift. The excitation and the fluorescence spectra of this tandem fluorochrome are also shown in Figure 11.7. A synthetic tandem fluorochrome is PE covalently linked to Cy5 whose excitation and florescence spectra are also shown in Figure 11.7. It can be excited at 488 nm by absorption into the PE unit which transfers energy to Cy5 emitting at 670 nm. For a

TABLE 11.4. List of Fluorochromes for Flow Cytometry

Fluorochrome	Excitation (nm)	Emission (nm)	Applications
Fluorescein (FITC)	495	520	Phenotyping
R-Phycoerythrin (PE)	480	575	Phenotyping
Tricolor	488	650	Phenotyping
PerCP	470	670	Phenotyping
TRITC (Tetramethyl rhodamine)	488	580	Phenotyping
Coumarin	357	460	Phenotyping
Allophycocyanin (APC)	650	660	Phenotyping
APCCy7	647	774	Phenotyping
Cascade blue	350	480	Phenotyping
Red 613	480	613	Phenotyping
Texas red	595	620	Phenotyping
Cy3	550	570	Phenotyping
Cy5	648	670	Phenotyping
Red 670	480	670	Phenotyping
Quantum red	480	670	Phenotyping
Hoechst 33342	350	470	DNA analysis/apoptosis
Hoechst 33258	350	475	DNA analysis/chromosome staining
DAPI	359	462	DNA staining, preferentially of AT sequences
Chromomycin A3	457	600	DNA analysis/chromosome staining
Propidium iodide	495	637	DNA analysis
Ethidium bromide	518	605	DNA analysis
TOPRO3	642	661	DNA analysis
Acridine orange	490	530/640	DNA, RNA staining
Sytox green	488	530	DNA
Fluorescein diacetate	488	530	Live/dead discrimination
SNARF1	488	530/640	pH measurement
Indo1	335	405/490	Calcium flux measurement
Fluo3	488	530	Calcium flux measurement
Rhodamine 123	515	525	Mitochondria staining
Monochlorobimane	380	461	Glutathione specific probe

Source: http//www.icnet.uk/axp/facs/davies/Flow.html.

three-color analysis, using 488 nm, the PECy5 tandem together with FITC and PE have been the frequent choice for phenotyping application.

A number of fluorochromes have been used for nucleic acid staining and analysis of DNA content. These fluorochromes have been described in Chapter 8. The most popular choice as a DNA fluorochrome for flow analysis is propidium iodide. This fluorochrome is not very specific because it stains

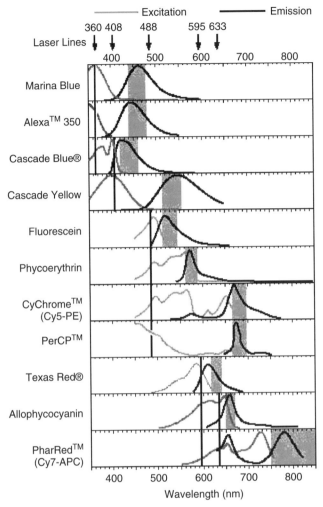

Figure 11.7. Excitation and emission (bold) spectra of some fluorochromes used in flow cytometry. (Reproduced with permission from BD Biosciences.)

the double-stranded regions of both DNA and RNA. Also, it is not able to penetrate an intact cell membrane (live cell). However, propidium iodide has the advantage of being efficiently excitable by the 488-nm line of an argon ion laser, whereas other DNA dyes such a DAPI, Hoechst and acridine orange require different wavelengths for efficient excitation. Table 11.4 lists the excitation wavelengths for these dyes. Both DAPI, which specifically stains the AT base pair, and Hoeshst, which stains the GC base pair on DNA, requires excitation in the UV. As discussed in Chapter 8, acridine orange shows an interesting feature in that it fluoresces red when bound to single-stranded or nonhelical nucleic acid such as RNA or denatured DNA, but fluoresces green when intercalated in the double-stranded helical nucleic acid of native DNA. This change in fluorescence color can be used to study the denaturability of DNA during the cell cycle.

11.4 DATA MANIPULATION AND PRESENTATION

This section describes the steps of data acquisition, storage, analysis, and display. Important steps in data acquisition also include calibration of the instrument, spectral compensation, and, sometimes, the use of gating. These topics are also discussed here.

Calibration. Often the performance of a flow cytometer is calibrated by using inert and stable standards that are nonbiological particles of dimensions comparable to biological cells. For this purpose, fluorescently labeled microbeads such as those made of polystyrene or latex are very popular and readily available from a number of commercial sources as listed in Section 11.7. These polystrene or latex microbeads can be conjugated to various fluorochromes and used for optical alignment and calibration. These microbeads can also be conjugated to antibodies for the purpose of calibrating the scale in terms of the number of binding sites (Givan, 2001). These beads also provide opportunity for multiplexed assays where the beads with capture molecules can be used to determine the concentration of soluble analytes.

For the purpose of calibration of fluorescence scales, the fluorescent beads serve as an external standard to which the fluorescence from a stained cell can be compared. Commercially available polystyrene beads, for example, have standardized fluorescence intensities that can be used for quantitation of fluorescence intensities of the stained cells. Two types of microbeads are available: (i) beads that have a fluorochrome such as fluorescein or PE bound to their surfaces and (ii) beads that have fluorochromes incorporated throughout the bead.

Beads are also available with known numbers of fluorochromes such as PE molecules on their surface. Using these beads, the background fluorescence of a control sample can be expressed as an equivalent number of fluorochrome molecules. This number can then be subtracted from the number of fluorochrome molecules of a stained sample to obtain information on the number of receptors or antigens on the surface of a cell.

Also available from commercial sources are beads with a known number of binding sites for immunoglobulin molecules. Therefore, they mimic cells and can be stained with a specific antibody. The fluorescence intensity of the beads with the known numbers of binding sites can, therefore, be used to calibrate the intensity scale, whereby the ADC channels can be converted to correspond to antibody binding sites per cell.

Beads can also be used for calibration of the volume flowing through a flow cytometer. This can be accomplished by the addition of a known number of beads to the sample. For multiplexed assays, each bead can be linked to a different capture molecule that can be a range of allergens, nucleotides, or antibodies. Each capture molecule can then bind to a specific soluble target analyte.

In summary, the microbeads play a significant role in calibration for flow cytometry.

Compensation. Compensation is the process applied to correct for an overlap between the fluorescence spectra of different fluorochromes. Even though bandpass filters are used to discriminate fluorescence in a given wavelength, derived from a specific fluorochrome, from that due to other fluorochromes, the bandpass filters have a finite bandwidth that may permit an overlapping fluorescence from a different fluorochrome to leak through. This crossover of the two fluorescence—for example, for the fluorochromes FITC and PE—is shown in Figure 11.8. This figure clearly shows crossover of the FITC fluorescence (darker region) detected by the PE-PMT and that of PE (shaded region) detected by the FITC-PMT. To correct for this crossover, flow cytometers use either an electronic or software compensation whereby a

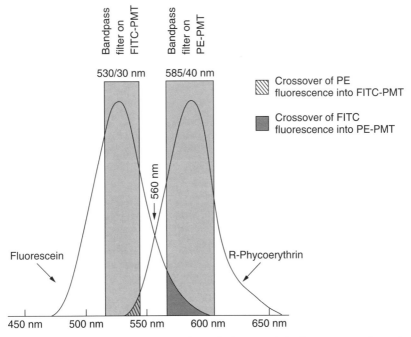

Figure 11.8. The crossover of fluorescein and phycoerythrin fluorescence through the filters on the "wrong" photo detectors. Excitation is at 488 nm. (Reproduced with permission from Givan, 2001.)

certain percentage of the signal from one photodetector is subtracted from the signal of another photodetector. These percentages are determined empirically. In electronic compensation, cells stained with only one fluorochrome at a time are run and signals are recorded on two photodetectors covering overlapping spectral regions. For example, in the case of FITC and PE labeled cells, first FITC stained cells are run, but the signal observed both with FITC and PE photodetectors are recorded. Then a percentage of FITC-PMT signal subtracted from the PE-PMT signal is varied until no signal above the background is recorded by the PE-PMT. The same procedure is repeated for cells stained with PE. The percentage determined thus provides the amount of compensation needed for each color detection.

The software compensation method permits one to apply compensation after data acquisition. However, even this type of compensation requires that files from single-stained samples be stored in order for the software compensation matrix to be generated. Therefore, single-stained control runs are always advisable.

Data Storage. A multiparameter flow cytometer generates a large amount of data. Software programs to store, manipulate, and present the data in useful form play an important role in flow cytometry. In most commercial flow cytometers the data are stored using a standard flow cytometry format called FCS. This standardization allows one to have the ability to write independent programs that can handle data acquired on any cytometer.

The data stored in the FCS format is usually as a "list mode" data file. In this mode, the multiparameters obtained for each cell are stored. For example, the data obtained for 10,000 cells on a four-parameter flow cytometer will consist of each set of four numbers describing each cell in the sequence in which the cell passes through the integration point (illumination region). Each cell can be analyzed again by retrieving the four parameters stored for it. Thus the intensity of each of the four signals for a specific cell can be known and correlated with each other as well as with the corresponding set for another cell.

Another type of data storage that has the advantage of requiring less storage space is a single-parameter data storage in which the intensity profiles for each parameter for the population of cells in a sample is stored separately. For example, the distribution of forward-scattered signal intensities for cells in the sample can be stored in one file. Similarly, another file records the distribution of green fluorescence signal intensities for the cells in the sample. The disadvantage of this mode of storage is that the information on any correlation between two parameters (for example, the forward-scatter signal and the green emission signal) for a cell is lost. Thus one may save storage capacity by using single-parameter data storage, but it is at a cost of severely limiting options for future analytes. Therefore, list-mode storage is the preferred option for storage of flow cytometry data.

In addition, a header to the file is generated that records any amount of desired textual information such as instrument settings, file name, and demographics about the experiment.

Gating. Gating in flow cytometry implies selection of cells of a specific set of characteristics for further analysis. Thus, gating is to place restriction on the flow cytometry data to be included for subsequent analysis. On the basis of the parameters monitored in flow cytometry, gating also defines regions that specify characteristics of a subset of cells. A region thus may cover all cells that produce a certain range of green fluorescence intensities or a certain set of forward- and side-scatter characteristics. A gate may involve one region or a combination of two or more individual regions. Gating can utilize two types of gates. A live gate restricts the data accepted by the data acquisition system for storage, based on the characteristics of cells needed to be fulfilled before the data are accepted for storage. However, in this mode of gating, data on all other cells are not stored, and information that may be valuable for subsequent analysis is also not stored. For this reason a live gate is not the preferred choice. An analysis gate selects cells with specified characteristics only after data from all cells in a heterogeneous population have been stored in a data file.

Data Display. The stored data can be analyzed by software to allow display formats that exhibit certain correlation of data. These correlations allow one to obtain profiles of a certain cell population which can also be used to identify certain diseases or to characterize certain cellular processes. These various formats of data display are discussed here.

Single-Parameter Histograms. In this mode of display, also known as *frequency histograms*, one plots the signal of one parameter (fluorescence from a specific labeling fluorochrome or a particular scatter light signal) against the number of events (number of cells registering this intensity level). As discussed in Section 11.2.2, an ADC assigns a certain intensity range to a certain channel number. Therefore, the histogram plot often substitutes the intensity range by the channel number that is represented on the horizontal axis. This histogram plot permits one to look at the intensity distribution from a large number of cells for a given fluorescence wavelength range. Figure 11.9 shows some typical histograms for a population of cells. In these, the y axis represents the number of cells (or cell counts) for each channel (defined light signal intensity) that is represented on the x axis. Software programs then allow one to obtain statistical analysis of the data, keeping in mind that histograms allow this analysis for one parameter (whose intensity is represented on the x axis) at a time.

The following types of statistical numbers are useful, and often the software of the flow cytometer automatically displays them.

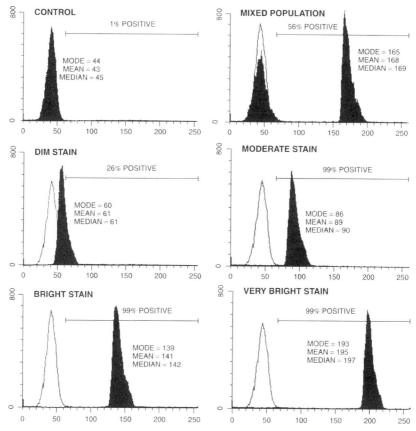

Figure 11.9. The histogram distribution of signal intensities from a population of cells. The plots show the number of cells on the vertical axis against channel numbers (related to scatter or fluorescence intensity) on the horizontal axis. (Reproduced with permission from Givan, 2001.)

- *Mean.* The mean represents the mean intensity channel for a group of cells. The two types of mean values represented are: (i) *arithmetic mean*, which is calculated by adding the intensities for all the cells and then dividing it by the number of cells (this type of mean is suited for data analysis that is collected on a linear scale) and (ii) *geometric mean*, which is calculated by multiplying the intensities of all the n cells and then taking the nth root of the product (this type of mean is more suited for data collected with a logarithmic amplifier).
- *Median.* The median value is used to describe the fluorescence intensity of a population of cells. It represents the midpoint in the sequence where all the cells are lined up in the order of increasing intensity. It is often the preferred statistical number for many analyses.

- *Mode.* The mode represents the most common intensity for a group of cells. The mode, therefore, represents the intensity (channel number) that describes the largest number of cells in a sample.
- *Coefficient of Variation (CV).* The CV is defined as the standard deviation of a series of values divided by the mean of those values. It is expressed as a percentage. For the histogram format of data display, the CV is represented by the width of the histogram curve. Often CV is used to calibrate the alignment of a flow cytometer. A narrower CV using microbead standards implies a better alignment and a higher-quality performance for a specific flow cytometer. In an actual analysis, the CV data can be used to measure the variation in characteristics within a cell population.

In the histogram display, markers can be used to define regions of intensity that are of interest. For example, a percentage of positively stained cells can be displayed by placing a marker at the position of intensity (channel), defined by the background fluorescence of unstained cells.

In Figure 11.9, the control cells (unstained) are indicated by the clear curve distribution overlayed with the black curve distribution for the stained cells. The positive intensity region is marked relative to the 1% level on an unstained control. The mixed population refers to a sample containing both the stained and unstained cells.

As seen from Figure 11.9, single-parameter histograms are generally multimodal and represent many separated peaks (and regions). Figure 11.9, for example, illustrates separated peaks corresponding to unstained and stained cells. Similarly, histograms of cellular DNA content exhibit a multimodal distribution containing contributions from one or more populations of cells in G_0/G_1, S, and G_2+M phases of the cell cycle as well as from nuclear fragments and other debris.

Dot Plots. In flow cytometric analysis of mixed cell populations, it is not uncommon to find a situation where frequency distributions of any one cellular parameter show considerable overlap from one cell type to another. In such a case a single-parameter histogram is not useful. However, correlation of one parameter with another can often be used to distinguish different types of cells. This is a bivariate plot in which the frequency distributions of two parameters are represented in a two-dimensional space. A frequently used bivariate display is a two-dimensional dot plot, also known as a *scattergram* or *bitmap*, in which the two axes represents the intensity channels for selected two parameters. Each cell then is represented as a dot on the plot according to the two intensity channels it registers in for the selected two parameters.

Figure 11.10 shows an example of the two one-parameter histograms for forward-scatter signal (FSC) and side-scatter signal (SSC) which have been

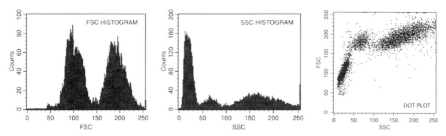

Figure 11.10. Two separate histograms (*left*) for FSC and SSC are converted into a two-dimensional dot plot (*right*). (Reproduced with permission from Givan, 2001.)

combined in a two-dimensional dot plot. A four-parameter data set will require six two-parameter correlations, each represented by a dot plot. Dot plots are frequently used where the fluorescence intensity associated with stained cells is bright enough that it is clearly separated from that of cells which are negative for each fluorescence marker. For the analysis of dot plots, one often defines four quadrants representing four types of cell populations:

- LL (lower left), which represents cells that are negative for each marker represented on the *x* and *y* axes
- UR (upper right), which represents cells that are positive for both markers
- UL (upper left), which represents cells that are positive only for marker represented on the *y* axis
- LR (lower right), which represents cells that are positive only for marker represented on the *x* axis.

Although dot plots are simple and frequently used, they suffer from blackout in the area where the number of registered cells is very dense.

Contour Plots. Contour plots are another type of two-dimensional bivariate plots that contain the same type of information on correlation between two parameters. They are specially useful if the number of registered cells in any region becomes too dense to produce a blackout.

In contour plots, lines are drawn to various levels of cell count. An example is shown in Figure 11.11, which also shows a three-dimensional correlation plot of the same data. Here the vertical axis represents the cell count for the two parameters represented on the *x* and *y* axis. In contour plots, each line represents a change in the number of events (cell counts).

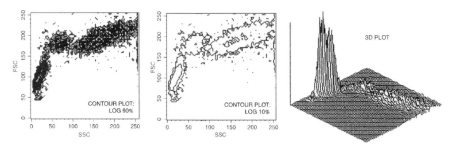

Figure 11.11. Contour plots for two scales, together with a three-dimensional plot. (Reproduced with permission from Givan, 2001.)

11.5 SELECTED EXAMPLES OF APPLICATIONS

11.5.1 Immunophenotyping

Immunophenotyping refers to identification of cells using fluorochrome-conjugated antibodies as probes for proteins (antigens) expressed by cells. The reasons for using the term immunophenotyping are twofold: (i) It relates to the activities of immunological species, namely, antibodies, and (ii) it is primarily used to identify lymphoid and hematopoietic cells, which are constituents of immune systems. Therefore, immunophenotyping basically deals with classification of normal or abnormal white blood cells according to their multiparameter surface antigen characteristics, which can then be used as a profile for a specific disease or malignancy. Immunophenotyping is one of the largest clinical applications of flow cytometry. Immunophenotyping is frequently used in evaluating malignancies of the hematopoietic system as well as in detecting various disease states. HIV immunophenotyping is another common clinical application of flow cytometry. The analysis utilizes multi-parameter flow cytometry of cells that combines the measurement of fluorescence from fluorochrome-labeled antibodies specific to expressed antigens, together with light-scattering properties of individual blood cells. Often, the bivariate dot plots, generated by plotting fluorescence, represent patterns of clusters which are called phenograms.

The various types of cells constituting blood are discussed in Chapter 3 and are summarized here in Table 11.5 for a normal human adult. Even the light-scattering properties of different types of cells generate different patterns in the forward light-scatter (FSC) and side-scatter (SSC) dot plots that reflect their relative sizes and granularity. Figure 11.12 represents the patterns corresponding to the various blood cells in a FSC versus SSC dot plot.

The pattern of light scatter can be used to gate (select) cells for fluorescence analysis using fluorochrome-labeled specific antibodies expressed by the different types of cells. The gating procedure as discussed above simply allows

TABLE 11.5. Cells in Normal Human Adult Peripheral Blood

Cells	Number per cm^3	Percent of WBC	Diameter (μm)
Platelets (thrombocytes)	$1–3 \times 10^8$		2–3
Erythrocytes (RBC)	$4–6 \times 10^9$		6–8
Leukocytes (WBC)	$3–10 \times 10^6$	100	
Granulocytes			
Neutrophils	$2–7 \times 10^6$	50–70	10–12
Eosinophils	$0.01–0.5 \times 10^6$	1–3	10–12
Basophils	$0–0.1 \times 10^6$	0–1	8–10
Lymphocytes	$1–4 \times 10^6$	20–40	6–12
Monocytes	$0.2–1.0 \times 10^6$	1–6	12–15

Values are from Lentner, C., ed. 1984. *Geigy Scientific Tables*, 8th edition, CIBA-Geigy, Basel; and from Diggs, L. W., et al. (1970). The *Morphology of Human Blood Cells*, 5th edition, Abbott Laboratories, Abbott Park, IL. (Reproduced with permission from Givan, 2001.)

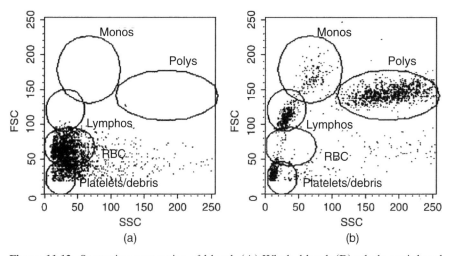

Figure 11.12. Scattering properties of blood. (A) Whole blood; (B) whole peripheral blood after erythrocyte lysis. The regions of five clusters representing the various leukocytes and platelets possess different level combinations of forward-scatter and side-scatter counts. (Reproduced with permission from Givan, 2001.)

one to select a specific cell subpopulation (e.g., corresponding to lymphocytes) based on its FSC and SSC characteristics and collects fluorescence signals only from this subpopulation. Table 11.6 lists some specific antigens (Fc receptors) expressed by various types of hematopoietic cells along with their side-scatter behavior.

The presence of nonmalignant abnormality (or infection) as well as hemolymphatic neoplasms such as leukemia and lymphoma is reflected in

TABLE 11.6. Fc Receptors on Hematopoietic Cells Together with the Side-Scattering Properties of These Cells

Cells	FcRI (CD64)	FcRII (CD32)	FcRIII (CD16)	Side Scatter
Erythroid cells	–	–	–	Low
Granulocytes				
Basophils		++	++	0 Low
Eosinophils	+/–	++	0	High
Neutrophils	++	++	[a]	High
Lymphocytes				Low
B cells	–	+	–	Low
Natural killer (NK) cells	+	–	–	Low
T cells	–	–	–	Low
Monocytes	[a]	++	+	Intermediate
Platelets	–	++	–	Intermediate

[a]None for resting cells, but up-regulated and variable during inflammation.

Source: Stewart and Mayers, 2000.

either (a) a change of the frequency of cells within a subset or (b) a change in the expression of surface antigens. For this purpose, morphologic information derived from FSC and SSC alone is not sufficient. Expression of antigens as identified by staining with the fluorochrome-labeled corresponding antibodies has been the approach to profile these diseases. Table 11.7 lists the antigenic expressions (using the CD designation) on the various normal mature blood cells.

Various kinds of leukemia (ALL, CLL, etc.) can be diagnosed on the basis of finding a deviation from normal behavior. The condition of abnormal combinations of antigens is often referred to as *lineage infidelity* or as *lineage promiscuity*. As an example, Figure 11.13 compares the dot plots of abnormal blood cell with that of a chronic lymphositic leukemia (CLL) cell. It utilizes staining CD19- and CD5-specific antibodies that have been respectively labeled by fluorochromes PE (hence CD19PE) and FITC (hence CD5 FITC). Usually, weakly expressed antigens such as CD19 are labeled with a bright fluorochrome such as PE, whereas strongly expressed antigens such as CD5 are labeled with a dim fluorochrome FITC. This avoids artifacts due to the nonlinearity of fluorescence compensation. In normal cells, the B-cell line does not coexpress both CD19 and CD5, but chronic lymphocyte leukemia cells strongly coexpress both, thus generating the pattern defined by the square block.

The HIV primary infection is characterized by a progressive loss of CD4-positive T cells and elevation of CD8-positive T cells. Therefore, CD4 enumeration can be used to monitor the disease course of HIV infection.

In profiling a disease, the immune response generated by a selected biological stimulus is of significant value. This immune response can also be

TABLE 11.7. Antigenic Expression on Normal Mature Blood Cells

CD No.	Lymphocyte			Monocyte	Neutrophil	Eosinophil	Basophil	Erythrocyte
	B	T	NK					
2	–	++	++	–	–	–	–	–
3	–	++	–	–	–	–	–	–
4	–	Sub	–	+	–	–	–	–
5	Sub	++	–	–	–	–	–	–
8	–	Sub	Sub	–	–	–	–	–
10	–	–	–	–	++	+	–	–
11b	–	–	+	+++	+++	+++	+	–
13	–	–	–	+++	+++	++	++	–
14	–	–	–	+++	–	–	–	–
15	–	–	–	+	++++	+	+	–
16	–	–	++	+\–	+++	–	–	–
19	++	–	–	–	–	–	–	–
20	+++	–	–	–	–	–	–	–
21	++	–	–	–	–	–	–	–
22	++	–	–	–	–	–	–	–
23	Sub	–	–	–	–	–	–	–
24	++	–	–	–	–	–	–	–
25	–	–	–	–	–	–	–	–
32	–	–	–	++	+++	+	+	–
33	–	–	–	+++	++	++	++	–
34	–	–	–	–	–	–	–	–
38	Sub	Sub	Sub	Sub	+	+	–	–
45	+++	+++	+++	+++	++	++	++	–
64	–	–	–	Sub	Sub	–	–	–
71	–	–	–	–	–	–	–	Sub

+, positive (first decade); ++, intermediate (second decade); +++, bright (third decade); ++++, very bright (fourth decade); Sub, subset.
Source: Loken and Wells (2000).

profiled using multiparameter flow cytometry. Figure 11.14 shows two-parameter dot plots of cells stimulated by LPS (lipopolysaccharide) and PMA (phorbal myristic acid).

11.5.2 DNA Analysis

DNA can be used to classify different types of cells. Cellular DNA content analysis has been clinically used to characterize solid tumors (Ross, 1996). Therefore, the measurements of the DNA content of cells is a major application of flow cytometry. The DNA content can also yield considerable information on the cell cycle as well as on the effect of added stimuli such as

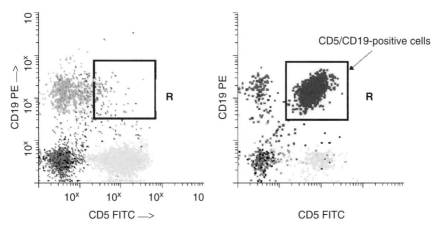

Figure 11.13. Comparison of the two-parameter dot plots for normal B cells (*left*) and chronic lymphositic leukemia B cells (*right*). The region R defining coexpression of both Cd19 and CD5 contains a substantial B cell subpopulation only in the case of chronic lymphositic leukemia cells. (Courtesy of C. Stewart/RPCI.)

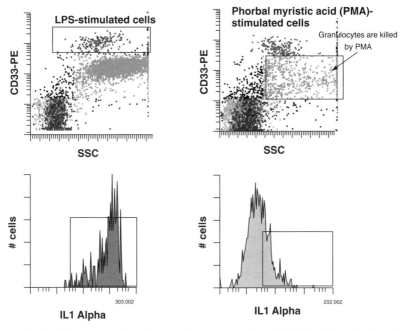

Figure 11.14. Response of cell populations to selected biological stimulus. *Left*: Stimulation with LPS (lipopolysaccharide). *Right*: Stimulation with PMA (phorbal myristic acid). (Courtesy of C. Stewart/RPCI.) (See color figure.)

transfected genes or drug action on the cell cycle (*source*: Cancer Research UK, FACS Laboratory: www.icnet.uk/axp/facs/davies/cycle.html). Furthermore, the measurement of DNA content can be combined with the quantification of an antigen in order to assess its expression during the cell cycle.

The amount of DNA in the nucleus of a cell (often called the 2C or diploid amount of DNA) is the same in all healthy cells of a specific organism. There are three major exceptions: (i) cells that have undergone meiosis, (ii) cells in preparation for cell division (mitosis), and (iii) cells that are undergoing apoptosis (Givan, 2001). Therefore, measurement of the DNA count can be used to identify cell abnormality the same way it can identify malignancy, which involves genetic changes.

As discussed in Section 11.3, a number of fluorochromes are used for DNA content analysis. They exhibit a large change in fluorescence intensity upon binding (interaction) with the nucleic acids. To measure the DNA content of the nucleus, one generally uses propidium iodide (with detergent or alcohol), which can permeate the outer membrane of normal cells. If the cells are treated with RNase, the fluorescence results only from the DNA content and the intensity of fluorescence (red in this case) is proportional to the DNA content. In the case of normal nondividing cells, a one-parameter histogram consists of a narrow peak showing all cells emitting nearly the same amount of red fluorescence and thus possessing the same DNA content. In the case of a malignant tissue, the histogram may consist of more than one peak (thus the abnormal cells are often referred to as *aneuploid*, or *DNA aneuploid*—in comparison to the normal cells, which are called *euploid* or *normal diploid*). Figure 11.15 compares the histogram plots for the normal cells with those for malignant breast tumors. It is worth pointing out that not all malignancies result from DNA changes. Further complications in the interpretation of data may arise from the presence of the cell division cycle, which will appear to increase the DNA content.

The information on cellular DNA content obtained by flow cytometry can readily be used for cell cycle analysis, which has been discussed in detail in Chapter 3. In the cell cycle, cells in the G0 phase are not cycling, while the ones in the G1 phase are either just recovering from division or preparing for another cycle. Their nuclei will contain 2C amount of DNA. Cells in the G2 phase have finished DNA synthesis, and cells in M phase are in mitosis—that is, undergoing chromosome condensation. Their nuclei contain twice as much (4C) DNA. Cells in the S phase are in the process of synthesizing new DNA. Hence, their nuclei will span the range of DNA content between 2C and 4C. A propidium iodide fluorescence histogram for cells that have been stimulated to divide is shown in Figure 11.16.

Flow cytometry, therefore, provides a rapid and simple method of studying cell proliferation. In actual analysis of the histograms, complications arise from certain widths associated with the Gausian shape of the G0/G1 and G2/M distributions. These widths, if they are large, may obscure the region covered by the S phase or even an aneuploid population. In such a case, various computer

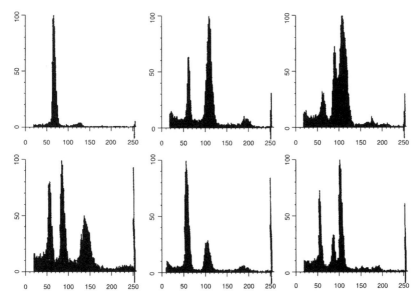

Figure 11.15. Propidium iodide fluorescence histograms from nuclei of cells aspirated from normal tissue (*upper left*) and malignant breast tumors (remaining histograms). Data courtesy of Colm Hennessy. (Reproduced with permission from Givan, 2001.)

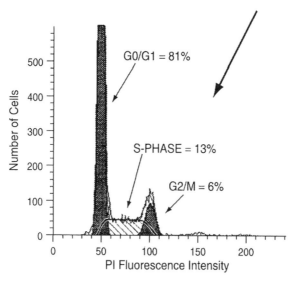

Figure 11.16. DNA histograms from lymphocytes stimulated to divide. (Reproduced with permission from Givan, 2001.)

algorithms have been proposed to estimate the proportion of S phase. For a brief introduction to these various algorithms, the readers are referred to the book by Givan (2001).

A more direct method in flow cytometry to measure the DNA synthesis process utilizes pulsing the cells with bromodeoxyuridine (abbreviated as BrdU or BrdUdr or BUDR), a thymidine analogue, which gets incorporated into the cell's DNA in place of thymidine during the synthesis cycle. The BrdU-incorporating DNA can then be stained with fluorescein-conjugated monoclonal antibodies, specific to BrdU, by partially denaturing the DNA to expose BrdU within the double helix. The denatured DNA can also be stained with propidium iodide at the same time. The two-parameter plot (dot or contour) can then be used to identify different phases of the cell cycle. Figure 11.17 shows the two-parameter contour plot. The cells in the middle region of the propidium iodide distribution, which represent the S phase, have all incorporated BrdU and exhibit strong green fluorescence from fluorescein (FITC).

Flow cytometry has also been used to study cell death by apoptosis and necrosis, which have been discussed in Chapter 3. Apoptosis is gene-directed cellular self-destruction, also called *programmed cell death*. Apoptotic cells can be recognized by flow cytometry by using morphological as well as biochemical changes. Some of the morphological changes are cell shrinkage and change in cell shape. A common event associated with apoptosis, which has been used in flow cytometry, is the flipping and stabilization of phosphatidylserine from the inner surface of the cytoplasmic membrane to the outer surface. Then staining of the intact cells with fluorochrome (e.g., FITC)-conjugated annexin

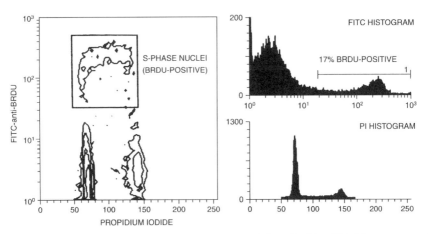

Figure 11.17. Fluorescein (FITC) histogram, propidium iodide (PI) histogram, and dual-color correlated contour plot of human keratinocytes cultured for 4 days, pulsed with BrdU, and then stained with FITCantiBrdU and PI. Data courtesy of Malcolm Reed. (Reproduced with permission from Givan, 2001.)

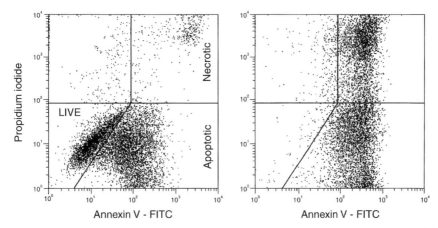

Figure 11.18. Cells stained with propidium iodide for permeable membranes and with annexin V for phosphatidylserine can distinguish live (double negative), apoptotic (annexin V-positive, propidium iodidenegative), and necrotic (double positive) cells. This figure (with data from Robert Wagner) shows cultured bovine aortic endothelial cells; the left plot displays data from adherent cells, and the right plot displays data from the "floaters." (Reproduced with permission from Givan, 2001.)

V, which binds to phosphatidylserine, will detect cells in the early stages of apoptosis.

Figure 11.18 shows a two-parameter dot plot of cells stained with propidium iodide (PI) and annexin V-FITC. It identifies three regions of the cell populations based on the strength of staining by these two colors. Unstained cells are alive; therefore, they do not have leaky membranes (for PI staining) and do not express phosphatidylserine on their surface (for annexin V-FITC staining). Cells that just stain with annexin V-FITC are apoptotic because they have not yet gone through the process that leads to membrane permeabilization. Cells that stain with both PI and annexin VFITC are dead (necrotic).

11.6 FUTURE DIRECTIONS

Flow cytometry is a rapidly expanding field worldwide where an enormous increase in its capability can be expected over the coming years. As indicated in the beginning of this chapter, there has been renewed interest in flow cytometry from the point of view of research where a major impetus is derived from its applications to genomics and proteomics. Recent advances in solid-state lasers, microfluidics, microarray technology, micro-optics, and miniaturized detectors provide challenging technological opportunities for developing small and compact flow cytometers with enhanced capabilities to simultaneously monitor many more parameters than currently possible. It is refueling the expectation that perhaps a flow cytometer-on-a-chip is not such a distant dream. New applications of flow cytometry are already emerging and will continue to expand the dimensions of flow cytometry. An excellent review on emerging technology and future development in flow cytometry is by Stewart et al. (2002). Some of these future directions are listed here.

Research. A newly emerging field is single-molecule flow cytometry, which takes advantage of a variety of techniques that have been developed during recent years to detect individual fluorochrome molecules in solutions (Keller et al., 1996; Goodwin et al., 1996; Nie and Zare, 1997). Single molecule flow cytometry offers tremendous prospects for molecular biology. The technique can be used for DNA fragment sizing (Castro et al., 1993; Goodwin et al., 1993; Huang et al., 1996) and DNA sequencing (Ambrose et al., 1993; Goodwin et al., 1997). In this method, the concentration and flow is adjusted so that each molecule (or fluorescently labeled DNA fragment) flows through the illumination zone of a flow cytometer individually, one at a time, at the same rate, and experiences the same light intensity. The DNA sequencing information can readily be obtained by using fluorescence *in situ* hybridization (FISH) involving a pool of fluorescently labled oligonucleotide probes. This technique of hybridization using single-stranded oligonucleotides synthesized with a known base sequence has been discussed in Chapter 9, which describes biosensors for identifying the specific DNA and mRNA sequences in individual cells. A major limitation of the single-molecule flow cytometry is the background count, which can be significantly larger than that produced by single-molecule fluorescence. Various methods are being pursued to overcome this limitation. A promising new method utilizes single-molecule fluorescence detection by two-photon excitation (Mertz et al., 1995; VanOrden et al., 1999). Efficient two-photon excitation, provided by using ultra-short (femtosecond) pulses of high intensity but with low average power, facilitates single-molecule detection, because the excitation wavelength (in near IR) and emission wavelength (in the visible) are well separated. Recent reports of fluorochromes with a considerably enhanced two-photon (Bhawalkar et al., 1996) and three-photon (He et al., 2002) absorption to produce even population inversion and resulting stimulated emission provides further promise to single molecule flow cytometry. This also offers considerable opportunities for chemists to synthesize highly efficient multiphoton excitable fluorochromes for fluorescence *in situ* hybridization to be used in molecular flow cytometry.

Another intriguing prospect is to couple laser capture microdissection (LCM) with flow cytometry. In this approach, precise cell-type-specific microdissection using LCM yields pure specimen for nucleic acid analysis (DiFrancesco et al., 2000).

Another area of research activities is functional assays that utilize various activation markers to stimulate a cellular process and monitor the progress on a flow cytometer in real time. Using a pulse laser source and a vertical time-resolved scan over a length of the flow, one can monitor the dynamics. For detailed information, one can couple the time resolution with spectroscopic resolution, where the fluorescence signal obtained as a function of time (transit time in the flow) can be dispersed in a spectrograph and collected using an array detector.

A strategy, originally introduced by Liu et al. (1989), that is receiving more attention recently is that of multiplexing antibodies. It combines multiple antibodies having the same color fluorochrome to resolve multiple subsets of cells in a single tube. This strategy will benefit from the production of high affinity (strongly binding) antibodies and development of site-directed fluorochrome conjugation.

Another area of future development involves nanotechnology to produce 10- to 30-nm size highly efficient up-converting nanophores. These nanoparticles, when appropriately functionalized, can be used to target specific cells, can permeate through the membrane because of small size, and can be used to target a specific organelle where they can be detected by up-converted emission using excitation by a 970-nm laser. An important class of up-converting nanoparticles are those containing rare-earth ions (Chen et al., 1999; Kapoor et al., 2000). A major issue to address in this case is the long lifetime (in hundreds of microseconds) of the emitting rare-earth ions.

Technology. A major future direction of development in the technology of flow cytometry is in the area of miniaturization and use of robotics. Important development in the area of micro-lasers, detectors, micro-electro-mechanical systems (MEMS), dense wavelength division multiplexing (DWDM), and micro-optics is taking place, driven by their application to optical information processing. A monolithic integration of these components is already envisioned to produce a photonic chip. Significant progress has also been made in the area of microfluidics. These two developments, coupled together, can provide a fertile ground to produce a flow cytometer-on-a-chip, which is more versatile and offers expanded scope. Use of robotics to implement an automated system for specimen processing with increased number of probes will greatly enhance the capabilities of a flow cytometer in data acquisition, significantly reduce the specimen processing time, and permit operations with smaller volumes.

Another area of technical development is their use to detect and probe microbial activities. Although laboratory demonstration of detection of bacteria has already been successful, currently available commercial flow cytometers fall short of achieving this goal. Current advances in biomedical optics and lasers will provide designs to focus the beam to dimensions compatible with bacteria. It may also be able to immunophenotype specific organelles such as mitochondria, endoplasmic reticulum vesicles, golgi, lysosomes, and so on.

An approach being introduced to provide high-throughput analyte analysis involves suspension array technology (SAT). The SAT method utilizes microsphere (beads) capable of quantifying up to 100 analytes in a single well, and one 96 well microtiter plate can be processed in less than one hour. The beads are functionalized with antibodies or genomic probes that can covalently couple to them to capture any analyte. Multiplexing in a single well, can be used to further enhance resolution. This assay format has recently been used for genomic evaluation (Iannone et al., 2001; Cai et al., 2000).

Applications. New applications of flow cytometry are constantly emerging. The future will see a considerable use of flow cytometers as a reliable research and clinical instrument for diagnosis and for monitoring the progress of a treatment.

Other new applications will be in the area of water and food quality control. Development of flow cytometers to detect microbial species will open up this vast application for continuous monitoring of water quality to detect any contaminants. Agriculture industry can also utilize flow cytometry to detect infection to plants and to develop new resistant species.

11.7 COMMERCIAL FLOW CYTOMETRY

Table 11.8 lists commercial flow cytometry providers and some vendors providing supplies and reagents (Source: http://flowcyt.salk.edu/comflow.html). Their websites are also provided.

HIGHLIGHTS OF THE CHAPTER

- A flow cytometer is a device that measures physical and/or chemical characteristics of one cell or biological assembly at a time. A suspension of cells or biological assemblies flows in a single file passing the point of interrogation defined by a highly focused light beam.

- The light scattered from cells or fluorescent light emitted from the staining fluorochrome at the point of interrogation is used to detect and characterize the cells.

- Applications of flow cytometry are very diverse, covering both basic research and clinical research areas, with the ability to provide molecular profiling of a disease in an early stage of progression.

- The basic steps of flow cytometry are (i) fluorescence labeling of biological substances in cells, (ii) hydrodynamic focusing to produce a laminar flow of cells in a single file at the interrogation point, (iii) laser illumination and collection of optical response, and (iv) data acquisition and processing.

TABLE 11.8. Commercial Sources for Flow Cytometers

Aber Instruments	Aberystwyth, UK	Cell Analyzer
Advanced Cytometry Instrumentation Systems	Amherst, NY	Prototype flow cytometer
Applied Cytometry Systems	Sheffield, UK	WinFCM for BD cytometers, EXPO32 for Coulter cytometers, HPDisk LIF to DOS converter, OptiMATE HP to PC link
Bangs Laboratories	Fishers, IN	QuantumPlex beads, microspheres, FCSC products
Becton Dickinson Immunocytometry Systems	San Jose CA	Instruments, reagents, software, source book
Biødesign International	Saco, ME	Catalog, antibody reference
Biømeda	Foster City, CA	Flow cytometry, fluorescence reagents, flow reagents, Mil Bio reagents
Clontech	Palo Alto, CA	Products, living colors fluorescent proteins, living colors Brochure, green fluorescent Protein (GFP), GFP flow cytometry
CompuCyte	Cambridge, MA	Laser scamming cytometer
Beckman Coulter	Miami, FL	Instruments, reagents, software
Cytek	Fremont, CA	Time zero module, time window system, automated micro-sampler, cell washer, 201 sheath tank, aerosol containment, sample filters, tube cooler
Cytomation	Fort Collins, CO	Cytometers and high-speed sorters, software supplies
Cytometry Research LLC	San Diego, CA	Flow cytometry services
Dako	Multinational	Reagents
Duke Scientific	Palo Alto, CA	Microspheres
Exalpha	Boston, MA	Reagents, CD4 beads
FCSPress	Cambridge, UK	FCS Press software, FCS assistant software
Flow Cytometry Standards Corp.	Jan Juan, PR	Acquired by Bangs Labs
Immune Source Immunotech	Los Altos, CA	Reagents, antibodies, links Reagents
Jackson Immunoresearch	West Grove, PA	Reagents, methods: technical information
Luminex	Austin, TX	Microspheres, FlowMetrix

TABLE 11.8. Continued

Martek Biosciences	Columbia, MD	Fluorescent products
Molecular Probes	Eugene, OR	Reagents
Omega Optical	Brattleboro, VT	Optical products, flow cytometry filters, fluorescence filters
Partec	Münster, Germany	Cytometers
PharMingen	San Diego, CA	Reagents, protocols
Phoenix Flow Systems	San Diego, CA	Software, support, multi-cycle, WinFCM for Fac-scan, apoptosis reagents and software
Polysciences	Warrington, PA	Reagents, microspheres
QBiogene	Multinational	Reagents
R&D Systems	Minneapolis, MN	Reagents
Research Diagnostics Inc.	Flanders, NJ	Antibodies
Riese Enterprises	Grass Valley, CA	Biosure reagents, standards, sheath buffer
Seradyn	Indianapolis, IN	Microspheres, monoclonals
SoftFlow	Burnsville, MN	Free FCAPList software, free HPtoMac
Spherotech	Libertyville, IL	Microspheres
Tree star	San Carlos, CA	FlowJo software, links/sites
Vector Laboratories	Burlingame, CA	Reagents
Verity Software House	Topshame, ME	Windows & Mac software, including Winlist and Modfit

- Some flow cytometers also provide cell-sorting capability to collect cells with one or more specific characteristics using an electrical or mechanical deflection method.
- Hydrodynamic focusing of a cell suspension in an appropriate buffer solution is achieved by flowing it as a core liquid surrounded by an outer concentric flowing saline solution, called the *sheath liquid*. The sheath liquid exerts hydrodynamic pressure on the core liquid to focus it and produce a single flow of cells.
- Cells are illuminated with one or more laser beams designed to selectively excite particular fluorochromes used to stain specific substances expressed by the cells.
- The illumination optics focuses the laser beam(s) to an elliptical spot coincident with the flowing suspension to illuminate each cell, one at time, as it passes through the illumination zone.

- The optical response detected by the collection optic system consists of forward-scattered count (FSC), side-scattered count (SSC), and the fluorescence intensities of the various fluorochromes used to stain the substances expressed by different cells.

- FSC and SSC, together with the number of fluorescent wavelengths detected, define the total number of parameters collected by a flow cytometer.

- The FSC signal, collected at 7–22° from the incident laser beam, is often used to measure cell size because it is proportional to the cross-sectional area of the particle. Generally, the larger the size of the cell, the greater the FSC signal.

- The SSC signal, generally collected at 90° from the excitation beam, provides information on the granularity of the cell (differences in refractive index).

- The fluorescence signals at different wavelengths are also detected at 90° from the excitation beam. They are separated from SSC and from each other generally by using sets of dichroic mirrors and wavelength selective filters.

- Important steps in data acquisition and analysis include calibration of the instruments, spectral compensation, and gating.

- Calibration is performed by using fluorescently labeled polystyrene or latex microbeads available from a number of sources.

- Compensation is the process used to correct for overlap between fluorescence spectra of different fluorochromes (fluorescence crossover). Electronic or software compensation is used for this purpose.

- Gating is an analytical process that utilizes selection criteria of cells of a specific set of characteristics for further analyzes. An acquisition gate refers to data selection by the acquisition system based on cell characteristics specification. An analysis gate refers to selection from a complete, stored data file.

- Flow cytometry generates a large amount of data representing multiparameter detection. They are generally stored as a "list mode" file in which multiple parameters obtained for each cell are stored.

- The stored data are analyzed by software and displayed in a variety of formats which exhibit certain correlations of data. These correlations allows one to profile a certain cell subpopulation as a signature for certain diseases or to characterize certain cellular processes.

- A single-parameter histogram is a plot of each channel in one mode of display which represents the relative signal height (channel number) of one parameter (FSC, SSC, or a specific fluorescence) against the event count (number of events) registering this intensity level.

- Dot plots, also referred to as bitmaps or scattergrams, are bivariate plots in which each cell is placed as a dot on the plot according to the two intensity channels it registers for the selected two parameters represented on the x and y axes. Three-dimensional plots in which a z axis is present can also be displayed.

- Dot plots are useful for correlation of one parameter with others and can be used to distinguish different types of cells.
- Immunophenotyping, which refers to identification of cells using fluorochrome-conjugated antibodies as immunosensitive probes for antigen proteins expressed by cells, is an important application for cellular analysis.
- Measurements of the DNA content of cell is another major application of flow cytometry and is used to characterize disease or a tumor based on the cellular content analysis.
- Future directions of research and development are in the areas of genomics and proteomics for rapid analysis.
- Single-molecule flow cytometry is another emerging area to benefit molecular biology. Here the concentration and the flow is adjusted to detect individual molecules such as fluorescently labeled DNA fragments for DNA sequencing.
- Areas of future technology development are (i) miniaturization and the use of robotics, (ii) detection of microbes, and (iii) use of suspension array technology (SAT) that utilizes microspheres functionalized with antibodies or genomic probes for high-throughput analysis.
- Some new areas of applications are water and food quality control.

REFERENCES

Ambrose, W. P., Goodwin, P. M., Jett, J. H., Johnson, M. E., Martin, J. C., Marrone, B. L., Schecker, J. A., Wilkerson, C. W., Keller, R. A., Haces, A., Shih, P. J., and Harding, J. D., Application of Single-Molecule Detection to DNA-Sequencing and Sizing, *Ber. BunsenGes. Phys. Chem.* **97**, 1535–1542 (1993).

Bhawalkar, J. D., He, G. S., and Prasad, P. N., Nonlinear Multiphoton Processes in Organic and Polymeric Materials, *Rep. Prog. Phys.* **59**, 1041–1070 (1996).

Buican, T. N., An Interferometer for Spectral Analysis in Flow, *Cytometry Suppl.* **1**, 1–10 (1987).

Buican, T., Real-Time Fourier Transform Spectroscopy for Fluorescence Imaging and Flow Cytometry, *Proc. SPIE* **1205**, 126–133 (1990).

Cai, H., White, P. S., Torney, D., Deshpande, A., Wang, Z., Keller, R. A., Marrone, B., and Nolan, J. P., Flow Cytometry-Based Mini-sequencing: A New Platform for High-Throughput Single-Nucleotide Polymorphism Scoring, *Genomics* **66**(2), 135–143 (2000).

Castro, A., Fairdield, F. R., and Shera, E. B., Fluorescence Detection and Size Measurement of Single DNA-Molecules, *Anal. Chem.* **65**, 849–852 (1993).

Chen, Y., Kalas, R. M., and Faris, G. W., Spectroscopic Properties of Upconverting Phosphor Reporters, *Proc. SPIE* **3000**, 151–154 (1999).

Darzynkiewicz, Z., Juan, G., Li, X., Gorczyca, W., Murakami, T., and Traganos F., Cytometry in Cell Necrobiology: Analysis of Apoptosis and Accidental Cell Death (Necrosis), *Cytometry* **27**, 1–20 (1997).

DiFrancesco, L. M., Murthy S. K., Luider J., and Demetrick, D. J., Laser Capture Microdissection-Guided Fluorescence In Situ Hybridization and Flow Cytometric Cell Cycle Analysis of Purified Nuclei from Paraffin Sections, *Modern Pathology* **13**(6), 705–711 (2000).

Givan, A. L., *Flow Cytometry: First Principles*, 2nd edition, Wiley-Liss, New York, 2001.

Goodwin, P. M., Ambrose, W. P., and Keller, R. A., Single-Molecule Detection in Liquids by Laser-Induced Fluorescence, *Acc. Chem. Res.* **29**, 607–613 (1996).

Goodwin, P. M., Cai, H., Jett, J. H., Ishaug-Riley, S. L., Machara, N. P., Semin, D. J., Van Orden, A., and Keller, R. A., Application of Single Molecule Detection to DNA Sequencing, *Nucleosides Nucleotides* **16**, 543–550 (1997).

Goodwin, P. M., Johnson, M. E., Martin, J. C., Ambrose, W. P., Marrone, B. L., Jett, J. H., and Keller, R. A., Rapid Sizing of Individual Fluorescently Stained DNA Fragments by Flow Cytometry, *Nucleic Acids Res.* **21,** 803–806 (1993).

He, G. S., Markowicz, P. P., Lin, T.-C., and Prasad, P. N., Observation of Stimulated Emission by Direct Three-Photon Excitation, *Nature*, **415**, 767–770 (2002).

Huang, Z. P., Petty, J. T., O'Quinn, B., Longmire, J. L., Brown, N. C., Jett, J. H., and Keller, R. A., Large DNA Fragment Sizing by Flow Cytometry: Application to the Characterization of P1 Artificial Chromosome (PAC) Clones, *Nucleic Acids Res.* **24**, 4202–4209 (1996).

Iannone, M. A., Consler, T. G., Pearce, K. H., Stimmel, J. B., Parks, D. J., and Gray, J. G., Multiplexed Molecular Interactions of Nuclear Receptors Using Fluorescent Microspheres, *Cytometry* **44**(4), 326–337 (2001).

Kapoor, R., Friend, C., Biswas, A., and Prasad, P. N., High Efficient Infrared-to-Visible Energy Upconversion in $Er^{3+}:Y_2O_3$, *Optics Lett.* **25**, 338–340 (2000).

Keller, R. A., Ambrose, W. P., Goodwin, P. M., Jett, J. H., Martin, J. C., and Wu, M., Single Molecule Fluorescence Analysis in Solution, *Appl. Spectrosc.* **50**, 12A–32A (1996).

Liu, C. M., Muirhead, K. A., George, S. E., and Landay, A. L., Flow Cytometric Monitoring of HIV-Infected Patients: Simultaneous Enumeration of Five Lymphocyte Subsets, *Am. J. Clin. Pathol.* **92**, 721–728 (1989).

Loken, M. R., and Wells, D. A., Normal Antigen Expression in Heamtopoiesis, in C. C., Stewart, and J. K. A., Nicholson, eds., *Immunophenotyping*, Wiley-Liss, New York, 2000, pp. 133–160.

Mertz, J., Xu, C., and Webb, W. W., Single Molecule Detection by Two-Photon Excited Fluorescence, *Opt. Lett.* **20**, 2532–2534 (1995).

Nie, S., and Zare, R. N., Optical-Detection of Single Molecules, in *Annual Reviews of Biophysics and Biomolecular Structure*, Vol. 26, R. M. Stroud, ed., Annual Reviews, Palo Alto, CA, 1997.

Nunez, R., *Flow Cytometry for Research Scientists: Principles and Applications*, Horizon Press, 2001.

Ormerod, M. G., ed., *Flow Cytometry: A Practical Approach*, Oxford University Press, Oxford, 2000.

Ross, J. S., *DNA Ploidy and Cell Cycle Analysis in Pathology*, Igaku Shtoin, New York, 1996.

Shapiro, H. M., *Practical Flow Cytometry*, 3rd edition, Wiley-Liss, New York, 1995.

Stewart, C. C., Goolsby C., and Shackney, S. E., Emerging Technology and Future Developments in Flow Cytometry, in R. Riley, ed., *Hematology Oncology–Clinics of North America*, W. B. Saunders, Philadelphia, 2002, pp. 477–495.

Stewart, C. C., and Mayers, G. L., Kinetics of Antibody Binding to Cells, in C. C. Stewart, and J. K. A. Nicholson, eds., *Immunophenotyping*, Wiley-Liss, New York, 2000.

Stewart, C. C., and Nicholson, J. K. A., eds., *Immunophenotyping*, Wiley-Liss, New York, 2000.

VanOrden, A., Cai, H., Goodwin, P. M., and Keller, R. A., Efficient Detection of Single DNA Fragments in Flow Sample Streams by Two-Photon Fluorescence Excitation, *Anal. Chem.* **71**, 2108–2116 (1999).

Vesey, G., Narai, J., Ashbolt, N., and Veal, D., Detection of Specific Microorganisms in Environmental Samples Using Flow Cytometry, *Methods: Cell Biol.* **42**, 489–522 (1994).

Light-Activated Therapy: Photodynamic Therapy

An important area of biophotonics is use of light for therapy and treatment. This chapter and Chapter 13 provide examples of the use of light for therapy and medical procedures. Chapter 12 covers the area of light-activated therapy, specifically the use of light to activate a photosensitizer that eventually leads to the destruction of cancer or a diseased cell. This procedure is called *photodynamic therapy* (abbreviated as PDT) and constitutes a multidisciplinary area that has witnessed considerable growth in activities worldwide. Treatment of certain types of cancer using photodynamic therapy is already approved in the United States by the Food and Drug Administration as well as in other countries by equivalent agencies. Therefore, this chapter can be useful not only for researchers but also for clinicians and practicing oncologists.

The basic principles utilized in photodynamic therapy are introduced. The nature of the photosensitizers, also called PDT drugs, plays an important role in determining the conditions and effectiveness of PDT. The various types of photosensitizers are described. Another section is devoted to the various applications of PDT, which are very diverse.

A very active area of investigation is the understanding of the mechanism of photodynamic action. This topic is covered in Section 12.4. Section 12.5 provides information on various light sources along with some examples of required light dosage and modes of light delivery for PDT. A new area of interest is the use of nonlinear optical techniques such as two-photon photodynamic therapy that show promise for the treatment of deeper tumors. This topic is covered in Section 12.6.

The chapter concludes with a discussion of current research and future directions in Section 12.7. This discussion is subjective, reflecting the views of this author. Nonetheless, it clearly illustrates that opportunities are manyfold and multidisciplinary: for chemists, physicists, engineers, biologists, and practicing clinicians.

For further reading, suggested general references are:

Introduction to Biophotonics, by Paras N. Prasad.
ISBN: 0-471-28770-9   Copyright © 2003 John Wiley & Sons, Inc.

Henderson and Doughtery (1992): Covers basic principles and clinical applications of PDT

Fisher et al. (1996): Covers clinical and preclinical PDT

12.1 PHOTODYNAMIC THERAPY: BASIC PRINCIPLES

Photodynamic therapy (PDT) has emerged as a promising treatment of cancer and other diseases utilizing activation of an external chemical agent, called a photosensitizer or PDT drug, by light. This drug is administered either intravenously or topically to the malignant site as in the case of certain skin cancers. Then light of a specific wavelength, which can be absorbed by the PDT photosensitizer, is applied. The PDT drug absorbs this light, producing reactive oxygen species that can destroy the tumor. This type of process induced by a photosensitizer was discussed briefly in Chapter 6.

The key steps involved in photodynamic therapy are shown in Figure 12.1. They are:

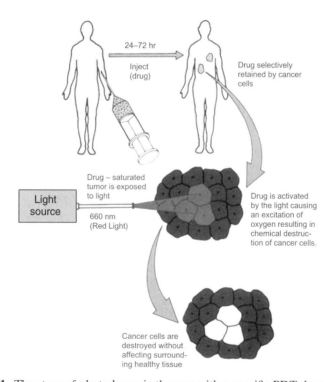

Figure 12.1. The steps of photodynamic therapy with a specific PDT drug.

- Administration of the PDT drug
- Selective longer retention of the PDT drug by the malignant tissue
- Delivery of light, generally laser light, to the malignant tissue site
- Light absorption by the PDT drug to produce highly reactive oxygen species that destroy cancer cells with minimal damage to surrounding healthy cells
- Clearing of the drug after PDT to reduce sunlight sensitivity

As indicated above, PDT relies on the greater affinity of the PDT drug for malignant cells. When a PDT drug is administered, both normal and malignant cells absorb the drug. However, after a certain waiting period ranging from hours to days, the concentration of the PDT drug in the normal cell is significantly reduced. Recent studies with tumor-targeting agents attached to the PDT drug have shown that their waiting period can be reduced to a matter of a few hours. In contrast, the malignant cells still retain this drug, thus producing a selective localization of this drug in the malignant tissue site. At this stage, light of an appropriate wavelength is applied to activate the PDT drug, which then leads to selective destruction of the malignant tissue by a photochemical mechanism (nonthermal, thus no significant local heating). In the case of cancer in an internal organ such as a lung, light is administered using a minimally invasive approach involving a flexible fiber-optic delivery endoscopic system. In the case of a superficial skin cancer, a direct illumination method can be used. Since coherence property of light is not required, any light source such as a lamp or a laser beam can be used. However, to achieve the desired power density at the required wavelength, a laser beam is often used as a convenient source for this treatment. The use of a laser beam also facilitates fiber-optic delivery.

The light activation process of a PDT drug is initiated by the absorption of light to produce an excited singlet state (S_1 or often written as $^1P^*$, where P* represents the excited photosensitizer), which then populates a long-lived triplet state T_1 (or $^3P^*$) by intersystem crossing. These terms and processes have been defined in Chapter 4. It is the long-lived triplet state that predominantly generates the reactive oxygen species. Two types of processes have been proposed to produce reactive species that oxidize the cellular components (hence, produce photooxidation) (Ochsner, 1997). These are described in Table 12.1.

A type I process generates reactive free radicals, peroxides, and superoxides by electron or hydrogen transfer reaction with water or with a biomolecule to produce a cytotoxic result. For the sake of simplicity, Table 12.1 only shows the generation of peroxides (H_2O_2) and the hydroxyl radical ($^{\bullet}OH$). In a type II process the excited triplet state of the photosensitizer reacts with the oxygen in the tissue and converts the oxygen molecule from the normal triplet state form to a highly reactive excited singlet-state form. It is the type II

TABLE 12.1. Mechanism of Photodynamic Photooxidation

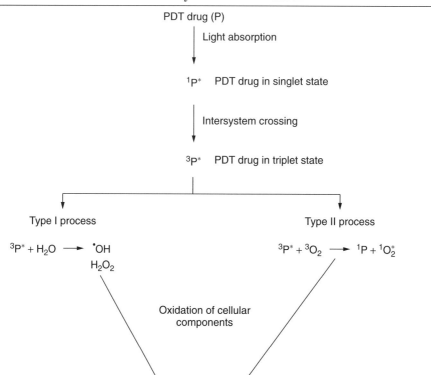

process that is generally accepted as the major pathway for photodynamic therapy—that is, the destruction of malignant cells.

The generation of singlet oxygen by an excited PDT photosensitizer can be detected spectroscopically or by chemical methods. The spectroscopic method utilizes the observation of phosphorescence emissions at ~1290 nm involving the transition from the excited singlet state of oxygen to its triplet ground state. Figure 12.2 shows the phosphorescence from singlet oxygen that is generated by a PDT drug, HPPH. This photosensitizer, discussed in Section 12.2, is being investigated at our Institute, in collaboration with the Roswell Park Cancer Institute (where it was developed).

The chemical method relies on the bleaching of absorption of a known singlet oxygen quencher such as 9,10-anthracenedipropionic acid (ADPA) (Bhawalkar et al., 1997). The absorption of ADPA at 400 nm is bleached (considerably reduced) by reaction with singlet oxygen.

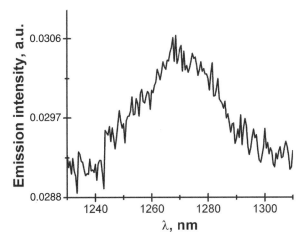

Figure 12.2. Phosphorescence from the lowest excited singlet state of oxygen generated by a photoexcited PDT drug, HPPH.

12.2 PHOTOSENSITIZERS FOR PHOTODYNAMIC THERAPY

A suitable choice of a PDT drug as a photosensitizer requires the following:

- The photosensitizer must have the ability to selectively accumulate in cancerous and precancerous tissues. In other words, while it is eliminated from normal tissue, it is retained in cancerous tissues and precancerous cells. Alternatively, the photosensitizer must target specific cancer cells.

- From the point of view of localization in tumors, the best sensitizers are those that are hydrophobic in order for them to penetrate cell membranes most readily. However, if they are to be administered intravenously, the sensitizers should be at least partially water-soluble and thus also hydrophilic to disperse in the bloodstream. Therefore, combining the two requirements, it is preferable to make the photosensitizers amphiphilic by chemically modifying a fundamentally hydrophobic PDT drug, by attaching polar residues such as amino acids, sugars, and nucleotides.

- The sensitizer should absorb significantly at a wavelength in the region of maximum transparency of biological tissues. This transparency region has been discussed in detail in Chapter 6. This choice will allow light to penetrate deeper in the tissue to activate a PDT drug, localized in malignant tissues which are deep, if the PDT drug absorbs at a long wavelength. However, wavelengths longer than 900 nm are energetically too low to provide the energy needed for excitation of triplet oxygen to its singlet state.

- The photosensitizer should exhibit minimum toxicity in the dark in order for light activation of the drug to produce maximum benefits without side effects derived from any inherent toxicity.
- The photosensitizer should have a high quantum yield of triplet-state formation and a long triplet lifetime. In other words, the nonradiative intersystem crossing from the excited singlet state of the photosensitizer to its excited triplet state should be efficient compared to the direct radiative transition (fluorescence) from the excited singlet. A longer triplet lifetime enhances the chance of producing a cytotoxic reagent or a cytotoxic reaction from this excited state.
- The photosensitizer should not aggregate since that can reduce the extinction coefficient and shorten the lifetime and quantum yield of the excited triplet state. Aggregated forms of the photosensitizer can also affect its pharmacokinetics and biodistribution.
- The photosensitizer should be rapidly excreted from the body. This will produce low systemic toxicity and will reduce sunlight sensitivity following PDT treatment.

12.2.1 Porphyrin Derivatives

The first group of photosensitizers used in clinical PDT were hematoporphyrin derivatives. The structure of hematoporphyrin is shown in Structure 12.1. Photofrin® (porfimer sodium), a PDT drug approved by the U.S. Food and Drug Administration as well as by other regulatory agencies throughout the world for the treatment of a variety of malignant tumors (see Figure 12.3), is obtained from hematoporphyrin by treatment with acids. Photofrin® actually is a complex mixture consisting of various derivatives as well as dimeric and

$\lambda_{abs}^{max} = 630$ nm

Structure 12.1. Structure of hematoporphyrin.

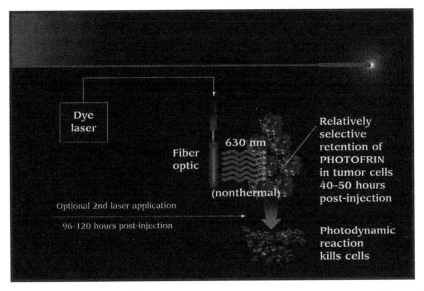

Figure 12.3. PDT with Photofrin® (courtsey of T. Mang). By lowering the dose of the drug, doctors can increase the laser intensity without damaging surrounding tissue.

oligomeric fractions. In commercial Photofrin®, the fractions are partly purified to be around 85% oligomeric material. Because Photofrin® is a complex mixture, there are still questions concerning the identity of the active components as well as the reproducibility of the synthetic process producing it. In clinical PDT, Photofrin® is excited with a red light at 630 nm (see Figure 12.3). At this wavelength, the penetration depth in the biological tissue is on the order of 1–2 mm. Thus while interstitial fibers can be used for thick cancers, Photofrin® is unsuitable for treatment of tumors extending more than about 4 mm from the source of illumination.

Photofrin® is a nontoxic drug; however, it is retained for some time by skin. For this reason, patients are required to avoid direct sunlight, very bright artificial lights, or strong residential indoor lighting for a period of 4–6 weeks after injection of the drug.

In order to prepare "second-generation" photosensitizers that consist of pure single components (as opposed to a mixture with Photofrin®) and that also absorb at a wavelength further in the red to provide deeper penetration in tissues, efforts have already led to many promising compounds. These include modified porphyrins, chlorins, bacteriochlorins, phthalocyanines, naphthocyanines, pheophorbides, and purpurins (Dougherty et al., 1998; Detty, 2001). Some of these are described here.

12.2.2 Chlorins and Bacteriochlorins

Chlorins are related to porphyrins because they are derived from porphyrins by hydrogenation of the *exo*-pyrrole double bonds of the porphyrin ring (Sharman et al., 1999; Sternberg et al., 1998). This derivatization produces a red-shifted intense absorption at wavelengths longer than 600 nm. An example is bonellin, a naturally occurring chlorin. Saturation of a second pyrrole double bond leads to bacteriochlorins, such as bacteriochlorin-a, which absorb at even longer wavelengths (750–800 nm). Their chemical structures are shown in Structure 12.2.

Mono-L-aspartyl chlorin e6

$\lambda_{abs}^{max} = 654$ nm

Bonellin

$\lambda_{abs}^{max} = 625$ nm

m-**Tetrahydroxyphenyl chlorin (mTHPC)**

$\lambda_{abs}^{max} = 652$ nm

Structure 12.2. Structures of three bacteriochlorins.

These PDT drugs are attractive because of their longer wavelength absorption. However, these classes of drugs undergo rearomatization of the pyrrole rings to produce porphyrins, which limits their lifetime *in vivo* as photosensitizers. None of these classes of materials are yet FDA approved for cancer treatment.

Efforts have also been made to produce chlorin derivatives that offer increased solubility. *Meta*-tetrahydroxyphenyl chlorin (*m*-THPC, see structure 12.2), also known as Foscan® or Temoporfin, is a promising photosensitizer that has a hydrophobic chlorin core and hydroxyphenyl groups at the meso position to increase its solubility. Clinical studies with *m*-THPC are in progress for the treatment of human mesothelioma as well as gynecological, respiratory, and head and neck cancers. *m*-THPC has been shown to be 200 times more effective than Photofrin®, resulting in a lower dose and shorter illumination times. Furthermore, it is a pure compound, in contrast to Photofrin®, which is a mixture, as described earlier. Also, *m*-THPC is excited at a longer wavelength of 652 nm (compared to photofrin at 630 nm), at which it has a molar extinction coefficient of $22,400 \, M^{-1} cm^{-1}$. Other advantages of *m*-THPC are its longer triplet lifetime, higher selectivity for tumors, and higher hydrophobicity than Photofrin® to provide increased cellular uptake. However, recent clinical trials have shown that some patients have experienced severe surface burns from sunlight post-treatment.

Because of its hydrophobicity, *m*-THPC is dissolved in polyethylene glycol 400 (PEG):ethanol:water (3:2:5 by volume) for clinical studies (Hornung et al., 1999). A water-soluble chlorin derivative is mono-L-aspartyl chlorin e6 (NPe6 or MACE, see structure 12.2), which has an extinction coefficient of $40,000 \, M^{-1} cm^{-1}$ at 654 nm. It has shown increased retention in tumors and efficient photodynamic effect with little skin photoxicity. Another chlorin is tin etiopurpurin (SnET2, see structure 12.5) with a molar extinction coefficient of $28,000 \, M^{-1} cm^{-1}$ at 660 nm.

A related group of compounds is pyropheophorbides, an example being the hexyl ether derivative, known as HPPH (see structure 12.5), which absorbs at ~670 nm (Pandey et al., 1996). HPPH is currently in Phase I clinical trials for the treatment of basal cell carcinoma.

12.2.3 Benzoporphyrin Derivatives

Benzoporphyrins are fused-ring chlorin derivatives produced by cycloaddition reactions (see Structure 12.3). The mono-acid benzoporphyrin derivative, called veteroporfin (also labeled BPD-MA), absorbs at 690 nm with a molar extinction coefficient of $35,000 \, M^{-1} cm^{-1}$ (Sternberg et al., 1998). Phase I and II clinical trials of this photosensitizer show rapid tumor accumulation and reduced skin photosensitivity. A benzoporphyrin derivative has been approved for the treatment of age-related macular degeneration (ARMD), where a rapid clearance is more desirable.

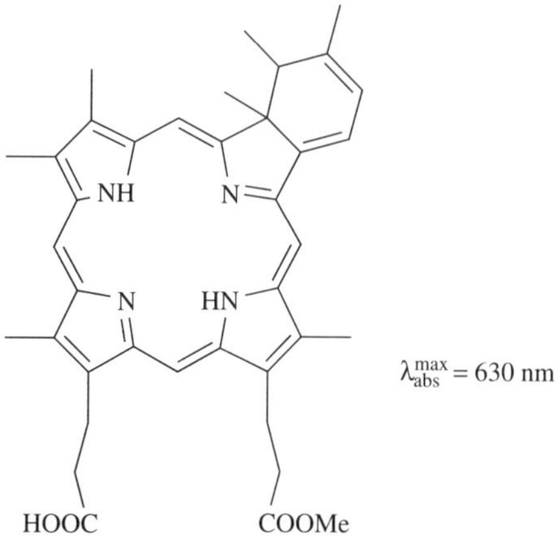

$\lambda_{abs}^{max} = 630$ nm

Structure 12.3. Benzoporphyrin derivative.

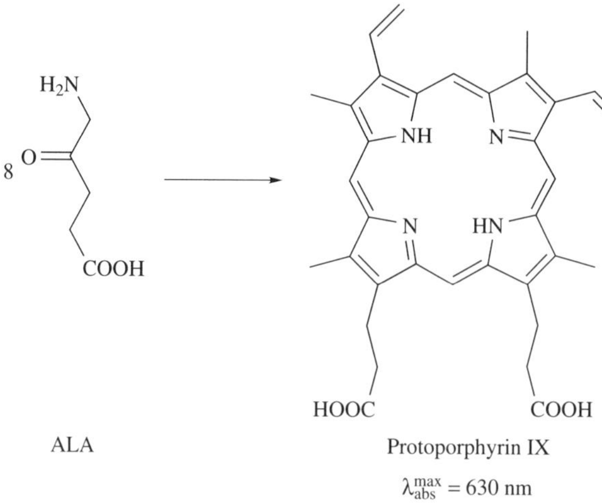

ALA

Protoporphyrin IX

$\lambda_{abs}^{max} = 630$ nm

Structure 12.4. The conversion of ALA to protoporphyrin.

12.2.4 5-Aminolaevulinic Acid (ALA)

ALA is a metabolic precursor in the biosynthesis of hemotoporphyrin, which generates endogenously an effective PDT sensitizer protoporphyrin IX, as shown in Structure 12.4. It thus provides an attractive alternative to the administration of an exogenous photosensitizer. Even though ALA can be endogenously generated from glycine and succinyl CoA, exogenous administration of ALA is chosen for a controlled buildup of protoporphyrin IX (PpIX). The advantages offered by ALA-induced PpIX over Photofrin® are:

- Ability to reach optimum therapeutic ratio in 4–6 hours
- Rapid systemic clearance of the photosensitizer within 24 hours, thus not only eliminating prolonged skin photosensitivity, but also allowing repeated treatment every 24 hours
- Accurate analysis of sensitizer levels by *in situ* monitoring of its fluorescence

PpIX has a molar extinction coefficient of only $<5000\,M^{-1}\,cm^{-1}$ at 635 nm. It undergoes photobleaching. ALA can be administered both systemically and topically. However, for local treatment of superficial skin lesions, it is often used in the form of a cream for topical application. Another limitation of ALA stems from its hydrophilic nature, which restricts its penetration through the keratinous lesion of normal skin. For this reason, lipophilic ALA esters may be preferable because they can penetrate cells more readily. ALA-induced PDT application has recently been accepted for the clinical treatment of actinic keratosis (Sharman et al., 1999).

12.2.5 Texaphyrins

Texaphyrins are related to porphyrins, except that they have five nitrogen atoms in the central core (Sessler and Miller, 2000). Lutetium texaphyrin (Lu-Tex) is currently been used in Phase II clinical trials for recurrent breast cancer (see Structure 12.5). A major advantage offered by Lu-Tex is its ability to be photoactivated at a much longer wavelength. It has a molar extinction coefficient of $42,000\,M^{-1}\,cm^{-1}$ at 732 nm. Lu-Tex shows minimal skin photosensitivity as it rapidly clears, providing only a narrow treatment window of 4–6 hours after injection. Lu-Tex and other derivatives are also being tested for ocular disorders such as ARMD.

12.2.6 Phthalocyanines and Naphthalocyanines

Phthalocyanines and naphthalocyanines are another class of promising PDT photosynthesizers that absorb in the long-wavelength region 670–780 nm and exhibit high molar extinction coefficients ($100,000\,M^{-1}\,cm^{-1}$) (Sharman et al., 1999). In phthalocyanine structures (as shown in Structure 12.6), the pyrrole groups are conjugated (fused) to the benzene rings and bridged by aza nitrogens.

The incorporation of a diamagnetic metal (M) such as Zn or Al in the center of the ring yields a longer-lived triplet state when compared to porphyrins, a necessary requirement for efficient photosensitization. The phthalocyanines and naphthalocyanines, being hydrophobic, exhibit limited solubility. The solubility can be enhanced by attaching sulfonic acid, carboxylic acid, or amino

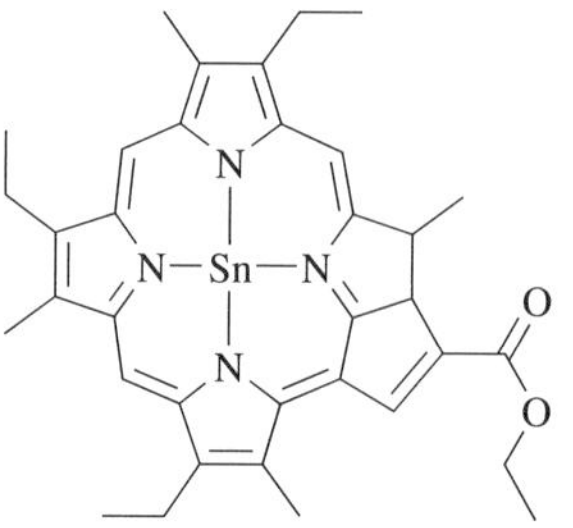

SnET2

$\lambda_{abs}^{max} = 660$ nm

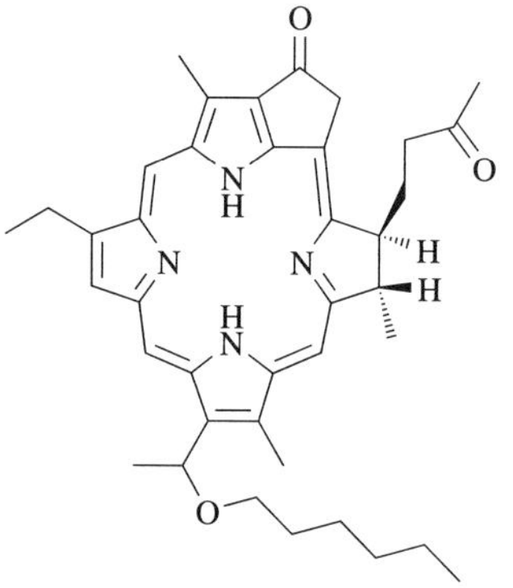

(1-hexyloxyalkyl)-pyropheophorbide-a
derivative (HPPH)

$\lambda_{abs}^{max} = \sim 665$ nm

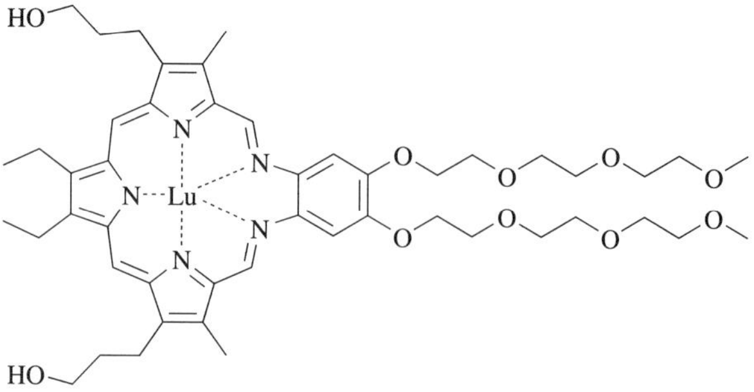

Lutelium texaphyrin

$\lambda_{abs}^{max} = 732$ nm

Structure 12.5. Structure of texaphyrins (Detty, 2001).

groups to the ring. A particular sulfonated compound, the chloroaluminum sulfonated phthalocyanine (AlPcS), has received attention recently because it also exhibits selective retention in some tumors. Clinical evaluation of AlPcS for PDT has been further motivated by its negligible dark toxicity, its minimal

$\lambda_{abs}^{max} = 674$ nm

Structure 12.6. Structure of phthalocyanines.

skin photosensitivity, and its ability to be photoactivated at a much longer wavelength. The phthalocyanines and naphthalocyanines are already in the early stages of preclinical and clinical evaluations (Colussi et al., 1999). A problem encountered with these compounds is their tendency to aggregate in aqueous media at relatively low concentration, resulting in a loss of their photoactivity.

12.2.7 Cationic Photosensitizers

This class of photosensitizers carries a positive charge on the heteroatom of the ring structure. Some representative dyes of this class are shown in Structure 12.7 (Detty, 2001). These cationic PDT photosensitizers tend to be bound intracellularly. Another distinction is that some of these dyes (e.g., rhodamine 123, abbreviated as Rh-123) are selectively taken up by the mitochondria of living cells (Johnson et al., 1981, see chapter 8). This appears to be responsible for the selective uptake of Rh-123 in cells enhancing its cytoplasmic concentration. However, the quantum yield for singlet oxygen generation by Rh-123 photosensitization is rather poor. Heavy atom derivatives of Rh-123 in which a core oxygen or nitrogen is replaced with a tellurium or selenium atom produce higher concentrations of triplet by increased intersystem crossing due to the heavy atom effect, as discussed in Chapter 4. This increased triplet yield can be expected to increase the efficiency of singlet oxygen generation and thus the efficiency of PDT action. Methylene blue is one cationic photosensitizer which is currently in clinical use.

12.2.8 Dendritic Photosensitizers

Dendrimers are highly branched structures using chemical units that provide multiple branching points (sites for chemical attachment). They are sequentially linked to build different layers (also called generation) of growth. These dendrimers or dendritic structures provide multiple sites for covalent linking

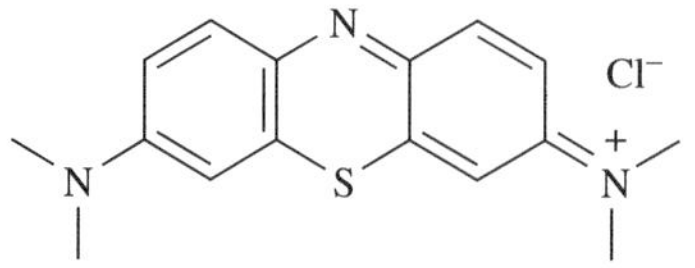

Methylene blue

$\lambda_{abs}^{max} = 660\,nm$

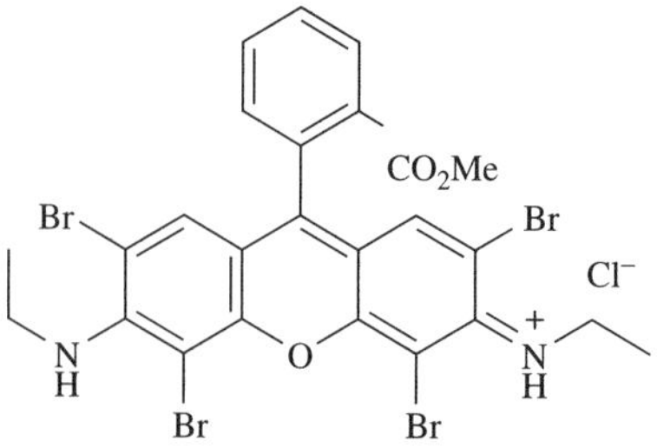

Tetrabromo rhodamine

$\lambda_{abs}^{max} = 540\,nm$

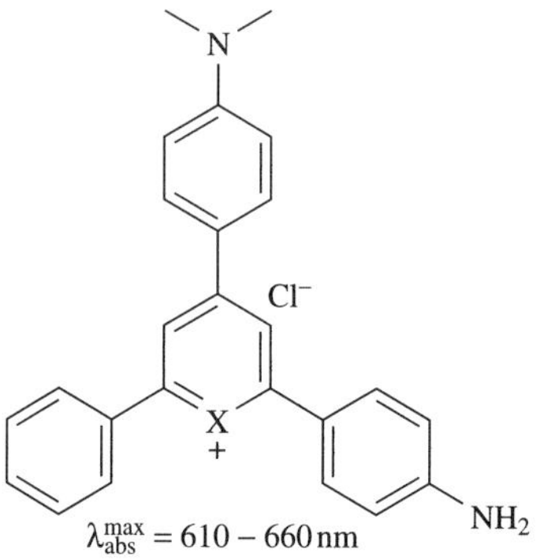

$\lambda_{abs}^{max} = 610 - 660\,nm$

Structure 12.7. Bottom: Structure of chalcogenopyrylium analogue, X = S, Se, Te.

of photosensitizers. Furthermore, they provide the opportunity to incorporate multiple photosensitizers by linking them to different arms of a dendrimer, which can then be delivered to a tumor site. The use of a dendrimer containing multiple photosensitizers permits different modes of actions and different wavelengths of activation for different photosensitizers. The development of PDT dendrimers as highly specific vehicles for targeted therapy and delivery of multiple photosensitizers is a very exciting, novel approach (Vogtel, 1998; Fisher and Vogtel, 1999). However, dendrimers can cause antibody responses to their introduction, which may preclude their distribution intraveneously. Their large size may also prove problematic in being able to penetrate cells.

Recent reports include dendrimers containing pheophorbide, a photosensitizer (Hackbarth et al., 2001). The structure of this type of dendrimer is shown in Structure 12.8.

Another report is of dendrimers containing 5-aminolevalinic acid (Battah et al., 2001). Even though these dendrimers contain only one type of photosensitizer, they serve as carriers with increased density of photosensitizers to increase their local concentration.

R= H or pheophorbide-*a* moiety

Pheophorbide-*a* (Pheo):

Structure 12.8. Pheophorbide-a-containing dendrimer (Hackbarth et al., 2001).

12.3 APPLICATIONS OF PHOTODYNAMIC THERAPY

Potential Cancer Therapy

- Microinvasive (early) endobronchial non-small-cell lung cancer
- Other endobronchial lung tumors
- Advanced, partially, or totally obstructing cancer of the esophagus

- Other lung tumors, including mesothelioma
- Early-stage esophageal cancer with Barrett's esophagus
- Skin cancers
- Breast cancer
- Brain tumors
- Colorectal tumors
- Gynecologic malignancies

The above applications *have not* yet been approved by the FDA. While PDT studies are currently being pursued for these *potential* applications, most have not yet completed their Phase II trials. Until PDT drugs complete Phase III, they cannot be used outside of an approved clinical trial, nor can they be marketed or distributed for medical use. There are only two currently approved drugs: Photofrin® and verteroporfin.

In all cases, Photofrin® has been used as a photosensitizer that is activated at 630 nm (see Figure 12.4). This PDT drug is administered by an intravenous injection. In the case of solar keratosis, a common premalignant skin lesion in light-skinned people, ALA is used as a photosensitizer that is administered as a topical cream. It is activated at 635 nm. HPPH (~665 nm) is utilized for a number of cancers, including breast cancer, because it appears to avoid the long-lasting photosensitivity of Photofrin®. Both ALA and HPPH are still in clinical trials and are not FDA-approved.

PDT for Other Diseases

- Cardiovascular (e.g., alternative to angioplasty)
- Chronic skin diseases [e.g., psoriasis (in development)]
- Autoimmune (e.g., rheumatoid arthriritis)
- Macular degeneration
- Antibacterial (wound healing, oral cavity)
- Antiviral (blood products, warts)
- Vaccines—especially anticancer vaccines
- Endometriosis
- *Precancerous conditions*: carcinoma in-situ and severe dysplasia in Barrett's; actinic keratoses (AK); cervical dysplasia; and so on

A phase I clinical trail at Stanford University is focusing on PDT treatment of arterosclerosis, which is the narrowing of the arteries caused by plaque accumulation. A PDT drug with a trademark ANTRIN®, when photoactivated, dissolves plaque in blood vessels with little or no damage to the surrounding

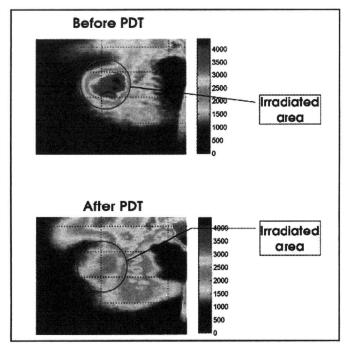

Figure 12.4. Fluorescence images showing the effectiveness of PDT treatment of a tumor (Reproduced with permission from Pifferi et al., 2000). Fluorescence images show a tumor on the back before and after irradiation with PDT light. Drug dose: 5 mg/kg b.w.; uptake time: 12 hr; excitation light: 660 nm.

healthy blood vessel walls. This new approach shows the promise to remove plaque over long segments of arteries, minimize damage to artery walls, and preclude artery re-closure.

Another nononcologic application of PDT that has attracted a great deal of attention is for the treatment of age-related macular degeneration, often abbreviated as AMD or ARMD (Schmidt-Erfurth and Hasan, 2000). ARMD is a degenerative eye disease that creates severe irreversible loss of vision among adults over 60 years of age. There are two types of ARMD: neovascular (also called wet) and non-neovascular (also called dry). In the neovascular (wet) form of ARMD, leaky blood vessels grow under the center of vision. The vascular ingrowth causes destruction of photoreceptors with visual distortion. In more advanced forms, bleeding with extensive loss of vision ensues. Until recently, laser photocoagulation has been used to destroy the vascular growth beneath the retina. However, a major drawback of this procedure is nonselective necrotic damage to the adjacent normal retina where the laser beam is applied, thus creating additional loss of vision. Furthermore, thermal damage in the subfocal area can stimulate recurrence of the neovascular tissue even after it is destroyed.

In photodynamic therapy for wet ARMD, a new photosensitizer verteporfin—a benzoporphyrin derivative—is injected, which accumulates in the leaky blood vessels in the eye. Fifteen minutes later a low-intensity laser beam of wavelength 680–695 nm is directed into the eye. This wavelength light can penetrate blood, melanin, and fibrotic tissue. The photodynamic action closes off the leaky blood vessels while minimizing damage to the adjacent tissues. This treatment is now preferred over laser photocoagulation (discussed in Chapter 13) for the majority of treatments for this disease. However, a disadvantage of PDT treatment for wet ARMD is that the closure can be temporary, resulting in the frequent need for more than one treatment. This treatment has been approved by the FDA for limited applications, has been approved for use in almost all countries (2001), and is currently being marketed worldwide by Novartis Ophthalmics under the trade name Visudyne®.

12.4   MECHANISM OF PHOTODYNAMIC ACTION

In order to improve the efficacy of photodynamic therapy with minimal side effects, it is important to understand the mechanism of photodynamic action at the cellular and tissue levels. This is a very active area of current research that will continue to attract attention. There are several excellent reviews that focus on the mechanism of photodynamic action, identifying the tissue/cellular sites for localization of photosensitizer and the nature of photodamage. They are Henderson and Dougherty (1992), Dougherty et al. (1998), Morgan and Oseroff (2001), and Schmidt-Erfurth and Hasan (2000).

Although the exact mechanism by which photodynamic therapy produces destruction of cells and tissues is still a subject of debate, three principal mechanisms have been suggested. These mechanisms are described in Table 12.2.

It is also thought that a combination of all three mechanisms may produce the best long-term response. However, the relative role of each mechanism may be determined by the characteristics of the photosensitizer, the nature of the tumor tissue and its microvasculation, the subcellular and tissue distribution of the photosensitizer, the type and duration of inflammatory and immune responses produced, and, finally, the treatment parameters used. It is also thought that each mechanism can influence the others.

Each of these principal mechanisms is discussed briefly in the following subsections.

Cellular Targeting by PDT. Cellular damage produced by photodynamic therapy is now believed to involve targeting of specific subcellular sites or

TABLE 12.2. Three Principal Mechanisms of Photodynamic Therapy

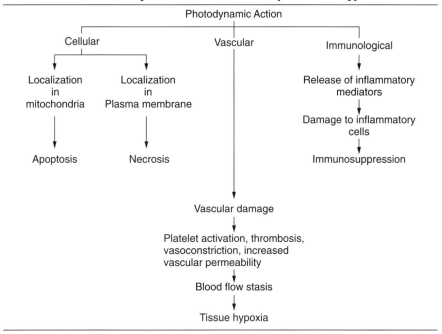

organelles by a particular photosensitizer. This type of cell damage is produced by the action of singlet oxygen generated by the photosensitizer. Since the singlet oxygen ($^1O_2$) has a very short lifetime (microseconds), the photodamage can be expected to be within a very short radius ($<0.02\,\mu$m) of the subcellular component that is targeted by the photosensitizer (i.e., the organelle in which the photosensitizer localizes because of its chemical affinity). The principal subcellular sites are mitochondria, plasma or internal membranes, and lysosomes.

Photosensitizers such as Photofrin® localize in mitochondria. 5-Aminolevalinic acid (ALA)-induced photoporphyrin IX is generated in mitochondria (Dougherty et al., 1998; Morgan and Oseroff, 2001). Photosensitizers that localize in mitochondria are now believed to cause photodamage by involving the process of apoptosis. As discussed in Chapter 3, apoptosis produces cell death by activation of a series of cellular enzymes that lead to fragmentation of nuclear DNA and disruption of the cell into membrane-bound particles that are eventually to be engulfed by nearby cells. Another class of photosensitizers that also targets mitochondria are the cationic photosensitizers. They accumulate in mitochondria along the membrane potential gradient.

Photosensitizers such as phthalocyanines localize in plasma membranes and are believed to cause necrosis. Necrosis has been discussed in Chapter 3. The photodamage produced by plasma membrane localization includes a variety of manifestations as outlined by Doughtery et al. (1998). Oxidative degradation of membrane lipids, caused by either a type I or type II process, as discussed in Section 12.1, can produce the loss of membrane integrity resulting in impairment of membrane transport, rupturing of membrane, and increased permeability. Another possible effect is the cross-linking of membrane-associated polypeptides, which may lead to inactivation of enzymes, receptors, and ion channels. The net result is necrosis by inhibition of energy production through glycolosis or oxidative phosphorylation.

Chlorin, benzoporphyrin, and phthalocyanine photosensitizers have been shown to cause damage to lysozymes, producing hydrolytic enzyme leakage. A major current emphasis is on the design of new PDT drugs and on conjugating a photosensitizer to carriers such as low-density lipoproteins, antitumor monoclinic antibodies (MAb), dextran, and so on, whereby a specific organelle such as the mitochondria or plasma membrane can be more effectively targeted (Konan et al., 2002).

Vascular Damage.   There are many experimental studies which indicate that vascular damage induced by PDT action plays a major role in the destruction of tumors. The Porphyrin®-induced PDT effect produces a rapid onset of vascular stasis, vascular hemorrhage, and both direct and hypoxia/anoxia-induced tumor cell death (Fisher et al., 1996). Vascular injury also contributes to cell death derived from oxygen and nutrient deprivation. The exact nature of vascular photodamage differs greatly from one photosensitizer to another. Also, the extent of vascular damage and blood flow stasis appears to be directly related to the level of circulating photosensitizer at the time of irradiation.

An important first step in vascular damage may involve the damage of endothelial cells causing a rearrangement of the cytoskeletal structure and leading to the shrinkage of endothelial cells away from each other. As a result, the vascular basement membranes are exposed, triggering platelet binding and aggregation at the sites of damage. The activated platelets subsequently release vasoactive mediators such as thromboxane, histamine, or tumor necrosis factor (abbreviated as TNF-α), triggering a multitude of events such as amplification of platelet activation, thrombosis, vasoconstriction, and increased permeability. These events subsequently produce blood flow stasis (stopping) culminating in tissue hypoxia (deficiency of oxygen) and eventual shutdown of the vasculature.

Immunological Response.   Another PDT-induced response is a strong inflammatory reaction leading to tumor destruction. These inflammatory processes release a wide variety of inflammatory mediators such as cytokines, which contribute to tumor destruction. The inflammatory response is often accompanied by immunosuppressive effects that aid in long-term tumor control.

The role of inflammation in the host's response to PDT seems to vary with different photosensitizers (e.g., Photofrin® versus HPPH, being of more importance in the former). The inflammation processes are involved in the concentration of the immune response in the area of need. Inflammation occurs in response to infection or tissue damage. As a result, collateral damage to cells by the inflammatory processes has an additive effect in tumor destruction. This collateral damage has a significant role in tumor destruction in PDT using Photofrin®, but plays less of a role in treatment with HPPH. Despite activation of many common components of inflammation (e.g., neutrophil and vascular adhesion molecules activation), the mode of cell death in PDT (necrosis, apoptosis, mixed response) may alter the role played by inflammation. In other models, the effects of mediators of inflammation [e.g., interleukin 6 (IL-6)] may either enhance, inhibit, or have no effect on the PDT response, depending on the model studied. IL-6 has a wide range of effects on many organ systems in regard to the initiation of inflammatory processes. PDT stimulation of the release of this cytokine would significantly enhance the role of inflammation in the destruction of the tumor and/or the immune response to specific tumor antigens.

12.5 LIGHT IRRADIATION FOR PHOTODYNAMIC THERAPY

12.5.1 Light Source

As discussed in Section 12.1, the PDT treatment using photoexcitation of the photosensitizer by linear absorption (as opposed to excitation by a nonlinear, two-photon absorption) does not require a high-peak power or a coherent light source. For this reason, incandescent filament lamps (tungsten) and arc lamps (xenon or mercury) were used in early clinical studies and continue to play a useful role. However, lasers are becoming more of a standard light source for most PDT studies and clinical applications (Fisher et al., 1996). The two practical advantages offered by a laser PDT are:

- The laser's ability to serve as a monochromatic source for selective and efficient excitation of a specific photosensitizer
- The efficiency and ease of coupling of a laser beam into fibers, making it ideal for insertion in flexible endoscopes and for interstitial use

For the excitation of PDT drugs photosensitized at 665 nm (such as Photofrin® and HPPH), a popular source has been a dye laser with rhodamine B as the lasing medium. A dye laser can be pumped by an argon-ion laser (a gas laser) or an intracavity KTP-doubled Nd:Vanadate laser (a solid-state laser), both producing a CW dye laser output in the range of 1–4 W. This is the typical power range requirement for most PDT applications. Pulse laser sources providing high repetition rates in the kilohertz range have also been used. These are gold vapor lasers, copper vapor laser-pumped dye lasers, and quasi-CW Q-switched Nd:YAG laser-pumped dye lasers.

Solid-state diode lasers are, perhaps, the choice of the future. These lasers already produce CW and quasi-CW powers in the range of 1–4 W with a single emitter source in the range of 780–850 nm. A diode bar containing an array of diode emitters can produce powers in excess of 100 W. Diode lasers are thus ready to serve as a suitable light source with new PDT drugs that can be activated at these longer wavelengths. However, the available diode lasers operating at 630 nm which meet the PDT power requirement are expensive. An additional consideration is that it is likely that future PDT applications will utilize near-infrared laser sources to treat subcutaneous cancerous tumors. These are areas of future development of laser technology for PDT applications.

Other laser sources for PDT applications at longer wavelengths are tunable solid-state lasers, such as (i) the Ti:sapphire laser (Ti:Al$_2$O$_3$), which covers the wavelength range of 690–1100 nm, and (ii) the Alexandrite lasers covering the range 720–800 nm. These solid-state lasers were discussed in Chapter 5.

Another new source for future applications may utilize optical frequency generation by optical parametric oscillation (OPO) and parametric amplification. This approach also provides a broad-band tunability and was discussed in Chapter 5.

A few studies have been reported that compare the efficacies of CW laser sources with those of pulse laser sources for PDT. It appears that both types of lasers can be used for PDT. As long as the peak power is not too high (as encountered with a short-pulse, high-intensity laser), as in the case for a high repetition rate quasi-CW laser source, the results obtained from this type of pulse laser and a CW laser source are biologically equivalent for the same irradiance and power densities.

12.5.2 Laser Dosimetry

The appropriate light dose for a specific PDT treatment is determined by the size, location, and type of tumor. The light dose requirements for Photofrin® using 630-nm excitation are as follows (Fisher et al., 1996; Dougherty et al., 1998):

- *Radiant Exposure*: 25–500 J/cm^2 for surface treatment; 100–400 J/cm^2 for interstitial applications
- *Maximum Irradiance*: 200 mW/cm^2 for surface treatment; 400 mw/cm^2 for interstitial applications
- *Typical Output Power*: 1–2 W

The power requirements are not expected to be less for the new generation of PDT photosensitizers, because the primary focus has been on increasing the wavelength of excitation to achieve deeper penetration.

12.5.3 Light Delivery

In determining an appropriate light delivery mode, some of the considerations are as follows (Fisher et al., 1996):

- Compatibility of the light source with other clinical instrumentations such as endoscopes and stereotactic devices
- Ability to continuously monitor light output and light dosimetry
- Ability to tailor spatial distribution of light to match the tumor shape and size in individual patients

In an optical fiber delivery system, a more uniform irradiation is obtained by fitting a microlens to the fiber for forward surface illumination (Doiron, 1991). For an interstitial laser irradiation needed for treating thicker lesions and tumors, the fiber can be directly inserted into the tumor mass, either by point insertion or inside a needle using a flat-cut fiber tip. Typically, the fiber will have a spherical or cylindrical diffusing tip. If more than one site needs to be irradiated, one can surgically implant translucent nylon catheters for subsequent laser treatments. For treatment of a tumor in an internal organ such as a lung, light is delivered through a flexible bronchoscope.

A major focus of current PDT studies from a light source perspective is the incorporation of light monitoring and dosimetry instruments into clinical delivery systems to gain information from each patient. This will also provide real-time information during the PDT procedure. Direct measurement of the PDT drug concentration, for example, can be obtained from quantitative fluorometry or reflectance spectrophotometry. Studies are also being conducted to provide methods for *in vivo* measurements of singlet oxygen production, which is generally accepted as a mechanism for destruction of a tumor. One such method utilizes the emission of singlet oxygen at 1270 nm (Gormann and Rodgers, 1992, see Figure 12.2). However, a severe limitation to this approach of detection is the relatively short lifetime (microseconds) of singlet oxygen.

12.6 TWO-PHOTON PHOTODYNAMIC THERAPY

As discussed in Section 12.2 on photosensitizers, currently most photosensitizers in clinical applications are being photoactivated using a light source in the range 630–690 nm. At this wavelength, the tissue penetration (defined by $1/e$ loss of intensity; see Chapter 6) is about 2–4 mm and the photodynamic effect is generally seen up to 2–3 times deeper than that. As a result, the largest attainable depth of PDT-induced cellular changes could reach up to 15 mm, but in most cases it is much less than half of that. For this reason, the increase of light penetration is considered to be an important factor in increasing the clinical efficacy of PDT. This is one of the focuses of current research. One approach is the design of new photosensitizers that absorb at longer wavelengths, as discussed in Section 12.2. Another approach is the use of two-photon absorption to photoactivate the photosensitizer. This two-photon process was discussed in Chapter 5. However, even two-photon PDT may prove ineffective beyond 1 to 2-cm penetration due to the large amount of scattering in some tissue types.

The spectral window for transmission through tissue lies around 800–1000 nm, which is in the near-infrared region. Such wavelengths may be used for excitation of the photosensitizer by using two-photon absorption. The idea of using two-photon excitation for PDT has been proposed by many investigators (Stiel et al., 1994; Lenz, 1994; Bhawalkar et al., 1997). However, the two-photon absorption cross-section of existing photosensitizers has been too small to be of practical significance until very recently (Karotki et al., 2001). The intensities required for direct two-photon excitation of these photosensitizers may cause damage to healthy tissues. Using efficient two-photon pumped upconverting dyes in conjunction with a PDT photosensitizer, at the Institute of Lasers, Photonics, and Biophotonics we have proposed a novel approach to PDT using infrared laser light for treatment (Bhawalkar et al., 1997). In this approach, an efficient two-photon absorbing dye is excited by short laser pulses. The dye molecule transfers the energy to the photosensitizer that is in proximity to it (or covalently bonded to it). The photosensitizer is thus excited to the singlet state from which the same sequence of energy transfer occurs as described earlier to produce the singlet oxygen.

The initial two-photon absorption of the dye molecules requires high-intensity IR laser pulses. These can be easily generated by ultra-short pulse lasers even with relatively low pulse energy. An example of such a laser is a typical mode-locked Ti:sapphire laser, which can produce 4-nJ pulses of about 70-fsec duration. This corresponds to a peak power of $5\,MW\,cm^{-2}$ in a 2-mm-diameter beam. The low pulse energy is highly desirable because it minimizes thermal side effects. An added advantage of using two-photon absorption arises from the quadratic dependence of the efficiency of such a process on the incident light. Therefore, the photodynamic effect is restricted to a small area around the focal point. Such spatial selectivity is important in many treatments such as PDT of brain cancers.

Preliminary studies were conducted at the Institute of Lasers, Photonics, and Biophotonics (Bhawalkar et al., 1997) as an initial assessment of the potential value of two-photon-induced PDT. To test the concept of the cascading energy transfer process in the photochemical generation of singlet oxygen, a new two-photon absorbing red dye, ASPS (*trans*-4-[*P*-(*N*-ethyl-*N*-hydroxyethylamino)styryl]-*N*-butansulfonpyridinum), and a well-known porphyrin photosensitizer (TPPS$_4$, obtained from Logan, UT) were selected. The fluorescence emission from three cuvettes irradiated with 1064-nm laser pulses from a Q-switched Nd:YAG laser was monitored in a fluorometer. One cuvette contained a solution of the dye alone, the second contained a solution of TPPS$_4$, and the third contained a mixture of both ASPS and TPPS$_4$ in solution. The dye solution showed a strong two-photon-induced fluorescence with a peak at around 610 nm while the TPPS$_4$ solution showed no detectable two-photon-induced fluorescence. The mixture of the two compounds showed, in addition to the characteristic emission spectrum of the dye, a new fluorescence peak at 653 nm which is the characteristic peak of TPPS$_4$. This is evidence of an energy transfer from the dye to the porphyrin. To further determine if the excited photosensitizer could generate singlet oxygen in the presence of atmospheric oxygen, a singlet oxygen-detecting compound was used. ADPA (9,10-anthracenedipropionic acid) is an excellent singlet oxygen quencher and a reaction with singlet oxygen leads to a bleaching of its 400-nm absorption band. This compound was added to the three cuvettes, and the solutions were exposed to IR pulses for several hours. Every hour, a sample from each cuvette was removed and its absorbance was measured. The cuvettes containing ADPA and the dye did not show any bleaching, nor did the ADPA and the porphyrin. However, in the cuvette containing the mixture, the absorbance at 400 nm was found to be steadily decreasing with each sample. This clearly indicated an increasing concentration of singlet oxygen during the exposure period. On repeating the observations with argon bubbled into the mixture, the bleaching was significantly lower.

A preliminary test of the feasibility of the two-photon process as an *in vivo* light source (at 500 nm) was performed in DBA (strain designation) mice with auxiliary SMT-F tumors, in collaboration with the PDT Center headed by Dr. Thomas J. Dougherty at the Roswell Park Cancer Institute in Buffalo. The treatment included APSS as the two-photon absorbing dye and Photofrin® as the photosensitizer. Immediately upon administering the mixture, the tumor area was illuminated with an unfocused train of 800-nm pulses from a mode-locked Ti:sapphire laser oscillator. The pulse duration was 90 fsec and the average power in the beam was 300 mW. The tumor was flat at 24 hours post-illumination, while the control group of Photofrin® plus light showed some hemorrhage and the light-alone control showed some edema (abnormal accumulation of serous fluid in the body). Additional unpublished studies at our Institute showed that APSS was nontoxic to mice.

Two-photon photodynamic therapy is currently an active area of both *in vitro* and *in vivo* studies, however, at the current time there have been no FDA-approved two-photon PDT protocols for cancer treatment.

12.7 CURRENT RESEARCH AND FUTURE DIRECTIONS

The field of photodynamic therapy is truly multidisciplinary, providing exciting opportunities for biomedical researchers, chemists, physicists, engineers, and practicing oncologists. Some areas of current activities offering prospects for future research are listed here. The selection of these areas is based partly on the author's personal views and partly on ongoing activities at the Photodynamic Therapy Center at Roswell Park Cancer Institute (courtesy of Dr. Janet Morgan). These selected areas are discussed in the following sections.

Molecular and Cellular Mechanisms of PDT. The fundamental nature of the photosensitizer structure and its photoactivity and the importance of its subcellular drug localization and photoaction are topics that are still not fully understood and are under intensive investigation. Various chemical, analytical, and spectroscopic probes are being used to understand the molecular and cellular mechanisms of PDT.

Effect of PDT on Cytokine Gene Expression and Immune Response. The subjects of intensive studies include (i) immune suppression after cutaneous PDT, (ii) immune potentiation after tumor PDT on other tumors, (iii) molecular mechanisms of regulation of some of the cytokines involved in potentiation, and (iv) different gene expression models, with different photosensitizers.

Tissue Oxygen Level Limitation. An important limitation of PDT utilizing photosensitizers that act by a type II process (Section 12.1) producing singlet oxygen is that the oxygen level is depleted both by consumption of singlet oxygen in a photoinduced chemical reaction and by vascular damage, leading to the shrinkage of its radius. This effect limits further therapy for producing direct tumor cell killing. This limitation is being addressed in a number of ways:

1. Adjusting the light fluence rate to slow oxygen consumption sufficiently so that the tumor tissue oxygen level can be maintained at the necessary level. A useful method has been the delivery of light in fractions, such as very short (20–50 sec) light and dark intervals, which allows reoxygenation during dark periods.
2. Providing PDT treatment in oxygen-enriched conditions (such as in a hyperbaric oxygen chamber)
3. Developing oxygen-independent photosensitizers that utilize free radicals (such as hydroxyl groups) as the active agent. However, these photosensitizers are not very efficient because one can only use each photosensitizer molecule once.

New Photosensitizers. Further acceptance of photodynamic therapy, increasing its efficacy, reducing side effects, and broadening the scope of its applications are crucially dependent upon the development of new photosensitizers. This provides unique opportunities for chemists. Some areas of opportunities are:

- One-photon PDT sensitizers that operate in the near IR ($\lambda > 800\,$nm)
- Efficient multiphoton-absorbing photosensitizers
- Dendrimers carrying multiple photosensitizers
- Targeting photosensitizers that carry an antibody, small proteins or peptides, sugars, and so on
- Oxygen-independent sensitizers
- Amphiphilic photosensitizers

The benefits of these types of photosensitizers have been discussed at various sections in this chapter. For example, it has recently been shown that porphyrins can be designed and synthesized with dramatically enhanced two-photon cross sections (up to two orders of magnitude enhancement) (Karotki et al., 2001). These new materials have also exhibited very efficient singlet oxygen production in *in vitro* studies.

Enhanced Transport of PDT Drugs. The more efficient transport of a photosensitizer into a tumor tissue can increase the efficacy of PDT treatment and shorten the waiting period. A highly active area of research is the use of various methods as well as chemical conjugation with various carrying units to enhance the transport of the sensitizer (Konan et al., 2002). For example, transdermal transport of amino levulinic acid, ALA, a PDT pro-drug for protoporphyrin IX, can be enhanced severalfold by electroporation as compared to topical application. Electroporation is a technique whereby pulsed electrical stimulation of the skin results in the opening of the interdermal spaces (spaces between the cells), allowing for more efficient transport of the sensitizer into the tissue. Another approach is to attach an imaging reagent conjugated to a small peptide that can bind to over-expressed receptor sites on the tumor.

Enhanced Drug Delivery to Tumors by Low-Dose PDT. Subcurative PDT for tumors can make the tumor vasculature highly permeable to large molecules. The subclinical dose disrupts the tumor vasculature as a result of cell destruction and/or activation of inflammatory processes. The result is increased permeability to large molecules, toxic drugs such as doxorubicin that are encapsulated and delivered locally after application of PDT (Henderson and Dougherty, 1992).

New Light Sources. In order for PDT to gain wide acceptance by the medical community, there is a need for lasers that are compact, low cost, user-friendly, and relatively maintenance-free. Furthermore, the need to activate more than one photosensitizer requires a multiwavelength laser source. New-generation diode lasers, other solid-state lasers, and optical parametric oscillators offer great opportunities for laser physicists and engineers. Looking futuristically, one can even think of implantable high-fluence diode light sources and low-fluence attachable device "patches" for long treatment.

Real-Time Monitoring of PDT.   There is a real need for further development of techniques that will allow real-time monitoring of the parameters that determine PDT action. Some of these parameters are photosensitizer tissue concentration, photobleaching rates, blood flow, and oxygen pressure in tissue (pO_2). These types of studies will provide insights into ways to enhance treatment effectiveness and selectivity.

HIGHLIGHTS OF THE CHAPTER

- Photodynamic therapy, abbreviated as PDT, utilizes light, often laser light, to destroy cancerous or diseased cells and tissues.
- Photodynamic therapy involves selective light absorption by an external chemical agent, called a *photosensitizer* or a *PDT drug.*
- The PDT drug, when administered either intravenously or topically, has the property of producing selective longer retention by the malignant (or diseased) tissue.
- The mechanisms of PDT action can involve either of two processes, often labeled as Type I and Type II. They both involve the formation of excited triplet state of the PDT drug by intersystem crossing from the excited singlet state generated by light absorption.
- In the Type I process, the excited triplet state of the PDT drug generates highly reactive radicals, peroxides, and superoxides by photochemistry, which then destroy the cancer cells by oxidation.
- In the Type II process, the PDT drug in its excited triplet state interacts with an oxygen molecule in its ground triplet state to produce a highly reactive excited singlet form of oxygen, which is a powerful oxidant that destroys the cancer cell. The Type II process is believed to be the major pathway for PDT.
- Treatment of certain types of cancers using PDT is already in clinical practice.
- Most of the PDT drugs are porphyrin derivatives. Photofrin®, a complex mixture of various porphyrin derivatives and containing dimeric and oligomeric fractions, is FDA-approved and being used for treatment of a variety of malignant tumors.
- Other PDT drugs being investigated are phthalocyanines, naphthocyanines, chlorins, and tetraphyrins that absorb at longer wavelengths (in the red), providing a better penetration in tissues to allow for treatment of deeper tumors. This is an area of intense research activity.

- Multibranched dendritic photosensitizers provide the opportunity to utilize different modes of actions and different wavelength of activation by simultaneously incorporating multiple sets of photosensitizers.
- Besides cancer treatment, PDT also is useful for the treatment of a number of diseases such as cardiovascular disease, psoriasis, rheumatoid arthritis, and age-related macular degeneration.
- The three principal mechanisms proposed for the destruction of cells and tissues by photodynemic therapy are (i) cell damage by targeting of a specific organelle by a particular photosensitizer, (ii) vascular damage induced by PDT action, and (iii) PDT-induced immunological response.
- Even though PDT does not require a coherent light source, a CW laser source provides a convenient source of light with concentrated energy at the wavelength of absorption of the PDT drug. Also, a laser source can readily be coupled with a wide variety of light delivery systems and endoscopic devices.
- Two-photon photodynamic therapy is a new area where the light activation of a PDT drug is achieved by two-photon absorption of near-IR photons using a short pulse laser source. This approach shows the promise of treating deeper tumors using greater tissue penetration by near IR light.
- Some areas of intense current research and future directions are (i) improving the understanding of molecular and cellular mechanisms of PDT and (ii) developing new photosensitizers, activatable by linear (one-photon) absorption in the near IR and those with the ability to be efficiently excited by two-photon absorption.
- Some other areas of future direction are (i) development of carriers conjugated to PDT drug for enhanced transport and efficient targeting of specific sites (or organelles) and (ii) development of *in vivo* techniques for real-time monitoring of PDT action.

REFERENCES

Battah, S. H., Chee, C.-E., Nakanishi, H., Gerscher, S., MacRobert, A. J., and Edwards, C., Synthesis and Biological Studies of 5-Aminolevulinic Acid-Containing Dendrimers for Photodynamic Therapy, *Bioconjugate Chem.* **12**, 980–988 (2001).

Bhawalkar, J. D., Kumar, N. D., Zhao, C.-F., and Prasad, P. N., Two-Photon Photodynamic Therapy, *J. Clin. Laser Med. Surg.* **15**, 201–204 (1997).

Bonnett, R., White, R. D., Winfield, V. J., and Berenbaum, M. C., Hydroporphyrins of the *meso*-tetra(hydroxyphenyl)porphyrin Series as Tumor Photosensitizers, *Biochem. J.* **261**, 277–280 (1989).

Colussi, V. C., Feyes, D. K., Mulivhill, J. W. et al., Phthalocyanine 4 (Pc4) Photodynamic Therapy of Human OVCAR-3 Tumor Xenografts, *Photochem. Photobiol.* **69**, 236–241 (1999).

Detty, M. R., Photosensitizers for the Photodynamic Therapy of Cancer and Other Diseases, *Expert Opin. Ther. Patents* **11**(12), 1849–1860 (2001).

Doiron, D. R., Instrumentation for Photodynamic Therapy, in A. N. Chester, S. Martelluci, and A. M. Scheggi, eds., *Laser Systems for Photobiology and Photomedicine*, NATO ASI Series, NY, Plenum (1991), 229–230.

Dougherty, T. J., Gomer, C. J., Henderson, B. W., et al., Photodynamic Therapy, *J. Natl. Cancer Inst.* **32**, 889–905 (1998).

Fisher, A. M. R., Murphree, A. L., and Gomer, C. J., Clinical and Preclinical Photodynamic Therapy, in C. A. Puliafito, ed., *Laser Surgery and Medicine*, Wiley-Liss, New York, 1996, pp. 339–368.

Fisher, M., and Vogtel, F., Dendrimers: From Design to Application—A Progress Report, *Angew. Chem. Int. Ed. Engl.* **38**, 884–905 (1999).

Gormann, A. A., and Rodgers, M. A. J., Current Perspectives of Singlet Oxygen Detection in Biological Environments, *J. Photochem. Photobiol. B:Biol.* **14**, 159–176 (1992).

Hackbarth, S., Horneffer, V., Wiehe, A., Hillenkamp, F., and Röder, B., Photophysical Properties of Pheophorbide-a-substituted Diaminobutane Poly-propylene-imine Dendrimer, *Chem. Physics* **269**, 339–346 (2001).

Henderson, B., and Dougherty, T., How Does Photodynamic Therapy Work?, *J. Photochem. Photobiol. B: Biol.* **55**, 145–157 (1992).

Hornung, R., Fehr, M. K., and Montifrayne, J., Highly Selective Targeting of Ovarian Cancer with the Photosensitizer PEG-*m*-THPC in a Rat Model, *Photochem. Photobiol.* **70**, 624–629 (1999).

Johnson, L. V., Walsh, M. L., Bochus, B. J., and Chen, L. B., Monitoring of Relative Mitochondrial Membrane Potential in Living Cells by Fluorescence Microscopy, *J. Cell. Biol.* **88**, 526–535 (1981).

Karotki, A., Kruk, M., Drobizhev, M., Rebane, A., Nickel, E., and Spangler, C. W., Efficient Singlet Oxygen Generation Upon Two-Photon Excitation of New Pophyrin with Enhanced Nonlinear Absorption, *IEEE J. Sel. Top. Quantum Electron.* **7**, 971–975 (2001).

Konan, Y. N., Gurny, R., and Allemann, E., State of the Art in the Delivery of Photosensitizers for Photodynamic Therapy, *J. Photochem. Photobiol. B: Biol.* **66**, 89–106 (2001).

Lenz, Z. P., Nonlinear Optical Effects in PDT, *J. Physi. IV* **C4**, 237–240 (1994).

Morgan, A. R., Garbo, G. M., Keck, R. W., and Selman, S. H., New Photosensitizers for Photodynamic Therapy: Combined Effect of Metallopyrin Derivatives and Light on Transplantable Bladder Tumors, *Cancer Res.* **48**, 194–198 (1988).

Morgan, J., and Oseroff, A. R., Mitochondria-Based Photodynamic Anti-cancer Therapy, *Adv. Drug Delivery Rev.* **49**, 71–86 (2001).

Ochsner, M., Photophysical and Photobiological Processes in the Photodynamic Therapy of Tumors, *J. Photochem. Photobiol. B: Biol.* **39**, 1–18 (1997).

Pandey, R. K., Sumlin, A. B., Constantine, S., Aouda, M., Potter, W. R., Henderson, B. W., Rodgers, M. A., and Dougherty, T. J., Alkyl Ether Analogs of Chlorophyll-a Derivatives, Part 1: Synthesis, Photophysical Properties and Photodynamic Efficacy, *Photochem. Photobiol.* **64**, 194–204 (1996).

Pifferi, A., Taroni, P., Torricelli, A., Valentini, G., Comelli, D., D'Andrea, C., Angelini, V., and Canti, G., Fluorescence Imaging During Photodynamic Therapy of Experimental Tumors in Mice Sensitized with Disulfonated Aluminum Phthalocyanine, *Photochem. Photobiol.* **72**, 690–695 (2000).

Schmidt-Erfurth, U., and Hasan, T., Mechanisms of Action of Photodynamic Therapy with Verteporfin for the Treatment of Age-Related Macular Degeneration, *Surv. Ophthalmol.* **45**, 195–214 (2000).

Sessler, J. L., and Miller, R. A., Texaphyrins—New Drugs with Diverse Clinical Applications in Radiation and Photodynamic Therapy, *Biochem. Pharmacol.* **59**, 733–739 (2000).

Sharman, W. M., Allen, C. M., and Van Lier, Jr., J. E., Photodynamic Therapeutics, *Drug Discovery Today* **4**, 507–517 (1999).

Sternberg, E. D., Dolphin, D., and Brockner, C., "Porphyrin-Based Photosensitizers for Use in Photodynamic Therapy," *Tetrahedron* **54**, 4151–4202 (1998).

Stiel, H., Teuchner, K., Paul, A., Freyer, W., and Leupold, D., Two-Photon Excitation of Alkyl-Substituted Magnesium Phthalocyanine: Radical Formation via Higher Excited States, *J. Photochem. Photobiol. A* **80**, 289–298 (1994).

Vogtel, F., ed., Dendrimers, in Topics in Current Chemistry, Vol. 197, Springer, Berlin, Germany, 1998.

Tissue Engineering with Light

Lasers have emerged as powerful tools for tissue engineering. Tissue engineering with light utilizes various types of light–tissue interactions discussed in Chapter 6. Consequently, some readers may find it helpful to revisit Chapter 6. Chapter 13 also has sufficient medical focus to be useful to medical practitioners as well.

This chapter covers three main types of laser-based tissue engineering: (i) tissue contouring and restructuring, (ii) tissue welding, and (iii) tissue regeneration. Two specific examples of tissue contouring and restructuring covered in this chapter are used in dermatology and ophthalmology. Dermatological applications discussed here are (i) the treatment of vascular malformations, such as port-wine stains, (ii) the removal of pigment lesions and tattoos, (iii) skin resurfacing (wrinkle removal), and (iv) hair removal. Appropriate lasers used for these applications are presented.

The ophthalmic applications covered are (i) repair of blockage, leaky blood vessels, or tears in the retina using photocoagulation, (ii) refractory surgery to reshape the cornea for vision correction using the procedures of photorefractive keratectomy (PRK), laser *in situ* keratomileusis (LASIK), and laser thermal keratoplasty (LTK), and (iii) photodisruptive cutting during posterior capsulotomy in post-cataract surgery. These procedures are defined, and there is a discussion of their respective underlying principles of laser–tissue interactions. Lasers commonly used for these procedures also are described.

The section on laser welding of tissues discusses how lasers are used to join or bond tissues. Also described are the three types of welding: (i) direct welding, (ii) laser soldering, and (iii) dye-enhanced laser soldering.

Laser tissue regeneration is a relatively new area; recent work suggests that laser treatment can effect tissue regeneration to repair tissue damage in an injury. Some results from studies in this area conducted at our Institute are presented.

A major impetus to the area of laser-based tissue engineering has been provided by developments in laser technology. Wide availability of ultra-short pulsed lasers (e.g., Ti:sapphire lasers discussed in Chapter 5) from a number

Introduction to Biophotonics, by Paras N. Prasad.
ISBN: 0-471-28770-9   Copyright © 2003 John Wiley & Sons, Inc.

of commercial sources has opened up new opportunity for more precise laser surgery with very little collateral damage. Hence, an emerging field is "femtolaser surgery," which employs femtosecond pulses to cut or ablate tissues.

This chapter concludes with a brief discussion of future directions. This section provides examples of the author's views on multidisciplinary opportunities that exist for future research and development in the area of tissue engineering with light.

The following references are suggested for further reading on the topics covered in this chapter:

- Puliafito, C. A., ed., *Laser Surgery and Medicine: Principles and Practice*, Wiley-Liss, New York, 1996.
- Alster, T. S., *Manual of Cutaneous Laser Technique*, Lippincott-Raven, Philadelphia, 1997.
- Goldman, M. P., and Fitzpatrick, R. D., *Cutaneous Laser Surgery*, Mosby, St. Louis, 1994.
- Reiss, S. M., Laser Tissue Welding: The Leap from the Lab to the Clinical Setting, *Biophotonics International*, **March**, 36–41 (2001).
- Talmor, M., et al., Laser–Tissue Welding, *Archives of Facial Plastic Surgery* **3**, 207–213 (2001).

13.1 TISSUE ENGINEERING AND LIGHT ACTIVATION

Tissue engineering is a field of bioengineering that recently has seen an immense amount of growth. It covers a broad spectrum including biocompatible artificial implants, tissue regeneration, tissue welding and soldering, and tissue restructuring and contouring. It is a multidisciplinary field that has resulted in the development of materials by chemists and material scientists, fabrication of engineering tools by engineers, determination of biocompatibility and reduced risk of dysfunction by biomedical research, and skills of implementation by surgeons. A vast number of approaches and procedures are being applied to tissue engineering.

The objective of this chapter, however, is significantly focused. It deals only with tissue engineering that utilizes light, which is generally produced by a laser. Lasers have emerged as promising tools for tissue engineering. The principles that drive these applications utilize various types of laser–tissue interactions discussed in Chapter 6 on photobiology. The scope of applications of lasers for tissue engineering is outlined in Table 13.1.

Lasers also are commonly used in general and other surgeries. Although these applications also can fall under the broad definition of tissue restructuring and tissue engineering, they will not be covered here. A good general reference covering many aspects of tissue engineering using lasers may be found in the book *Laser Surgery and Medicine: Principles and Practice*, edited

TABLE 13.1. Lasers Applied for Various Types of Tissue Engineering

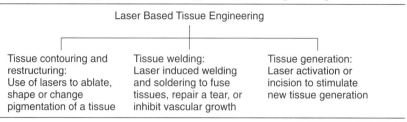

Tissue contouring and restructuring: Use of lasers to ablate, shape or change pigmentation of a tissue	Tissue welding: Laser induced welding and soldering to fuse tissues, repair a tear, or inhibit vascular growth	Tissue generation: Laser activation or incision to stimulate new tissue generation

by Puliafito (1996). The two main areas of laser activated tissue contouring and restructuring briefly discussed here deal with (i) dermatological applications in plastic and cosmetic surgeries and (ii) ophthalmic applications. These applications in current practice and represent a rapidly growing market (Alora and Anderson, 2000). They are covered in Section 13.2. The development of new, compact and cost-effective solid-state lasers, advancements in new protocols, and pre- and post-treatment regimens will lead to further demand of these laser-based plastic, cosmetic, and ophthalmic applications by both physicians and patients.

Another active area falling within the general scope of tissue restructuring is laser angioplasty (a cardiac procedure that dilates and unblocks atherosclerotic plaque from the walls of arterial vessels and often involves the placement of a mesh stent to prevent the vessels from closing again) (Deckelbaum, 1996). Fiber optics can be utilized to transmit laser radiation anywhere in the cardiovascular system accessible by an optical fiber. The laser is then used to vaporize obstructing atherosclerotic plaque (the thickening of arterial vessels with cholesterol buildup). Another approach is laser balloon angioplasty. With laser angioplasty, the laser beam heats the vessel wall during balloon angioplasty to improve the vessel remodeling induced by balloon dilation. Laser angioplasty may be particularly useful for treating chronic coronary artery occlusions and diffuse atheroscleric disease. The lasers used for this are a pulsed xenon chloride eximer laser operating in the UV at 308 nm or a pulsed holmium laser emitting in the infrared at 2.0–2.1 μm.

Some other applications of laser ablations include:

- *Otolaryngology*: A CO_2 laser is often used to create intense localized heating of the target tissue to vaporize both extra- and intracellular water, producing coagulative necrosis and soft tissue retraction or fusion.
- *Dentistry*: Lasers have been used for ablation of both soft and hard tissues. Soft tissue procedures have focused on incising and excising materials from the mucosa and gingiva in the oral cavity using a variety of lasers such as CO_2, Nd:Yag, Ho:Yag, and argon lasers. Er lasers with a wavelength in the 2.79- to 2.94-μm range have been used for cutting

dental tissues (drilling and preparation of cavities) as well as for removing dental materials.

Laser tissue welding is a developing biotechnology that looks promising for applications in practically all surgeries (Bass and Treat, 1996). Laser tissue welding utilizes the energy from the laser beam to join or bond tissues. The absorbed laser energy can produce alterations in the molecular structure of the tissues to induce bonding between neighboring tissue structures. Since the laser tissue-welding process is a noncontact and nonmechanical method, it is ideally suited for cases where suturing and stapling is difficult. The surgical requirements for tissue welding are to produce stronger welding strength while minimizing tissue thermal injury. To achieve these goals, current efforts are focused on developing new techniques using low laser energy and reduced energy absorption to produce localized thermal transmissions. The following approaches are being used (Xie et al., 2001):

- Use of a short pulse laser and thermal feedback to limit energy output
- Selection of laser wavelength to limit absorption in the tissue
- Application of solders and chromophores activated by lasers to increase bonding strength

However, clinical acceptability of laser tissue welding is limited by concerns about the stability of the weld strength (tensile strength, burst strength) and the difficulties in controlling the process. Laser tissue welding is covered in Section 13.3.

Recent studies at our Institute also show some promise in using lasers to promote the generation of new tissues in incisions. This topic is covered in Section 13.4.

Ultra-short pulse lasers promote nonthermal laser tissue interactions, primarily by the mechanism of photodisruption (discussed in Chapter 6), thus reducing the undesirable effect of collateral damage by a thermal mechanism. Interest in the use of these lasers has grown rapidly with the availability of femtosecond pulsed lasers, giving rise to a new field of femtolaser surgery. This topic is covered in Section 13.5.

13.2 LASER TISSUE CONTOURING AND RESTRUCTURING

The two specific applications discussed here are for the use of lasers in dermatologic and ophthalmologic procedures. The theory of selective photothermolysis, introduced by Anderson and Parrish in 1981, is the basis for much advancement in dermatological lasers (Anderson and Parrish, 1983). It allows for highly localized destruction of light absorbing "targets" in skin, with minimal damage to the surrounding tissue. To achieve selective photothermolysis, an appropriate wavelength, exposure duration, and sufficient fluence are necessary. Various targets absorb at different wavelengths, and the wavelength of the laser should be absorbed more by the target structure than by the surrounding structures. Light absorbed in the target structure is converted to heat, which begins to diffuse away immediately. In general, the exposure duration should be shorter than or about equal to the thermal relaxation time of the target. Clinically, selective photothermolysis involves ensuring that a maximum tissue-damaging temperature occurs only in the desired tissue targets. When treating dermal targets (blood vessels, tattoos, hair, etc.), light must pass through the epidermis. Epidermal injury is the most frequent side effect in these settings.

Some of the dermatological applications include:

1. *The Treatment of Vascular Malformations (e.g., Cutaneous Port-Wine Stains of Sturge–Weber Syndrome).* Here, the target chromophore is oxyhemoglobin. Laser light is absorbed by hemoglobin and is converted into heat, which damages the endothelium and the surrounding vessel wall. This is followed by thrombosis (a blockage of a blood vessel) and vasculitis (an inflammatory disease of the vessels). As the removal of the abnormal venules (or small veins that serve as collecting channels for adjacent capillaries) occurs, the lesion regresses into a more normally colored skin area.

2. *Removal of Pigmented Lesions and Tattoos.* In this case the target chromophore is melanin or tattoo pigment. Laser light causes extremely rapid heating of melanin or tattoo pigment granules. This fractures these submicrometer particles and kills the cells that contain them. Figure 13.1 illustrates the clinical results of tattoo removal using a laser beam.

3. *Resurfacing.* The target chromophore here is water. A superficial layer of skin is ablated in wrinkle removal. The laser deposits energy in the

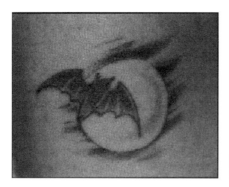 

Figure 13.1. Tattoo removal using laser technology. Four treatments with Q-switched frequency doubled Nd:YAG laser (532-nm green) removed the tattoo. (Reproduced with permission from Hogan, 2000.)

upper 1 μm (Er:YAG laser) or 20 μm (CO_2 laser) skin because of the strong absorption of energy by water. This typically leaves 0.05–1 mm of residual thermal damage, which also achieves hemostasis. Lasers have also been used effectively for ablation of warts, actinic cheilitis, and other benign epidermal lesions.

4. *Hair Removal.* The target chromophore is follicular melanin. Selective photothermolysis of the hair follicles is achieved without damaging the skin. It is unknown at present whether the bulge, dermal papilla (nonvascular core elevations of tissues associated with irritation or immunological challenge), or both have to be destroyed to achieve permanent hair removal. Also, it is currently debatable if the hair removal achieved is permanent.

Table 13.2 lists the dermatological applications of skin resurfacing (a more popular form being wrinkle removal), hair removal, and tattoo removal. The lasers and their parameters used for these procedures are also listed.

Ophthalmic applications of lasers are some of the oldest medical applications going back more than three decades. New laser applications and techniques are being implemented in an exciting fashion and cover a broad range of ophthalmic problems. Ophthalmic applications utilize a number of laser–tissue interaction mechanisms discussed in Chapter 6, where the structure and function of the human eye is also discussed. The ophthalmic applications that correct medical conditions fall into two categories:

1. *Use of Visible or Near-Visible Infrared Laser Wavelengths to Treat Retinal Disease or Glaucoma.* Examples are: (i) diabetic retinopathy associated with capillary nonperfusion or swelling caused by leaking microaneurysms, (ii) retinal vein occlusions that block ocular blood drainage causing retinal hemorrhage, ischemia, and swelling, (iii) age-related macular degeneration (discussed in Chapter 12, which discusses photodynamic therapy), which, in the wet-type neovascular tissue, invades normal retina, producing macular edema and hemorrhage, (iv) retinal tears, which can occur as a part of aging, as a complication following cataract surgery or from an eye injury (tears allow vitreous liquids to leak beneath the retina, lifting the retinal photoreceptors away from their vascular blood supply and supporting eye structures), and (v) glaucoma, which may be treated by producing a channel in iris structures or shrinkage of drainage tissues in order to facilitate lowering of eye pressure.

2. *Use of Nonvisible Wavelengths for Refractive Surgery to Reshape the Cornea for Vision Correction.* Lasers are now routinely used to correct for myopia (near-sightedness) with two techniques: photorefractive keratectomy (PRK) and laser *in situ* keratomileusis (LASIK). In these procedures, a pulsed laser beam flattens the cornea by removing more tissue from the center of the cornea than from its midzone. The result of

TABLE 13.2. Dermatological Applications of Lasers

Procedure	Skin Resurfacing		Hair removal				Tattoo Removal	
	CO₂ laser	Er:YAG laser	Alexandrite laser	Diode laser	Nd:YAG laser	Ruby laser	Q-switched frequency-doubled Nd:YAG laser	Q-switched alexandrite laser
Commonly used lasers								
Wavelength	$10.6\,\mu m$	$2.94\,\mu m$	$0.755\,\mu m$	$0.81\,\mu m$	$1.064\,\mu m$	$0.694\,\mu m$	$0.532\,\mu m$	$0.752\,\mu m$
Pulse duration	800 μsec	0.3–10 msec	2–20 msec	0.2–1 sec	10–50 msec	3 msec	10–80 nsec	50 nsec
Fluence (energy)	3.5–$6.5\,J/cm^2$ $(0.250$–$0.4\,J)$	5–$8\,J/cm^2$ $(1$–$1.5\,J)$	25–$40\,J/cm^2$	23–$115\,J/cm^2$	90–$187\,J/cm^2$	10–$60\,J/cm^2$	6–$10\,J/cm^2$	2.5–$6\,J/cm^2$
General references and websites	1, 4		2–4				4–6	

1. http://www.lasersurgery.com/laser_resurfacing_aging_and_scars.html.
2. Goldberg, D. J., Unwanted Hair: Evaluation and Treatment with Lasers and Light Source Technology, *Adv. Dermatol.* **14**, 223–248 (1999).
3. Goldberg, D. J., ed., *Laser Hair Removal*, Dunitz, London, 2000.
4. Alora, M. B. T., and Anderson, R. R., Recent Developments in Cutaneous Lasers, *Lasers Surg. Med.* **26**, 108–118 (2000).
5. http://www.bli.uci.edu/medical/laserskinsurfacing.html.
6. http://www.bmezine.com/tattoo/tr/iqp1.html.

flattening of the cornea is that the focus of the eye moves farther back toward its desired spot on the retina and corrects the vision for distance. As discussed below, PRK and LASIK use the same laser system and the same interaction mechanism to achieve the same goal. However, there is a major difference. In PRK, the epithelial (outer) layer of the cornea first is removed by a mechanical (soft brush) or chemical (alcohol) means or even by using a laser beam (transepithelial ablation). The laser beam then is used to ablate and reshape the cornea. A soft contact lens is used as a bandage and is placed over the eye to help the epithelial layer grow back. This generally takes 3–5 days. In LASIK, the ophthalmologist creates a hinged flap of the cornea approximately 125 μm in thickness using a specialized cutting blade mounted on a vacuum device. The cutting tool, known as a microkeratome, is then removed, thereby exposing the underlying corneal tissue to ultraviolet ablation of the desired degree. Finally, the corneal flap is returned to its original position. PRK and LASIK have also been used to a much lesser extent for hyperopia (far-sightedness). A new method, also now approved by the Food and Drug Administration in the United States, for the treatment of hyperopia and presbyopia (loss of near-focusing ability due to aging), which affect many people over 40 years of age, is laser thermal keratoplasty (LTK). With LTK, the laser is utilized to shrink the cornea, causing its central part to become steeper. Unlike PRK and LASIK, LTK does not involve ablation of any corneal tissue. It utilizes the application of concentric rings of laser energy to gently heat the cornea and steepen its curvature.

Other ophthalmic applications of the laser are for posterior capsulotomy in post-cataract surgery or cutting strands of vitreous in the posterior segment of the eye. In capsulotomy, a laser beam is used to open a hole in the membrane to correct for the opacity of the membrane, which may occur after cataract surgery. As stated above, a number of laser–tissue interaction mechanisms play a role in these treatments. Table 13.3 lists these mechanisms.

TABLE 13.3. Various Laser-Tissue Interaction Mechanisms for Ophthalmic Applications

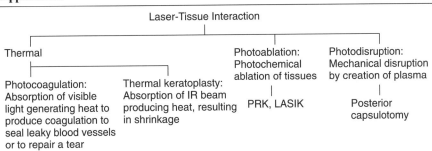

Table 13.4 provides the information on the types of lasers and their characteristics for some of these treatments.

13.3 LASER TISSUE WELDING

Laser tissue welding employs the process of using laser energy to join or bond tissues. Currently, the tissues welded by this technique are soft tissues. The approaches to join or bond tissues are listed in Table 13.5.

Laser tissue welding was first demonstrated by Jain and Gorisch (1979), who used Nd:YAG laser light to seal rat arteries. Subsequent studies suggested that laser interaction could be used to heat a tissue sufficiently to denature proteins (collagens) in the tissue surfaces to form new connecting structures (Jain, 1984; Schober et al., 1986). Most early studies of laser tissue welding employed CO_2 lasers. The use of the CO_2 laser relied on water, the largest constituent of most tissues, absorbing strongly at its wavelength ($10.6\,\mu m$). This strong absorption leads to a shorter optical penetration depth (~$13\,\mu m$), limiting its use to extremely thin tissues. Also, under a CW laser exposure, lateral spread of heating produces a large zone of injury.

Other lasers employed for laser tissue welding are argon-ion and Nd:YAG, which produce deeper and more uniform tissue heating than that achieved by using a CO_2 laser. In the case of the Nd:YAG laser, the 1.320-μm laser output has been used, because at this wavelength both water and hemoglobin absorb. Pulsed lasers have the appeal that they can minimize collateral thermal damage. However, the choice of laser wavelength and exposure parameters (energy, pulse duration, etc.) is clearly dependent on the tissue absorption, optical penetration depth, and the thermal relaxation time in the tissues to be welded. The optical penetration depth clearly has to be matched with the extent of the thickness to be welded to provide uniform heating.

The laser soldering technique utilizes laser light to fuse a proteineous solder to the tissue surface, thereby providing greater bond strength with less collateral damage compared to direct welding. Blood was the first material used as a solder. Subsequently, egg-white albumin followed by other proteins such as those derived from blood fibrinogen and other albumins were used a solder substitutes.

Dye-enhanced soldering was introduced to take advantage of the strong absorption of light by the selected dye and the efficient conversion of light into heat by the dye dispersed in the solder. This method also provided the benefit that an appropriate dye can be selected to match its absorption peak with the particular laser wavelength utilized. This method has allowed the ability to use the more common and relatively inexpensive 808-nm diode laser with the help of a biocompatible dye, indocyanine green (ICG) (Oz et al., 1990; Chivers, 2000). In yet another approach, a polymer scaffold doped with serum albumin and ICG was used (McNally et al., 2000). They found that the

TABLE 13.4. Ophthalmic Laser Procedures and Lasers Used with the Appropriate Specifications

Procedure	Laser Photocoagulation			Laser Thermal Keratoplasty (LTK)	Laser-Assisted In Situ Keratomileusis (LASIK)
Commonly used lasers	Argon ion laser	Krypton ion laser	Laser diode	Ho:YAG laser	ArF excimer laser
Wavelength	514.5 nm	647 nm	810 nm	2.1 μm	193 nm
Operation regime (pulse duration)	CW (0.1–1.0 sec)	CW (up to 10 sec)	CW (up to 2 sec)	Pulse (0.25–1 sec)	Pulse (15–25 nsec)
Power (energy)	0.05–0.2 W	0.3–0.5 W	2 W	20 mJ	50–250 mJ
General references and websites	1, 2			3–5	6–8

1. http://www.eyecenters.com/brochures/frames/laserfr.htm.
2. De Roo-Merritt, L., Lasers in Medicine: Treatment of Retinopathy of Prematurity, *Neonatal New.* **19**(1):21–26 (2000).
3. http://www.emedicine.com/oph/topic660.htm.
4. http://www.eyemdlink.com/EyeProcedure.asp?EyeProcedureID = 76.
5. Bower, K. S., Weichel, E. D., and Kim, T. I., Overview of Refractive Surgery, *Am. Fam. Physician* **64**:1183–90 (2001).
6. http://www.fda.gov/cdrh/LASIK/what.htm.
7. Roger, F. S., and Shamik, B., II., PRK and LASIK Are the Treatments of Choice, *Surv. Ophthalmol.* **43**(2):157–159 (1998).
8. http://www.lasik1.com.

TABLE 13.5. The Approaches for Tissue Bonding

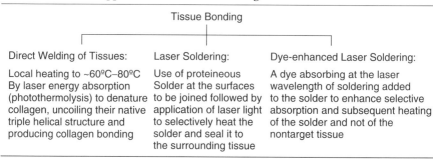

Direct Welding of Tissues:	Laser Soldering:	Dye-enhanced Laser Soldering:
Local heating to ~60ºC–80ºC By laser energy absorption (photothermolysis) to denature collagen, uncoiling their native triple helical structure and producing collagen bonding	Use of proteineous Solder at the surfaces to be joined followed by application of laser light to selectively heat the solder and seal it to the surrounding tissue	A dye absorbing at the laser wavelength of soldering added to the solder to enhance selective absorption and subsequent heating of the solder and not of the nontarget tissue

addition of the polymer membrane improved the weld strength and provided better flexibility compared to the use of albumin protein solder alone. The polymer scaffold makes the solder sufficiently flexible, allowing it to wrap around the tissue. Solders can be used for applications other than tissue bonding. Laser-assisted tissue sealing (LATS) can be used to seal bleeding surfaces for hemostasis (blood clotting). Anastomoses (sites where blood vessels have been rejoined surgically) that leak can be sealed and made impermeable.

Laser welding or soldering can be used endoscopically and laparoscopically to extend the range of its applications to cases where sutures or staples cannot be used. Other advantages are:

- Microsurgery
- Reduced inflammation
- Faster healing
- Watertight seal
- Ease and speed of application

Applications of laser welding and soldering have been diverse (Bass and Treat, 1996):

- *Cardiovascular Surgery*: Primary vascular anastomosis; sealing to reduce blood loss in vascular surgery
- *Thoracic Surgery*: Sealing of air leaks after lung biopsy or wedge resection; sealing of the bronchial stump
- *Dermatology*: Skin closure with improved cosmesis and faster healing
- *Gynecology*: Repair of fallopian tubes
- *Neurosurgery*: Welding and repair of peripheral nerves
- *Ophthalmology*: Laser solder closure of incisions in the sclera and cornea
- *Urology*: Closure of ureter, ureteroneocystostomy, urethra, and bladder. [Most urinary tract closures must be watertight to prevent leakage of

urine, reducing the subsequent development of infection or fistula (i.e., blind sac) formation.]

13.4 LASER TISSUE REGENERATION

Laser-induced tissue regeneration is an exciting prospect to repair tissue damage after an injury. Since the early report of low-level light therapy for wound healing (Mester et al., 1971), there have been numerous reports of effects of light on wound healing and tissue regeneration (Basford, 1996). Many investigators report visible and IR radiation at relatively low fluences (irradiation densities) of 1–4 J/cm^2 stimulates capillary growth and granulation of tissue formation (Basford, 1986). However, these reports have not gone uncontested. Variability of experimental models, fluences, wavelengths, and other parameters have compounded the problem and lead to seemingly contradictory results.

At our own Institute, studies have focused on the prospect of laser-induced tissue regeneration. The following hypotheses were used to explore the prospect of laser-induced tissue regeneration:

- Postoperative wound healing begins with blood clot formation.
- Blood clot directs scar tissue formation.
- There is an absence of blood clot formation after laser ablation.
- Absence of clot may allow for regeneration of native tissue.

The following method was used to study any tissue regeneration:

- Bilateral surgical defects (3 mm × 3 mm × 3 mm) were created in the gluteal muscles of hamsters (*Mesocricetus auratus*).
- Each subject received one laser wound and a contralateral scalpel wound.
- Subjects were injected with BrDU (800 mg/kg) throughout the postoperative phase.
- Subjects were sacrificed and the wounds harvested for both histological and immunohistochemical analysis.

The results obtained yielded the following observations:

17 DAYS AFTER SURGERY

Laser	Scalpel
Disorganized myotubules	Fibrous (scar) tissue
BrDU incorporated into myoblasts	No BrDU incorporation

The incorporation of BrDU clearly suggests the growth of fresh tissue, thus providing the exciting prospect of laser-induced tissue regeneration. Further

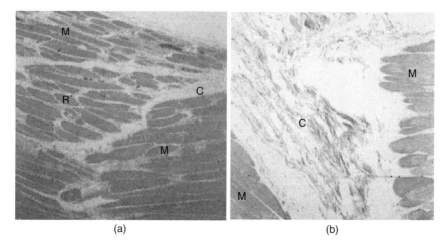

(a)	(b)

Figure 13.2. Muscle regeneration following laser excision of a 3 × 3 × 3-mm section of tissue; C, connective (scar) tissue; R, regenerated muscle tissue; and M, muscle tissue. (a) Regeneration of muscle tissue after laser excision; (b) scar formation after scalpel excision (Kingsbury, Liebow, Bergey, and Prasad, unpublished results).

studies are warranted to firmly establish tissue regeneration using this procedure. This conclusion was confirmed histologically, as is shown in Figure 13.2. After H&E (hematoxylin & eosin) staining, regenerated muscle cells are clearly evident in the laser excised region. The contralateral component of muscle, excised with a scalpel, shows only connective or scar tissue in the excised region.

13.5 FEMTOLASER SURGERY

An area of growing interest is the use of ultra-short pulsed (femtoseconds) lasers for surgery and tissue ablation (Juhasz et al., 2002). The advantages offered by these ultra-short pulses are that cuts or ablations can be made more precisely, with very little collateral damage. The mechanism of laser–tissue interactions that occur using ultra-short laser pulses is also different from the photothermal and photoablation mechanisms discussed above because they pertain to tissue contouring and welding. The high peak power of the ultra-short pulses lead to photodisruption, discussed in Chapter 6. The mechanism of photodisruption involves laser-induced optical breakdown (LIOB), in which a strongly focused short-duration pulse generates a high-intensity electric field and leads to efficient multiphoton ionization and subsequent avalanche ionization to produce a hot microplasma. This hot microplasma expands with supersonic velocity, displacing (ablating or cutting) the surrounding tissue. Since the displacement is adiabatic (i.e., it occurs on a time scale short compared to the local thermal diffusion time), the effect of abla-

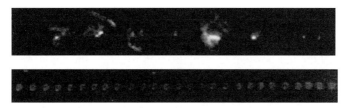

Figure 13.3. Laser tissue ablation using lasers of two different pulse widths. Top: pulse width of 200 ps; bottom: pulse width of 80 fs. (Reproduced with permission from http://www.eecs.umich.edu/CUOS/Medical/Photodisruption.html.)

tion or cutting is spatially confined and any spread due to thermal damage is also confined. The wide availability of mode-locked Ti:sapphire lasers producing ~100-fesec pulses at ~800 nm has provided much of the impetus for using them for femtolaser surgery or tissue ablation. Figure 13.3 shows the results of tissue ablation produced by two different sources: (a) a laser with 200-psec pulse width and (b) a laser with 80-fsec pulse width. These results are from the Center of Ultrafast Optical Science at the University of Michigan.

For these studies, the beams were focused to a circular spot and scanned across the sample. The spot separation is the same in both cases, but the ablation process takes place in an uncontrolled way in the case of the picosecond pulses, and the ablated domains are large. The femtosecond laser pulses, on the other hand, produce reproducible cuts that are spatially confined.

An advantage of using the photodisruption mechanism resulting from femtosecond pulses is that no specific absorbing target such as a pigment or a dye is required. Thus, a tissue that is totally transparent at the wavelength of the laser can be cut or ablated at any specific location in 3-D space. For this reason, a major application has been for refractive surgery involving the cornea. Four surgical procedures using laser surgery techniques approved by the Food and Drug Administration in the United States include (Juhasz et al., 2002):

- Corneal flap creation for LASIK
- Anterior lamellar corneal transplantation
- Keratomileusis
- Channel creation for corneal implants

A microkeratome has traditionally been used in LASIK for cutting a corneal flap to expose the internal corneal layers (stroma) for subsequent excimer laser ablation, as discussed in Section 13.2. However, flap creation with microkeratomes also produces a majority of the intraoperative and postoperative LASIK complications. Using femtosecond laser pulses, a flap is created by scanning a spiral pattern of laser pulses at the appropriate depth. This provides a greater control of precision and reliability as well as improved safety and performance. The process also provides a highly precise control of flap parameters such as flap thickness, diameter, hinge position and angle, and entry cut angle.

Anterior lamellar corneal transplantation involves the replacement of diseased or damaged superficial corneal tissue using tissue obtained from a cadaver donor eye. The femtolaser surgery allows the recipient and donor corneas to be cut with a high degree of accuracy, ensuring a proper fit of the donor corneal graft in the recipient bed. In addition, femtolasers provide local tissue sealing to improve stability and healing at the donor and recipient tissue interfaces.

Other ophthalmic applications of femtolaser surgery currently being investigated include treatment of glaucoma; preparation of corneal tissue for LASIK surgery precuts for the placement of implants for presbyopia; and photodisrutpion of the lens for cataract surgery. Nonophthalmic applications include dermatological and neurosurgical procedures.

13.6   FUTURE DIRECTIONS

Some examples of future directions of tissue engineering with light include:

Computer-Aided Tissue Engineering. Development of appropriate hardware and software to control the precision of tissue ablation or welding and surgery will build the confidence of both the doctor and the patient. The computer-aided systems will also provide a monitoring and feedback mechanism to achieve the desired result with minimal collateral damage. Introduction of robotics is a promising opportunity in this area. Laser safety issues may also be addressed using computer-aided systems and robotics, making these systems more user-friendly. Therefore, this area is definitely projected for a future growth opportunity (Sun and Lal, 2002).

New Laser Solders and Dyes to Assist Soldering. New biocompatible materials for tissue bonding will broaden the scope and applicability of tissue bonding. A major emphasis is to use light activation at the wavelengths and outputs provided by inexpensive diode lasers. For example, McNally et al. (1999) have reported the use of solid protein solder strips containing indocyanine green dye that strongly absorbs at the commonly available GaAlAs diode laser wavelength of ~800 nm for peripheral nerve repair.

Mechanism of Tissue Ablation and Welding. Even though there is a general consensus on the primary mechanisms of the procedures presented above, there appears to be suggestions that other molecular processes are occurring during laser–tissue interactions. Improved techniques to monitor molecular changes in real time will be of significant value and thus enhance the efficacy of a given treatment.

Femtolaser Technology. As is discussed in Section 13.5, femtolasers will emerge as a powerful surgical and tissue engineering tool. Femtolasers are still at a level of technology and require great care and maintenance from a highly skilled technician. Furthermore, they are expensive. The technology is, however, rapidly developing with a major motivation derived from the potential applications of a femtosecond fiber laser to telecommunications. These 1.55-μm femtosecond fiber lasers, when frequency doubled to produce ~777.5-nm output, will be suitable for the frequency range provided by the current mode-locked Ti:sapphire lasers. The development of applications in telecommunication may also bring down the price of a fiber-based femtolaser, while providing a convenient and flexible laser source for integration with other medical instruments at the same time.

HIGHLIGHTS OF THE CHAPTER

- Laser light provides a new dimension for tissue engineering, covering a broad spectrum of usage, such as (i) tissue contouring and restructuring, (ii) tissue welding and soldering, and (iii) tissue regeneration.
- Tissue contouring and restructuring utilize lasers to ablate or shape a tissue or change the pigmentation of tissue.
- One major application of tissue contouring and restructuring is in dermatology. Here, lasers are now routinely used for (i) treatment of vascular malformation, such as port-wine stain, (ii) removal of pigmented lesions and tattoos, (iii) skin resurfacing (wrinkle removal), and (iv) hair removal.
- Dermatological applications use the process of selective photothermolysis, which utilizes highly localized distribution of light absorbing "targets" in the skin, with minimal damage to the surrounding tissue.
- The second major application of tissue contouring and restructuring is in ophthalmology. Some principal examples are (i) repair of blockage, leaky blood vessels, or tears in the retina, (ii) refractive surgery to reshape the cornea for vision correction, and (iii) posterior capsulotomy in post-cataract surgery.
- Photocoagulation is used to repair blockage of leaky blood vessels or tears in the retina. This is accomplished by using the heat generated by light absorption to produce coagulation.
- Photorefractive keratectomy (PRK) used to correct near-sightedness involves (i) removing the outer layer of the cornea and (ii) ablating the cornea with a laser beam to appropriately reshape it. The process of photochemical ablation with ultra-short laser pulses in the UV is utilized.
- Laser *in situ* keratomileusis (LASIK) involves the opening of a flap of the top layer of the cornea, laser ablating the underlying tissue to reshape

the cornea, and returning the corneal flap to its original position. Again, the process of photochemical ablation with a UV laser is utilized.

- Laser thermal keratoplasty (LTK) is a new procedure that utilizes heat produced by the absorption of an IR laser beam to shrink the cornea in such a way as to cause steepening of the central part of the cornea. This type of restructuring corrects for the loss of near focusing due to aging.
- Another ophthalmic application is posterior capsulotomy in post-cataract surgery, where to correct opacity that may occur after cataract surgery a photodisruption mechanism is used to open a hole in the membrane that has formed on the implanted lens.
- The three approaches used for tissue welding are (i) direct welding of tissues, (ii) laser soldering, and (iii) dye-enhanced soldering.
- Direct welding of tissues uses lasers to locally heat tissue to a temperature that denatures collagen and forms a collagen bond.
- Laser soldering utilizes a proteineous solder at the surfaces to be joined. Laser light selectively heats the solder, sealing it to the surrounding tissue.
- The dye-enhanced soldering procedure adds a dye, with enhanced absorption at the laser wavelength used for soldering, to the solder to enhance selective heating at the soldering point.
- Laser tissue regeneration, an area in an early stage of development, deals with the prospect of using lasers to effect tissue regeneration for repairing tissue damage from an injury.
- Preliminary studies conducted at our Institute provide indications of tissue regeneration in a tissue operated on (cut) with laser surgery.
- Femtolaser surgery is a new field that utilizes femtosecond laser pulses to cut or ablate a tissue segment with great precision and with very little collateral damage.
- The field of tissue engineering with light offers potential for further research and development through the continued exploration of the underlying mechanisms of laser engineering, computer-aided tissue engineering and the search for new types of laser soldering materials.

REFERENCES

Alora, M. B. T., and Anderson, R. R., Recent Developments in Cutaneous Lasers, *Lasers Surg. Med.* **26**, 108–118 (2000).

Alster, T. S., *Manual of Cutaneous Laser Technique*, Lippincott-Raven, Philadelphia, 1997.

Anderson, R. R., and Parrish, J. A., Selective Photothermolysis: Precise Microsurgery by Selective Absorption of Pulsed Radiation, *Science* **220**, 524–527 (1983).

Basford, J. R., Low-Energy Laser Treatment of Pain and Wounds: Hype, Hope, or Hokum?, *Mayo Clinic. Proc.* **61**, 671–675 (1986).

Basford, J. R., Low Intensity Laser Therapy: Still Not an Established Clinical Tool, in C. A. Puliafito, ed., *Laser Surgery and Medicine: Principles and Practice*, Wiley-Liss, New York 1996, pp. 195–206.

Bass, L. S., and Treat, M. R., Laser Tissue Welding: A Comprehensive Review of Current and Future Clinical Applications, in C. A. Puliafito, ed., *Laser Surgery and Medicine. Principles and Practice*, Wiley-Liss, New York, 1996, pp. 381–415.

Chivers, R., *In Vitro* Tissue Welding Using Albumin Solder: Bond Strengths and Bonding Temperatures, *Int. J. Adhes. Adhes.* **20**, 179–187 (2000).

Deckelbaum, L. I., Cardiovascular Applications of Laser Technology, in C. Puliafito, ed., *Laser Surgery and Medicine*, Wiley-Liss, New York, 1996, pp. 1–27.

Goldman, M. P., and Fitzpatrick, R. D., *Cutaneous Laser Surgery*, Mosby, St. Louis, 1994.

Hogan, H., Technology is Being Fine-Tuned for Erasing Body Art, *Biophotonics Int.* **7**, 62–64 (2000).

Jain, K. K., and Gorisch, W., Repair of Small Blood Vessels with the Neodimium–YAG Laser: A Preliminary Report, *Surgery* **85**, 684–688 (1979).

Jain, K. K., Sutureless Extra-Intracranial Anastomoses by Laser, *Lancet* **8046**, 817–817 (1984).

Juhasz, T., Kuztz, R., Horvath, C., Suarez, C., Raksi, F., and Spooner, G., The Femtosecond Blade: Applications to Corneal Surgery, *Opt. Photonics News* **13**, 24–29 (2002).

McNally, K. M., Dawes J. M., Parker, A. E., Lauto, A, Piper, J. A., and Owen, E. R., Laser-Activated Solid Protein Solder for Nerve Repair: *In Vito* Studies of Tensile Strength and Solder/Tissue Temperature, *Laser Med. Sci.* **14**, 228–237 (1999).

McNally, K. M., Song, B. S., Hammer, D. X., Heintzelman, D. L., Hodges, D. E., and Welch, A. J., Improved Laser-Assisted Vascular Tissue Fusion Using Solder-Doped Polymer Membranes on a Cyanine Model, *Proc. SPIE* **3907**, 65–73 (2000).

Mester, E., Spiry, T., Szende, B., and Tota, J. G., Effect of Laser Rays on Wound Healing, *Am. J. Surg.* **122**, 532–535 (1971).

Oz, M. C., Chuck, R. S., Johnson, J. P., Parangi, S., Bass, L. S., Nowygrod, R., and Treat, M. R., Indocyanine Green Dye Enhanced Vascular Welding with the Near Infrared Diode Laser, *Vasc. Surg.* **24**, 564–570 (1990).

Puliafito, C. A., ed., *Laser Surgery and Medicine: Principles and Practice*, Wiley-Liss, New York, 1996.

Reiss, S. M., Laser Tissue Welding: The Leap from the Lab to the Clinical Setting, *Biophotonics Int.* **March**, 36–41 (2001).

Schober, R., Ulrich, F., Sander, T., Dürsclen, H., and Hessel, S., Laser Induced Alteration of Collagen Substructure Allows Microsurgical Tissue Welding, *Science* **232**, 1421–1422 (1986).

Sun, W., and Lal, P., Recent Development on Computer Aided Tissue Engineering: A Review, *Computer Methods and Programs in Biomedicine* **67**, 85–103 (2002).

Talmor, M., Bleustein, C. B., Poppas, D. P., Laser Tissue Welding: A Biotechnological Advance for the Future, *Archives of Facial Plastic Surgery*, **3**, 207–213 (2001).

Xie, H., Buckley, L., Prath, S., Schaffer, B., and Gregory, K., Thermal Damage Control of Dye-Assisted Laser Tissue Welding: Effects of Dye Concentration, *SPIE Proc.* **4244**, 189–192 (2001).

CHAPTER 14

Laser Tweezers and Laser Scissors

Lasers are useful tools for micromanipulation of biological specimens. This chapter covers two types of laser micromanipulations: laser tweezers for optical trapping and laser scissors for microdissection.

The principle of laser optical trapping using a laser beam has been explained using minimal amounts of theoretical discussion. Readers finding the concept still difficult to grasp may simply assume that submicron to micron size objects can be trapped in a focused laser beam spot, and then they can move on to appreciate the various biological applications of laser tweezers. These applications span a large number of areas.

This chapter also provides a detailed discussion of the design of a laser tweezer. Readers interested in building their own laser tweezers will find this section quite useful. Also presented are variations on laser tweezer techniques, such as using them as optical stretchers or as tools for the simultaneous, multiple trapping of many biological species.

Laser scissors function on the principle of photoablation, which is discussed in Chapter 6. They can be used to punch a hole in a cell membrane to allow the injection of a drug or genetic material. A more popular application is microdisscetion, used to excise a single cell or pure cell population from a tissue specimen. The two approaches used to capture the dissected portion— laser pressure catapulting (LPC) and laser capture microdissection (LCM)— are discussed in this chapter.

A vast, diverse number of applications in fundamental research cover the understanding of single biomolecule (DNA and protein) structure, function, and interactions (e.g., protein–protein interactions). Examples of these applications are provided. Selected practical applications of laser tweezers and scissors to genomics, proteomics, plant biology, and reproductive medicine are presented.

The chapter also includes a discussion of future directions of research and applications. A list of some commercial sources of these laser microtools is also given.

Introduction to Biophotonics, by Paras N. Prasad.
ISBN: 0-471-28770-9   Copyright © 2003 John Wiley & Sons, Inc.

A highly recommended book for further reading is by Greulich (1999). Some websites on laser micromanipulation are:

Harvard University: http://www.lightforce.harvard.edu/tweezer
UMEA, Sweden: http://www.phys.umu.se/laser/
Beckman Laser Institute: www.bli.uci.edu

14.1 NEW BIOLOGICAL TOOLS FOR MICROMANIPULATION BY LIGHT

Imagine the following:

- Grasping a biological cell, noninvasively, by using a focused laser beam, holding it in place, and moving or stretching it.
- Holding an egg by one laser beam and bringing a sperm trapped in another beam for fertilizing the egg.
- Drilling a microhole in a cell to inject molecules for manipulation and control of intracellular activities, without permanently damaging the plasma membrane which then seals within a fraction of a second.
- Performing microsurgery using a laser as a scalpel to cut a portion of the intracellular structure (an organelle or a DNA segment) and to modify the structure and function of a cell without affecting the cell viability.

It may have appeared as science fiction at one time. These types of micromanipulation are now routinely conducted in many laboratories around the world. Laser tweezers and laser scissors are two different micromanipulation tools that can be independently used or used in combination. As discussed in Chapter 2, light as photon particles carries momentum, a property that is utilized for the operation of laser tweezers. Light is also a carrier of energy as energy packets called *quanta*; it is the energy aspect of light that is used in laser scissor action. When an electromagnetic wave interacts with a small particle, it can exchange energy and momentum with the particle. The force exerted on the particle is equal to the momentum transferred per unit time. The force exerted by an optical tweezer is on the order of piconewtons (10^{-12} N). It is too weak to manipulate macroscopic-sized objects but is large enough to manipulate individual particles on a cellular level. The force is distributed over most of the area of the particle, so fragile and delicate objects can be manipulated without causing damage. Near-infrared laser beams can manipulate cells without damaging them, because cells do not absorb at these wavelengths.

Laser tweezers, also known as *optical tweezers* or *optical traps*, utilize the principle of trapping small particles/biological cells in the waist of a focused continuous-wave (CW) laser beam based on the gradient force derived from a change in the momentum of light. The wavelength of the laser beam (usually 1064 nm) is chosen from the region of optical transparency of the particle so that no exchange of energy (absorption of light) occurs. The particle/cell thus trapped in the optical beam can be moved around by moving the laser focal spot, hence the name optical tweezers as if the particle is picked up by a tweezer to manipulate its position. The development of optical tweezers is credited to the pioneering work of Ashkin et al. (Ashkin and Dziedzic, 1985; Ashkin et al., 1986). The first report of trapping and manipulating a living biological cell in a laser beam without harming it was also by Ashkin et al. (1987). Since then, optical tweezers have come a long way to be recognized as an important tool for biological micromanipulation. Current applications range from basic studies of biophysics and biochemistry at the single cell level to medical applications in blood cell analysis and *in vitro* fertilization (Greulich, 1999; Berns, 1998; Mehta et al., 1998; Strick et al., 2001). There is even a suggestion of the use of optical tweezers in early detection of cancer based on changes in viscoelastic properties of cells. Laser tweezers have provided much of the impetus for the study of single molecule biophysics, an area of considerable current interest.

A laser tweezer offers a number of benefits over a traditional mechanical micromanipulator. Some of these are:

- It does not involve any mechanical contact that can introduce a risk of contamination.
- It is a noninvasive method of manipulation that does not cause any damage to living cells. Thus a living cell can be optically trapped and manipulated without affecting its survivability.
- Subcellular organelles in a living cell can be manipulated (repositioned) without opening the membrane as required by other biological methods.
- Optical trapping has provided unprecedented capabilities to measure different forces in biology, down to the level of piconewtons, thus permitting one to correlate these forces with specific biological functions.
- Ability to use laser tweezers to mechanically unzip DNA can provide important applications to genomics by speeding up the sequencing of nucleotides.

An extension of laser tweezers or optical tweezers is an optical stretcher that utilizes placing of an object (e.g., biological cell) between two opposed, nonfocused laser beams to produce stretching of the cell along the axis of the beam (Guck et al., 2001).

The history of laser scissors is even older. Berns and Round (1970) showed that lasers can be used to microdissect cells. Other terms used for laser scissors are laser microscapel, laser microbeam, and laser microdissection unit. Laser scissors are convenient microtools for performing microscopic surgery on tissue specimen, cells, and molecules.

In contrast to laser tweezers with focused IR continuous wave laser beam, laser scissors utilize short pulses of high irradiance at a wavelength at which a tissue specimen or its cellular component absorbs. Typically, it can be a nanosecond, subnanosecond, or even a femtosecond solid-state laser with the output in a visible or UV spectral range. Often it is a UV laser source such as a nitrogen laser (337 nm). The absorbed energy produces the scissor action by photoablation to conduct delicate microsurgery on a tissue, a cell, or its organelle. In a more general sense, the laser scissor action has been used to include a broad range of action from pricking a hole in a cell, to ablating a portion of it. As explained in Chapter 6, the photoablation process involves a photochemical process of breaking of chemical bonds without generating heat. However, photodisruption involving a mechanical disruption produced by microplasma-induced shockwaves has also been used for microdissection. Therefore, the use of the term *laser scissors*, which literally implies a cutting action, may be confusing. This is why alternate terms such as laser microbeams, laser microdissection, and optoinjection are also used to represent different laser functions.

Laser scissors provide precision and selectivity compared to an invasive mechanical device. Compared to a regular scalpel, laser scissors provide the ability to act on dimensions as small as 0.25 μm in diameter. It can, therefore, be used to produce changes in a chromosome by cutting a portion of it while it is still deep within a living cell.

Laser scissors can be used to cut a micron-size hole in a membrane that seals within a fraction of a second. Exogenous species can be inserted in a cell through these holes without permanently damaging the membranes. This feature provides a convenient approach for genetic manipulation of cells.

The term *laser microdissection* is often used to refer to excise a portion of a tissue specimen to obtain clean (uncontaminated) tissue samples, or to separate tumor cells from precancerous neoplasm and supporting stroma.

Another term is *optoinjection*, which refers to a process in which a pulse laser beam pricks a hole in a cell to load it with exogenous molecules, without any visible damage to the cell and with high survival rates (Tsukakoshi et al., 1984; Krasieva et al., 1998; Rink et al., 1996). Yet another term used is *optoporation*, which implies pore production through optical means (Berns, 1998; Krasieva et al., 1998; Lee et al., 1997; Soughayer et al., 2000). This process refers to laser-induced transient permeabilization of a membrane to again allow entrance of exogenous species into selected cells.

The research groups of Gruelich and Berns pioneered applications involving the combined powers of laser tweezers and laser scissors. They utilized an Nd:YAG laser tweezer to bring two human myeloma cells close together, then used a pulsed UV nitrogen laser scissor to cut the adjoining membranes to fuse the two cells (Gruerlich, 1999; Berns, 1998). The two cells merged into a single hybrid cell containing the genomes of both. Greulich's group utilized a UV laser scissor and an optical tweezer combination for manipulation of pieces of chromosomes for gene isolation (Seeger et al., 1991).

Over the past decade the usage of laser tweezers and laser microbeams (scissors) have expanded considerably at a rapid pace. It is expected that new developments in laser micromanipulation as well as new applications will continue to emerge. The range of applications, already demonstrated, covers both fundamental research at single cell and subcellular level and in biotechnology. Some of these are listed here (*source*: www.PALM-microlaser.com). Selected examples of applications will be presented in somewhat more detail in a later section.

AREAS OF APPLICATIONS OF LASER MICROMANIPULATION

Biology

Microsurgery	Basic studies in cell biology
Cell fusion	Plant breeding
Force measurements	Food engineering
Cell sorting	Patch-clamp studies
Bacteria separation	Cloning

Genetics

Fetal cell capture	Cytogenetic analysis
Chromosome preparation, microinjection	Genetic engineering
Prenatal diagnosis	Gene therapy

Neuroscience

Single-neuron capture	Analysis of neuronal disorders
Microinjection	Study of nerve stimulation processes
Expanding artificial dendrites	Patch-clamp studies
Microsurgery	

Molecular Medicine

Single-cell capture	Diagnosis of diseases
Microinjection	Gene therapy
Cell fusion	Immunology
Analysis of cancer	

Pharmacy

Preparation of pure samples	DNA array
Trapping of cells	Chip technology
Laser microinjection	Gene therapy
Capture of living cells	Genetic engineering
Drug screening and design	Tumor banking

Biotechnology	
Separation of single cells, yeast, bacteria	Genetic engineering
Gene analysis	Cloning studies

Reproductive Medicine	
Laser zona drilling	*In vitro* fertilization
Sperm trapping	Embryo hatching
Polarbody extrusion	Preimplantation diagnosis
Blastomere biopsy	Embryo development
Blastomere fusion	

Forensic	
Selective isolation of suspect material	Fingerprinting Analysis
DNA isolation	Genetic database of suspects

14.2 PRINCIPLE OF LASER TWEEZER ACTION

Laser tweezers utilize trapping of small particles in a focused laser beam. The principle of optical trapping of small particles by laser is based on the forces arising from a change in the momentum of the light itself. With lasers one can make these forces large enough to accelerate, decelerate, deflect, guide, and even trap small particles. This is a direct consequence of the high intensities and high-intensity gradients achievable with continuous-wave (CW) coherent light beams. Laser trapping and manipulation techniques apply to particles as diverse as atoms, large molecules, and small dielectric spheres in size ranges of tens of nanometers, and they even apply to biological particles such as viruses, single living cells, and organelles within cells.

A satisfactory explanation for large-particle trapping can be obtained using geometrical optics. For this, let us place a spherical particle with refractive index greater than of surrounding medium in the laser beam with a wavelength λ, much smaller than the radius r of the particle. Additionally, the laser beam is focused to a spot with a diameter comparable to the wavelength.

Optical trapping is based on the fact that photons have linear momentum. It changes when a photon changes direction, as when crossing an interface between two media of different refractive index. Since the total momentum is conserved, the difference between the initial and the final momentum of a photon is transferred to the particle and is responsible for appearance of the force acting on the sphere. The force equation is

$$F = \frac{\Delta P}{\Delta t} \tag{14.1}$$

where F is the force, ΔP is the change in momentum, and Δt is the change in time.

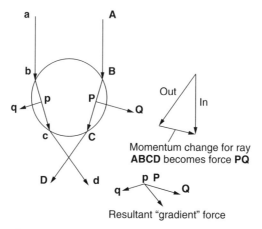

Figure 14.1. Force diagram for a sphere in a beam containing a power density gradient represented by two rays of unequal power.

The description provided here is based on an article by Ulanowski (2001). In Figure 14.1, a photon traveling along the path **abcd** imparts momentum to a spherical particle at **b** and **c**. The result is shown as a vector **pq**. Similarly, two photons in the path **ABCD** of a stronger ray transfer momentum at **B** and **C**, the resultant being the twice-longer vector **PQ**. From this we see that the sphere is forced toward the region of more intense light in the beam axis. This force is called the *gradient force*. Figure 14.2 shows the balance of force vectors for a particle positioned below the focal point of a focused beam directed downwards. We can see that the resultant force draws the particle upwards along the direction of propagation of the beam. This is the unexpected result, since a particle can be pulled toward the source of light against radiation pressure.

There is another force (not shown in Figure 14.1) present in such an experimental setup. This force appears since some light is reflected off the particle and is often called *scattering force*. This force is one reason why trapping is carried out on particles suspended not in air but in a liquid such as water so that the difference in the refractive indices between the particle and the immediate surround is less, thus resulting in smaller reflectivity. In this case the particle undergoes a smaller scattering force.

There is also the gravity force acting on the particle which, together with the scattering force, makes the sphere reside in equilibrium, a little beyond the focal point.

According to the size of the particle compared to the wavelength, different models of trapping interactions are used. For the $r \ll \lambda$ the Rayleigh model, for $r < \lambda$ the electromagnetic (EM) model, and for $r > \lambda$ the Ray-Optics (RO) model can be used. Since most biological cells are in the RO regime, the RO

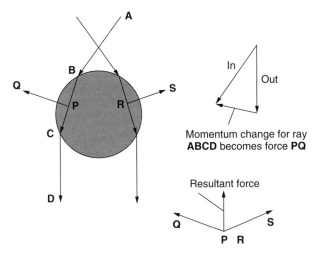

Figure 14.2. Balance of force vectors for a spherical particle positioned below the focal point of a focused beam. The resultant force pulls the particle upward.

model can be used to describe trapping of cells. In this case the forces F_{scatter} and F_{gradient} are given by (Ashkin, 1992)

$$F = \frac{\Delta P}{\Delta t} = F_{\text{scatter}} = n_m P Q_s / c$$

$$= n_m P / c \left\{ 1 + R \cos 2\Theta - \frac{T^2 [\cos(2\Theta - 2\varepsilon) + R \cos 2\Theta]}{1 + R^2 + 2R \cos 2\varepsilon} \right\} \quad (14.2)$$

and

$$F = \frac{\Delta P}{\Delta t} = F_{\text{gradient}} = n_m P Q_g / c$$

$$= n_m P / c \left\{ R \sin 2\Theta - \frac{T^2 [\sin(2\Theta - 2\varepsilon) + R \sin 2\Theta]}{1 + R^2 + 2R \cos 2\varepsilon} \right\} \quad (14.3)$$

where n_m, P, c, θ, and ε are refractive index of the medium, power, speed of light, angle of incidence, and angle of refraction, respectively. R and T are the Fresnel coefficients of reflection and refraction. Q is a dimensionless angle-dependent factor, different for both the scattering and the gradient forces.

Using the above formula and the Fresnel coefficients, the scattering and the gradient forces can be calculated for a beam coming perpendicular at a spherical polystyrene particle in water. They are: $F_{\text{scattering}} = 2.45 \times 10^{-12}\,\text{N}$ and $F_{\text{gradient}} = 3.18 \times 10^{-12}\,\text{N}$. For this calculation it has been assumed that $\lambda = 1.064\,\mu\text{m}$, $P = 100\,\text{mW}$, and the diameter r of the polystyrene sphere is $5\,\mu\text{m}$. Such a size is typical of many living cells. The force diagram for polystyrene

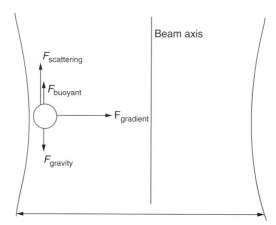

Figure 14.3. The force diagram for polystyrene particle. The arrow lengths are proportional to the magnitude of the forces.

particle is shown in Figure 14.3. The restoring gradient force is larger than the scattering force, and the gravitational force is balanced by the buoyancy force. Therefore, the polystyrene particle of radius 5 μm can be trapped by the laser beam.

14.3 DESIGN OF A LASER TWEEZER

Block (1998) provides a good description of the construction of optical tweezers. The single-beam optical trapping system used at our Institute for Lasers, Photonics, and Biophotonics is described here as an example of a basic design for a laser tweezer. The optic layout of this laser trap unit is shown in Figure 14.4. The main components of an optical trap used in this design are as follows:

1. *A Microscope (Either Upright or Inverted) with a High-Numerical-Aperture Objective Lens.* Some typical specifications are: NA 1.25–1.40, magnification 40–100×. The displayed configuration uses a Nikon TE200 inverted microscope with a 1.30-NA oil-immersion, 100× magnification objective lens. The high-numerical aperture allows for a tight focusing of the laser beam to generate a high-intensity gradient (and, thus, a large gradient force). Furthermore, for studies utilizing fluorescence imaging, this microscope can be either an epi-fluorescence or a confocal microscope system. To combine the optical trap and the epifluorescence mode in our case, the Nikon TE200 microscope design is modified by inserting mirror M_3, which permits us to introduce the trapping laser beam. The beam is introduced as a parallel beam by the use of lens L_2, reflected by mirrors M_3 and M_4 and, subsequently, focused on the specimen by lens L_3. The mirrors M_3 and M_4 were designed to transmit and to

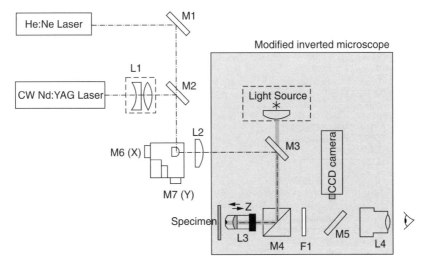

Figure 14.4. Single-beam optical trapping system.

reflect respectively the light from a mercury lamp (labeled as light source in Figure 14.4). The mercury lamp source can be used either for reflection mode (viewing) or for luminescence mode (imaging). All microscope functions are computer-controlled. The option of a CCD-TV camera port permits video viewing of microscope images and monitoring of optical trapping.

2. *A CW Laser Source that Provides Wavelengths at Which Biological Samples Are Transparent.* Based on the optical transparencies of cells; the near-IR region covering 700–1300 nm are used for optical trapping. CW lasers with powers in the range of several hundred milliwatts to several watts are utilized which can provide intensities in the range of 10^6–10^8 W/cm^2. Suitable choices are Nd:YAG at 1064 nm, Nd:YLF at 1047 nm, Ti:sapphire in the range 695–1100 nm, and various diode lasers, generally in the range 800–900 nm (where they are available with highest power). In order to produce the steepest gradient force, a laser beam with the TEM$_{00}$ mode is used. Such high-quality mode structures can easily be achieved from diode bar-pumped Nd:YAG lasers and Nd:YAG laser second-harmonic-pumped Ti:sapphire lasers. However, in the case of diode lasers, which produce elliptical beams, special optical beam correction is required to make the beam circular. A diode-pumped continuous-wave Nd:YAG TEM$_{00}$ laser was used for optical trapping in the configuration represented by Figure 14.4. A coincident red beam from a low-power He:Ne laser was used as the aiming beam.

3. *Beam Steering to Realize a Movable Trap.* A number of methods used to realize a movable trap for manipulation of trapped particles are shown in Figure 14.5. Figure 14.5 also shows the scheme of Figure 14.4, which uses an x–y galvano-head and a microscope objective lens on a movable mount. In

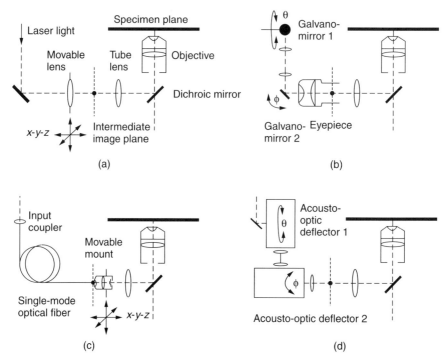

Figure 14.5. Four ways to scan the position of the laser spot in an optical trap in the object plane: (a) Translating the movable lens; (b) rotating galvanometer mirrors; (c) translating the end of an optical fiber; and (d) deflecting the beam with an acousto-optic deflector (AOD). (Reproduced with permission from Svoboda and Block, 1994.)

Figure 14.4, the in-plane x–y position of the laser focus (hence the optical trap) is controlled by the use of deflection from a set of galvano-mirrors M_6 and M_7. The z position of the laser focus is adjusted by piezoelectric displacement of microscope objective lens L_3.

Certain experiments require simultaneous use of more than one optical trap (Fallman and Axner, 1997). For two traps, a single laser beam can be split in two, utilizing a polarizing beam splitter (Misawa et al., 1992). A more flexible scheme to produce multiple optical traps utilizes time-sharing of the same beam among a set of positions in the specimen plane. This feature is achieved by rapidly scanning the beam focus position back and forth among the desired set of positions (Visscher et al., 1993). When the light is scanned sufficiently quickly, such as by using galvo-deflection or an acousto-optic deflector, the optical traps formed behave similar to what would be formed under steady illumination.

Figure 14.6 shows the schematics of a dual-beam trap produced by splitting a laser beam (Visscher et al., 1996). This design utilizes a polarized beam from

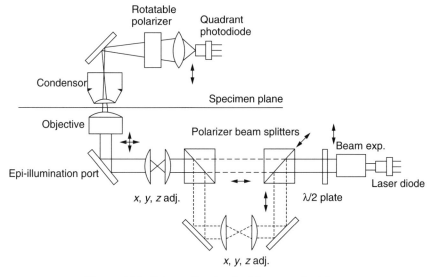

Figure 14.6. Schematic of a dual-beam optical trap.

an 830-nm diode laser which is expanded three times and passed through a half waveplate. The rotation of this half wave plate is used to alter the power split between the two orthogonally polarized trapping beams that are pro-duced by a polarizing beam-splitting cube. The beam reflected by the first polarizing cube then passes through a pair of lenses forming a 1:1 telescope in which the first lens is movable in the x, y, and z directions. This allows adjust-ment of the position of the optical trap, formed by this beam, with respect to the trap formed by the horizontally polarized beam which is transmitted by the first cube. Both beams are subsequently combined together using a second polarizing beam-splitting cube and then pass through a second 1:1 telescope consisting of a movable first lens which is used to jointly adjust the position of both optical traps.

Another design of a dual-beam optical trap is shown in Figure 14.7. The unit utilizes the optical trap shown in Figure 14.4. Here thin-film polarizers P_1–P_3 are used to split and to combine the beam of a Nd:YAG laser. One beam, with selected s polarization, is reflected from thin-film polarizers P_1, P_2, and P_3. Subsequently, it is focused on the specimen by a combination of optical components L_2, M_7, M_9, and L_3. This beam forms a fixed x–y plane optical trap because it dos not incorporate the galvano-mirrors M_{10} and M_{11}. Another beam, with selected p polarization, passes through polarizer P_1 and is reflected from mirror M_4 as well as from galvano-mirrors M_{10} and M_{11}. Then it passes through polarizer P_3 and is subsequently focused onto the specimen, where it forms an optical trap movable in the x–y plane. Both trapping beams can be controlled in the z direction. This dual-trap arrangement permits the flexibil-

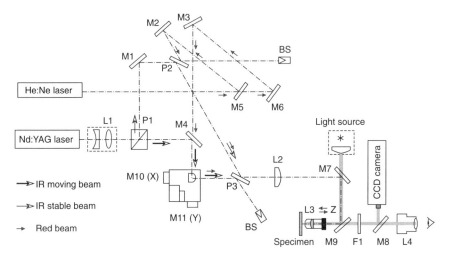

Figure 14.7. Optic layout of dual beam optical trap.

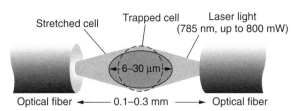

Figure 14.8. Schematic of the stretching of a cell trapped in an optical stretcher. (Reproduced with permission from Guck et al., 2001.)

ities of keeping one biological object in a fixed trap and manipulating another object using the movable trap.

A variation of optical trapping is the concept of an *optical stretcher* for micromanipulation of cells (Guck et al., 2001). An optical stretcher utilizes trapping of a cell between two opposed, nonfocused laser beams. The schematic is shown in Figure 14.8. This arrangement utilizes counterpropagating nonfocused laser beams and generates additive surface forces that produce stretching of a trapped cell along the axis of the beams. This optical stretcher can be used to measure viscoelastic properties of cells, with sensitivity sufficient to distinguish between different individual cytoskeletal phenotypes. Guck et al. (2001) used this type of optical stretcher to deform human erythrocytes and mouse fibroblasts.

14.4 OPTICAL TRAPPING USING NON-GAUSSIAN BEAMS

Optical traps, as discussed in Section 14.3, generally utilize a Gaussian beam (TEM$_{00}$), which is tightly focused with a high-numerical-aperture lens. A Gaussian beam, however, has certain limitations. First, it diverges as it propagates beyond a distance called the *Rayleigh range*. Second, it cannot trap particles more than a few micrometers apart in the propagation direction. The reason is that the original Gaussian beam gets distorted when passing through the trapped particle.

Recently, there has been considerable interest in the use of a non-Gaussian beam such as a Bessel beam (Arlt et al., 2001; MacDonald et al., 2002; Garcies-Chavez et al., 2002). A Bessel beam consists of light waves arranged in a cone. Thus its transverse profile (in a plane perpendicular to propagation) consists of a bright spot in the center of the beam, with a set of concentric bright rings around it. Some unique features of a Bessel beam are as follows: (i) The center bright beam propagates over several Rayleigh ranges without appreciable divergence. Hence, it is also sometimes referred to as "nondiffracting," or propagation invariant. This central beam thus can act as a rod of light. (ii) If the central bright spot is obstructed or distorted by a particle in the path, such as a trapped particle, the beam reconstructs itself to the original shape after a characteristic propagation distance (Bouchal et al., 1998). This effect is produced by the parts of light waves, far removed from the center, that move past the particle unhindered and recreate (as if self-healing) the beam center at some distance beyond the particle. These two features allow the use of a Bessel beam to be used as a two-dimensional laser tweezer, which can trap at multiple locations in the rod light. However, there is no confining force in the beam propagation direction, but an appropriate geometry (providing radiation pressure) can be used to push a sample down against the microscope slide. The Bessel beam can be used to trap particles in different sample cells, which are apart even by millimeters. A simple way to convert a Gaussian beam to a nearly Bessel beam is by using a conical-shaped optical element, called *axicon*.

Figure 14.9 shows micromanipulation with a Bessel light beam as utilized by Arlt et al. (2001). An axicon having an opening angle of 1° was illuminated by the expanded Gaussian output beam from a Nd:YVO$_4$ laser (1 W at 1064 nm). A telescope was used to reduce the size of the central maximum of the generated Bessel beam and to obtain a suitable propagation distance. The Bessel beam with the central maximum of 6–10 μm in diameter could be used for the manipulation. Both the central maximum size and the overall propagation distance was able to be varied by a judicious choice of optics and beam parameters.

To demonstrate that a Bessel beam can be used as single-beam line tweezers, several 5-μm spheres were trapped and maneuvered in unison. This function was performed by trapping the sphere at a power level at which it was guided upwards by the Bessel light beam along its propagation direc-

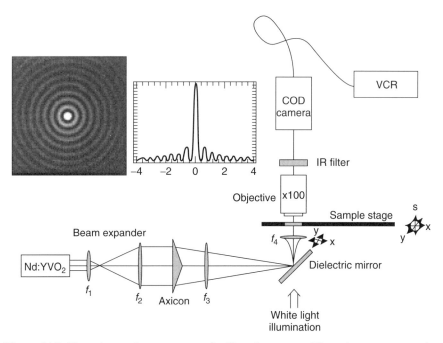

Figure 14.9. Experimental arrangement for Bessel tweezers. The axicon generates the Bessel beam. Lens f_4 is adjusted to manipulate the particles. The inset shows a picture and cross-sectional profile of the Bessel light beam that propagates in the vertical direction. (Reproduced with permission from http://www.st-andrews.ac.uk/~atomtrap/Research/BesselManip.htm.)

tion, and then translating the beam in the x–y plane over another particle which was then also trapped. This action was repeated to trap up to nine particles in a single tweezer setup. The stack of spheres can be translated as a whole in the transverse plane as shown in Figure 14.10.

Bessel beam tweezers can also be used to align elongated particles. Several biological specimens have an elongated or rod-like form. Figures 14.11a and 14.11b show the rotational orientation of *E. coli* bacteria and a 50-μm-long fragment of a glass fiber, respectively, captured by Bessel beam tweezers. Vertical alignment and subsequent manipulation could allow one to readily isolate and transfer elongated samples from one sample chamber to another.

14.5 DYNAMIC HOLOGRAPHIC OPTICAL TWEEZERS

Recently a University of Chicago group, in collaboration with Arrix Inc. (Chicago, Illinois, USA), has produced computer-generated dynamic holographic optical tweezers (abbreviated as HOT). The details can be found on the following websites: http://mrsec.uchicago.edu/Nuggets/Holographic_Optical_Tweezers/ and http://www.arryx.com/.

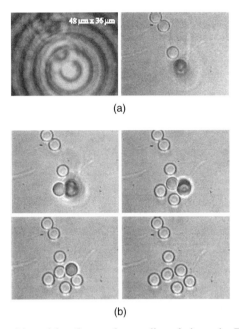

(a)

(b)

Figure 14.10. The stacking of five 5-μm spheres aligned along the Bessel beam: (a) The stack is held vertically by Bessel beam and can be translated in the transverse plane; (b) a sequence of frames showing the collapse of the five stacked spheres after blocking the beam. (Reproduced with permission from http://www.st-andrews.ac.uk/~atomtrap/ Research/BesselManip.htm.)

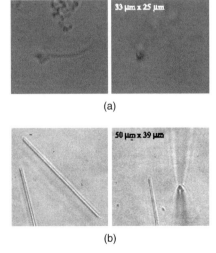

(a)

(b)

Figure 14.11. Rotational orientation through 90° of (a) an *E. coli* bacterium (the second frame is focused about 15 μm higher, at the top of the upright bacterium) and (b) a 50-μm fragment of a glass fiber. Both samples could subsequently be manipulated once in the upright position. (Reproduced with permission from http://www. st-andrews.ac.uk/~atomtrap/Research/BesselManip.htm.)

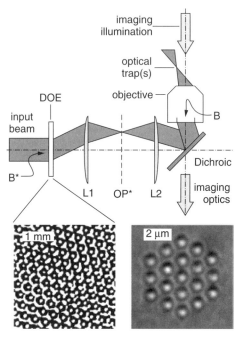

Figure 14.12. Schematic of a typical holographic optical tweezer array. (Reproduced with permission from Dufresne et al., 2001.)

Holographic optical tweezers utilize a specially designed, computer-generated diffractive optical element (DOE) to split a single laser beam into many beams (Dufresne et al., 2001). Each of these beams is then focused to create an optical trap. Furthermore, this computer-generated DOE can be reconfigured to manipulate the configurations of multiple optical trap patterns. For this reason they are also referred to as *dynamic holographic optical tweezers*. A schematic of a holographic optical tweezer is shown in Figure 14.12. This design utilizes a computer-controlled 512×512 pixel liquid crystal spatial light modulator (SLM) (Igasaki et al., 1999) as a reconfigurable DOE, together with a CW green (532 nm) Nd:YAG laser and a Nicon TE200 inverted light microscope. A liquid crystal SLM consists of a series of computer-controllable pixilated patterns. By adjusting the orientation (birefringence) of the liquid crystal at each pixel (using applied electric field), one can vary the effective optical path length and introduce a corresponding phase shift. This phase shift then produces a corresponding intensity modulation at the plane of optical trapping. Hence the light reflected from the SLM creates a pattern of bright spots, each capable of trapping an individual particle. Computer-controlled SLM allows manipulation of thousands of particles in the sample. This approach can be generalized to three-dimensional arrangements of trapping patterns.

14.6 LASER SCISSORS

The laser scissor action utilizes photoablation by a pulsed UV laser. A suitable UV laser, frequently used, is a pulsed nitrogen laser that provides a 3-nsec pulse output of energy 120–300 μJ at 337 nm. When focused through a high-numerical-aperture objective, it can produce a high photon density in a spot, less than 1 μm. This high photon density of UV radiation produces photoablation by the photochemical decomposition of cellular and/or tissue structures (primarily melanin). This process has been discussed in Chapter 6. Since it is a photochemical process, little heat is produced and thus it is often called a *cold ablative process*. Using this process, the laser scissor can cut a spot size less than 1 μm, without damaging the neighboring subcellular/cellular structures. The laser beam can be introduced into a standard inverted microscope through the epi-fluorescence port. Substage robotics can introduce flexibility that enables cutting of any shape and size. The use of a spatially localized photochemical decomposition without any significant thermal damage allows one to perform microsurgery on live cells in tissue cultures, ova, and sperm without affecting the cell viability. The cells and chromosomal fragments, dissected using a UV laser beam, have exhibited no evidence of UV-induced damage to collected or adjacent specimens. This advantage has led to applications such as cutting of cytoplasmic fragments, fusion of live cells in culture, blastomere fusion, and cutting of flagella or sperm tails (Schütze et al., 1997).

The same principle of photoablation can be used to punch holes in the cell membrane, which rapidly close. This process is also sometimes referred to as *optoinjection* because drugs or genetic materials dissolved in the surrounding medium then can be injected into the cell. As indicated in Section 14.1, combining a laser scissor with a laser tweezer opens up new doors for a number of applications. Some of these applications are discussed in Section 14.7. Figure 14.13 shows the schematics of a combination of an optical scissor and an optical tweezer. This design combines a UV laser beam, used for scissor action, with an IR beam that is used for optical trapping.

Another type of laser scissor operation, more popularly known as *laser microdissection*, is to excise a single cell or pure cell population from a tissue specimen (Willingham, 2002; Roberts, 2002). It involves a coordinated use of microscopic, laser, and robotic techniques to localize and dissect. The laser microdissection technique can be used to select specific cells such as tumor cells from a malignant tissue or prostate-specific antigen expressing cells from prostate. Integration of a capturing mechanism to collect cells, microdissected from a tissue element, enables one to conduct studies on pure populations of cells for analysis of molecular functions. The two different innovations providing both laser microdissection and capture of the dissected cells or tissue fragments are "laser pressure catapulting" (LPC) and "laser capture microdissection" (LCM). Both of these methods are noncontact and allow extraction of a single cell or a small number of cells from a tissue. These two techniques are described below.

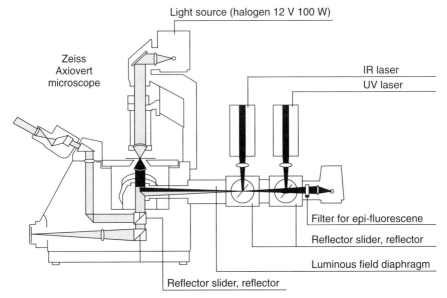

Figure 14.13. Schematic of a combination of a laser scissor with a laser tweezer. In this system the fluorescence illumination lamp is still present and the two lasers are combined using a dielectric mirror. (Reproduced with permission from Greulich, 1999.)

14.6.1 Laser Pressure Catapulting (LPC)

This type of capture, along with laser microdissection, is used in the system sold by Palm GmbH in Germany. In the first step, a nitrogen laser beam is used for microdissection, as described above. Once the cell is dissected, the same laser with approximately twice the energy level is focussed from below the tissue specimen. Depending on the size of the excised sample, one or a few laser shots produce enough laser pressure force, due to the extremely high photon density of the focused beam, to catapult (eject) the sample into a collector cap. A schematic of this approach is shown in Figure 14.14.

This capture thus involves a completely noncontact procedure. The use of a 100× objective lens can generate a laser beam spot of 600-nm diameter, allowing one to dissect even an individual nucleus from cells or single chromosomes that may be used for microcloning and construction of splice-specific libraries. The sample can be a tissue specimen mounted directly on a glass slide, or on a 6-μm membrane and placed on a glass slide. In the latter case, the UV laser microbeam burns a rim of the membrane as well as producing microdissection of the tissue, which is then catapulted. Figure 14.15 shows excision and collection of cells using laser microdissection and laser pressure catapulting. This figure is taken from the website of PALM Microlaser Technologies. The top picture shows single cells and cell clusters from the glass-mounted tissue.

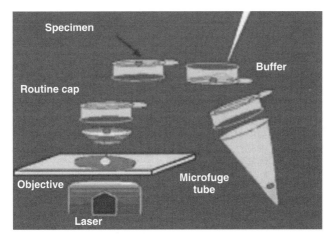

Figure 14.14. Schematic of a laser pressure catapulting system. (Reproduced with permission from www.palm-microlaser/com.)

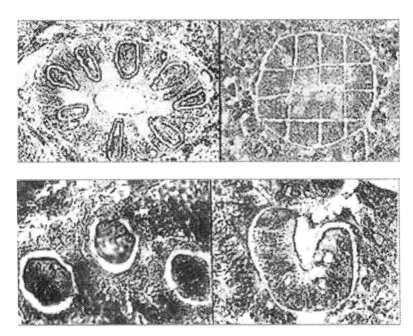

Figure 14.15. Laser microdissection and laser pressure catapulting of single cells and cell clusters from glass-mounted tissue (*top*); single cells and homogeneous cell area from membrane-mounted tissue (*bottom*). (Reproduced with permission from http://www.palm-mikrolaser.com/about_us_technology_microdissection.html.)

The bottom picture shows single cells and the homogeneous cell area from the membrane-mounted tissue.

14.6.2 Laser Capture Microdissection (LCM)

This approach was developed at the National Institutes of Health (Emmert-Buck et al., 1996; Best and Emmert-Buck, 2001) and is now commercialized as a product, PixCell II by Arcturus Engineering of Mountain View, California (Roberts, 2002). LCM is performed using a tissue section of 5- to 10-μm thickness that has been preserved either by freezing or by fixation followed by paraffin embedding. A 100-μm-thick ethylene vinyl acetate (EVA) film, which has been impregnated with an infrared absorbing dye, is attached to a rigid 6-mm laser cap. It is lowered exactly opposite to the area of tissue section to be harvested and acts as a transfer film. This schematic is shown in Figure 14.16. A pulsed near-infrared beam, usually of duration less than 5 msec, is directed through the cap. The membrane absorbs the energy from the IR beam due to the presence of the IR absorbing dye, raising its temperature momentarily to 90°C and consequently melting it when it adheres to the underlying tissue. By adjusting the laser beam spot size (for example, between 7.5, 15, and 30 μm), one can select a single cell or a group of cells with one laser pulse. Furthermore, by moving the laser spot around on the tissue with the help of a joystick, one can select multiple sites of the same tissue with the same cap. When the cell selection is finished, the cap can be lifted off the tissue pulling off the cells attached to the membrane. The EVA films can be dissolved under the effect of lysis buffer to release the cells. Since the duration of the pulse is short, there is minimal transfer of thermal energy to the tissue, thereby reducing the danger of damage to the tissue and extracted cells.

14.7 SELECTED EXAMPLES OF APPLICATIONS

14.7.1 Manipulation of Single DNA Molecules

The use of laser tweezers has provided valuable insight into determining the forces that keep the DNA molecule in a randomly coiled configuration. Laser tweezers have also been successfully used to study DNA–protein interaction, gene transcription, and enzymatic degradation of DNA, all at the single molecule level. In these studies, one or both ends of a fluorescently labeled DNA chain is attached to a polystyrene microbead. For this purpose, one couples a biotin group at the end of the DNA chain which is then conjugated to commercially available avidin-coated polystyrene microbeads (typically of diameter ~1 μm). In the case where only one end of DNA is attached to the microbead, one can apply a viscous drag force using a hydrodynamic flow to stretch the DNA chain. If both ends of DNA have a microbead attached, one can use laser tweezers at each end to pull the DNA simultaneously from both

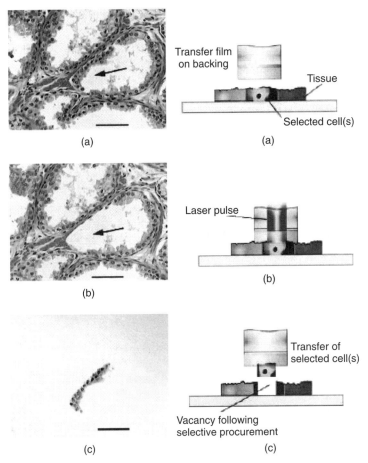

Figure 14.16. Schematic of laser capture microdissection (LCM). *Right*: Fixed, stained, microscopic tissue cells of interest are selected using LCM and transferred onto the area of the polymer surface activated by the laser beam. *Left*: Visualization of LCM-procured cells. Target region designated by arrow. (a) Before LCM. (b) Tissue after LCM; two shots. Note the vacancy left by the removal of selected cells. The bar represents 30 μm. (c) Epithelial cells transferred to cap surface. (Reproduced with permission from Simone at al., 2000.)

sides. The group of Chu (Perkins et al., 1995) used a laser tweezer to hold one end of a DNA molecule and used the hydrodynamic force from a flowing liquid to extend a 64.5-μm-long DNA molecule with increasing fluid velocity of flows. Figure 14.17 shows this extension at a number of flow velocities. These are fluorescence microscopic images.

The experimental result appeared to fit the force field elongation theory of Schurr and Smith (1990). The unstressed end (not attached to the bead) exhibits a disorder (k) expected at the free end. As the hydrodynamic force

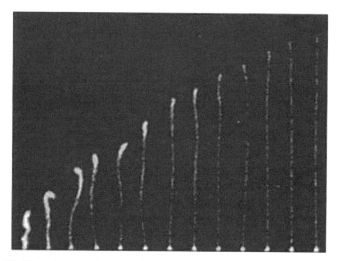

Figure 14.17. Use of hydrodynamic flow to stretch a DNA molecule held at the end coupled to a microbead using a laser tweezer. (Reproduced with permission from Perkins et al., 1995.)

field is increased, it increases elongation and decreases the size of k. When the flow is stopped, the DNA molecule relaxes to revert back to its condensed, more globular form, allowing one to follow this motion by time-resolved fluorescence microscopy (Perkins et al., 1994). Their result reveals that the relaxation is exponential, with a relaxation time dependent on the total length of the molecule.

Optical tweezers can be used as picotensiometers where displacement from the trapping center with nanometer accuracy can be calibrated in terms of force with piconewton accuracy for force measurements. This method has been used to measure forces involved in DNA stretching (Smith et al., 1996; Wang et al., 1997). Meiners and Quake (2000) have extended the range of force measurements to femtonewton sensitivity, using a dual optical-trap-based force spectroscopic technique. They have used this method to study thermal fluctuations of a single DNA molecule.

An important application of laser tweezers has been in the study of bio-chemistry at the single-molecule limit. An example is in the study of the force exerted by a single molecule of the enzyme, RNA polymerase, during gene transcription (Yin et al., 1995). For this investigation, the RNA polymerase was attached to the surface of a cover glass of a flow cell. In the presence of DNA and other components essential for *in vitro* transcription, the RNA poly-merase catches a DNA molecule, pulls it through its active site, and synthe-sizes RNA with a sequence, complementary to that of the segment of DNA just read. The RNA falls off the polymerase as the transcription is completed when the DNA signals a stop codon. One end of the DNA molecule is coupled

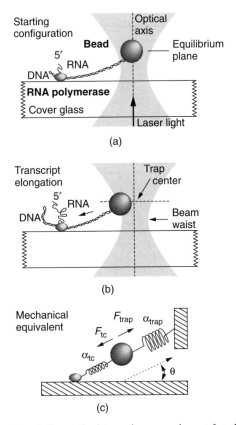

Figure 14.18. Schematic of the optical trapping experiment for the RNA polymerase produced gene transcription. (Reproduced with permission from Yin et al., 1995.)

to a polystyrene microbead (diameter ~0.5 μm). In the experiment by Yin et al. (1995), when the RNA polymerase catches the DNA molecule during transcription, the bead is pulled. The microbead is held in the center of the beam of an optical tweezer (using relatively low laser power of 25 mW at 1.06 μm) which has been calibrated to measure forces. Thus the pull exerted by RNA polymerase on the DNA from the optical trapping region can be measured by the displacement of the microbead and the stiffness of the optical trap. The position of the bead was measured with resolution in nanometers using interferometry. A schematic of this process is shown in Figure 14.18. The result of this measurement showed that *E. coli* RNA polymerase can provide a force up to 14 pN. Compared to it, the maximum force exerted by a typical motor protein is only 6 pN. The obvious conclusion is that enzyme transcription through DNA is a more stringent process than proteins driving a muscle. This optical trapping study of gene transcription also yielded information on the efficiency with which chemical energy is converted into mechanical energy.

The mechanical energy produced during the insertion of each nucleotide into the nascent RNA is simply force multiplied by the distance, the latter being 0.34 nm. The chemical energy released by one reaction step is known from other measurements. The measurement of mechanical energy using the optical trapping study yielded a chemical-to-mechanical energy conversion rate of up to 42%, which compares quite favorably with that of up to 60% for motor proteins. Another interesting observation was that DNA does not run smoothly through the active sites of the enzyme, exhibiting a frictional behavior.

Baumann et al. (2000) used optical tweezers to study the elastic response of single plasmid and lambda phase DNA molecules in the presence of various concentrations of trivalent cations which provoke DNA condensation in the bulk. They investigated the dependence of a single-molecule condensation on ionic conditions and the extent of stretching. The facilitation of DNA condensation in the presence of a certain multivalent ion is thought to arise from attractive lateral interactions between the adjacent helices produced upon binding a critical amount of multivalent cations. Their finding that intramolecular condensation occurs only when the DNA molecule is sufficiently relaxed to form intramolecular loops provides support for a lateral interaction rather than an elastic buckling mechanism.

Hirano et al. (2002) reported another approach for manipulation of single-coiled DNA molecules. This method does not require any prior chemical modification of DNA (biotin–avidin coupling) to attach it to a microbead. In their approach, a bead cluster is formed using laser trapping that can then be manipulated to capture a single native DNA molecule. The bead cluster was then used to drag the end of a single DNA molecule.

14.7.2 Molecular Motors

Molecular motors are special enzymes that catalyze a chemical reaction such as hydrolysis of ATP, capture the free energy released by the reaction, and use it to perform a mechanical work such as muscle contraction. An example of such a motor enzyme is kinesin, which binds to subcellular organelles such as chromosomes and transports them through the cytoplasm by pulling them along microtubules. Optical trapping has been used to study the process of movement of kinosin from site to site on the microtubule lattice (Visscher et al., 1999). In this work, they used a molecular force clamp method utilizing a feedback-driven optical trap, capable of maintaining a constant load (force) on a single kinosin molecule. The kinosin molecule is composed of two heavy chains, each consisting of a force generating a globular domain head (hence double-headed), a long α-helical coil, and a tail portion that is a small globular C-terminal domain. Microtubules are cylinders comprised of parallel protofilaments that are linear polymers of α- and β-tubulin dimers.

Visscher et al. (1999) used a kinosin-coated silica bead (diameter ~0.5 μm) that trapped in a focused 1064-nm beam from a Nd:YVO$_4$ laser using the objective lens of an inverted microscope. The trap position within the speci-

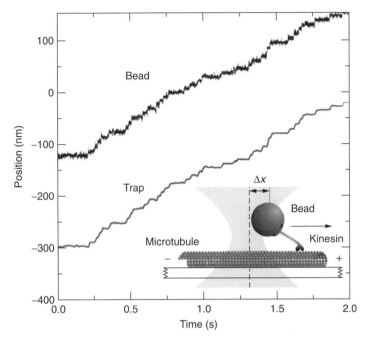

Figure 14.19. Schematic of the experimental geometry, which includes position measurement for kinosin-driven bead movement and the corresponding optical trap displacements at an ATP concentration of 2 mM. (Reproduced with permission from Visscher et al., 1999.)

men plane was specified using two digitally computer-controlled acousto-optic deflectors. The bead positions were determined by focusing a low-power He–Ne laser beam onto the optically trapped kinosin-coated silica bead and measuring the deflected light in a plane conjugate to the back focal plane of the microscope condenser, using a quadrant photodiode arrangement (Visscher et al., 1996).

The schematics of the experimental geometry and the results are shown in Figure 14.19. The results indicate kinosin stepping tightly coupled to ATP hydrolysis over a wide range of forces. A single hydrolysis produces kinosin movement along a microtubule with an 8-nm step that coincides with the α, β-tubulin dimer repeat unit. The progressive movement of kinosin was explained by a hand-over-hand mechanism in which one head remains bound to the microtubule, while the other detaches and moves forward.

14.7.3 Protein–Protein Interactions

Optical trapping has also been used to characterize individual intermolecular bonds in proteins and thus investigate protein–protein interactions (Stout, 2001). Figure 14.20 shows the schematics of the experimental arrangement

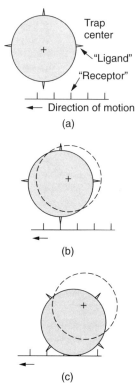

Figure 14.20. (a) A microbead coated with a small number of IgG held a short distance above a moving substrate coated with receptor molecules; (b) the binding of the bead with the surface produces bead movement away from the optical trap center with the same velocity as the moving surface; (c) as the bead moves away from the trap center it pivots, making contact with the surface which produces an additional normal force. (Reproduced with permission from Stout, 2001.)

used by Stout. A polystyrene microsphere, coated with immunoglobulin G (IgG), is held in a force-calibrated optical trap. A flat substrate is sparsely coated with a receptor protein (staphylococcus protein A [SpA]). The IgG-coated microsphere, as a probe, is held in contact with the substrate using a stationary optical trap to allow for protein–substrate binding. Then the substrate is scanned. This movement produces a lateral displacement of the microbead due to the pull exerted on the surface coated protein, now bound to the substrate, through protein–protein interaction. This displacement of the probe is monitored through an $x–y$ position detector. The bound probe (microsphere) moves with the substrate until the force applied by the optical trap overcomes the bond between the probe and the substrate. The rupture of this bond allows the optical trap to pull back the probe to the trap center. This geometry of utilizing interaction between the probe and the surface allows a fivefold enhancement of the force applied by the optical trap due to the substrate acting as a lever. Thus, this optical trapping method allows a laser tweezer to be used to access rupture forces up to 200 pN, as opposed to the regular upper limit of 50 pN. The experiment yielded a median single-bond rupture force from 25 to 44 pN for IgG from four mammalian species, which is in general agreement with predictions based on free energies of association obtained from solution equilibrium constants.

14.7.4 Laser Microbeams for Genomics and Proteomics

Laser microtools can be of significant value for genomics and proteomics in molecular profiling of cancer and other genetically based diseases. As discussed above, laser microbeam microdissection (LMM) coupled with laser pressure catapulting (LPC) or laser capture microdissection (LCM) allows isolation of a single cell, as well as a small number of specific cells from an archival tissue in a noncontact mode. Thus laser microdissection can be used to extract specific cell populations such as normal cells, precancerous cells, and invasive cancer cells. The purity of these specific cells then can permit one to compare and identify tumor suppressor genes as well as novel transcriptions and proteins that change in neoplastic cells (Best and Emmert-Buck, 2001; Maitra et al., 2002).

Genetic changes manifested in multistep progression of cancer can involve gain of mutation in dominant oncogenes, or loss of a function by delection, mutation, or methylation in repressive tumor suppressor genes. This loss of supressor gene function in a tumor is called *loss of heterozygosity* (LOH) (Gillespie et al., 2000). Laser microdissection has made a significant contribution to applications of LOH analysis to cancer studies because virtually pure populations of tumor cells or preneoplastic foci necessary for LOH analysis can be isolated without contamination even by a few unwanted cells. The LOH analysis has proved valuable in the mapping of tumor suppressor genes, localization of putative chromosomal "hot spots," and the study of sequential genetic changes in preneoplastic lesions. LCM in conjunction with fluorescence *in situ* hybridization (FISH, discussed in Chapter 8) demonstrated LOH on chromosome sp21 in prostate cancer. Loss of the dematin gene was observed, leading to dysregulation of cell shape (Lutchman et al., 2000). Study of preneoplastic lesions has revealed that genetic alterations in cancers actually starts in histologically "benign" tissue.

LCM used in conjunction with polymerase chain reaction (PCR) such as reverse transcriptase-PCR (RTPCR) provides an opportunity to study only a few hundred cells. The advantage is that even microscopic preneoplastic lesions can be studied. In addition to LOH analysis, other studies have been performed using laser microdissection. They include X-chromosome inactivation analysis to access clonality, single-strand conformation polymorphism (SSCP) analysis for mutations in critical genes, comparative genomic hybridization (CGH), and the analysis of promoter hypermethylation.

Microdissected cells have been used to obtain differential gene expression which is a useful parameter to differentiate tumors from their normal cells. Methods to study gene expression include expressed sequence tag (EST) sequencing, differential display, subtractive hybridization, serial analysis of gene expression (SAGE), and cDNA microarray technique. The microarray technology has been discussed in Chapter 10. LCM has been used to generate cDNA libraries for a number of cancers. These data can be accessed at the NIH Cancer Genomic Anatomy Project (CGAP) website (http://cgap.nci.nih.gov/). The cDNA libraries can be used for identification of novel genes that are either overexpressed or underexpressed in the multistage pathogenesis of cancer, which can eventually lead to genetic profiling of individual patient samples to customize treatment on an individual basis.

Laser microdissection for molecular profiling of global protein patterns will play an important role because it is crucial for protein analysis and differentiation to obtain pure populations of tumor cells and their preneoplastic lesions. Identification of proteins dysregulated during cancer progression will be valuable in formulating treatment and developing intervention strategies.

LCM has been used in conjunction with high-resolution two-dimensional polyacrylamide gel electrophoresis (2-D PAGE), a technique used to analyze populations of proteins in different cell types to resolve more than 600 proteins or their isoforms and identify dysregulated products in cancer cells. Sequencing of the altered peptide products unique to the tumor population can be used to identify novel tumor-specific alterations. For example, proteomic analysis of microdissected prostate cancers and benign prostatic epithelium revealed six differentially expressed proteins (Ornstein et al., 2000). Microdissected specimen of colon cancer has shown increased levels of gelatinase and cathepsin B, both implicated in cancer invasion and metastasis (Emmert-Buck et al., 1994).

To conclude, it can be envisioned that the use of a rapid microdissection technique, together with biomolecule amplification protocols, can provide more sensitive detection and database integration which can become a standard practice for cancer diagnostics. The molecular profile information on DNA, RNA, and protein alterations can lead to diseases management as well as to design of optimal, low-risk, and patient-tailored treatment.

14.7.5 Laser Manipulation in Plant Biology

Laser manipulation may hold promise for plant breeding. Many plant cells are transparent, thereby permitting the use of laser microtools to access subcellular structures such as mitochondria and chloroplasts. Furthermore, plant cells such as rapeseed cells contain subcellular organelles which are mobile and thus can be pulled through the cell with a high spatial control (Greulich, 1999).

Laser microinjection of genes is particularly suited for plant cells. The reason is that for plant cells, such as rapeseed cells, glass capillaries used for microinjection are either too thick and damage the cell or are sufficiently fine but too fragile to penetrate the rigid plant cells. Laser microinjection has been utilized to inject foreign genes into individual cells in suspension. Bacterial glucuronidase (GUS) reporter gene is a promoter of cauliflower mosaic virus. It has been shown that when GUS DNA is laser-microinjected into some selected cells of the embryo, an appearance of a blue color indicates success-ful expression of GUS. Thus the gene becomes active in its new host cell.

Photosynthesis in plants take place in the organelle's chloroplasts, which are 5–10 µm in diameter (see Chapter 6). Foreign genes have succesfully been injected into the chloroplast by opening a hole in the membrane using a single laser shot, which would close one second after the laser treatment (Weber et al., 1990).

14.7.6   Laser Micromanipulation for Reproduction Medicine

Laser micromanipulation may one day provide benefits in assisting *in vitro* fertilization at fertility clinics. Human infertility can often be overcome by simply performing sperm–egg fusion externally in a reaction tube. The problem of infertility is often mechanical, derived from the surrounding of a mammalian egg by a highly viscous envelope called *zona pellucida*, which can not be penetrated by the sperm cells. By laser zona drilling (LZD) to produce a micron-sized hole in the zona pellucida, this viscous barrier can be opened for a sperm to penetrate and fertilize the egg. This process can occur in the normal course, or it can be assisted by using a laser tweezer to trap the sperm and lead it to the site of the hole in the egg produced by LZD. The procedure thus is completely noncontact and has been used by a number of researchers (Clement-Sengewald et al., 1997). Figure 14.21 schematically represents this process. This procedure has been shown to produce significant improvement in fertilization when the sperm density is low.

Using 50- to 60 4 µJ pulses (337 nm) of 3-nsec duration, a straight channel could be driven into the zona pellucida. For this operation, the egg cell did not have to be fixed by a micromanipulator; that is, it was essentially a suspension procedure. An insemination was judged to be successful when the egg cell divided. Table 14.1 summarizes the major results of this study. At high sperm density, there is no significant difference between conventional *in vitro* fertil-ization and the laser zona drilling technique. However, at low sperm density, the effect is significant, with a 58% success rate while using the laser-supported technique as compared to 33% for the nonlaser technique. The very low success rate (18%) of subzonal insemination is not explained.

Another occasional cause of infertility is the low mobility of a sperm cell. Again, the use of laser tweezers can overcome the problem of low mobility. A major concern in this case is whether there is any adverse effect on the veloc-ity of a sperm after optically trapping it for some time. The results of a study (König et al., 1995) show that it is dependent on the wavelength as well as on the trapping period. The 760-nm light appeared to cause much higher damage

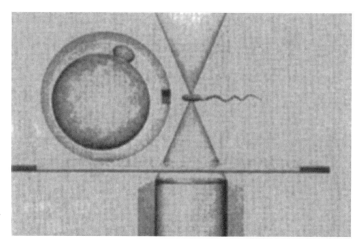

Figure 14.21. Schematic of laser zona drilling combined with laser tweezer capture of the sperm cell to bring it to the hole in the zona. (Reproduced with permission from Clement-Sengewald et al., 1997.)

TABLE 14.1. Success Rates in Mouse Gametes Using Different Types of Fertilization Techniques

Technique	Sperm Density	Cells Treated	Successfully Fertilized	Percent Success
Nonlaser IVF	Normal	85	45	53
Laser zona drilling	Normal	124	74	60
Nonlaser IVF	Low	63	21	33
Laser zona drilling	Low	40	23	58
Subzonal insemination	Low	22	4	18

than that produced by the 800-nm light. The longer trapping duration appears to increase the sperm cell damage.

14.8   FUTURE DIRECTIONS

Laser micromanipulation will continue to receive worldwide attention from fundamental researchers trying to understand single-molecule biochemistry and biofunction. It is receiving increasing attention for genomics and proteomics. Clinical interest in utilizing microdissection for cancer diagnostics and combined use of microdissection and optical trapping for genetic manipulation and fertilization may also see a growth. This field may also receive an impetus from laser developments, making them more efficient, user-friendly, reliable, cost-effective, and compact. Some of the current activities defining future directions are outlined here.

14.8.1 Technology of Laser Manipulation

New physical/chemical processes for more efficient laser trapping and more effective laser microdissection are likely to emerge. Agayan et al. (2002) have shown theoretically that by selecting an optical trapping wavelength near the resonance absorption, a 50-fold enhancement in trapping forces can be realized together with increased specificity.

A new approach to produce a stable optical trap is by Zemanek et al. (1999), who demonstrated optical trapping of nanoparticles and microparticles by a Gaussian standing wave. The standing wave was produced under a microscope objective as a result of interference between an incoming laser beam and a beam reflected from a microscope slide coated with reflective dielectric layers. Three-dimensional trapping of nanoparticles (100-nm polystyrene spheres) and one or several vertically aligned micro-objects (5-μm polystyrene spheres, yeast cells) was achieved by use of even highly aberrated beams or objectives with low numerical apertures.

The modern combination of optical multitrap, microdissection, and far-field confocal (Raman) microscope function in one unit could open up new possibilities for biophotonics.

For laser microdissection, the use of ultra-short femtosecond laser pulses is being investigated. The femtosecond pulses can enhance the probability of multiphonon processes as very intense fields but at very low average powers. These pulses could produce a much precise microdissection with considerably reduced collateral damage. As the prices for femtosecond laser systems come down, they would become affordable to be widely utilized for laser microdissection.

14.8.2 Single Molecule Biofunctions

The use of optical trapping to manipulate single molecules and perform biophysical and biochemical studies will remain an area of growing interest. As sensitive spectroscopic detection techniques are emerging for single-molecule analysis, their application will lead to the study of DNA–protein and protein–protein interactions as well as to the monitoring of subcellular functions at a single biomolecule level. Xie et al. (2002) used a low-powered diode laser at 785 nm to both trap and excite Raman spectra of single biological cells in solution. As mentioned above, a combination of the advantages of NIR Raman spectroscopy and optical tweezers for the characterization of single biological cells with a low-power semiconductor laser provides high sensitivity, making it possible to obtain Raman spectra from single living red blood cells (RBCs) or yeast cells placed in an optical trap.

TABLE 14.2. Commercial Sources of Laser Microtools

Company	Product	Function	Website
Arturus Engineering, Hercules, CA, USA	Pixcell II	Laser capture microdissection	www.arcture.com
Bio-Rad Laboratories, Hercules, CA, USA	Clonis	IR laser microdissection	www.microscopy.bio-rad.com
Cell Robotics, Albuquerque, NM, USA	Pro 300	Laser microdissection using UV laser	www.cellrobotics.com
	Laser Tweezers©	IR or NIR for trappingr	
Leica Microsystems, Wetzler, Germany	Leica AS LMD	Laser microdissection using UV beam	www.leica-microsysystems.com
P.A.L.M. Microlaser Technologies, Bernried, Germany	PALM Microbeam	Laser pressure catapulting	www.palm-microlaser.com
	PALM Microtweezers	IR or NIR for trapping	
MMI AG Heidelberg, Germany	μ-CUT	Laser microdissection using UV laser	www.mmi-micro.com
ARRYX, Inc., Chicago, IL, USA	ARRYX BioRyx™ 200 SYSTEM	Holographic optical trapping (multiple beam) system to independently manipulate large numbers of microobjects simultaneously	www.arryx.com

14.9 COMMERCIALLY AVAILABLE LASER MICROTOOLS

A number of companies sell laser microdissection systems and laser tweezers. Table 14.2 lists some of them. However, it should be kept in mind that they exist at the time of writing of this book. It is very likely that this list can change significantly quickly, making Table 14.2 rather obsolete. Nonetheless, it may serve as a starting point for those looking to acquire a commercial system. It also is a demonstration that the activities and interests in the use of laser microtools are sufficiently advanced to recognize business opportunities in commercializing them.

HIGHLIGHTS OF THE CHAPTER

- Laser tweezers are microtools used to trap biological cells or micron-sized particles in the focused laser beam spot of a continuous-wave laser.

- Laser tweezers accomplish optical trapping by using the net force derived from a change in the momentum of photons, created by their refraction to the particles.
- A particle must have a refractive index higher than the surrounding medium to experience a net attractive force at the focus spot of the laser beam.
- Using appropriate optics to break up a laser beam into many spots allows one to trap many particles simultaneously.
- An optical stretcher is a variation of a laser tweezer. It traps a cell between two opposed, nonfocused laser beams, stretching the cell along the axis of the beams.
- Laser scissors are used to cut a hole into the cell membrane for injection of a drug or genetic material; this process is often called *optoinjection*. They also are used to excise a single cell or extract a pure cell population from a tissue specimen, an operation popularly known as *laser microdissection*.
- Laser scissors operate on the principle of photoablation produced by a pulsed UV beam.
- The two different approaches used to provide both laser microdissection and the capture of the dissected cells or tissue fragments are "laser pressure catapulting" (LPC) and "laser capture microdissection" (LCM).
- In LPC, the same UV laser used for the scissor action, but now with twice the energy, is focused from below the tissue element to eject the sample from the matrix by the laser pressure, which is then collected in a reservoir.
- LCM utilizes an infrared beam to heat and melt a film, containing an infrared dye and coated on the bottom of a cap. When the cap is lowered, the melted film adheres to the tissue. Then as the cap is pulled away, it lifts the tissue.
- Laser tweezers have found important applications in the study of the structure and dynamics of single biomolecules such as in the manipulation of a single DNA chain.
- Other areas of basic research benefiting from optical tweezers are molecular motor proteins and protein–protein interactions.
- Laser scissor manipulation has benefited genomics and proteomics by using laser microdissection to extract specific cell populations, such as normal cells, precancerous cells, and invasive cancer cells, from a tissue specimen.
- Laser manipulation holds promise for a number of applications such as plant breeding and *in vitro* fertilization.
- Future opportunities for research and development include research on single molecule biofunctions as well as the development of new technology for laser manipulation and new applications for gene manipulation.

REFERENCES

Agayan, R. R., Gittes, F., Kopelman, R., and Schmidt, C. F., Optical Trapping Near Resonance Absorption, *Appl. Opt.* **41**, 2318–2327 (2002).

Arlt, J., Garces-Chavez, V., Sibbett, W., and Dholakia, K., Opical Micro Manipulation Using a Bessel Light Beam, *Opt. Commun.* **197**, 239–245 (2001).

Ashkin, A., Forces of a Single-beam Gradient Laser Trap on a Dielectric sphere in the Ray Optics Regime, *Biophys. J.* **61**, 569–582 (1992).

Ashkin, A., and Dziedzic, J. M., Observation of Radiation Pressure Trapping of Particles by Alternating Laser Beams, *Phys. Rev. Lett.* **54**, 1245–1248 (1985).

Ashkin, A., Dziedzic, J. M., Bjorkholm, J. E., and Chu, S., Observation of a Single-Beam Gradient Force Optical Trap for Dielectric Particles, *Opt. Lett.* **11**, 288–290 (1986).

Ashkin, A., Dziedzic, J. M., and Yamane, T., Optical Trapping and Manipulation of Single Cells Using Infrared Laser Beams, *Nature* **330**, 769–771 (1987).

Baumann, C. G., Bloomfield, V. A., Smith, S. B., Bustamante, C., Wang, M. D., and Block, S. M., Stretching of Single Collapsed DNA Molecules, *Biophys. J.* **78**, 1965–1978 (2000).

Berns, M. W., Laser Scissors and Tweezers, *Sci. Am.* **278**, 62–67 (1998).

Berns, M. W., and Round, D. E., Cell Surgery by Lasers, *Sci. Am.* **222**, 98–103 (1970).

Best, C. J. M., and Emmert-Buck, M. R., Molecular Profiling of Tissue Samples Using Laser Capture Microdissection, *Expert. Rev. Mol. Diagn.* **1**, 53–60 (2001).

Block, S. M., Construction of Optical Tweezers, in D. L. Spector, R. Goldman, and L. Leinward, eds., *Cell Biology: A Laboratory Manual*, Cold Spring Harbor Press, Cold Spring Harbor, NY, 1998.

Bouchal, Z., Wagner, J., and Chlup, M., Self-reconstruction of a distorted non-diffracting beam, *Opt. Commun.* **151**, 207–211 (1998).

Clement-Sengewald, A., Schütze, K., Ashkin, A., Palma, G. A., Kerlen, G., and Brem, G., Fertilization of Bovine Oocytes Induced Solely with Combined Laser Microbeam and Optical Tweezers, *J. Assist. Reprod. Genet.* **13**, 259–265 (1997).

Dufresne, E., Spalding, G., Dearing, M., Sheets, S., and Grier, D., Computer-Generated Holographic Optical Tweezer Arrays, *Rev. Sci. Instrum.* **72**, 1810–1816 (2001).

Emmert-Buck, M. R., Bonner, R. F., Smith, P. D. et al., Laser Capture Microdissection, *Science* **274**, 998–1001 (1996).

Emmert-Buck, M. R., Gillespie, J. W., Paweletz, C. P., Ornstein, D. K., Basrur, V., Appella, E., et al., An Approach to Proteomic Analysis of Human Tumors, *Mol. Carcinog.* **27**, 158–165 (2000).

Emmert-Buck, M. R., Roth, M. J., Zhuang, Z., Campo, E., Rozhin, J., Sloane, B. F., et al., Increased Gelatinase A (MWP-2) and Cathepsin B Activity in Invasive Tumor Regions of Human Colon Cancer Samples, *Am. J. Pathol.* **145**, 1285–1290 (1994).

Fallman, E., and Axner, O., Design for Fully Steerable Dual-Trap Optical Tweezers, *Appl. Opt.* **36**, 2107–2113 (1997).

Garcies-Chavez, V., McGloin, D., Melville, H., and Dholakia, K., Simultaneous Micromanipulation in Multiple Planes Using a Self-Reconstructing Light Beam, *Nature* **419**, 145–147 (2002).

Gillespie, J. W., Nasir, A., and Kaiser, H. E., Loss of Heterozygosity in Papillary and Follicular Thyroid Carcinoma: A Mini Review, *In Vivo* **14**, 139–140 (2000).

Greulich, K. O., *Micromanipulation by Light in Biology and Medicine*, Birkhäuser Verlag, Germany, 1999.

Guck, J., Ananthakrishnan, R., Mahmood, H., Moon, T. J., Cunningham, C. C., and Käs, J., The Optical Stretcher: A Novel Laser Tool to Micromanipulate Cells, *Biophys. J.* **81**, 767–784 (2001).

Hirano, K., Baba, Y., Matsuzawa, Y., and Mizuno, A., Manipulation of Single Coiled DNA Molecules by Laser Clustering of Microparticles, *Appl. Phys. Lett.* **80**, 515–517 (2002).

Igasaki, Y., Li, F., Yoshida, N., Toyoda, H., Inoue T., Mukohzaka, N., Kobayashi, Y., and Hara, T., High Efficiency Electrically Addressable Phase-Only Spatial Light Modulator, *Opt. Rev.* **6**, 339–334 (1999).

König, K., Liang, H., Berns, M. W., and Tromberg, B. J., Cell Damage by Near-IR Microbeams, *Nature* **377**, 20–21 (1995).

Krasieva, T. B., Chapman, C. F., Lamomte, V. J., Venugopalen, V., Berns, M. W., and Tromber, B. J., Mechanisms of Cell Permeabilization by Laser Microirradiation, *Proc. Soc. Photo-Opt. Instrum. Eng.* **3260**, 38–44 (1998).

Lee, S., McAuliffe, D. J., Zhang, H., Xu, Z., Taitelbaum, J., Flotte, T. J., and Doukas, A. P., Stress-Wave-Induced Membrane Permeation of Red Blood Cells Is Facilitated by Aquaporins, *Ultrasound Med. Biol.* **23**, 1089–1094 (1997).

Lutchman, M., Park, S., Kim, A. C., et al., Loss of Heterozygosity on Spin Prostate Cancer Implicates a Role for Dematin in Tumor Progression, *Cancer Genet. Cytogenet.* **115**, 65–69 (2000).

MacDonald, M. P., Volke-Sepulveda, K., Paterson, L., Arlt, J., Sibetti, W., and Dholakia, K., Revolving Interference Patterns for the Rotation of Optically Trapped Particles, *Opt. Commun.* **201**, 21–28 (2002).

Maitra, A., Gazdar, A. F., Moore, T. O., and Moore, A. Y., Loss of Heterozygosity Analysis of Cutaneous Melanoma and Benign Melanocytic Nevi: Laser Capture Microdissection Demonstrates Clonal Genetic Changes in Acquired Nevocellular Nevi, *Hum. Pathol.* **33**, 191–197 (2002).

Mehta, A. D., Pullen, K. A., and Spudich, J. A., Single Molecule Biochemistry Using Optical Tweezers, *FEBS (Fed. Eur. Biochem. Soc.) Lett.* **430**, 23–27 (1998).

Meiners, J.-C., and Quake, S. R., Femtonewton Force Spectroscopy of Single Extended DNA Molecules, *Phys. Rev. Lett.* **84**, 5014–5017 (2000).

Misawa, H., Sasaki, K., Koshioka, M., Kitamura, N., and Masuhara, H., Multibeam Laser Manipulation and Fixation of Microparticles, *Appl. Phys. Lett.* **60**, 310–312 (1992).

Ornstein, D. K., Gillespie, J. W., Paweletz, C. P., Duray, P. H., Herring, J., Vocke, C. D., et al., Proteomic Analysis of Laser Capture Microdissected Prostate Cancer and *In Vitro* Cell Lines, *Electrophoresis* **21**, 2235–2242 (2000).

Perkins, T. T., Quake, S. R., Smith, D. E., and Chu, S., Relaxation of a Single DNA Molecule Observed by Optical Microscopy, *Science* **264**, 822–826 (1994).

Perkins, T. T., Smith, D. E., Larson, R. G., and Chu, S., Stretching of a Single Tethered Polymer in a Uniform Flow, *Science* **268**, 83–87 (1995).

Rink, K., Delacretaz, G., Salathe, R. P., Sejin, A., Mocera, D., Germond, M., DeGrandi, P., and Fakan, S., Non-Contact Microdrilling of Mouse Zona Pellucida with an Objective-Delivered 1.48-μm Diode Laser, *Lasers Surg. Med.* **18**, 52–62 (1996).

Roberts, J. P., The Cutting Edge in Laser Microdissection, *Biophotonics Int.* **9,** 50–53 (2002).

Schurr, J. M., and Smith, S. B., Theory for the Extension of a Linear Polyelectrolyte Attached at One End in an Electric Field, *Biopolymers* **29**, 1161–1165 (1990).

Schütze, K., Becker, I., Becker, K. F., Thalhammer, S., Stark, R., Hecl, W. M., Böhm, M., and Pösl, H., Cut Out or Poke In—The Key to the World of Single Genes: Laser Manipulation as a Valuable Tool on the Lookout for the Origin of Disease, *Genet. Anal.* **14**, 1–8 (1997).

Seeger, S., Manojembaski, S., Hutter, K. J., Futterman, G., Welfrum, J., and Greulich, K. O., Application of Laser Optical Tweezers in Immunology and Molecular Genetics, *Cytometry* **12**, 497–504 (1991).

Simone, N. L., Remaley, A. T., Charboneau, L., Petricoin III, E. F., Glickman, J. W., Emmert-Buck, M. R., Fleisher, T. A., and Liotta, L. A., Sensitive Immunoassay of Tissue Cell Proteins Procured by Laser Capture Microdissection, *Am. J. Path.* **96**, 445–452 (2000).

Smith, S. B., Cui, Y., and Bustamante, C., Overstretching B-DNA: The Elastic Response of Individual Double Stranded and Single Stranded DNA Molecules, *Science* **271**, 795–799 (1996).

Soughayer, J. S., Krasieve, T., Jacobson, S. C., Ramsey, J. M., Tromberg, B. J., and Allbritton, N. L., Characterization of Cellular Optoporation with Distance, *Anal. Chem.* **72**, 1342–1347 (2000).

Stout, A. L., Detection and Characterization of Individual Intermolecular Bonds Using Optical Tweezers, *Biophys. J.* **80**, 2976–2986 (2001).

Strick, T., Allemand, J.-F., Croquette, V., and Bensimon, D., The Manipulation of Single Biomolecules, *Phys. Today* **54**, 46–51 (2001).

Svoboda, K., and Block, S. M., Biological Applications of Optical Forces, *Ann. Rev. Biophys. Biomol. Struct.* **23**, 247–285 (1994).

Tsukakoshi, M., Ksata, S., Nomiya, Y., and Katsuya, Y., A Novel Method of DNA Transfection by Laser Microbeam Cell Surgery, *Appl. Phys. B* **B35**, 135–140 (1984).

Ulanowski, E., Optical Tweezers Principles and Applications, *Proc. RMS* **36**, 7–14 (2001).

Visscher, K., Brakenhoff, G. J., and Krol, J. J., Micromanipulation by Multiple Optical Traps Created by a Single Fast Scanning Trap Integrated with the Bilateral Confocal Scanning Laser Microscope, *Cytometry* **14**, 105–114 (1993).

Visscher, K., Gross, S. P., and Block, S. M., Construction of Multiple-Beam Optical Traps with Nanometer-Resolution Position Sensing, *IEEE J. Quantum Electronics* **2**, 1066–1076 (1996).

Visscher, K., Schnitzer, M. J., and Block, S. M., Single Kinosin Molecules Studied with a Molecular Force Clamp, *Nature* **400**, 184–189 (1999).

Wang, M. D., Yin, H., Landick, R., Gelles, J., and Block, S. M., Stretching DNA with Optical Tweezers, *Biophys. J.* **72**, 1335–1346 (1997).

Weber, G., Monajembashi, S., Greulich, K. O., and Wolfrum, J., Genetic Changes Induced in Higher Plants by a UV Laser Microbeam, *Israel J. Botany* **40**, 115–122 (1990).

Willingham, E., Laser Microdissection Systems, *The Scientist* **16**, 42–44 (2002).

Xie, C., Dinno, M. A., and Li, Y.-Q., Near-Infrared Raman Spectroscopy of Single Optically Trapped Biological Cells, *Opt. Lett.* **27**, 249–251 (2002).

Yin, H., Wang, M. D., Svoboda, K., Landik, R., Block, S. M., and Gelles, J., Transcription Against an Applied Force, *Science* **270**, 1653–1657 (1995).

Zemanek, P., Jonas, L., Sramek, L., and Liska, M., Optical Trapping of Nanoparticles and Microparticles by a Gaussian Standing Wave, *Opt. Lett.* **24**, 1448–1450 (1999).

CHAPTER 15

Nanotechnology for Biophotonics: Bionanophotonics

Some describe us as living in an era of Nanomania where there is a general euphoria about nanoscale science and technology. The fusion of nanoscience and nanotechnology with biomedical research has also broadly impacted biotechnology. The subject covered in this chapter, however, is more focused, dealing with the interface between biomedical science and technology and nanophotonics, hence the term *bionanophotonics*. Nanophotonics is an emerging field that describes nanoscale optical science and technology.

Specifically, this chapter discusses the use of nanoparticles for optical bioimaging, optical diagnostics, and light-guided and activated therapy. Section 15.2 describes the power of nanochemistry to produce the various nanoparticles and tailor their structures and functions for biomedical applications. Specific examples provided for bioimaging are two classes of nanoparticle emitters. One consists of semiconductor nanoparticles, also known as *quantum dots*, whose luminescence wavelength is dependent on the size and the nature of the semiconductors. These nanoparticle emitters can be judiciously selected to cover the visible to the IR spectral range. They can also be surface-functionalized to be dispersable in biological media as well as to be conjugated to various biomolecules.

Another class of nanoparticle emitters for bioimaging consists of up-converting nanophores comprised of rare-earth ions in a crystalline host. They convert near-IR and IR radiation, which can penetrate deeper into a tissue, to emissions in the visible range by utilizing the process of sequential multiphoton absorption. In addition to bioimaging, the up-converting nanophores can also allow treatment of deeper tumors by using them for multiphoton photodynamic therapy described in Chapter 12. The use of metallic nanoparticles and nanorods for biosensing is described in Section 15.5.

The next two sections, 15.6 and 15.7, describe the use of a nanoparticles platform, for intracellular diagnostic and targeted drug delivery. Section 15.6 discusses the PEBBLE nanosensors approach for monitoring intercellular

Introduction to Biophotonics, by Paras N. Prasad.
ISBN: 0-471-28770-9   Copyright © 2003 John Wiley & Sons, Inc.

activities. Section 15.7 discusses the use of nanoclinics, which are thin silica shells (packaging various probes for diagnostics and agents for external activation) and are surface-functionalized with carrier groups to target specific biological sites such as cancer cells.

The chapter concludes with a discussion of future directions of research and development in bionanophotonics. For further reading, the following reviews are recommended:

Shen et al. (2000): A feature review article on nanophotonics

Murray et al. (2002): A review on synthesis and characterization of nanocrystals

15.1 THE INTERFACE OF BIOSCIENCE, NANOTECHNOLOGY, AND PHOTONICS

Imagine nanosubmarines navigating through our bloodstreams and destroying nasty viruses and bacteria. Imagine nanorobots hunting for cancer cells throughout our body, finding them, then reprogramming or destroying them. A subject of science fiction at one time has now been transformed into a future vision showing promise to materialize. The fusion of nanoscience and nanotechnology into biomedical research has brought in a true revolution that is broadly impacting biotechnology. New terms such as *nanobioscience*, *nanobiotechnology*, and *nanomedicine* have come into existence and gained wide acceptance.

Table 15.1 lists some nanotechnology frontiers in bioscience. The content of this chapter, however, is more focused on the applications of nanophotonics in biomedical science and technology. Nanophotonics is an emerging field that deals with optical interactions on a scale much smaller than the wavelength of light used (Shen et al., 2000). The three major areas of nanophotonics are shown in Table 15.2.

Nanoscale confinement of radiation is achieved in a near-field geometry. This allows one to break diffraction barriers and obtain optical resolution to less than 100 nm. Near-field microscopy, discussed in Chapter 7, is becoming a powerful biomedical research tool to probe structure and functions of submicron dimension biological species such as bacteria. Nanoscale confinement of matter is achieved by producing nanoparticles, nanomers, nanodomains, and nanocomposites. The nanosize manipulation of molecular architecture and morphology provides a powerful approach to control the electronic and optical properties of a material. An example is a semiconductor quantum dot, a nanoparticle whose electronic band gap and thus the emission wavelength are strongly dependent on its size. Nanoscale control of the local structure in a nanocomposite, consisting of many domains and separated only on the nanometer scale, provides an opportunity to manipulate excited-state dynamics and electronic energy transfer from one domain to another.

TABLE 15.1. Nanotechnology Frontiers in Bioscience

Biophysics	**Structural Biology**
Nanomechanics	Protein folding, design
Optical traps	"Rational" drug/ligand design
Flexible-probe methods	Novel and improved methods
Single-molecule methods	**Computational Biology**
FRET	Protein folding, design
New labels	"Rational" drug/ligand design
New reporters	Bioinformatic design, regulation
New imaging, microscopies	**Biotronics—Biomolecules on Chips**
Scanned-probe	DNA and protein nanotrays
Combinations	Sensors, detectors, diagnostics
	Labs-on-a-chip
Biochemistry	**Biofabrication**
Single-molecule enzymology	Nanoparticle delivery systems
Single-molecule kinetics	Biomaterials, tissue engineering
Single-molecule sequencing	Implants, prosthetics

Source: Steven Block from http:/grants.nih.gov/grants/becon/becon_symposia.htm.

TABLE 15.2. Three Major Areas of Nanophotonics

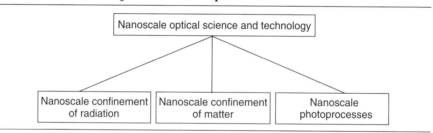

Such nanostructured materials can provide significant benefits in FRET imaging and in flow cytometry, the topics already covered in Chapters 8 and 11, respectively.

Nanoscale photoprocesses such as photopolymerization provide opportunities for nanoscale photofabrication. Near-field lithography can be used to produce nanoarrays for DNA or protein detection. The advantage over the microarray technology, discussed in Chapter 10, is the higher density of arrays obtainable using near-field lithography, thus allowing one to use small quantities of samples. This is a tremendous benefit for protein analysis in the case when the amount of protein produced is very minute and, as discussed in Chapter 10, there is no equivalent of DNA PCR amplification for proteins to enhance the detection.

The applications of nanophotonics to biomedical research and biotechnology range from biosensing, to optical diagnostics, to light activated therapy. Nanoparticles provide a highly useful platform for intracellular optical diagnostics and targeted therapy. The area of usage of nanoparticles for drug delivery has seen considerable growth. This chapter presents some selected examples of nanotechnology and applications to biophotonics.

A great deal of information can be obtained from visiting various websites on the Internet. Some selected examples of these websites are:

- National Nanotechnology Initiative—http://www.nano.gov
- Engines of Creation—http://www.foresight.org/EOC
- Stanford Nanofabrication Facility—http://www-snf.stanford.edu
- Cornell Nanofabrication Facility—http://www.cnf.cornell.edu
- NIH conference on nanotechnology and biomedicine—
  http://www.masimax.com/becon/index.html
- University at Buffalo Biophotonics and Nanophotonics Program—
  http://www.biophotonics.buffalo.edu

15.2 NANOCHEMISTRY

Nanochemistry is an active new field that deals with confinement of chemical reactions on nanometer length scale to produce chemical products that are of nanometer dimensions (generally in the range of 1–100 nm) (Murray et al., 2000). The challenge is to be able to use chemical approaches that would reproducibly provide a precise control of composition, size, and shape of the nano-objects formed. These nanomaterials exhibit new electronic, optical, and other physical properties that depend on their composition, size, and shape. Nanoscale chemistry also provides an opportunity to design and fabricate hierarchically built multilayer nanostructures to incorporate multifunctionality at nanoscale.

Nanochemistry offers the following capabilities:

- Preparation of nanoparticles of a wide range of metals, semiconductors, glasses, and polymers
- Preparation of multilayer, core-shell-type nanoparticles
- Nanopatterning of surfaces, surface functionalization, and self-assembling of structures on this patterned template
- Organization of nanoparticles into periodic or aperiodic functional structures
- *In situ* fabrication of nanoscale probes, sensors, and devices

TABLE 15.3. Overview of Nanochemistry

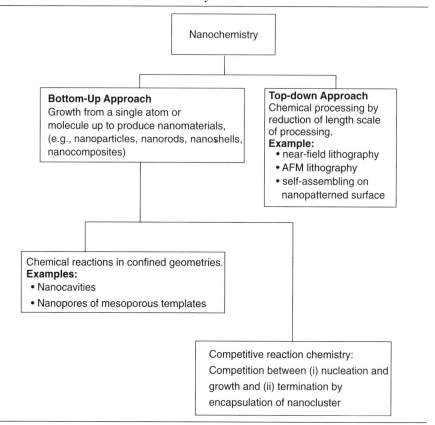

A number of approaches provide nanoscale control of chemical reactions. Some of these are shown in Table 15.3.

An example of a reaction in a confined geometry is the synthesis of nanoparticles such as CdS (quantum dots to be discussed in a subsequent section) in a reverse micelle nanoreactor. Figure 15.1 shows a schematic representation of the reverse micelle chemistry.

The reverse micellar system is generally composed of two immiscible liquids, water and oil, where the aqueous phase is dispersed as nanosize water droplets encapsulated by a monolayer film of surfactant molecules in a continuous nonpolar organic solvent such as a hydrocarbon oil. The continuous oil phase generally consists of isooctane; sodium bis(2-ethylhexyl) sulfosuccinate (AOT) serves as the surfactant. In addition to water, aqueous solutions containing a variety of dissolved salts, including cadmium acetate and sodium sulfide, can be solubilized within the reverse micelles (Masui et al., 1997). The size of the micelle, and subsequently the volume of the aqueous pool contained

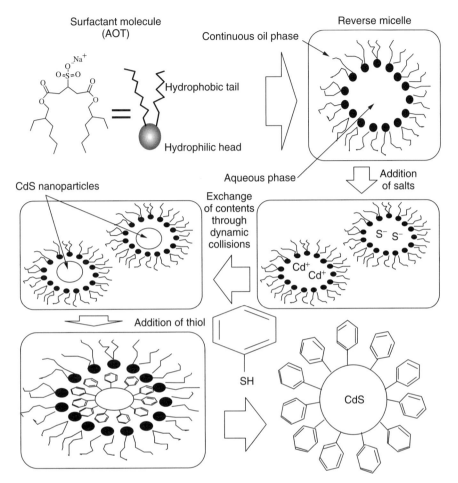

Figure 15.1. Synthesis of CdS nanocrystals in a reverse micelle nanocavity (J. Winiarz, Ph.D. dissertation, Chemistry, SUNY at Buffalo, 2002).

within the micelle, is governed by the water to surfactant ratio, also termed W_0, where $W_0 = [H_2O]/[\text{surfactant}]$ (Pileni, 1993). Continuous exchange of the micellar contents through dynamic collisions enables the reaction to proceed. However, since the reaction is confined within the cavity of the micelle, growth of the nanocrystal beyond the dimensions of the cavity is inhibited. In the final stage of this synthesis, the passivating capping reagent, p-thiocresol, is added to the continuous oil phase. This species is then able to enter the aqueous phase as an RS^- anion and bond to the surface of the contained nanocrystal, eventually rendering the surface of the nanocrystal hydrophobic and inducing precipitation of the capped CdS nanoparticles.

The reverse micelle chemistry also lends itself readily to a multistep synthesis of a multilayered nanoparticle such as a core-shell structure. An

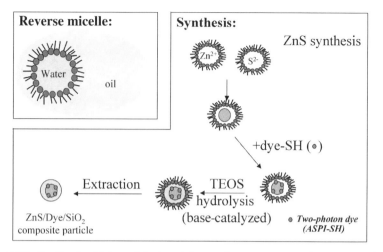

Figure 15.2. Synthesis of hierarchical multilayered nanostructures in reverse micelles.

example is that of a silica nanobubble containing an organic dye attached to a ZnS nanocrystal core (Lal et al., 2000). A schematic of this synthesis is shown in Figure 15.2.

Using reverse-micelles-mediated synthesis, Lal et al. (2000) prepared ZnS/dye/SiO$_2$ heterostructured nanoparticles through a multistep reaction. Specifically, by subsequent reaction and chemical processing within the cavity, inorganic–organic particles were made containing a zinc sulfide core coupled via the thiol group to a two-photon dye, ASPI-SH (1-methyl-4-(E)-2-{4-[methyl(2-sulfanylethyl)amino]-phenyl}ethenyl)pyridinium iodide). ASPI-SH belongs to a general family of hemicyanine dyes (containing short alkyl chain) such as amino styryl pyridine derivatives exhibiting a very high two-photon absorption cross section at 1.06 μm (wavelength of Nd:YAG laser) and producing efficient up-converted emission, useful for bioimaging. Zinc sulfide was chosen as a semiconductor because its band gap is well separated from that of the chromophore. In other words, there is no spectral overlap between zinc sulfide and the dye. During the reaction, the nucleus of zinc sulfide grows to a desired/required size and then the surface of these nanocrystallites is passivated or capped through covalent addition of the dye (ASPI-SH) thiol where the thiol group acts as a growth moderator (Swayambunathan et al., 1990). This property stems from the ability of thiolate ions to bind to the metal ions on the semiconductor surface, thereby effectively inhibiting the growth of the semiconductor nanoparticles. This process was followed by the introduction of sol–gel silica precursor, tetraethoxy orthosilicate (TEOS), which undergoes hydrolysis in the aqueous core forming a silica shell around the dye-capped zinc sulfide particles. The sol–gel processing has been discussed in Chapter 9 on biosensors. The advantage of using TEOS as the silica precursor is its relatively slow and controllable rate of reaction. Thus, the reverse micelle

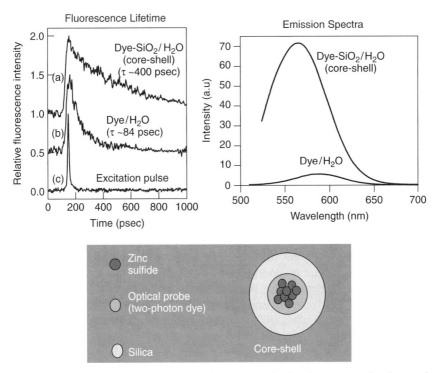

Figure 15.3. Comparison of the lifetime (as measured using femtosecond pulses and a 2-psec resolution streak camera) and the fluorescence efficiency of a dye solution and dispersion of the same dye when encapsulated. (Reproduced with permission from Lal et al., 2000.)

technique is versatile and can be used for the synthesis of other kinds of hybrid materials.

There are several advantages to using silica shells as stabilizers. Unlike polymers, they are not subject to microbial attack and there is no swelling or porosity change occurring in these particles with the change of pH (Jain et al., 1998). Silica is chemically inert and, therefore, does not affect the redox reactions at the core surface (Markowitz et al., 1999). The shell is optically transparent; furthermore, the shell prevents coagulation during chemical reactions and concentrated dispersions of nanosized semiconductors can be made. Also, the silica shell acts as a stabilizer, limiting the effect of the outside environment on the core particles. This is particularly important for dyes, which are sensitive to certain solvents, especially water, and which quench the emission due to certain nonradiative decay processes. Figure 15.3 shows a dramatic increase in the lifetime and the emission efficiency of the two-photon ionic dye (ASPI-SH) when deposited on a core and encapsulated in a silica nanobubble. The dispersion medium is water. Most important, however, is the ease of synthesis because no special conditions (e.g., initiator, temperature) are required for the synthesis.

Photobleaching and thermally induced degradation are the problems commonly encountered in laser dyes that reduce the operational lifetime of a dye. By encapsulation of a dye within the silica shell, where silica is chemically and thermally inert, photobleaching and photodegradation of the dye can also be minimized. This advantage has been demonstrated in our work. Another advantage of using silica is the introduction of a specific surface functionality, which can be obtained by modifying the surface hydroxyls on the silica surface with amines, thiols, carboxyls, and methacrylate. This modification can facilitate the incorporation of these nanoparticles into nonpolar solvents, glasses, and polymeric matrixes.

Competitive reaction chemistry (CRC) has also been utilized to prepare various sulfide and selenide nanocrystals (e.g., CdS, CdSe) (Herron et al., 1990). In the initiation phase of the synthesis, a solution containing cadmium ions, generated from cadmium acetate, is introduced into a solution containing S^{2-} and RS^- (RS^- represents an organic thiol anion) in the form of sodium sulfide and p-thiocresol, respectively, to create small nanocrystals of CdS. Once they are formed, a propagation step of nanocrystal growth competes with the growth terminating reaction of the thiolate with the surface of the nanocrystal. It has also been shown that although being covalently bonded to the surface of CdS nanocrystals, the thiocresol species are dislocated by additional sulfide ions, allowing for further growth of the cluster. However, once a sulfide ion has been incorporated into a given cluster, it cannot be replaced by a thiolate ion. Through this process the nanocrystals are allowed to grow until the supply of S^{2-} has been exhausted.

The above examples of nanocrystal formation are, clearly, examples of the "bottom-up" approach—that is, building naonobjects from smaller objects (molecules). On the other hand, examples of a "top-down" approach may be found in two-photon-induced photochemistry. Using near-field propagation of a femtosecond pulse laser beam at 800 nm (pulse with very high peak power to induce efficient two-photon excitation), Shen et al. (2000) successfully achieved two-photon induced photochemistry to produce structures of the dimension of 70 nm. The high spatial localization using two-photon excitation reduces the diameter of photofabrications to 70 nm, whereas single-photon excitation leads to 120-nm-size photopolymerized structures. Conventional photolithographic structures are much larger than this.

15.3 SEMICONDUCTOR QUANTUM DOTS FOR BIOIMAGING

Quantum dots (also frequently abbreviated as Qdots) are nanocrystals of semiconductors that exhibit quantum confinement effects, once their dimensions get smaller than a characteristic length, called the *Bohr's radius*. This Bohr's radius is a specific property of an individual semiconductor and can be equated with the electron–hole distance in an exciton that might be formed in the bulk semiconductor. For example, it is 2.5 nm for CdS. Below this length scale the band gap (the gap between the electron occupied energy level, similar to HOMO, and the empty level, similar to LUMO, which are discussed in Chapter 2) is size-dependent. The physical picture can be visualized in terms of the simple concept of particle in a box, discussed in Chapter 2. As the length of the box (quantum confinement size of the particle) decreases, the band gap (the energy level separation) increases. In other words, as the particle size decreases below the Bohr's radius, the absorption, and, subsequently, the emission wavelengths of the nanoparticles shift to a shorter wavelength (toward UV). The quantum dots, therefore, offer themselves as fluorophores where the emission wavelength can be tuned by selecting appropriate-size nanocrystals (Bruchez et al., 1998; Chan and Nie, 1998). By appropriate selection of the materials (e.g., CdS, CdSe, etc.) and the size of their nanocrystals, a wide spectral range of emission can be covered for bioimaging. Also, a significantly broad range of emission covered by many sizes of nanocrystals of a given material can be excited at the same wavelength. The typical line widths are 20–30 nm, thus relatively narrow, which helps if one wants to use the quantum dots more effectively for multispectral imaging. Compared to organic fluorophores, the major advantages offered by quantum dots for bioimaging are:

- Quantum dot emissions are considerably narrower compared to organic fluorophores, which exhibit broad emissions. Thus, the complication in simultaneous quantitative multichannel detection posed by cross-talks between different detection channels, derived from spectral overlap, is significantly reduced.
- The lifetime of emission is longer (hundreds of nanoseconds) compared to that of organic fluorophores, thus allowing one to utilize time-gated detection to suppress autofluorescence, which has a considerably shorter lifetime.
- The quantum dots do not readily photobleach.
- They are not subject to microbial attack.

A major problem in the use of quantum dots for bioimaging is the reduced emission efficiency due to the high surface area of the nanocrystal. A number of groups as well as new start-up companies are addressing this issue.

Alivisatos and co-workers (Bruchez et al., 1998) used a core-shell structure in which a shell of another semiconductor (ZnS) with a larger band gap encapsulated the core of a narrower band-gap semiconductor (CdSe). This encapsulation produced confinement of the excitation to the core and eliminated the surface-induced nonradiative relaxation pathways to enhance the emission efficiency of the core quantum dot.

Figure 15.4 illustrates the different emission colors obtainable from the quantum dots of a number of materials of different sizes. It illustrates the spectral tunability as well as the narrow line width of luminescence obtainable from quantum dots.

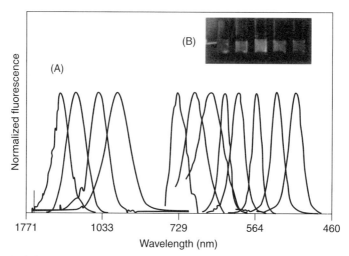

Figure 15.4. (A) Size- and material-dependent emission spectra of several surfactant-coated semiconductor nanocrystals in a variety of sizes. The first five from right represent different sizes of CdSe nanocrystals with diameters of 2.1, 2.4, 3.1, 3.6, and 4.6 nm (from right to left). The next three from right is of InP nanocrystals with diameters of 3.0, 3.5, and 4.6 nm. The IR emitters are InAs nanocrystals with diameters of 2.8, 3.6, 4.6, and 6.0 nm. (B) A true-color image of a series of silica-coated core (CdSe)-shell (ZnS or CdS) nanocrystal probes in aqueous buffer, all illuminated simultaneously with a handheld ultraviolet lamp. (Reproduced with permission from Bruchez et al., 1998.)

To make the quantum dots water-dispersable, Alivisatos and co-workers (Bruchez et al., 1998) added a layer of silica onto the core-shell structure. The silica encapsulated core-shell of nanocrystals were soluble and stable in water or buffered solutions. They also exhibited a fair fluorescence quantum yield (up to 21%). Alivisatos' group demonstrated these nanocrystals for biological staining by fluorescently labeling 3T3 mouse fibroblast cells using two different-sized CdSe–CdS core-shell quantum dots encapsulated in a silica cell.

Nie's group (Chan and Nie, 1998) covalently bonded the quantum dots to biomolecules (such as proteins) for use in ultrasensitive biological detection. Their approach utilized coupling to mercaptoacetic acid through sulfur binding, which also solubilizes the quantum dots in an aqueous medium. Then the acid group is attached to a protein through an amide linkage (Chapter 3). The mercaptoacetic acid layer is also expected to reduce passive protein adsorption on the quantum dots.

A schematic of the ZnS-capped CdSe quantum dots covalently coupled to a protein by mercaptoacetic acid is shown in Figure 15.5. The work of Nie's group showed that the optical properties of the quantum dots did not change after conjugations and solubilization. They also reported that the Qdot emission was 100 times as stable as that of the common organic dye rhodamine 6G

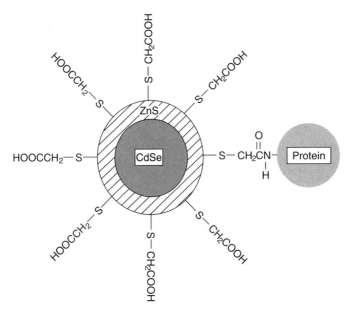

Figure 15.5. Schematics of a ZnS-capped CdSe quantum dot covalently coupled to a protein by mercaptoacetic acid. (Reproduced with permission from Chan and Nie, 1998.)

against photobleaching. They also demonstrated that the protein-attached Qdots were biocompatible *in vitro* as well as with living cells. For this they used transferrin-Qdot bioconjugates. Cultured HeLa cells were incubated with mercapto-Qdots as control and with transferrin-QD bioconjugates. Only the transferrin-Qdot bioconjugates were transported into the cell, as evidenced by emission from the stained cells, indicating receptor-mediated endocytosis. This result was taken as evidence that the attached transferrin molecules were still active and were recognized by the receptors on the cell surface.

Akerman et al. (2002) showed that ZnS-capped CdSe Qdots coated with a lung-targeting peptide accumulate in the lungs of mice after intravenous injection, whereas two other peptides directed Qdots to blood vessels or lymphatic vessels in tumors.

Bawendi and co-workers (Mattoussi et al., 2000) utilized a chimeric fusion protein to electrostatically bind it to the oppositely charged surface of capped colloidal core-shell-type CdSe–ZnS quantum dots to produce a bioconjugate. They suggested that this approach provided all the advantages of lipoic acid capped quantum dots (such as photochemical stability, size-dependent emission covering a broad spectral range, and aqueous compatibility) and at the same time yielded a facile electrostatic conjugation of a bioactive protein. In their approach, they capped the CdSe–ZnS nanocrystals, first with alkyl-COOH capping reagents. Adjusting the pH (basic) of an aqueous dispersion produces the quantum dots with negative charges ($-COO^-$ groups). The use of an engineered bifunctional recombinant protein consisting of a positive-charge domain leads to the formation of a bioconjugate by electrostatic attraction.

Thiol-terminated DNA segments (20–25 mers) have also been immobilized on mercaptopropionic acid capped CdSe–ZnS nanocrystals (Mitchell et al., 1999).

Nie's group (Han et al., 2001) has proposed the use of a porous microbead of polystyrene to capture quantum dots in specific quantities and in a wide range of colors and intensities. They demonstrated the application to DNA analysis by preparing microbeads of three different colors and attaching them to strips of genetic materials, each color corresponding with a specific DNA sequence. They then were used to probe complementary pieces of genetic material in a DNA mixture. The basics of this approach have already been discussed in Chapter 10.

15.4   METALLIC NANOPARTICLES AND NANORODS FOR BIOSENSING

Other types of materials used for biosensing are in the form of metal nanoparticles and nanorods. Storhoff and Mirkin (1999) linked a single-stranded DNA, modified with a thiol group at one terminal, to a gold nanoparticle ~15 nm in diameter via strong gold–sulfur interactions, discussed in Chapter 9. The 15-nm-diameter gold particles exhibit well-defined surface plasmon resonance, a topic also discussed in Chapter 9. Due to this resonance, the individual gold particles, even when attached to DNA, exhibit a burgundy-red color. When this DNA attached to the gold particle hybridizes with the complementary DNA in the test sample, the duplex formation leads to aggregation of the nanoparticles, shifting the surface plasmon resonance and, thus, the color to blue black. The reason for the shift is that the plasmon band is very sensitive to the interparticle distance as well as to the aggregate size.

15.5   UP-CONVERTING NANOPHORES

Another group of nanoparticles useful for bioimaging as well as for light activation of therapy is that of rare-earth-ion-doped oxide nanoparticles (Holm et al., 2002). The rare-earth ions are well known to produce IR to visible up-conversion by a number of mechanisms as shown in Figure 15.6. These up-conversion processes in rare-earth ions, like the two-photon absorption in organics, discussed in Chapter 5 and in Chapters 7 and 8 (two-photon bioimaging), are quadratically dependent on the excitation intensity. Thus, they provide better spatial resolution. They produce background-free (practically no autofluorescence) detection, because the excitation source is in the near-IR (generally 974-nm laser diodes). An advantage offered by these nanopar-

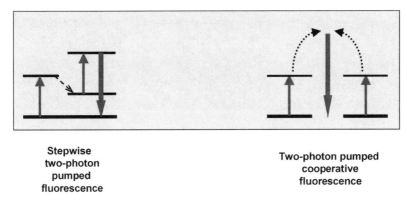

Stepwise
two-photon
pumped
fluorescence

Two-photon pumped
cooperative
fluorescence

Figure 15.6. Various up-conversion processes exhibited by rare-earth ions.

ticles over the two-photon dye is that the up-conversion process in the rare-earth nanoparticles is by sequential absorption through real states and is thus considerably stronger. Therefore, one can use a low-power continuous-wave diode laser at 974 nm (which is also very inexpensive and readily available) to excite the up-converted emission. By contrast, the two-photon absorption in organic dyes is a direct (simultaneous) two-photon absorption through a virtual state (see Chapter 5) that requires a high-peak-power pulse laser source. However, the emission from the rare-earth ion is a phosphorescence with a lifetime typically in milliseconds, compared to a dye fluorescence with a lifetime in nanoseconds. The concepts of phosphorescence and fluorescence have been discussed in Chapter 4. Therefore, applications that require short-lived fluorescence cannot use the phosphorescence from these up-converting nanoparticles, also referred to as *nanophores* or *nanophosphors*.

A considerable amount of work on up-converting nanophores and their applications was originally done by SRI (Chen et al., 1999). More recently, our group at the Institute for Lasers, Photonics, and Biophotonics has produced rare-earth-doped yttria (Y_2O_3) nanoparticles and coated them with silica to produce nanophores of size ~25 nm (Holm et al., 2002). These silica-coated nanophores are water-dispersable and extremely stable and exhibit no photobleaching. The size of these nanophores is still small enough for them to penetrate the cell by endocytosis or by functionalizing the surface of the silica coating with a carrier group.

These nanophores are prepared using the reverse micelle chemistry described in Section 15.2 (Kapoor et al., 2000). Salts of Y and Er are used to form functionalized surfactants by replacing the cation, Na^+, in the surfactant bis-2-ethylhexyl sulfosuccinate sodium salt (often abbreviated as Na-AOT). The nanophosphors are synthesized by dissolving appropriate amounts of the dried functionalized surfactant in isooctane. Particles of varying sizes can be synthesized by altering the water to surfactant ratio W_0 (see Section 15.2). Then, a Na-AOT reverse micelle solution of equivalent W_0 containing ammo-

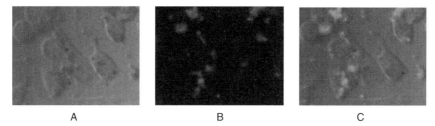

A   B   C

Figure 15.7. Bioimaging using up-converting nanoparticles on oral epithelial carcinoma cells (KB). KB cells were incubated with nanoparticles consisting of Er-doped Y_2O_3 nanophosphors in silica shell. Figure A represents the light transmission image of the KB cells. Figure B is the fluorescence emission after excitation with 974 nm. Figure C is the composite of Figures A and B.

nium hydroxide is added to precipitate the hydroxide precursor nanoparticles. Encapsulation and functionalization, for subsequent ligand coupling, of nanophosphors is accomplished by addition of silica shell. The targeting ligand is then coupled to the —COOH groups or NH_2 groups of the spacer arms by using carbodimides. The same procedure is followed to synthesize Er/Yb co-doped Y_2O_3 and Tm/Yb co-doped Y_2O_3. The up-converted emission of these nanoparticles is red (640 nm) for the Er/Yb co-doped Y_2O_3 particles, green (550 nm) for Er-doped Y_2O_3 particles, and blue (480 nm) for Tm/Yb co-doped Y_2O_3 particles. These wavelengths of light are readily detected with standard CCD arrays and/or a CCD-coupled spectrograph. The use of an IR laser drastically reduces the problems associated with the use of a UV excitation source.

Moreover, the ability to tailor the emission wavelength coupled with our ability to surface functionalize these nanoparticles allows for a number of unique applications of these materials. Our initial studies were conducted in the KB cells (Holm et al., 2002). As can be seen in Figure 15.7, the infrared excitation wavelength does not induce autofluorescence in the target cells. Only the fluorescence emission of the nanophores can be seen (Figure B). This signal-to-noise ratio reduction is of great benefit in the visualization of low-level fluorescent signals in biological systems.

At our Institute for Lasers, Photonics, and Biophotonics, the silica-encapsulated rare-earth-doped Y_2O_3 nanoparticles are also being investigated for multiphoton photodynamic therapy (Roy et al., 2003). The basic concept is similar to the one discussed in the section on two-photon photodynamic therapy in Chapter 12. However, here one utilizes the IR-to-visible up-conversion in these nanophores and not a two-photon active dye, discussed in Chapter 12. The benefits are again greater penetration into a tissue offered by the use of a near-IR (974-nm) excitation source.

For this purpose, we used a well-established photodynamic photosensitizer (PDT drug) HPPH, discussed in Chapter 12, to test the ability of the nanophosphors to excite HPPH. The following study was performed. Sintered nanopar-

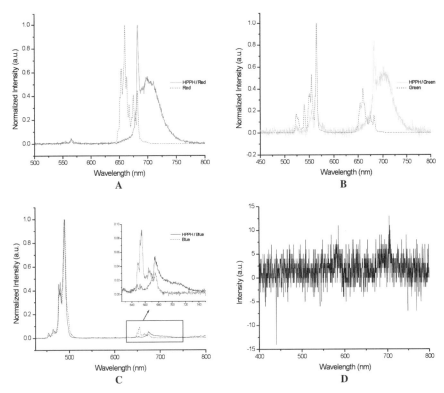

Figure 15.8. Nanophosphor excitation of HPPH. (A) Red emitting nanophosphors, (B) green emitting nanophosphors, (C) blue nanophosphors, (D) HPPH excitation by 974 nm.

ticles were dispersed in DMSO to obtain a translucent colloidal dispersion of nanophosphors. Equal volumes of the nanophosphor solution and 1 mM of HPPH in DMSO were mixed in individual cuvettes. Identical solutions containing only HPPH or the nanoparticles were also placed into cuvettes. Each cuvette was pumped with a 974-nm CW diode laser, and the emission spectra were collected at 90° to the excitation laser with a fiber-coupled CCD spectroscope. The data were normalized to the maximum peak intensity and plotted with identical nanophosphor blank solutions.

It is clearly seen in Figure 15.8 that within the experimental parameters, both the green and red emitting nanophosphors are capable of exciting HPPH. Coupling of the emission of the nanophosphor with HPPH is shown by loss of emission by particle and appearance of the HPPH emission. The blue-emitting nanophosphors, however, did not demonstrate any significant coupling with HPPH. This lack of fluorescent emission from HPPH is due to absence of overlap of emission of the blue nanophosphors with the absorption region of the HPPH.

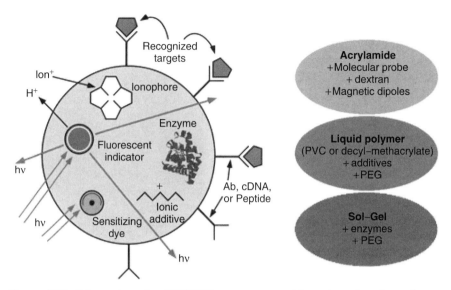

Figure 15.9. Schematics of a PEBBLE nanosensor, with various functions shown. Current matrix materials are presented on the right. (Reproduced with permission from http://www.umich.edu/ ~koplab/research2/analytical/EnterPEBBLEs.html.)

15.6 PEBBLE NANOSENSORS FOR *IN VITRO* BIOANALYSIS

A probe encapsulated by biologically localized embedding (PEBBLE), introduced by Kopelman and co-workers (Clark et al., 1999), enables optical measurement of changes in intracellular calcium levels and pH. It provides a major advancement in the field of nanoprobes and nanomedicine. PEBBLEs are nanoscale spherical devices consisting of sensor molecules entrapped in a chemically inert matrix. Figure 15.9 shows a schematic diagram of a PEBBLE nanosensor that can provide many functions. The matrix materials used for production of PEBBLES are also shown in the figure. The three matrix media used for PEBBLE technology are polyacrylamide hydrogel, sol–gel silica, and cross-linked decyl mathacrylate. These matrices have been used by Kopelman's group to fabricate sensors for H^+, Ca^{2+}, Na^+, Mg^{2+}, Zn^{2+}, Cl^-, NO_2^-, O_2, NO, and glucose. The PEBBLE size ranges from 30 to 600 nm. In the case of polyacrylamide (PAA), a nano-emulsion technique (Daubresse, 1994) similar to the reverse micelle method discussed in Section 15.2 is used to polymerize the monomer (acrylamide), which may contain a hydrophilic dye (selective for the analyte of interest) and an appropriate cross-linker (*N,N*-methylene bis(acrylamide)). The polymerization is initiated with 10% ammonium persulfate. The matrix porosity allows entrapment and sensing of the analyte.

The sol–gel processing has already been discussed in Chapter 9. The decyl methacrylate PEBBLE is made utilizing decyl methacrylate, hexane diol-

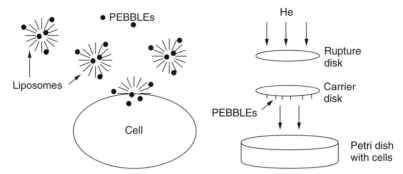

Figure 15.10. Various delivery methods for intracellular delivery of PEBBLE nanosensors. (*Left*) Liposome delivery, (*right*) gene gun delivery. (Reproduced with permission from http://www.umich.edu/ ~koplab/research2/analytical/DeliveryGeneGun.html.)

methacrylate, and doctyl sebacate to produce spherical particles. This PEBBLE matrix provides a hydrophobic environment.

A number of methods have been used by Kopelman and co-workers for intracellular delivery of these PEBBLEs. They are described in Figure 15.10. They include gene gun (using blast of helium shoots) and liposomal delivery.

An example of a PEBBLE nanosensor is the calcium PEBBLE that utilizes calcium Green-1 and sulforhodamine dyes as sensing components. The calcium green fluorescence intensity increases with increasing calcium concentrations, while the sulforhodamine fluorescence intensity is unaffected. Thus, the ratio of the calcium green intensity to the sulforhodamine intensity can be used to measure cellular calcium levels.

According to Kopelman and co-workers, PEBBLE technology offers the following benefits:

- It protects the cells from any toxicity associated with the sensing dye.
- It provides an opportunity to combine multiple sensing components (dyes, ionphores, etc.) and create complex sensing schemes.
- It insulates the indicator dyes from cellular interferences such as protein binding.

15.7 NANOCLINICS FOR OPTICAL DIAGNOSTICS AND TARGETED THERAPY

Our Institute for Lasers, Photonics, and Biophotonics has developed the concept of a nanoclinic, a complex surface functionalized silica nanoshell containing various probes for diagnostics and drugs for targeted delivery (Levy et al., 2002). Nanoclinics provide a new dimension to targeted diagnostics and therapy. These nanoclinics are produced by multistep nanochemistry in a

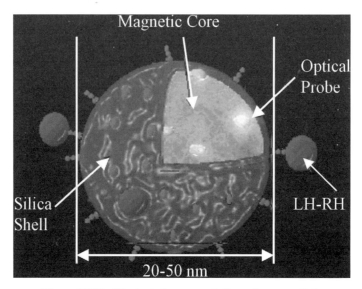

Figure 15.11. Illustrated representation of a nanoclinic

reverse micelle nanoreactor, a method discussed in detail in Section 15.2. An illustrated representation of a nanoclinic is shown in Figure 15.11.

These nanoparticles are subsequently surface-functionalized to target specific cells for biological sites. These nanoclinics are ~30-nm silica shells that can encapsulate various optical, magnetic, or electrical probes as well as platforms containing externally activatable drugs or therapeutic agents (see Figure 15.11). The size of these nanoclinics is small enough to enter the cell in order for them to function from within the cell. Through the development of nanoclinics (functionalized nanometer-sized particles that can serve as carriers), new therapeutic approaches to disease can be accomplished from within the cell. At our Institute, integration of the ferrofluid, nanotechnology, and peptide hormone targeting has resulted in the fabrication of multifunctional nanoclinics. One example of a nanoclinic is a multilayered nanosized structure consisting of an iron oxide core, a two-photon optical probe, and a silica shell with a LH–RH targeting hormone analogue, covalently coupled to the surface of the shell. This protocol can produce nanoclinics with a tunable size from 5 to 40 nm in diameter. They are small enough to be able to diffuse into the tissue and enter the cells (by endocytotic processes) and are large enough to respond to the applied magnetic field at 37°C. High-resolution transmission electron microscopy shows that the structure of the nanoparticle is composed of a crystalline core corresponding to Fe_2O_3 and one amorphous silica layer (bubble). The same crystalline/amorphous structure was obtained by electron diffraction of the particle and also confirmed by x-ray diffraction.

The two-photon dye is able to absorb one photon (with a wavelength of 400 nm) or two photons (with a wavelength of 800 nm) by direct two-photon absorption. The selective interaction and internalization of these nanoclinics with cells was visualized using two-photon laser scanning microscopy, allowing for real-time observation of the uptake of nanoclinics (Bergey et al., 2003). Two different types of particles were used in this study: LH–RH-positive (surface-coupled) and LH–RH-negative (spacer arm only). A suspension of nanoclinics was added to adherent (KB) oral epithelial carcinoma cells (LH–RH receptor positive), and uptake was observed using laser scanning microscopy. The time-dependent uptake of the LH–RH-positive nanoclinics by LH–RH receptor bearing cells was identified. A similar accumulation was not observed in LH–RH-negative nanoclinics studies or LH–RH-positive nanoclinics incubated with receptor-negative cells (UCI-107). Thus, targeting of LH–RH receptor-specific cancer cells and the specific effects of the nanoclinics were demonstrated.

The multifunctional nanoclinics containing the magnetic Fe_2O_3 nanoparticles also produced a new discovery for targeted therapy, a new effect that to our knowledge has not previously been reported, that being the selective lysing of cancer cells in a dc magnetic field using magnetic nanoclinics. Magnetic probes or particles have been investigated as a potential alternative treatment for cancer. Studies have demonstrated that the hyperthermic effect generated by magnetic particles coupled to a high-frequency ac magnetic field (requiring tremendous power) could be used as an alternate or adjuvant to current therapeutic approaches for cancer treatment. This hyperthermic effect (heat produced by the relaxation of magnetic energy of the magnetic material) was shown to effectively destroy tumor tissue surrounding the probes or particles. This approach resulted in reduction of the tumor size by hyperthermic effect when the particles were directly injected into the tissue and were exposed to an alternating magnetic field. However, no targeted therapy using a dc magnetic field has been reported previously, to our knowledge. Our work demonstrated the use of a dc magnetic field at a strength typically achievable by magnetic resonance imaging (MRI) systems for selectively destroying cancer cells. AFM studies together with a detailed study of magnetization behavior suggest mechanical disruption of the cellular structure by alignment of the nanoclinic.

15.8 FUTURE DIRECTIONS

The nanotechnology field is undergoing phenomenal growth. A primary impetus has been a major increase in funding for this field worldwide, such as the National Nanotechnology Initiative in the United States. It is beyond the limited scope of this book to cover all the new directions being pursued or to project all the prospects. Therefore, only some examples of future directions are presented here.

New Nanocrystals: Silicon Nanoparticles. The development of new semiconductor nanoparticles as efficient, inexpensive, stable, and tunable luminescent probes for biological staining and diagnostics is one future direction. In this regard, silicon and germanium nanoparticles appear to be promising.

Silicon nanoparticles had been produced by both top-down and bottom-up approaches. Nanoscale silicon has been extensively studied since 1990, when visible luminescence from porous silicon was first reported (Canham, 1990). Porous silicon contains a skeleton of crystalline nanostructures and is produced by electrochemically etching bulk silicon (which does not luminesce). Nanoparticles have been produced in a top-down fashion by many groups by using ultrasound to disperse porous silicon into various solvents, as first suggested by Heinrich et al. (1992). Bottom-up production methods have included both liquid phase methods (Holmes et al., 2001) and laser-induced decomposition of gas-phase species (Ehbrecht et al., 1995). At our Institute, such production methods are being used to generate silicon nanoparticles that emit at wavelengths throughout the visible spectrum.

An advantage of using silicon is that, through a controlled oxidation process, a thin shell of silica can be created on a silicon nanocrystal. This silica shell can then be functionalized to attach to DNA or to target specific biospecies as discussed earlier in this chapter.

Up-Converting Nanophores for Photodynamic Therapy. This subject, already discussed in Section 15.5, holds considerable promise for the treatment of tumors. The challenges are many. First, the up-conversion efficiency of these nanophores still needs to be improved. There appears to be an inverse relation between the efficiency of up-conversion and the size of nanocrystals. For more efficient up-conversion, one thus needs larger particles. However, larger particles cannot enter the cell through endocytosis. An appropriate balance of these two factors, together with the development of new host media for rare-earth ions to increase their up-conversion efficiency, has to be found.

Another area of investigation is real-time imaging and spectroscopy to determine the efficacy of photodynamic therapy. In the case of PDT drugs operating with singlet oxygen production, the singlet oxygen production can be monitored by its emission at ~900 nm.

In Vivo *Studies.* There are very few *in vivo* studies reported with the application of nanoparticles. This is an area that will attract a great deal of attention. The biocompatibility of the nanoparticles and nanoprobes, as well as their long-term toxicity, has to be studied.

Nanoarrays. The development of nanoarray technology for DNA and proteins is another future direction. Nanoarrays show promise for high-density analysis as well as for work with minute quantities of specimen. The challenges for biophotonics will be the use of optical methods to fabricate nanoarrays and to be able to use fluorescence detection for DNA or protein binding at nanodomains. Near-field lithography and fluorescence detection using simultaneous control of multiple-fiber probes need to be developed. This offers an opportunity for engineering development. Vo-Dinh et al. (2001) have reported nanosensors and biochips for single-cell analysis.

BIONEMS. The most active areas emerging from the fusion of biomedical technology with nanotechnology are nanoelectromechanical systems (NEMS) and nanofluidics. The NEMS devices for biotechnology are also sometimes labeled as BIONEMS. The NEMS devices are nanoscale analogues of micro-electro-mechanical systems (MEMS). MEMS and NEMS act to convert mechanical energy to electrical or optical signals, and vice versa. The mechanical–optical signal interconversion devices are also sometimes called optical MEMS and optical NEMS. In the case of biotechnology, MEMS and NEMS have been used in a broader sense to include micro- and nanosize motors, actuators, and even sensors. Although most of the NEMS devices utilize the well-developed fabrication process for semiconductors, plastics MEMS and NEMS provide future opportunities because of the structural flexibility offered by plastics.

HIGHLIGHTS OF THE CHAPTER

- Bionanophotonics refers to research and applications that involve both biomedical sciences and nanophotonics.
- Nanophotonics involves light–matter interactions on nanoscale. It is another exciting frontier dealing with nanoscale optical science and technology.
- Nanoscale light–matter interactions can be manifested in two ways: (i) by confining the light on nanoscale with the use of a near-field geometry such as that in near-field microscopy, discussed in Chapters 7 and 8, and (ii) by confining the matter on nanoscale by using nanoparticles and nanodomains.
- Nanochemistry involves the use of confined chemical reactions to produce nanoscale materials, such as nanoparticles and nanostructures.
- Nanoparticles can be produced by nanochemistry, by confining a chemical reaction within a reverse micelle. It consists of molecules with hydrophilic heads and hydrophobic tails that are self-organized around a water droplet.
- Multilayered particles can be produced subsequent to nanoparticle formation by additional multiple steps invoking various appropriate chemistries.
- Competitive reaction chemistry (CRC) is another nanochemistry approach to produce nanoparticles; here, conditions are chosen for the

reactants to initially combine and form particles, but particle growth is limited by some competing reaction.

- Quantum dots are nanoparticles in which electrons and holes are confined in a semiconductor whose size is smaller than a characteristic length called the *Bohr radius*.

- The luminescence band width in a quantum dot is very narrow and the wavelength of the peak emission depends strongly on the size of the nanoparticle.

- Three advantages of typical quantum dots over dyes in bioimaging applications are that (1) they exhibit longer lifetimes (hence their emission can be separated from any autofluorescence), (2) they do not readily photobleach, and (3) they are insensitive to microbial attack.

- Because of their large surface-to-volume ratios, the optical and chemical properties of quantum dots depend strongly on their surface characteristics.

- Semiconductor nanoparticles capped with shells as silica or other semiconductors have been used for biological labeling and imaging.

- Metallic nanoparticles and nanorods have also been used in biosensing.

- Oxide nanoparticles doped with rare-earth ions exhibit emission that generally is long-lived phosphorescence. Hence they are sometimes referred to as *nanophores*.

- Up-converting nanophores are those that produce up-converted visible emission when excited by an IR radiation. The up-conversion involves sequential absorption of multiphotons; hence a continuous-wave IR laser can induce visible emission.

- These nanophores are useful for bioimaging and also show promise for use in multiphoton photodynamic therapy, to reach deep tumors.

- PEBBLE is an acronym for probe encapsulated by biologically localized embedding and refers to sensor molecules entrapped in an inert nanoparticle. These devices are advantageous because cells and the indicator dyes are protected from each other. Also, multiple sensing mechanisms can be combined onto one particle.

- Nanoclinics are surface functionalized silica nanoshells that encapsulate probes as well as externally activatable drugs or therapeutic agents. They have shown to be capable of targeting specific cancer cells.

- Magnetic nanoclinics appear to be capable of destroying cancer cells in the presence of a dc magnetic field.

- Future work in the field of bionanophotonics will include the development of new nanoparticles, the usage of up-converting nanophores in photodynamic therapy, the conduction of nanoparticle-based *in-vivo* studies, the development of nanoarrays that might replace modern-day microarrays; and the fabrication of plastic-based bionanoelectromechanical devices (BioNEMS).

REFERENCES

Akerman, M. E., Chen, W. C. W., Laakkonen, P., Bhatia, S. N., and Ruoslalti, Nanocrystals Targeting *In Vivo*, *Proc. Natl. Acad. Sci.* **99**, 12617–12621 (2002).

Badley, R. D., Warren, T. F., McEnroe, F. J., and Assink, R. A., Surface Modification of Colloidal Silica, *Langmuir* **6**, 792–801 (1990).

Bergey, E. J., Levy, L., Wang, X., Krebs, L. J., Lal, M., Kim, K.-S., Pakatchi, S., Liebow, C., and Prasad, P. N., DC Magnetic Field Induced Magnetocytolysis of Cancer Cells Targeted by LH–RH Magnetic Nanoparticles *In Vitro*, *Biomed. Microdevices*, **4**, 293–299 (2002).

Bruchez, M., Jr., Moronne, M., Gin, P., Weiss, S., and Alivisatos, A. P., Semiconductor Nanocrystals as Fluorescent Biological Labels, *Science* **281**, 2013–2016 (1998).

Canham, L. T., Silicon Quantum Wire Array Fabrication by Electrochemical and Chemical Dissolution of Wafers, *Appl. Phys. Lett.* **57**, 1046–1048 (1990).

Chan, W. C., and Nie, S., Quantum Dot Bioconjugates for Ultrasensitive Nonisotopic Detection, *Science* **281**, 2016–2018 (1998).

Chen, Y., Kalas, R. M., and Faris, W., Spectroscopic Properties of Up-Converting Phosphor Reporters, *SPIE Proceedings* **3600**, 151–154 (1999).

Clark, H. A., Hoyer, M., Philbert, M. A., and Kopelman, R., Optical Nanosensors for Chemical Analysis Inside Single Living Cells. 1. Fabrication, Characterization, and Methods for Intracellular Delivery of PEBBLE Sensors, *Anal. Chem.* **71**, 4831–4836 (1999).

Daubresse, C., Granfilo, C., Jerome, R., and Teyssie, P., Enzyme Immobilization In Nanoparticles produced by Inverse Microemulsion Polymerization, *J. Colloid Interface Sci.* **168**, 222–229 (1994).

Ehbrecht, M., Ferkel, H., Smirnov, V. V., Stelmakh, O. M., Zhang, W., and Huisken, F., Laser-Driven Flow Reactor as a Cluster Beam Source, *Rev. Sci. Instrum.* **66**, 3833–3837 (1995).

Fischer, C. H., and Henglein, A., Photochemistry of Colloid Semiconductors: Preparation and Photolysis of CdS in Organic Solvents, *J. Phys. Chem.* **93**, 5578–5581 (1989).

Han, M., Gao, X., Su, J. Z., and Nie, S., Quantum-Dot Tagged Microbeads for Multiplexed Optical Coding of Biomolecules, *Nat. Biotechnol.* **19**, 631–635 (2001).

Heinrich, J. L., Curtis, C. L., Credo, G. M., Kavanagh; K. L., and Sailor, M. J., Luminescent Colloidal Silicon Suspensions from Porous Silicon, *Science* **255**, 66–68 (1992).

Herron, N., Wang, Y., and Eckert, H., Synthesis and Characterization of Surface-Capped, Size-Quantized CDS Clusters—Chemical Control of Cluster Size, *J. Am. Chem. Soc.* **112**, 1322–1326 (1990).

Holm, B. A., Bergey, E. J., De, T., Rodman, D. J., Kapoor, R., Levy, L., Friend, C. S., and Prasad, P. N., Nanotechnology in Biomedical Applications *Mol. Cryst. Liq. Cryst.* **34**, 589–598 (2002).

Holmes, J. D., Ziegler, K. J., Doty, R. C., Pell, L. E., Johnston, K. P., and Korgel, B. A., Highly Luminescent Silicon Nanocrystals with Discrete Optical Transitions, *J. Am. Chem. Soc.* **123**, 3743–3748 (2001).

Jain, T. K., Roy, I., De, T. K., and Maitra, A. N., Nanometer Silica Particles Encapsulating Active Compounds: A New Ceramic Drug Carrier, *J. Am. Chem. Soc.* **120**, 11092–11095 (1998).

Kapoor, R., Friend, C., Biswas A., and Prasad, P. N., Highly Efficient Infrared-to-Visible Energy Upconversion in Er^{3+}:Y_2O_3, *Opt. Lett.* **25**, 338–340 (2000).

Lal, M., Levy, L., Kim, K. S., He, G. S., Wang, X., Min, Y. H., Pakatchi, S., and Prasad, P. N., Silica Nanobubbles Containing an Organic Dye in a Multilayered Organic/Inorganic Heterostructure with Enhanced Luminescence, *Chem. Mater.* **12**, 2632–2639 (2000).

Levy, L., Sahoo, Y., Kim, K.-S., Bergey, E. J., and Prasad, P. N., Nanochemistry: Synthesis and Characterization of Multifunctional Nanoclinics for Biological Applications, *Chem. Mater.* **14**, 3715–3721 (2002).

Markowitz, M. A., Schoen, P. E., Kust, P., and Gaber, B. P., Surface Acidity and Basicity Functionalized Silica Particles, *Colloids Surf.* **150**, 85–94 (1999).

Masui, T., Fujwara, K., Machida, K., Adachi, G., Sakata, T., and Mori, H., Characterization of Cerium (IV) oxide Ultrafine Particles Prepared Using Reverse Micelles, *Chem. Mater.* **9**, 2197–2204 (1997).

Mattoussi, H., Mauro, J. M., Goldman, E. R., Anderson, G. P., Sundor, V. C., Mikulec, F. V., and Bawendi, M. G., Self-Assembly of CdSe–ZnS Quantum Dot Bioconjugates Using an Engineered Recombinant Protein, *J. Am. Chem. Soc.* **122**, 12142–12150 (2000).

Mitchell, G. P., Mirkin, C. A., and Letsinger, R. L., Programmed Assembly of DNA Functionlized Quantum Dots, *J. Am. Chem. Soc.* **121**, 8122–8123 (1999).

Murray, C. B., Kagan, C. R., and Bawandi, M. G., Synthesis and Characterization of Monodisperse Nanocrystals and Closed-Packed Nanocrystals Assemblies, *Annu. Rev. Mater. Sci.* **30**, 545–610 (2002).

Pileni, M. P., Reverse Micelles as Microreactors, *J. Phys. Chem.* **97**, 6961–6973 (1993).

Roy, I., Ohulchanskyy, T. Y., Pudavar, H. E., Bergey, E. J., and Prasad, P. N., Ceramic-based Nanoparticles Entrapping Water-Insoluble Photosensitizing Anticancer Drugs: A Novel Drug-Carrier System for Photodynamic Therapy (PDT), submitted (2003).

Shen, Y., Friend, C. S., Jiang, Y., Jakubczyk, D., Swiatkiewicz, J., and Prasad, P. N., Nanophotonics: Interactions, Materials and Applications, *J. Phys. Chem.* **104**, 7577–7587 (2000).

Storhoff, J. J., and Mirkin, C. A., Programmed Materials Syntheses with DNA, *Chem. Rev.* **99**, 1849–1862 (1999).

Swayambunathan, V., Hayes, D., Schmidt, K. H., Liao, Y. X., and Miesel, D., Thiol Surface Complexation on Growing CdS Clusters, *J. Am. Chem. Soc.* **112**, 3831–3837 (1990).

vanBlaaderen, A., and Vrij, A., Synthesis and Characterization of Monodisperse Colloidal Organosilica Spheres, *J. Colloid Interface Sci.* **156**, 1–18 (1993).

Vo-Dinh, T., Cullum, B. M., and Stokes, D. L., Nanosensors and Biochips: Frontiers in Biomoleular Diagnostics, *Sensors and Actuators B* **74**, 2–11 (2001).

CHAPTER 16

Biomaterials for Photonics

Photonics, which utilizes light–matter interactions for information processing, transmission, data storage, and display, is being hailed as the dominant technology for the 21st century. The continued development of photonics technology is crucially dependent on the availability of suitable optical materials. Biomaterials are emerging as an important class of materials for a variety of photonics applications. This chapter describes potential applications of various types of biomaterials for photonics.

The four types of biomaterials that hold promise for photonics applications are (i) bioderived materials, naturally occurring or their chemical modifications, (ii) bioinspired materials, synthesized based on guiding principles of biological systems, (iii) biotemplates for self-assembling of photonic active structures, and (iv) bacteria bioreactors for producing photonic polymers. These biomaterials are discussed in this chapter.

A wide range of photonics applications using these biomaterials are being explored. They include efficient harvesting of solar energy, low-threshold lasing, high-density data storage, optical switching, and filtering. The chapter discusses these applications of biomaterials. The most extensively investigated photonics applications are of bacteriorhodopsin for holographic data storage. This application is described in detail, together with the coverage of its current status.

Finally the chapter concludes with a presentation of some areas of future directions for research and development.

Suggested general reading materials are:

Saleh and Teich (1991): *on Fundamentals of Photonics*
Birge et al. (1999): *on Biomaterials for Holographic Optical Memory*

16.1 PHOTONICS AND BIOMATERIALS

The previous chapters have dealt with the applications of interactions of light with biological materials for optical diagnostics and light activated therapy. In

Introduction to Biophotonics, by Paras N. Prasad.
ISBN: 0-471-28770-9   Copyright © 2003 John Wiley & Sons, Inc.

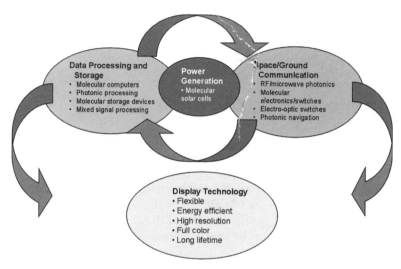

Figure 16.1. Interconnection of the various key areas to produce paradigms for new generation information technology.

other words, the focus has been how photonics benefits biotechnology. The other side of the coin is how biomaterials have benefited photonics. In this case, one can exploit interaction of light with naturally occurring biological matter, or materials that may be produced using the same fundamental principles that produce hierarchically built biological assemblies. The latter class of materials is often called *bioinspired materials*.

Photonics is expected to revolutionize many aspects of data collection, processing, transmission, interpretation, display, and storage. It is the dominant part of information technology for the 21st century and, in its more comprehensive scope, is presented in Figure 16.1.

Availability and future development of new multifunctional materials that can dramatically improve speed and encryption, as well as provide terabit data storage and large-area high-resolution display, are of vital importance for implementation of the full scope of new-generation information technology. Biological systems have provided researchers with a fertile ground with regard to materials enabling new technologies that cover a wide range, from disease therapy, to sensory systems, to computing, and to photonics. In Nature, bioprocesses yield structures that are nearly flawless in composition, stereospecific in structure, flexible, and ultimately biodegradable. Compounds of biological origin can spontaneously organize into complex structures and function as systems possessing long range and hierarchical order. Biological systems also lend themselves to modifications to enhance a specific functionality using both chemical modification and genetic engineering. An important area of application of biomaterial is DNA computing. Although this application has drawn considerable attention, it really does not fall under photonics applications. Photonics applications can utilize a number of diverse groups

TABLE 16.1. Biomaterials for Photonics

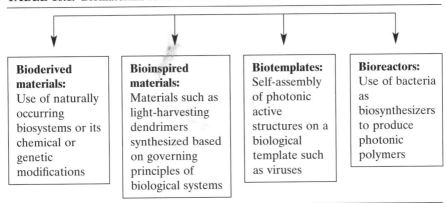

Bioderived materials:	Bioinspired materials:	Biotemplates:	Bioreactors:
Use of naturally occurring biosystems or its chemical or genetic modifications	Materials such as light-harvesting dendrimers synthesized based on governing principles of biological systems	Self-assembly of photonic active structures on a biological template such as viruses	Use of bacteria as biosynthesizers to produce photonic polymers

of biomaterials for a variety of active and passive functions. Table 16.1 lists biomaterials for photonics.

An example of bioderived material for photonics is green fluorescent protein (GFP) in its wild and mutant forms, which have attracted a great deal of interest as biological fluorescent markers for *in vivo* imaging and fluorescence energy transfer imaging (FRET) to study protein–protein and DNA–protein interactions. This subject has already been discussed in Chapter 8. Other photonics applications of GFP have also been proposed. Another widely investigated bioderived material for photonics is bacteriorhodopsin (Birge et al., 1999). The main focus has been to utilize its excited-state properties and associated photochemistry for high-density holographic data storage. In addition, a number of other applications have been proposed which are listed in Section 16.2. More recently, native DNA has been proposed as photonic media for optical waveguide and host for laser dyes (Kawabe et al., 2000). Another example is biocolloids which consist of highly structured and complex, discrete biological particles that can be organized into close-packed arrays via surface-directed assembly to form photonics crystals (discussed in Chapter 9). These are some examples of naturally occurring biomaterials for photonics which are described in some detail in Section 16.2.

Bioinspired materials are synthetic materials produced by mimicking natural processes of synthesis of biological materials. A growing field is biomimicry with a strong focus on producing multifunctional hierarchical materials and morphologies that mimic Nature. An example of this category is a light-harvesting photonic material that will be presented in Section 16.3.

Biotemplates refer to natural microstructures with appropriate morphologies and surface interactions to serve as templates for creating multiscale and multicomponent photonics materials. The biotemplates can be naturally occurring biomaterials or a chemically modified, bioderived material. Examples are viruses with organized structures of varied morphologies. This topic is discussed in Section 16.4.

Bioreactors refer to the naturally occurring biosynthetic machinery that can be manipulated to produce a family of helical polymers having a wide range of optical properties. An example is a bacterial reactor that can be used to synthesize customized polymeric structures for photonics applications. This topic is discussed in Section 16.5.

16.2 BIODERIVED MATERIALS

This section presents selected examples of naturally occurring biomaterials or their chemically derivatized forms that have been investigated for photonics. Among those are:

- Bacteriorhodopsin for holographic memory
- Green fluorescent proteins for photosensitization
- DNA as host for laser dyes
- Biocolloids for photonics crystal media

These examples are discussed below.

Bacteriorhodopsin. Bacteriorhodopsin (often abbreviated as bR) grows in the purple membrane of a salt marsh bacterium known as *Halobacterium salinarium* or *Halobacterium halobium* (Birge et al., 1999). A broad range of photonics applications for this naturally occurring protein has been proposed, taking advantage of its robustness, ease of processing into optical quality films, suitable photophysics and photochemistry of the excited state, and flexibility for chemical and genetic modifications. The photonics applications include random access thin-film memories (Birge et al., 1989), photon counters and photovoltaic converters (Marwan et al., 1988; Sasabe et al., 1989; Hong, 1994), spatial light modulators (Song et al., 1993), reversible holographic media (Vsevolodov et al., 1989; Hampp et al., 1990), artificial retinas (Miyasaka et al., 1992; Chen and Birge, 1993), two-photon volumetric memories (Birge, 1992), and pattern recognition systems (Hampp et al., 1994).

Bacteriorhodopsin consists of seven *trans*-membrane α-helices which form the secondary structure of this protein. The light absorption by the light-adapted form of this protein, often labeled as bR, is due to the chromophore, called all-*trans*-retinal. This chromophore is also involved in the process of vision as discussed in Section 6.4.1 of Chapter 6; here it is covalently bound to Lys-216 via a protonated Schiff base linkage. Light absorbed by this chromophore induces an all-*trans* to 13-*cis* photoisomerization in its structure, which is followed by a series of protein intermediates exhibiting different absorption spectra and vectoral proton transport. Ultimately, the reisomerization of the chromophore leads to regeneration of the protein's original (resting) state.

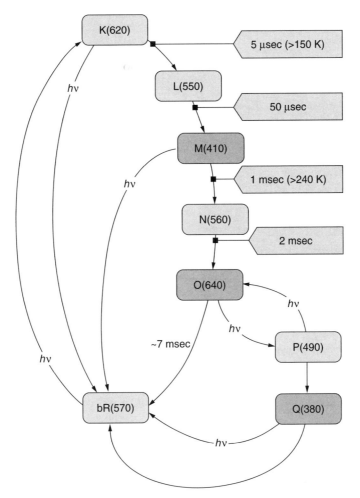

Figure 16.2. A simplified model of the photocycle of the light adapted bacteri-orhodopsin. The height of the symbols is representative of the relative free energy of the intermediates. (Reproduced with permission from Birge et al., 1999.)

Figure 16.2 shows this photoinduced cycle. The absorption maxima (in nanometers) for each excited state, labeled by a capital letter, in this cycle are shown in parentheses. The indicated lifetime and temperatures apply to the wild type only. Bacteriorhodopsin can undergo these photocycles a large number of times (10^6 or more) without any degradation.

Not all the intermediates are shown in Figure 16.2. Although there are two species of M states (M_1 and M_2), only one is shown in Figure 16.2 for the sake of simplicity. Figure 16.3 shows the absorption spectra of the resting state, bR, along with those of the various excited states, K to Q.

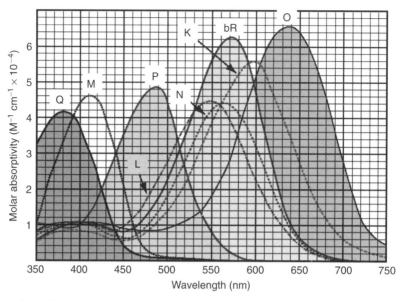

Figure 16.3. The electronic absorption spectra of selected intermediates in the photocycle, together with that of the resting state (bR). (Reproduced with permission from Birge et al., 1999.)

The principle of data storage or optical correlation using a volume hologram utilizes the change in refractive index of a medium when a particular excited state is created. In the case of bacteriorhodopsin, the excitation to a particular intermediate level such as the M state produces a change in the absorption spectrum and thus a change in absorbance at a given frequency (ω in angular unit), denoted as $\Delta\alpha(\omega)$. The Kramers–Kronig relation of optics relates the change in absorbance, $\Delta\alpha(\omega)$, to the corresponding change in refractive index, $\Delta n(\omega)$, as follows (Finlayson et al., 1989):

$$\Delta n(\omega) = \frac{c}{\pi}\text{p.v.}\int_0^\infty \frac{d\omega'\Delta\alpha(\omega')}{\omega'^2 - \omega^2} \tag{16.1}$$

where p.v. stands for the principal value of the integral. To write a hologram, two monochromatic beams (an object beam and a reference beam) of wavelength λ are crossed at an angle in a holographic medium as shown in Figure 16.4. Their interference produces an intensity modulation with alternate bright and dark stripes; the separation between them, the fringe spacing Λ, is given by

$$\Lambda = \lambda/(2n\sin\theta/2) \tag{16.2}$$

for a transmission grating, where θ is the angle of crossing and n is the refractive index of the medium. In the bright areas, the action of light is to induce

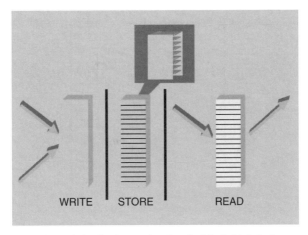

WRITE   STORE   READ

Figure 16.4. Illustration of the basic mechanism behind thick hologram. Courtsey of R. Burzynski.

the change in absorption, $\Delta\alpha$, due to the population of the excited state, which creates a change in refractive index, Δn, given by the above Kramers–Kronig relation. The result is a refractive index grating (periodic variation in refractive index) that constitutes the hologram. This hologram can be read by a weak probe (readout) beam that is diffracted by the refractive index grating. Thus, when the hologram is illuminated by the reference beam, a bright diffraction beam is produced in the direction of the object beam. This reproduction process is very sensitive to angle and wavelength, due to Bragg selectivity of thick gratings (Kogelnik, 1969).

The diffraction efficiency η is related to the refractive index change Δn for a symmetric grating as

$$\eta = \sin^2\left(\frac{\pi\Delta n d}{\lambda\cos\theta/2}\right) \tag{16.3}$$

where d is the thickness of the sample. A large Δn produces a large diffraction efficiency, until the sine function reaches its maximum value of 1.

The advantage of using stored holograms for memory application is that in the same space (volume element) many different holograms (thousands) can be recorded by changing the angle of the writing incident beams. This process is called *angular multiplexing* and is shown in Figure 16.5.

Consider first the storage of a single image. This is achieved by passing the object beam through an image plate, for example. Next, the reference beam is rotated, but the direction of the object beam remains unchanged. However, a new image plate is inserted in front of the object beam. This process of angular multiplexing can be repeated many times, storing a separate image each time. Figure 16.6 illustrates the process of recalling these images one at

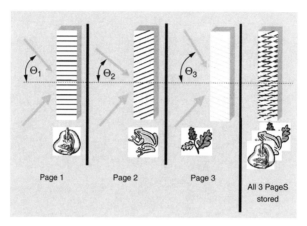

Figure 16.5. Illustration of the mechanism for storing multiple images in a hologram. (Courtesy of R. Burzynski.)

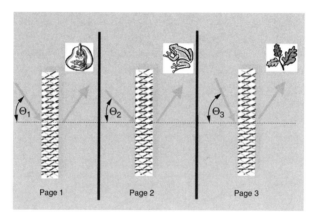

Figure 16.6. Illustration of the mechanism for restoring individual images from a hologram. (Courtesy of R. Burzynski.)

a time, by simply changing the angles of the probe beam to coincide each time with the angles of the reference beam. Note that all the images come out in the same direction.

This idea can be used in reverse to identify an image, a process termed *optical correlation*. Briefly, assume that a set of images have been stored in a medium as a hologram, as in Figure 16.5. Now, we are given a random image, which may or may not have been stored in this medium as a hologram. Our goal is to determine (i) if the image is in there and (ii) if it is in there, which one it is. To accomplish this goal, we pass the object beam (which had a fixed angle during writing) through the image and illuminate the holographic medium in which multiple holograms are recorded. If the image matches one that was stored, a bright diffracted spot will be generated in the direction corresponding to the direction of the reference beam used to store that image. By determining the angle at which bright beam was diffracted, we can determine which one of the stored images we have matched. The absence of such a spot indicates no match.

The magnitude of photoinduced refractive index modulation can be used as a direct measure of the material's storage capacity. A large index modulation allows a large number of holograms to be recorded with good diffraction efficiency. This property, represented by a $M_\#$ parameter (M number), is used to describe the dynamic range of the material and is defined as (Mok et al., 1996)

$$\sqrt{\eta} = (M_\#)/M \tag{16.4}$$

where M is the number of holograms recorded in the same volume, and η is the diffraction efficiency of each hologram. Diffraction efficiency η is defined as the ratio of intensities of the diffracted beam to the incident beam. In some materials with large index modulation value and/or of large thickness, η can reach values close to unity (or 100%) while $M_\#$ can be as high as 10 or more.

Most holographic applications of bacteriorhodopsin utilize the optical change ($\Delta\alpha$ and subsequently Δn) in going from the ground state, bR, to the M intermediate state. As shown in Figure 16.7, the Δn calculated using the Kramers–Kronig transformation [equation (16.1)] is large because of the wide wavelength separation between these two states. The quantum yield of the bR to M conversion is also high. A primary limitation, however, is the short lifetime of the M state. However, chemical and genetic manipulation have provided much improved bacteriorhodopsin analogues for long-term storage.

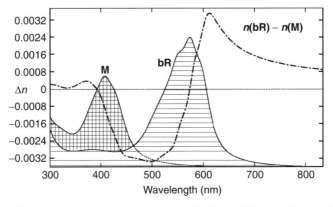

Figure 16.7. The refractive index change $\Delta n = n(\text{bR}) - n(\text{M})$ for a 30-μm film of bacteriorhodopsin with an optical density maximum of ~6. The absorption spectra of bR and M are also shown. (Reproduced with permission from Birge et al., 1999.)

TABLE 16.2. Key Properties of Bacteriorhodopsin Films

Spectral range	400–650 nm
Achievable resolution	~5000 lines/mm
Light sensitivity	1–80 mJ/cm^2 (B-type recording)
30 mJ/cm^2 (M-type recording)	
Reversibility	>10^6 write/erase cycles
Photochemical bleaching	>95% in selected films
Diffraction efficiency	1–3%
Polarization recording	possible

The transition from the bR state to the M state is photochemically induced with a yellow light (λ_{max} for the bR form is at ~570 nm) after going through the intermediate K and L states. An inverse approach using M-type recording is effected with the yellow light to control the population of the M state, and with the blue light (λ_{max} for the M state is at 410 nm) to write information. Munich Innovative Biomaterials (abbreviated as MIB; website: www.mib-biotech.de/products_films. htm), which commercializes the bacteriorhodopsin films, quotes the key parameters for holographic storage in their films as listed in Table 16.2.

The return of the M state back to the bR state occurs through a sequence of nonradiatively formed intermediates. The time associated with these thermal relaxation processes depends on the type of bacteriorhodopsin (wild type or variant), pH value of the film matrix, and the film temperature.

Recent interest has focused on the use of branched photocycles originating from the O state (Figure 16.2) which contains long-lived P and Q states, thus forming the basis for long-term storage. Illumination of bacteriorhodopsin in the O state, with red light, produces a small amount of P state that converts thermally to the Q state. The chromophore configuration in both the P and the Q states is 9-*cis* (Popp et al., 1993). The P and Q intermediates appear to differ in that the 9-*cis* chromophore exists in a trapped but unbound state in Q, while it is still bound in P. The exposure to blue light regenerates the bR state. A current limitation is the extremely low quantum yield of the P and Q states.

Green Fluorescent Protein (GFP). The green fluorescent protein (GFP) has been discussed in Chapter 8. By now, a number of recombinant and mutant forms of GFP have been produced which exhibit different excitation and emission profiles. Some variants are blue FP, cyan FP, yellow FP, and red FP (see Section 8.2). They are being widely used as fluorescent markers in the determination of gene expression, protein localization, and protein–protein interactions (through FRET, discussed in Chapter 8). A number of other photonics applications have been proposed for GFP utilizing a number of properties exhibited by these molecules. Some of these are:

- Absorption in two bands at $\sim$395 nm (extinction coefficient 30,000 L mol^{-1}cm^{-1}) and 475 nm (extinction coefficient 7000 L mol^{-1}cm^{-1}) covers a broad range of UV and visible regions. This feature has led to the application of GFP as a photosensitizer, which is discussed below.
- The two absorption bands attributed to the presence of two resonant forms, a neutral and an anionic, of the same chromophore, p-hydroxybenzylidene-imidazolidone, can be interconverted in the excited state by proton transfer (Chattoraj et al., 1996).
- The relative stabilities of these two forms can be manipulated by the appropriate choice of the close environment surrounding the chromophore. For example, an alteration of the Ser 65 (serine, S) favors the anionic form which absorbs at 475 nm and thus can readily be pumped by widely available green lasers (Cubitt et al., 1995; Yang et al., 1996).
- Single molecules of GFP mutants, when immobilized in aerated aqueous polymer gels and excited by 488-nm light, exhibit an unusual repeated cycle of fluorescence emission (on/off blinking) on a time scale of several seconds (Dickson et al., 1997). This behavior is not observed in bulk studies (ensemble, averaged over many molecules). Dickson et al. have suggested a possible application of this phenomenon for molecular photonic switches or optical storage elements, addressable on the single-molecule level.
- GFP also exhibits efficient two-photon excitation when excited at 800 nm (Kirkpatrick et al., 2001). Two-photon excitation has successfully been used to produce up-conversion lasing in GFP (Pikas et al., 2002). This is the first report of two-photon pumped lasing in a biological system.

In addition, a very attractive feature of GFP is its environmental stability. Because its 3-D structure insulates the chromophore from the external environment, GFP fluorescence is insensitive to oxygen quenching and is stable in a variety of harsh environments (temperatures up to 70°C, pH 6–12; detergents, proteolysis). This protein is highly resistant to denaturation by heat. GFP requires heating to 90°C with 6 M guanidine hydrochloride, or a pH outside the range of 4–12 to denature. Furthermore, renaturation can be achieved to restore the optical properties of GFP by reversing the conditions of denaturation.

A photonic application presented here is that of making a molecular photodiode of sandwich configuration S(electron sensitizer)/M(mediator)/A(electron acceptor). Nature exhibits a very efficient photoinduced electron transfer such as those found for photoelectric conversion in retina and long-range electron transfer in photosynthetic organisms (Chapter 6). Furthermore, these electron transfers are unidirectional (Deisenhofer et al., 1985). A number of artificial molecular devices have been fabricated where the electron transport functions of biological photosynthesis have been mimicked.

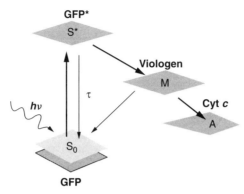

Figure 16.8. Schematics of energy diagrams for the GFP photosensitizer (S)/viologen media(M)/cytochrome C acceptor(A) system. (Reproduced with permission from Choi et al., 2001.)

The approach adopted by Choi et al. (2001) is schematically illustrated in Figure 16.8. GFP was chosen as a photosensitizer because of its high fluorescence quantum yield of 80% and excitation covering a major portion of the UV to blue-green wavelength regions. The M/I/M (metal/insulator/metal) device structure utilized the GFP (S)/viologen(M)/cytochrome C(A) heterofilms. The viologen and the cytochrome films were deposited onto an ITO-coated glass by the Langmuir–Blodgett technique, on top of which the GFP molecules were adsorbed by dipping the M/A heterostructure into the GFP solution. Finally, an Al electrode was deposited to produce Al/GFP/viologen/cytochrome C/ITO, as an M/I/M device. A photoinduced unidirectional flow of electrons, detected as photocurrents in this M/I/M device, was achieved by irradiation of the structure with a 460-nm light from a 500-W xenon lamp. Thus, the rectification function of a photodiode was demonstrated.

Naturally Occurring DNA. Ogata and co-workers from Chitose Institute of Science and Technology have shown that naturally occurring DNA from salmon can be used as photonic medium (Kawabe et al., 2000; Wang et al., 2001). They have shown that good-optical-quality films of waveguiding quality can be fabricated using salmon DNA. They doped a laser dye into a DNA–surfactant complex film to achieve amplified spontaneous emission.

As discussed in Chapter 8 (Section 8.5.4), many fluorescent dyes can readily be intercalated into the helices of DNA. For a number of dyes, the fluorescence intensity is greatly enhanced (Jacobsen et al., 1995; Spielmann, 1998). Ogata and co-workers found that the intercalated dye molecules can be well-aligned and stabilized. Dyes can be intercalated in high concentration without showing any concentration quenching effect derived from aggregation.

The procedure used by them to dope a laser dye (P-mehemicyanine), a nonlinear optical molecule (disperse red 13), or a photochromic dye (spiropyran) into a DNA-surfactant complex is shown in Figure 16.9. This procedure was

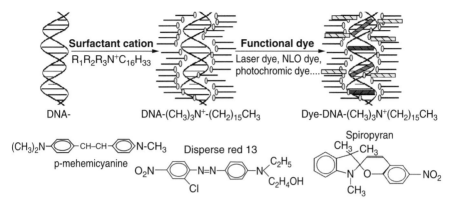

Figure 16.9. Schematics representation for the preparation method and a possible structure of a dye-doped DNA-surfactant film. (Courtesy of N. Ogata.)

utilized to make rhodamine 6G-doped film for lasing. First, the DNA–surfactant complex was prepared by mixing a salmon DNA solution with hexadecyltrimethylammonium chloride (surfactant) aqueous solution to which rhodamine 6G was added. Then a rhodamine 6G-doped DNA–surfactant film was cast from an ethanol solution (with a DNA base pair to dye ratio of 25:1) on a glass slide in a closed chamber (55% humidity), and a solid film was formed by slow evaporation. Above a certain threshold energy (20 μJ) and power density (300 kW/cm^2) of the pump beam (532 nm; 7 nsec; 10 Hz), a narrowing of the lineshape, together with a superlinear dependence on the pump intensity, was taken as an indication of light amplification by stimulated emission.

Kawabe et al. (2000) suggested that rhodamine 6G may not be intercalating DNA structure in the strict sense because of its chemical structure. But even then, no aggregation of the dye takes place because the film shows strong fluorescence and stimulated emission even at a high dye concentration of 1.36 weight %.

Bioobjects and Biocolloids. Nature exhibits many unique forms of bioobjects that have highly precise shapes and are of monodisperse size in 1, 2, and 3 dimensions (plates, rods, icosahedral, etc.). Examples of these bioobjects are viruses, sponges, sea urchin needles, platelets from abalone shell, and so on. Furthermore, the surface chemistry of these bioobjects is heterogeneous and precise. For example, virus particles are comprised of a capsid consisting of arranged protein subunits that form a hollow particle (with diameter in the range 20–300 nm) that encloses the genome. The genomic material in the core of a virus particle can be replaced by other functional interiors to produce novel photonic functions. In addition, using appropriate protein chemistry, surfaces of virus particles are being exploited for various applications, by the research group of E. L. Thomas at MIT.

One exciting prospect is the use of these monodispersed bioobjects as building blocks for photonics crystals. The topic of photonics crystals has been discussed in Chapter 9 (Section 9.9). In a traditional approach, monodispersed colloidal crystals of silica or polystyrene of appropriate diameters are close-packed in a face-centered cubic (fcc) periodic array to produce a photonic crystal that reflects light of a specific wavelength, determined by the periodicity (the size of the colloidal particles and the respective refractive indices).

The biological objects such as virus particles of sizes 100–300 nm and varied shapes (icosahedral, rod, etc.) enable one to assemble them in both fcc and non-fcc packing to produce a wide range of self-assembled photonics crystals. When dispersed in an appropriate solvent media, these bioobjects form biocolloids that can self-assemble into a close-packed structure exhibiting a photonic crystal behavior. The research group at the Polymer Branch of the Air Force Research Laboratory (United States; R.Vaia) in collaboration with those at MIT (United States; E. L. Thomas) and Otago University (New Zealand; V. K. Ward) are using this approach to produce novel photonics crystals.

In their first approach, they have used iridovirus consisting of an icosahedral capsid with a diameter of ~200 nm to produce a photonic crystal (Vaia et al., 2002). They applied a strong gravitational force of ~11,000 G from a centrifuge for 15 minutes to an aqueous viral solution containing 4% formaldehyde to cross-link the sedimented particles together. Figure 16.10 shows the local regions of the fcc packing of the iridovirus particles, very similar to the packing of the colloidal particles presented in Chapter 9 (Figure 9.26). Furthermore, McPherson's group (Yu et al., 2000) has shown that viruses can be packed not only in an fcc structure, but also in other lattices such as orthorhombic and monoclinic systems.

Parker et al. (2001) produced a spine from the sea mouse aphrodita sp. (polychaeta:aphroditidae) which is an example of close-packed bioobjects, but now not in an fcc arrangement. The spine normally exhibits a deep red color; but when light is incident perpendicular to the axis of the spine, different colors are seen as stripes running parallel to the axis of the spine. They report that, even over a range of smaller incident angles, the complete visible spectrum is reflected with a reflectivity of 100%. The electron micrograph of a section of the spine reveals a close-packed array of hollow cylinders, with the long axis of the cylinders along the spine and each cylinder having six nearest neighbors.

Another potential application is offered by the ability to include other materials such as high-refractive-index nanoparticles within the capsid of a virus to manipulate its refractive index, which, in turn, can be utilized to enhance the dielectric contrast of a photonic crystal. Douglas and Young (1998) have shown that under appropriate conditions, the viral capsid can be temporarily opened to allow the transport of various substances inside it, which can then be trapped within the capsid by closing it again.

Figure 16.10. SEM micrograph of close packing of iridoviruses in a periodic structure. (Courtesy of R. Vaia.)

16.3 BIOINSPIRED MATERIALS

An understanding of biorecognition, multilevel processing, self-assembly, and templating in biological systems can provide valuable methodology and tools to design a new class of bioinspired materials for electronics and photonics. An example of a bioinspired material is a new class of light-harvesting dendrimer, developed by Frechet and co-workers (Adronov et al., 2000), which has been modeled after a naturally occurring photosynthetic system, a chlorophyll assembly. The photosynthetic system consists of a large array of chlorophyll molecules that surround a reaction center, as discussed in Chapter 6 (see Section 6.4). This chlorophyll array acts as an efficient light-harvesting antenna to capture photons from the sun and transfer the absorbed energy to the reaction center. The reaction center utilizes this energy to produce charge separation, eventually forming ATP and NADPH (Section 6.4.3).

Frechet and co-workers prepared light-harvesting dendrimers that consist of a number of nanometer-size antennas in a hyperbranch arrangement. Figure 16.11 shows schematic representation of two dendritic systems. Figure 16.11A exhibits a nanoscale light-harvesting dendrimer. It consists of multiple peripheral sites (light absorbing chromophores, represented by spheres) that

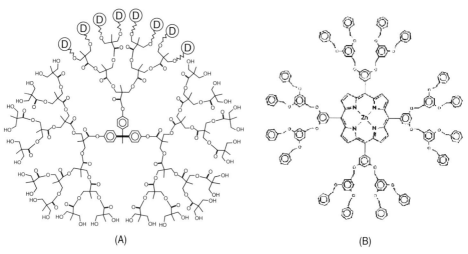

Figure 16.11. (A) Light-harvesting antenna-based dendritic structure. (B) light-harvesting dendrimer with a photocatalytic center. (Courtesy of J. M. J. Frechet).

can absorb and thus harvest sunlight. The absorbed energy is eventually transferred (funneled) quantitatively by a Förster-type energy transfer, discussed in Chapters 4 and 7, to an excitation energy acceptor at the center of the dendrimer where it can be "reprocessed" into a monochromatic light of a different wavelength (by emission from the core), or converted into electrical or chemical energy. Figure 16.11A represents a case where the dendritic antenna is based on a single molecule about 3 nm in size. Figure 16.11B represents the schematics of a porphyrin-base catalytic site located at the center of the dendrimer. The light absorbed by the antenna system in the case of the dendritic structure of Figure 16.11B produces a photocatalytic product at the center.

Another example of energy transfer dendrimer is provided by the work of Kopelman's group (Swallen et al., 2000), as shown in Figure 16.12.

Recently, Frechet, Prasad, and co-workers have demonstrated two-photon excited efficient light harvesting in novel dendrite systems (Brousmiche et al., 2003; He et al., 2003). Here the antennas are efficient two-photon absorbers (and green emitters) that absorb near-IR photons at ~800 nm and transfer the excitation energy quantitatively to the core molecule (a red emitter).

16.4 BIOTEMPLATES

The biotemplate approach is a top-down approach (discussed in Chapter 15) for building a nanostructured photonic material. Biotemplates refer to natural microstructures that can be used, in the pristine form or a surface functionalized form of a natural or bioinspired material, as templates to produce multi-

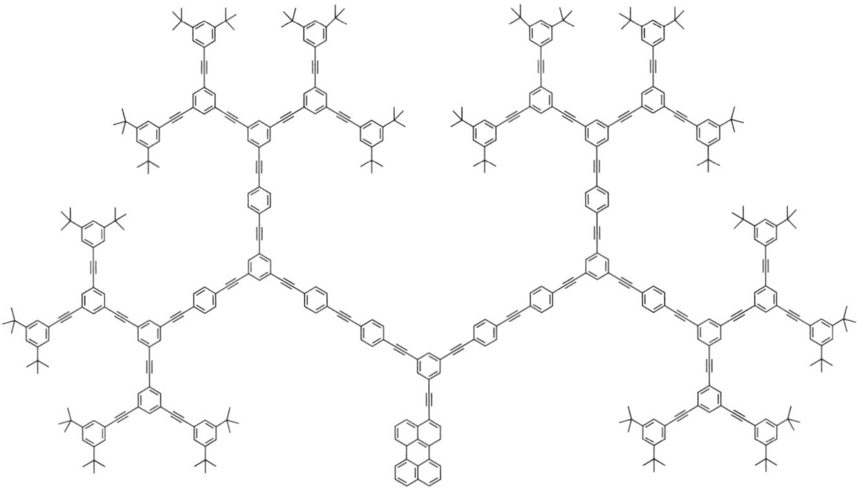

Figure 16.12. The UV and blue photons absorbed by the dendrimeric antenna form excitons that migrate within picoseconds to the perylenic "reaction center," forming excimers that emit orange-red light with 99% photon efficiency. Furthermore, the dendrimeric antenna enhances the perylenic photostability by several orders of magnitude. (Reproduced with permission from Swallen et al., 2000.)

scale, multicomponent materials through iterative mesophase synthesis and processing. Biotemplates can be used for the development of new assembling and processing techniques to produce periodic, aperiodic, and other engineering architectures on nanoscale for photonics applications. This approach can provide new aspects of achieving cooperative amplification of a photonic function, producing synergism between various electronic and photonic functions as well as creating new manifestations, together with a broad spectral response. Two specific examples are provided here: (i) DNA and (ii) virus as templates.

DNA as a Template. A great deal of research has been reported utilizing DNA as a template to grow inorganic quantum confined structures (quantum dots, quantum wires, metallic nanoparticles as discussed in Chapter 15) and to organize nonbiological building blocks into extended hybrid materials. An excellent review on this topic is by Storhoff and Mirkin (1999). Some also describe the DNA template as a "smart glue" for assembling nanoscale building blocks (Mbindyo et al., 2001). Some of the DNAs used are naturally occurring, while others have been synthesized with appropriate length and base (nucleotide) sequence. In these approaches, a major advantage derived from the use of DNA is the ability of complementary DNA strands to hybridize selectively. This feature has been discussed in this book at several places (see Chapters 8 and 10). Coffer and co-workers were the first to utilize DNA as a

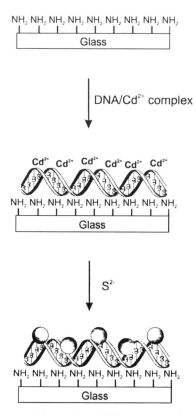

Figure 16.13. The scheme used to produce CdS nanoparticle arrays. (Reproduced with permission from Coffer, 1997.)

template for CdS nanoparticles (Coffer et al., 1992; Coffer, 1997). In their initial approach, CdS nanoparticles were formed by first mixing an aqueous solution of calf thymus DNA with Cd^{2+} ions, followed by the addition of one molar equivalent of Na_2S. The formed nanoparticles of CdS exhibited the optical properties of a CdS quantum dot of approximate size of 5.6 nm. A subsequent study by Bigham and Coffer (1995) demonstrated that the DNA base sequence and, particularly, the content of the adenine base had a significant influence on the size of the CdS nanoparticles, which clearly established the influence of a DNA template. In an improved strategy to produce a well-defined quantum dot array, Coffer and co-workers used a new strategy to bind a DNA template to a solid substrate (Coffer et al., 1996). The scheme used by them is shown in Figure 16.13. They used a circular plasmid DNA and formed a plasmid DNA/Cd^{2+} complex by adding Cd^{2+} ions. This complex was then bound to a polylysine-coated glass surface and subsequently exposed to produce CdS nanostructures. Tour and co-workers used a similar approach to assemble DNA/fullerene (C_{60}) material (Cassell et al., 1998). Braun et al. (1998) used DNA as a template for nanoscale silver wires.

Mirkin and co-workers assembled 13-nm colloidal Au nanoparticles, modified with thiolated single-stranded DNA, for calorimetric DNA sensor application (Mirkin et al., 1996; Mucic et al., 1998). They have also assembled hybrid materials composed of Au and CdSe nanoparticles (Mitchell et al., 1999). Alivisatos' group (Loweth et al., 1999) used single-stranded DNA as a template for the directed self-assembly of nanoparticles, modified with single-stranded DNA that is complementary to a particular section of a DNA template. Mallouk and co-workers have reported the DNA-directed assembly of a long Au nanowire, up to 6 μm in length and 0.2 μm in diameter (Mbindyo et al., 2001). At our Institute (Suga, Prasad, and co-workers), the focus is on 2-D and 3-D DNA periodic arrays that can be synthesized by producing an intermolecular bridge between two strands as shown in Figure 16.14 (H. Suga,

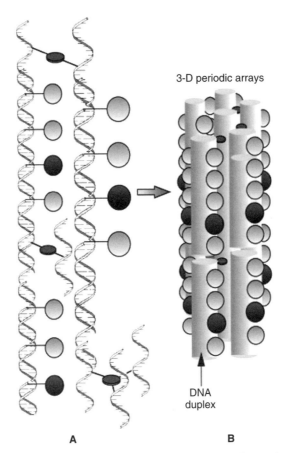

A   **B**

Figure 16.14. Schematic representation of strategy for DNA-directed nanostructure of photonic and electronic materials. (Courtesy of H. Suga.)

unpublished work). We refer to this type of DNA as bridger DNA. The bridger DNA will have two functions (Figure 16.14A). First, it can bridge two strands in head-to-tail fashion. Second, it can bridge two strands in parallel fashion. To incorporate the first function into the bridger DNA, the sequence can be designed to be complementary to the 5′ and 3′ ends of the antisense DNA template. For the second function, two or three bridger DNAs can be cross-linked with each other by using dimeric or trimeric cross-linkers. These bridger DNAs can arrange each unit to program 2-D and 3-D periodic arrays (Figure 16.14B).

An example of this technique is the fabrication of simple metal nanocrystal arrays (e.g., linear arrays of Au nanocrystals of different size and with different separation). These nanocrystals can form metal–insulator–metal tunnel structures, in which the width of the tunneling barrier can be adjusted by the crystallite size and spacing of the sites along the chains. They can then be followed by semiconductor (e.g., CdSe or CdTe) barrier arrays. Another possibility is making p–n nanojunction structures and semiconductor heterostructures. One can even think of fabricating 3-D arrays of metals, metal–semiconductor structures, and semiconductor heterostructures.

Virus as a Template. Viruses with well-defined morphology, flexible microstructures, and surfaces that can be modified relatively easily can serve as suitable templates for producing novel photonic materials. An example is the use of cowpea mosaic virus (CPMV) particles, which are 30-nm-diameter icosahedra, as a template to attach dye molecules or gold nanoparticles (Wang et al., 2002). In this report, functionalized mutant CPMV particles were prepared and reacted with dye using a dye-maleimide reagent to attach up to 60 dye molecules per CPMV particle. Similarly, by using monomaleimido-nanogold, Wang et al. were able to attach gold nanoclusters on the surface of the CPMV particles. This approach provides an opportunity to produce a high local concentration of the attached specie.

16.5 BACTERIA AS BIOSYNTHESIZERS FOR PHOTONIC POLYMERS

The biosynthetic behavior of bacteria can be harnessed to prepare unique polymers for photonics. An example is the production of a family of polyhydroxyalkanoic acid (PHA) polymers synthesized by the bacteria *Pseudomonas oleovorans*. This organism has the capacity to synthesize various PHAs containing C_6 to C_{14} hydroxyalkanoic acid dependent on the 3-hydroxyl alkanoate monomer present.

Poly(3-hydroxyalkanoates) are thermoplastic polyesters produced by bacteria as a carbon reserve in response to conditions of stress such as shortage of essential nutrient. PHAs are linear polyesters composed of 3-hydroxy fatty acid monomers; the basic structure is shown in Figure 16.15. This naturally occurring polymer (crystalline thermoplastic, resembling isostatic polypropy-

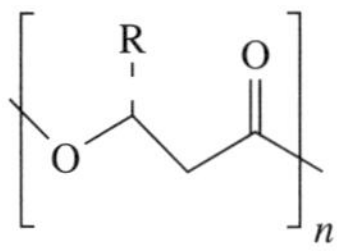

Figure 16.15. Repeating unit structure of PHA.

lene) consists of left-handed 2_1 helices ("all R" confirmation) grouped into crystals of one chirality with a 5.96-Å repeat structure. Isolation and crystallization of these polymers results in a multilaminate "lath" structure that may have significant applications in the development of new photonic materials (Nobes et al., 1999).

The molecular mass of PHAs are generally of the order of 50,000 to 1,000,000 Da (Madison and Huisman, 1999). The carboxyl group of one monomer forms an ester bond with the hydroxyl group of the neighboring monomer. The hydroxyl-substituted carbon atom is of the R stereoconformation. Different PHAs vary at the C-3 or β position. The most common PHA and probably the most abundant of this general class of optically active (chiral) microbial polyesters is poly(3-hydroxybutyrate) in which the alkyl group is a methyl group. Poly(3-hydroxyalkanoates) accumulate in bacteria cells, as discrete granules that can be stored at very high levels. For example, under certain conditions, the polymer may be accumulated at levels of >80% of the cell's dry weight.

Over 100 different PHAs have been isolated from various bacteria in which the 3-hydroxyalkanoate monomer units range from 3-hydroxypropionic acid to 3-hydroxyhexadecanoic acids (Steinbüchel and Valentin, 1995). There are PHAs in which unsaturated 3-hydroxyalkenoic acids occur with one or two double bonds in the R group. Other PHAs have 3-hydroxyhexanoic acids with a methyl group at various positions of the R group. In addition, there are 3HAs in which the R group contains various functional groups such as halogens (—Br, —Cl, —F), olefin, cyano, and hydroxy groups (Curley et al., 1996a, 1996b). This diversity is significant in the development of photonic polymers that possess the optical properties required for the next generation of optical systems. These sets of PHA polymers synthesized by bacteria provide backbone structures and chirality for novel self-ordered photonic polymers and provide unique properties that are difficult to obtain synthetically. They are more heat- and light-stable than other biopolymeric structures (i.e., protein and nucleic acids). Furthermore, the compositional variability provides a unique opportunity to investigate various polymer structure/functional relationships in these novel photonic materials. In order to produce functional polymers, a chromophore having an appropriate β-hydroxyalkanoic acid side chain can be fed to the bacteria under conditions that stimulate nutritional limitations (excess carbon, limited nitrogen). Many monomers can be used to produce PHAs, and there are numerous opportunities to introduce side groups to confer specific optical and mechanical properties. The native structure of the natural polymers is a helix with a strong tendency to form a beta sheet (Chapter 3). Finally, they are economically feasible to produce in relation to costs of synthetic production of specialty polymers. The long-range prospect of this approach poses unique opportunities to prepare nanoscopic and microscopic structures of interest. More specifically, these biosynthetically produced photonic polymers could provide new benefits such as those derived from structural modifications.

The helical and β-sheet structure of materials can be designed or modified to provide further opportunities for photonics. By incorporating cross-linking functionalities and using established electric field poling techniques, it will be possible to develop polymers with a well-defined and temporally and thermally stable noncentrosymmetric structure. Similar materials have been designed for nonlinear optical applications such as high-frequency E-O modulators (Prasad and Williams, 1991). Electroactive or photosensitive moieties could be introduced to effect structural changes in the polymers which could lead to applications in transducers, smart materials, and actuators.

To test the applicability of the bacteria synthesized PHA as thin-film optical media, a study was conducted at our Institute (Bergey and Prasad, unpublished work) to fabricate thin films of a bacteria-synthesized PHA polymer and evaluate its optical properties. Both dip-coating and spin-coating methods were successful in depositing optical quality films on a glass substrate. Also, an important chromophore, APSS (see Chapter 8, Figure 8.4), developed at our Institute, was successfully doped in this polymer film. This chromophore exhibits important nonlinear optical properties such as electro-optic activities (under electrically poled conditions) and two-photon excited up-converted fluorescence.

Figure 16.16a shows the PHA film containing APSS, laid on top of letter characters to demonstrate its optical quality. Figure 16.16b exhibits the one-photon (UV)-excited fluorescence from the same film. Figure 16.16c shows two-photon excited up-converted fluorescence using 800 nm femtosecond pulses from a mode-locked Ti-sapphire laser. This preliminary work clearly

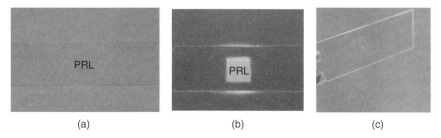

(a)   (b)   (c)

Figure 16.16. Thin film of dye-doped PHA: (a) Transmission, (b) UV-excited fluorescence, (c) two-photon-excited fluorescence.

reveals that good optical quality film structures for photonics applications can be fabricated with bacteria-synthesized PHA polymeric materials.

16.6   FUTURE DIRECTIONS

Photonics will continue to evolve and play a major role in the technological revolution. Future development of photonics requires new materials that are multifunctional and hierarchical and provide the ability to produce heterogeneous integration of interfaces with a 3-D architecture. New guiding principles and design criteria are needed for the development of multifunctional nanoscale building blocks, methods of organizing them and controlling the relevant interactions, and dynamics to achieve desired device functions. This need provides numerous challenges and truly unique cross-disciplinary opportunities at the interface of biology, chemistry, and physics. The very fertile area of biomaterials, which provides many attractive features as discussed in this chapter, together with biomaterials being environmentally friendly and biodegradable, is well poised to be at the center of attention. Some of the potential areas of future development are presented in this section.

Nanoscale Templates for Self-Assembly of Building Blocks. Taking advantage of specificity of interactions exhibited by molecular building blocks in biological systems, a wide use of biomaterials for self-assembly can readily be seen. One can envision the use of nucleotide base-pair specificity to pattern a 3-D optical and/or electronic design for self-assembling of components to produce these functions. Use of a biological template can provide a true opportunity for 3-D integration of heterogeneous components to produce a complex active/passive optoelectronic circuit with multifunctionality. Another advantage offered is that biomaterials can lend themselves for incorporating both inorganic and organic blocks to produce hybrid structures.

Control of Photophysics and Photochemistry with Mutant Biostructures. Use of mutation to produce variant structures to study the structure/photophysical and photochemical property relationship will permit a control of the photoprocesses and their time scales in order to meet the device application need. An example is the case of holographic data storage in bacteriorhodopsin, where the challenge is to either (a) increase the lifetime of the M state for bR to M transition for data storage or (b) increase the quantum yield of the O,P state photobranching for P to Q state transition to be used for data storage.

Surface Functionalization and Bulk Encapsulation. Development of new chemical or mutogenic approaches to surface functionalize a biotemplate or to incorporate a photonic active group in the bulk of a biomaterial will be of considerable value in producing multidomain nanocomposites where each domain may be called upon to perform a specific electronic/photonic function (Ruland et al., 1996). For example, in the case of a virus, one can use functionalization of the surface to graft various functional groups. In addition, one can introduce functional structures inside the capsid of a virus. In the case of DNA, one can intercalate or incorporate active nanoblocks within its duplex structure or attach them on the outer perimeter.

HIGHLIGHTS OF THE CHAPTER

- Photonics for information technology involves functions of data collection, processing, transmission, display, and storage.
- A full implementation of photonics technology is crucially dependent on the availability of suitable optical materials, which are still in the development stage for many of photonics applications.
- Biomaterials are emerging as potentially attractive multifunctional materials for many photonics applications.
- Important classes of biomaterials for photonics are (i) bioderived materials, (ii) bioinspired materials, (iii) biotemplates, and (iv) bioreactors.
- Bioderived materials are naturally occurring biomaterials or their chemically modified forms.
- Bioinspired materials are those synthesized on the basis of governing principles of biological systems.
- Biotemplates refer to natural microstructures that serve as suitable templates for self-assembling of photonic active structures.
- Bioreactors are live biological objects such as bacteria which can act as biosynthesizers to produce photonic polymers.
- Some examples of naturally occurring biomaterials for photonics are (i) bacteriorhodopsin for holographic memory, (ii) green fluorescent protein for photosensitization, (iii) DNA as the host media for laser dyes, and (iv) bio-objects and biocolloids for photonics crystal media.
- Bacteriorhodopsin is the most intensively investigated biomaterials for its application in high-capacity data storage using holographic principles. Even though they are being sold commercially, problems need to be worked out for their practical implementation.
- More recently, naturally occurring DNA has shown promise as a high-optical-quality host media for dyes to produce efficient lasing.
- Bio-object and biocolloids with highly precise shapes can serve as useful building blocks for photonic crystals.
- An example of bioinspired materials is provided by light-harvesting dendrimers, designed following the principles of the antenna effect in natural photosynthetic units.

- An example of a biotemplate is the use of DNA as a template to organize nonbiological blocks, such as organic dyes, semiconductor quantum dots, and metallic nanoparticles, into extended hybrid structures.
- Another example of biotemplate is provided by a virus that has a well-defined morphology, flexible microstructures, and easily modifiable surfaces.
- The biosynthetic ability of bacteria makes them suitable as bioreactors to prepare unique polymers for photonics.
- Some future directions for research and development to produce novel photonics materials are (i) nanoscale self-assembling on a biotemplate, (ii) control of photophysics and photochemistry with mutant biostructures, and (iii) functionalization of the virus surface and/or encapsulation inside the virus.

REFERENCES

Andronov, A., Gilat, S. L., Frechet, J. M. J., Ohta, K., Neuwahl, F. V. R., and Fleming, G. R., Light Harvesting and Energy Transfer in Laser Dye-Labeled Poly (Aryl Ether) Dendrimers, *J. Am. Chem. Soc.* **122**, 1175–1185 (2000).

Bigham, S. R., and Coffer, J. L., The Influence of Adenine Content on the Properties of Q-CdS Clusters Stabilized by Polynucleotides, *Colloids and Surfaces A* **95,** 211–219 (1995).

Birge, R. R., Protein-Based Optical Computing and Memories, *IEEE Comput.* **25**, 56–67 (1992).

Birge, R. R., Gillespie, N. B., Izaguirre, E. W., Kusnetzow, A., Lawrence, A. F., Singh, D., Song, Q. W., Schmidt, E., Stuart, J. A., Seetharaman, S., and Wise, K. J., Biomolecular Electronics: Protein-Based Associative Processors and Volumetric Memories, *J. Phys. Chem. B* **103**, 10746–10766 (1999).

Birge, R. R., Zhang, C. F., and Lawrence, A. F., Optical Random Access Memory Based on Bacteriorhodopsin, in F. Hong, ed., *Molecular Electronics*, Plenum, New York, 1989, pp. 369–379.

Braun, E., Eichon, Y., Sivan, V., and Ben-Yoseph, G., DNA-Templated Assembly and Electrode Attachment of a Conducting Silver Wire, *Nature* **391**, 775–778 (1998).

Brousmiche, D. W., Serin, J. M., Frechet, J. M. J., He, G. S., Lin, T.-C., Chung, S. J., and Prasad, P. N., Fluorescence Resonance Energy Transfer in a Novel Two-Photon Absorbing System, *J. Am. Chem. Soc.* (2003), in press.

Cassell, A. M., Scrivens, W. A., and Tour, J. M., Assembly of DNA/Fullerene Hybrid Materials, *Angew. Chem. Int. Ed. Engl.* **37**, 1528–1531 (1998).

Chattoraj, M., King, B. A., Bublitz, G. U., and Boxer, S. G., Ultra-Fast Excited State Dynamics in Green Fluorescent Protein: Multiple States and Proton Transfer, *Proc. Natl. Acad. Sci. USA* **93**, 8362–8367 (1996).

Chen, Z., and Birge, R. R., Protein-Based Artificial Retinas, *Trends Biotechnol.* **11**, 292–300 (1993).

Choi, J.-W., Nam, Y.-S., Park, S.-J., Lee, W.-H., Kim, D., and Fujihira, M., Rectified Photocurrent of Molecular Photodiode Consisting of Cytochrome C/GFP Heterothin Films, *Biosensors & Bioelectronics* **16**, 819–825 (2001).

Coffer, J. L., Approaches for Generating Mesoscale Patterns of Semiconductor Nanoclusters, *J. Cluster Sci.* **8**, 159–179 (1997).

Coffer, J. L., Bigham, S. R., Li, X., Pinizzotto, R. F., Rho, Y. G., Pirtle, R. M., and Pirtle, I. L., Dictation of the Shape of Mesoscale Semiconductor Nanoparticle Assemblies by Plasmid DNA, *Appl. Phys. Lett.* **69**, 3851–3853 (1996).

Coffer, J. L., Bigham, S. R., Pinizzotto, R. F., and Yang, H., Characterization of Quantum-Confined CdS Nanocrystallites Stabilized by Deoxyribonucleic Acid, *Nanotechnology* **3**, 69–76 (1992).

Cubitt, A. B., Heim, R., Adams, S. R., Boyd, A. E., Gross, L. A., and Tsien, R. Y., Understanding, Improving and Using Green Fluorescent Proteins, *Trends Biochem. Sci.* **20**, 448–455 (1995).

Curley, J. M., Hazer, B., Lenz, R. W., and Fuller, R. C., Production of Poly(3-hydroxyalkanoates) Containing Aromatic Substituents by *Pseudomonas oleovorans*, *Macromolecules* **29**, 1762–1766 (1996a).

Curley, J. M., Lenz, R. W., and Fuller, R. C., Sequential Production of Two Different Polyesters in the Inclusion Bodies of *Pseudomonas oleovorans*, *Int. J. Biol. Macromolecules* **19(1)**, 29–34 (1996b).

Deisenhofer, J., Epp, O., Miki, K., Huber, R., and Michel, H., Structure of the Protein Subunits in the Photosynthetic Reaction Center of *Rhodopseudomonas viridis* at 3× Resolution, *Nature* **318**, 618–624 (1985).

Dickson, R. M., Cubitt, A. B., Tslen, R. Y., and Moerner, W. E., On/Off Blinking and Switching Behavior of Single Molecules of Green Fluorescent Protein, *Nature* **388**, 355–358 (1997).

Douglas, T., and Young, M., Host–Guest Encapsulation of Materials by Assembled Virus Cages, *Nature* **393**, 152–155 (1998).

Finlayson, N., Banyai, W. C., Seaton, C. T., Stegeman, G. I., Neill, M., Cullen, T. J., and Ironside, C. N., Optical Nonlinearities in CdS_xSe_{1-x}-Doped Glass Wave-Guides, *J. Opt. Soc. Am. B* **6**, 675–684 (1989).

Hampp, N., Bräuchle, C., and Oesterhelt, D., Bacteriorhodopsin Wildtype and Variant Aspartate-96-Asparagine as Reversible Holographic Media, *Biophys. J.* **58**, 83–93 (1990).

Hampp, N., Thoma, R., Zeisel, D., and Bräuchle, C., Bacteriorhodopsin Variants for Holographic Pattern-Recognition, *Adv. Chem.* **240**, 511–526 (1994).

He, G. S., Lin, T.-C., Cui, Y., Prasad, P. N., Brousmiche, D. W., Serin, J. M., and Frechet, J. M. J., Two-Photon Excited Intramolecular Energy Transfer and Light Harvesting Effect in Novel Dendritic Systems, *Opt. Lett.* (in press).

Hong, F. T., Retinal Proteins in Photovoltaic Devices, *Adv. Chem.* **240**, 527–560 (1994).

Jacobsen, J. P., Pedersen, J. B., and Wemmer, D. E., Site Selective Bis-Intercalation of a Homodimeric Thiazole Orange Dye in DNA Oligonucleotides, *Nucl. Acid Res.* **23**, 753–760 (1995).

Kawabe, Y., Wang, L., Horinouchi, S., and Ogata, N., Amplified Spontaneous Emission from Fluorescent-Dye-Doped DNA-Surfactant Complex Films, *Adv. Mater.* **12**, 1281–1283 (2000).

Kirkpatrick, S. M., Naik, R. R., and Stone, M. O., Nonlinear Saturation and Determination of the Two-Photon Absorption Cross-Section of Green Fluorescent Protein, *J. Phys. Chem. B* **105**, 2867–2873 (2001).

Kogelnik, H., Coupled Wave Theory for Thick Hologram Grating, *The Bell System Tech. J.* **48**, 2909–2948 (1969).

Kuznetsov, Yu. G., Malkin, A. J., Lucas, R. W., and McPherson, A., Atomic Force Microscopy Studies of Icosahedral Virus Crystal Growth, *Colloids and Surfaces B: Biointerfaces* **19**, 333–346 (2000).

Loweth, C. J., Caldwell, W. B., Peng, X., Alivisatos, A. P., and Schultz, P. G., DNA-Based Assembly of Gold Nanocrystals, *Angew. Chem. Int. Ed.* **38**, 1808–1812 (1999).

Madison, L. L., and Huisman, G. W., Metabolic Engineering of Poly(3-Hydroxyalkanoates): From DNA to Plastic, *Microbiology and Molecular Biology Reviews: MMBR* **63(1)**, 21–53 (1999).

Marwan, W., Hegemann, P., and Oesterhelt, D., Single Photon Detection by an Archaebacterium, *J. Mol. Biol.* 663–664 (1988).

Mbindyo, J. K. N., Reiss, B. D., Martin, B. R., Keating, C. D., Natan, M. J., and Mallouk, T. E., DNA-Directed Assembly of Gold Nanowires on Complementary Surfaces, *Adv. Mater.* **13**, 249–254 (2001).

Mirkin, C. A., Letsinger, R. L., Mucic, R. C., and Storhoff, J. J., A DNA-Based Method for Rationally Assembling Nanoparticles into Macroscopic Materials, *Nature* **382**, 607–609 (1996).

Mitchell, G. P., Mirkin, C. A., and Letsinger, R. L., Programmed Assembly of DNA Functionalized Quantum Dots, *J. Am. Chem. Soc.* **121**, 8122–8123 (1999).

Miyasaka, T., Koyama, K., and Itoh, I., Quantum Conversion and Image Detection by a Bacteriorhodopsin-Based Artificial Photoreceptor, *Science* **255**, 342–344 (1992).

Mok, F. H., Burr, G. W., and Psaltis, D., System Metric for Holographic Memory Systems, *Opt. Lett.* **21**, 896–898 (1996).

Mucic, R. C., Storhoff, J. J., Mirkin, C. A., and Letsinger, R. L., DNA-Directed Synthesis of Binary Nanoparticle Network Materials, *J. Am. Chem. Soc.* **120**, 12674–12675 (1998).

Nobes, G. A. R., Marchessault, R. H., Chanzy, H., Briese, B. H., and Jendrossek, D., Splintering of Poly(3-Hydroxybutyrate) Single Crystals by PHB-Depolymerase A from *Pseudomonas lemoignei*, *Macromolecules* **29**, 8330–8333 (1996).

Parker, A. R., McPherson, R. C., McKenzie, D. R., Botten, L. C., and Nicorovici, N.-A. P., Aphrodite's Iridescence, *Nature* **409**, 36–37 (2001).

Pikas, D. J., Kirkpatrick, S. M., Tewksbury, E., Brott, L. L., Naik, R. R., and Stone, M. O., Nonlinear Saturation and Lasing Characteristics of Green Fluorescent Protein, *J. Phys. Chem. B* **106**, 4831–4837 (2002).

Popp, A., Wolperdinger, M., Hampp, N., Bräuchle, C., and Oesterhelt, D., Photochemical Conversion of the O-Intermediate to 9-*cis*-Retinal Containing Products in Bacteriorhodopsin Films, *Biophys. J.* **65**, 1449–1459 (1993).

Prasad, P. N., and Williams, D. J., Introduction to Nonlinear Optical Effects in Molecules and Polymers, Wiley, New York, 1991.

Ruland, G., Gvishi, R., and Prasad, P. N., Multiphasic Nanostructured Composites: Multi-dye Tunable Solid-State Laser, *J. Am. Chem. Soc.* **118**, 2985–2991 (1996).

Saleh, B. E. A., and Teich, M. C., *Fundamentals of Photonics*, Wiley-Interscience, New York, 1991.

Sasabe, H., Furuno, T., and Takimoto, K., Photovoltaics of Photoactive Protein Polypeptide LB Films, *Synth. Met.* **28**, C787–C792 (1989).

Song, Q. W., Zhang, C., Gross, R., and Birge, R. R., Optical Limiting by Chemically Enhanced Bacteriorhodopsin Films, *Opt. Lett.* **18**, 775–777 (1993).

Spielmann, H. P., Dynamics of a bis-Intercalator DNA Complex by H-1-Detected Natural Abundance C-13 NMR Spectroscopy, *Biochemistry* **37**, 16863–16876 (1998).

Steinbüchel, A., and Valentin, H. E., Diversity of Bacterial Polyhydroxyalkanoic Acids, *FEMS Microbiol. Lett.* **128(3)**, 219–228 (1995).

Storhoff, J. J., and Mirkin, C. A., Programmed Materials Synthesis with DNA, *Chem. Rev.* **99,** 1849–1862 (1999).

Swallen, S. F., Zhu, Z., Moore, J. S., and Kopelman, R., A Perylene Derivative of a Phenylacetylene Dendrimer, called "Nanostar," *J. Phys. Chem. B* **104**, 3988–3995 (2000).

Vaia, R., Farmer, B., and Thomas, E. L. (2002), private communications.

Vsevolodov, N. N., Druzhko, A. B., and Djukova, T. V., Actual Possibilities of Bacteriorhodopsin Application in Optoelectronics, in F. T. Hong, ed., *Molecular Electronics: Biosensors and Biocomputers*, Plenum Press, New York, 1989, pp. 381–384.

Wang, Q., Lin, T., Tang, L., Johnson, J. E., and Finn, M. G., Icosahedral Virus Particles as Addressable Nanoscale Building Blocks, *Angew. Chem. Int. Ed.* **41**, 459–462 (2002).

Wang, L., Yoshida, J., and Ogata, N., Self-Assembled Supramolecular Films Derived from Marine Deoxyribonucleic Acid (DNA)–Cationic Surfactant Complexes: Large Scale Preparation and Optical and Thermal Properties, *Chem. Mater.* **13**, 1273–1281 (2001).

Yang, F., Moss, L. G., and Phillips, J. G. N., The Molecular Structure of Green Fluorescent Protein, *Nat. Biotechnol.* **14**, 1246–1251 (1996).

Introduction to Biophotonics, by Paras N. Prasad
ISBN: 0-471-28770-9   Copyright © 2003 John Wiley & Sons, Inc.